RADIO FREQUENCY POWER IN PLASMAS

Previous Proceedings in the Series of Topical Conferences on Radio Frequency Power in Plasmas

Year	Conference	Publisher	ISBN
2001	14th	AIP Conf. Proceedings Vol. 595	0-7354-0038-5
1999	13th	AIP Conf. Proceedings Vol. 485	1-56396-861-4
1997	12th	AIP Conf. Proceedings Vol. 403	1-56396-709-X
1995	11th	AIP Conf. Proceedings Vol. 355	1-56396-536-4
1993	10th	AIP Conf. Proceedings Vol. 289	1-56396-264-0
1991	9th	AIP Conf. Proceedings Vol. 244	0-88318-937-2

Other Related Titles from AIP Conference Proceedings

692 Non-Neutral Plasma Physics V: Workshop on Non-Neutral Plasmas
Edited by Martin Schauer, Travis Mitchell, and Richard Nebel, December 2003, 0-7354-0165-9

691 High Energy Density and High Power RF: 6th Workshop on High Energy Density and High Power RF
Edited by Steven H. Gold and Gregory S. Nusinovich, November 2003, CD-ROM included, 0-7354-0164-0

669 Plasma Physics: 11th International Congress on Plasma Physics: ICPP2002
Edited by Ian S. Falconer, Robert L. Dewar, and Joe Khachan, June 2003,
Print: 0-7354-0133-0; CD-ROM: 0-7354-0134-9

649 Dusty Plasmas in the New Millennium: Third International Conference on the Physics of Dusty Plasmas
Edited by R. Bharuthram, M. A. Hellberg, P. K. Shukla, and F. Verheest, December 2002, 0-7354-0106-3

625 High Energy Density and High Power RF: 5th Workshop on High Energy Density and High Power RF
Edited by B. E. Carlsten, August 2002, 0-7354-0078-4

611 Superstrong Fields in Plasmas: Second International Conference on Superstrong Fields in Plasmas
Edited by Maurizio Lontano, Gérard Mourou, Orazio Svelto, and Toshiki Tajima, April 2002, 0-7354-0057-1

563 Plasma Physics: IX Latin American Workshop
Edited by Hernán Chuaqui and Mario Favre, May 2001, 1-56396-999-8

To learn more about these titles, or the AIP Conference Proceedings Series, please visit the webpage **http://proceedings.aip.org/proceedings**

RADIO FREQUENCY POWER IN PLASMAS

15th Topical Conference on Radio Frequency Power in Plasmas

Moran, Wyoming   19-21 May 2003

EDITOR
Cary B. Forest
University of Wisconsin
Madison, Wisconsin

SPONSORING ORGANIZATIONS
University of Wisconsin
The American Physical Society

Melville, New York, 2003
AIP CONFERENCE PROCEEDINGS ■ VOLUME 694

Editor:

Cary B. Forest
University of Wisconsin-Madison
3277 Chamberlain Hall
1150 University Avenue
Madison, WI 53706
USA

E-mail: cbforest@wisc.edu

Copyright in the following articles remains the property of the authors: pp. 41-49, 66-73, 74-81, 98-101, 118-121, 122-125, 130-133, 138-141, 219-226, 227-234, 255-258, 259-262, and 279-282.

The articles on pp. 146-149, 150-153, and 359-366 are covered by British Crown Copyright, EURATOM/UKAEA Fusion Association, Oxfordshire.

Authorization to photocopy items for internal or personal use, beyond the free copying permitted under the 1978 U.S. Copyright Law (see statement below), is granted by the American Institute of Physics for users registered with the Copyright Clearance Center (CCC) Transactional Reporting Service, provided that the base fee of $20.00 per copy is paid directly to CCC, 222 Rosewood Drive, Danvers, MA 01923. For those organizations that have been granted a photocopy license by CCC, a separate system of payment has been arranged. The fee code for users of the Transactional Reporting Service is: 0-7354-0158-6/03/$20.00.

© 2003 American Institute of Physics

Individual readers of this volume and nonprofit libraries, acting for them, are permitted to make fair use of the material in it, such as copying an article for use in teaching or research. Permission is granted to quote from this volume in scientific work with the customary acknowledgment of the source. To reprint a figure, table, or other excerpt requires the consent of one of the original authors and notification to AIP. Republication or systematic or multiple reproduction of any material in this volume is permitted only under license from AIP. Address inquiries to Office of Rights and Permissions, Suite 1NO1, 2 Huntington Quadrangle, Melville, N.Y. 11747-4502; phone: 516-576-2268; fax: 516-576-2450; e-mail: rights@aip.org.

L.C. Catalog Card No. 2003114582
ISBN 0-7354-0158-6
ISSN 0094-243X
Printed in the United States of America

CONTENTS

Preface xv

REVIEWS

Waves and Instabilities in Dusty Plasmas *(Review)* 3
    R. L. Merlino

Electron Bernstein wave heating in fusion plasmas *(Review)* 15
    H. P. Laqua, W7-AS Team, and ECRH-Group

Lower Hybrid Current Drive: An Overview of Simulation Models, Benchmarking with Experiment, and Predictions for Future Devices *(Review)* 24
    P. T. Bonoli, E. Barbato, R. W. Harvey, and F. Imbeaux

ION CYCLOTRON RANGE OF FREQUENCIES

Bulk Plasma Rotation in the Presence of Waves in the Ion Cyclotron Range of Frequencies *(Invited)* 41
    L.-G. Eriksson, J.-M. Noterdaeme, S. Assas, C. Giroud, J. DeGrassie,
    T. Hellsten, T. Johnson, V. G. Kiptily, K. Kirov, M. Mantsinen,
    K.-D. Zastrow, M. DeBaar, J. Brzozowski, R. Budny, R. Cesario, V. Chan,
    C. Fenzi-Bonizec, A. Gondhalekar, N. Hawkes, G. T. Hoang, P. Lamalle,
    A. Meigs, F. Meo, F. Nguyen, E. Righi, A. Staebler, D. Testa, A. Tuccillo,
    H. Weisen, and the JET-EFDA Contributors

Study of Ion Cyclotron Range of Frequencies Mode Conversion in the Alcator C-Mod Tokamak *(Invited)* 50
    Y. Lin, S. J. Wukitch, P. T. Bonoli, A. Mazurenko, E. Nelson-Melby,
    M. Porkolab, J. C. Wright, I. H. Hutchinson, E. S. Marmar,
    D. Mossessian, S. Wolfe, C. K. Phillips, G. Schilling, and P. Phillips

The Analysis of ICRF Heating Experiment in View of the Confinement of High-Energy Particles in LHD *(Invited)* 58
    K. Saito, R. Kumazawa, T. Mutoh, T. Seki, T. Watari, F. Shimpo,
    G. Nomura, A. Kato, M. Yokota, M. Isobe, T. Ozaki, M. Osakabe,
    M. Sasao, T. Saida, T. Yamamoto, Y. Torii, N. Takeuchi,
    and the LHD Experimental Group

Recent ^{3}He Radio Frequency Heating Experiments on JET *(Invited)* 66
    D. Van Eester, F. Imbeaux, P. Mantica, M. Mantsinen, M. de Baar,
    P. de Vries, L.-G. Eriksson, R. Felton, A. Figueiredo, J. Harling, E. Joffrin,
    K. Lawson, H. Leggate, X. Litaudon, V. Kiptily, J.-M. Noterdaeme,
    V. Pericoli, E. Rachlew, A. Tuccillo, K.-D. Zastrow,
    and the JET-EFDA Contributors

Local Improvement Confinement in the Ion Bernstein (IBW) Experiment on FTU (Frascati Tokamak Upgrade) *(Invited)* 74
    R. Cesario, A. Cardinali, C. Castaldo, M. Marinucci, G. Ravera,
    S. Bernabei, E. Giovannozzi, M. Leigheb, F. Paoletti, V. Pericoli-Ridolfini,

F. Zonca, G. Apruzzese, M. Borra, R. De Angelis, E. Giovannozzi,
L. Gabellieri, H. Kroegler, M. Leigheb, G. Mazzitelli, P. Micozzi,
L. Panaccione, P. Papitto, B. Angelini, M. L. Apicella, E. Barbato,
L. Bertalot, A. Bertocchi, G. Buceti, C. Centioli, V. Cocilovo, F. Crisanti,
F. De Marco, B. Esposito, G. Gatti, C. Gormezano, M. Grolli, F. Iannone,
G. Maddaluno, G. Monari, P. Orsitto, D. Pacella, M. Panella, L. Pieroni,
G. B. Righetti, F. Romanelli, E. Sternini, N. Tartoni, A. A. Tuccillo,
O. Tudisco, V. Vitale, G. Vlad, and M. Zerbini

Properties of Segmented Ion Cyclotron Antennas 82
G. Bosia and S. Bremond

**Interaction of Neutral Beam (NB) Injected Fast Ions with Ion
Cyclotron Resonance Frequency (ICRF) Waves** 86
M. Choi, V. S. Chan, S. C. Chiu, Y. A. Omelchenko, Y. Sentoku,
and H. E. St John

RF-Sheath Assessment of ICRF Antenna Geometry for Long Pulses 90
L. Colas, S. Heuraux, and S. Bremond

**ELM Resilient External Matching System for the ICRF System of
ITER: 2. Design of the Components and Implementation** 94
P. Dumortier, F. Durodié, A. Messiaen, and S. Brons

The ITER-like ICRH Launcher Project for JET 98
F. Durodié, G. Agarici, G. Amarante, F. W. Baity, B. Beaumont,
S. Brémond, P. Chappuis, C. Damiani, J. Fanthome, R. H. Goulding,
J. Hosea, G. H. Jones, A. Kaye, P. U. Lamalle, G. D. Loesser, A. Lorenz,
M. Mead, M. A. Messineo, I. Monakhov, B. Nelson, M. Nightingale,
J. Paméla, C. Portafaix, V. Riccardo, L. Semeraro, C. Talarico, E. Turker,
K. Vulliez, A. Walden, R. Walton, J. R. Wilson, and P. Wouters

**Initial Operation of the JET ITER-like High-Power Prototype
ICRF Antenna** 102
R. H. Goulding, F. W. Baity, F. Durodié, A. Fadnek, J. C. Hosea,
G. H. Jones, G. D. Loesser, B. E. Nelson, D. A. Rasmussen,
P. M. Ryan, D. O. Sparks, D. W. Swain, and R. Walton

**Measurements and Calculations of Electrical Properties
of ICRF Antennas** 106
D. A. Hartmann, R. Bilato, D. Birus, M. Brambilla, F. Braun,
J.-M. Noterdaeme, and F. Wesner

**The Effect of Weak Single Pass Damping on the Coupled ICRH
Power Spectrum** 110
T. Hellsten and M. Laxåback

**The Radial Mode Transition of Excited Fast Alfvén Waves in the
Mirror Plasmas** 114
H. Higaki, M. Ichimura, Y. Yamaguchi, S. Kakimoto, K. Horinouchi,
K. Ide, H. Utsumi, D. Inoue, H. Hojo, and K. Yatsu

**Radio-Frequency Matching Studies for the JET ITER-like
ICRF System** 118
P. U. Lamalle, F. Durodié, R. H. Goulding, I. Monakhov, M. Nightingale,
A. Walden, P. Wouters, and the EFDA JET Workprogramme Contributors

Three-Dimensional Electromagnetic Modelling of the JET ITER-like ICRF Antenna 122
    P. U. Lamalle, F. Durodié, A. Whitehurst, R. H. Goulding, P. M. Ryan, and the EFDA JET Workprogramme Contributors

Self-Consistent Modelling of Polychromatic ICRH in Tokamaks 126
    M. Laxåback, T. Johnson, T. Hellsten, and M. Mantsinen

Effect of ICRF Mode Conversion at the Ion-Ion Hybrid Resonance on Plasma Confinement in JET 130
    A. Lyssoivan, M. J. Mantsinen, D. Van Eester, R. Koch, A. Salmi, J.-M. Noterdaeme, I. Monakhov, and the JET EFDA Contributors

Efficient Self-Consistent 3D/1D Analysis of ICRF Antennas 134
    R. Maggiora, G. Vecchi, V. Lancellotti, and V. Kyrytsya

Comparison of Monochromatic and Polychromatic ICRH on JET 138
    M. J. Mantsinen, M. Laxåback, A. Salmi, V. Kiptily, D. Testa, Y. Baranov, R. Barnsley, P. Beaumont, S. Conroy, P. de Vries, C. Giroud, C. Gowers, T. Hellsten, L. C. Ingesson, T. Johnson, H. Leggate, M.-L. Mayoral, I. Monakhov, J.-M. Noterdaeme, S. Podda, S. Sharapov, A. A. Tuccillo, D. Van Eester, and the EFDA JET contributors

ELM Resilient External Matching System for the ICRF System of ITER: 1. Principle and Performances 142
    A. Messiaen, P. Dumortier, and F. Durodié

Studies of JET ICRH Antenna Coupling during ELMs. 146
    I. Monakhov, T. Blackman, M.-L. Mayoral, M. Nightingale, A. Walden, P. U. Lamalle, P. Wouters, and the JET EFDA Contributors

New Techniques for the Improvement of the ICRH System ELM Tolerance on JET 150
    I. Monakhov, T. Blackman, A. Walden, M. Nightingale, A. Whitehurst, F. Durodie, P. U. Lamalle, and JET EFDA Contributors

Status and Development of the ICRF Antennas on ASDEX Upgrade 154
    J.-M. Noterdaeme, W. Becker, V. Bobkov, F. Braun, D. Hartmann, and F. Wesner

Investigation of ICRF Coupling Resistance in Alcator C-Mod Tokamak 158
    A. Parisot, S. J. Wukitch, A. Ram, R. Parker, Y. Lin, J. W. Hughes, B. Labombard, P. Bonoli, and M. Porkolab

Optimized AT Target Plasma Studies in Alcator C-Mod with I_p-Ramp and Intense ICRH 162
    M. Porkolab, P. T. Bonoli, Y. Lin, S. J. Wukitch, C. Fiore, A. Hubbard, J. Irby, L. Lin, E. Marmar, R. Parker, J. Rice, G. Schilling, J. R. Wilson, S. Wolfe, H. Yuh, K. Zhurovich, and the C-Mod Team

Analysis of 4-Strap ICRF Antenna Performance in Alcator C-Mod 166
    G. Schilling, S. J. Wukitch, R. L. Boivin, J. A. Goetz, J. C. Hosea, J. H. Irby, Y. Lin, A. Parisot, M. Porkolab, J. R. Wilson, and the Alcator C-Mod Team

Mode Conversion of High-Field-Side-Launched Fast Waves at the Second Harmonic of Minority Hydrogen in Advanced Tokamak Reactors 170
    R. Sund and J. Scharer

Ion Heating in Field-Reversed Configuration by Radio-Frequency Waves 174
    V. A. Svidzinski and S. C. Prager

Overview of Alcator C-Mod ICRF Experiments 178
    S. J. Wukitch, P. T. Bonoli, C. Fiore, I. H. Hutchinson, Y. Lin, E. Marmar,
    A. Parisot, M. Porkolab, J. Rice, G. Schilling, and the Alcator C-Mod Team

FAST WAVE CURRENT DRIVE

Characterization of Fast Ion Absorption of the High Harmonic Fast Wave in the National Spherical Torus Experiment 1 *(Invited)* 185
    A. L. Rosenberg, J. E. Menard, J. R. Wilson, S. Medley, C. K. Phillips,
    R. Andre, D. S. Darrow, R. J. Dumont, B. P. LeBlanc, M. H. Redi,
    T. K. Mau, E. F. Jaeger, P. M. Ryan, D. W. Swain, R. W. Harvey, J. Egedal,
    and the NSTX Team

Observations of Anisotropic Ion Temperature during RF Heating in the NSTX Edge 193
    T. M. Biewer, R. E. Bell, D. S. Darrow, and J. R. Wilson

Fast Wave Current Drive in JET ITB-Plasmas 197
    T. Hellsten, M. Laxåback, T. Johnson, M. Mantsinen, G. Matthews,
    P. Beaumont, C. Challis, D. van Eester, E. Rachlew, T. Bergkvist,
    C. Giround, E. Joffrin, A. Huber, V. Kiptily, F. Nguyen, J.-M. Noterdaeme,
    J. Mailloux, M.-L. Mayoral, F. Meo, I. Monakhov, F. Sartori, A. Staebler,
    E. Tennfors, A. Tuccillo, A. Walden, B. Volodymyr,
    and the JET-EFDA Contributors

High-Harmonic Fast-Wave Driven H-Mode Plasmas in NSTX 201
    B. P. LeBlanc, R. E. Bell, S. I. Bernabei, K. Indireshkumar, S. M. Kaye,
    R. Maingi, T. K. Mau, D. W. Swain, G. Taylor, P. M. Ryan, J. B. Wilgen,
    and J. R. Wilson

Modeling of HHFW Heating and Current Drive on NSTX with Ray Tracing and Adjoint Techniques 205
    T. K. Mau, P. M. Ryan, M. D. Carter, E. F. Jaeger, D. W. Swain,
    C. K. Phillips, S. Kaye, B. P. LeBlanc, A. L. Rosenberg, J. R. Wilson,
    R. W. Harvey, P. Bonoli, and the NSTX Team

High Harmonic Fast Wave Current Drive Experiments on NSTX 209
    P. M. Ryan, D. W. Swain, J. R. Wilson, J. C. Hosea, B. P. LeBlanc,
    S. Bernabei, P. Bonoli, M. D. Carter, E. F. Jaeger, S. Kaye, T. K. Mau,
    C. K. Phillips, D. A. Rasmussen, J. B. Wilgen, and the NSTX Team

Investigations of Low and Moderate Harmonic Fast Wave Physics on CDX-U 213
    J. Spaleta, R. Majeski, C. K. Phillips, R. J. Dumont, R. Kaita,
    V. Soukhanovskii, and Z. Zakharov

LOWER HYBRID CURRENT DRIVE

Hybrid and Steady-State Operation on JET and Tore Supra *(Invited)* 219
    A. Bécoulet, the Tore Supra Team, and the Contributors to the EFDA-JET Workprogramme

Long Distance Coupling of Lower Hybrid Waves in ITER Relevant Edge Conditions in JET Reversed Shear Plasmas *(Invited)* 227
    A. Ekedahl, G. Granucci, J. Mailloux, V. Petrzilka, K. Rantamäki,
    Y. Baranov, K. Erents, M. Goniche, E. Joffrin, P. J. Lomas, M. Mantsinen,
    D. McDonald, J.-M. Noterdaeme, V. Pericoli, R. Sartori, C. Silva,
    M. Stamp, A. A. Tuccillo, F. Zacek, and the EFDA-JET Contributors

Profile Control and Plasma Start-up by RF Waves Towards Advanced Tokamak Operation in JT-60U *(Invited)* 235
    Y. Takase for the JT-60 Team

Control and Data Acquisition System for Lower Hybrid Current Drive in Alcator C-Mod 243
    N. P. Basse, J. Bosco, T. W. Fredian, M. Grimes, N. D. Kambouchev,
    Y. Lin, R. R. Parker, Y. I. Rokhman, J. A. Stillerman, D. R. Terry, and
    S. J. Wukitch

A Proposal for a Planar Lower Hybrid Launcher 247
    G. Bosia, S. Kuzikov, and P. Testoni

Lower Hybrid System Upgrade on Tore Supra 251
    B. Beaumont, A. Beunas, P. Bibet, and F. Kazarian

Modeling of Lower Hybrid Current Drive (LHCD) and Parametric Instability (PI) for High Performance Internal Transport Barriers (ITBs) 255
    R. Cesario, A. Cardinali, C. Castaldo, F. Paoletti, C. Challis,
    J. Mailloux, D. Mazon, and the EFDA-JET Contributors

Density Convection near Radiating ICRF Antennas and its Effect on the Coupling of Lower Hybrid Waves 259
    A. Ekedahl, L. Colas, M.-L. Mayoral, B. Beaumont, P. Bibet, S. Brémond,
    F. Kazarian, J. Mailloux, J.-M. Noterdaeme, A. A. Tuccillo,
    and the EFDA-JET Contributors

Lower Hybrid Experiments in MST 263
    J. A. Goetz, M. A. Thomas, P. K. Chattopadhyay, M. C. Kaufman,
    R. O'Connell, S. P. Oliva, and P. J. Weix

Effect of the Launched LH Spectrum on the Fast Electron Dynamics in the Plasma Core and Edge 267
    M. Goniche, V. Petržílka, A. Ekedahl, V. Fuchs, J. Laugier, Y. Peysson, and
    F. Záček

Absorption of Lower Hybrid Waves by Alpha Particles in ITER 271
    F. Imbeaux, Y. Peysson, and L. G. Eriksson

Separate and Combined Lower Hybrid and Electron Cyclotron Current Drive Experiments in Non-inductive Plasma Regimes on Tore Supra 275
    F. Imbeaux, J. F. Artaud, V. Basiuk, G. Berger-By, F. Bouquey,
    C. Bourdelle, J. Clary, C. Darbos, G. Giruzzi, G. T. Hoang,
    G. Huysmans, M. Lennholm, P. Maget, R. Magne, D. Mazon,
    Y. Peysson, J. L. Ségui, X. L. Zou, and the Tore Supra Team

Real-Time Control of the Current Profile in JET 279
    D. Moreau, F. Crisanti, X. Litaudon, D. Mazon, P. De Vries, R. Felton,
    E. Joffrin, L. Laborde, M. Lennholm, A. Murari, V. Pericoli-Ridolfini,
    M. Riva, T. Tala, G. Tresset, L. Zabeo, and K. D. Zastrow

Construction, Calibration, and Testing of a Multi-waveguide Array for the Application of LHCD on Alcator C-MOD 283
    J. R. Wilson, S. Bernabei, E. Fredd, N. Greenough, J. C. Hosea,
    C. C. Kung, and G. D. Loesser

ELECTRON CYCLOTRON HEATING AND CURRENT DRIVE

Optimization of Electron Cyclotron Heating in LHD *(Invited)* 289
    S. Kubo, T. Shimozuma, H. Idei, Y. Yoshimura, T. Notake,
    K. Ohkubo, Y. Mizuno, S. Ito, S. Kobayashi, Y. Takita,
    and the LHD Experimental Group

Physics Studies with ECH/CD in the TCV Tokamak *(Invited)* 297
    A. Pochelon, S. Alberti, C. Angioni, G. Arnoux, R. Behn, P. Blanchard,
    Y. Camenen, S. Coda, I. Condrea, T. P. Goodman, J. Graves,
    M. A. Henderson, J-P. Hogge, E. Nelson-Melby, P. Nikkola, L. Porte,
    O. Sauter, A. Scarabosio, M. Q. Tran, G. Zhuang, and the TCV Team

Electron Cyclotron Current Drive in DIII-D: Experiment and Theory *(Invited)* 305
    R. Prater, C. C. Petty, T. C. Luce, R. W. Harvey, M. Choi, R. J. La Haye,
    Y.-R. Lin-Liu, J. Lohr, M. Murakami, M. R. Wade, and K.-L. Wong

Toroidal Rotation in ECH H-Modes in DIII-D 313
    J. S. deGrassie, K. H. Burrell, D. R. Baker, L. R. Baylor, J. Lohr,
    R. I. Pinsker, and R. Prater

Neoclassical Tearing Mode Suppression with ECRH and Current Drive in ASDEX Upgrade 317
    G. Gantenbein, A. Keller, F. Leuterer, M. Maraschek, W. Suttrop, H. Zohm,
    and the ASDEX Upgrade-Team

Stabilization of Neoclassical Tearing Mode by Electron Cyclotron Wave Injection in JT-60U 321
    A. Isayama, K. Nagasaki, T. Ide, T. Fukuda, T. Suzuki, M. Seki,
    S. Moriyama, Y. Ikeda, and the JT-60 team

Launcher Performance in the DIII-D ECH System 325
    K. Kajiwara, C. B. Baxi, J. Lohr, Y. A. Gorelov, M. T. Green, D. Ponce,
    and R. W. Callis

Absorption of X-Wave at the Second Harmonic in HSX 331
    K. M. Likin, J. N. Talmadge, A. F. Almagri, D. T. Anderson,
    F. S. Anderson, C. Deng, S. P. Gerhardt, and K. Zhai

The 110 GHz Microwave Heating System on the DIII-D Tokamak 335
    J. Lohr, R. W. Callis, J. L. Doane, R. A. Ellis, Y. A. Gorelov, K. Kajiwara,
    D. Ponce, and R. Prater

Start-up of Spherical Tokamak by ECH on LATE 340
    T. Maekawa, H. Tanaka, M. Uchida, H. Igami, T. Yoshinaga, and K. Higaki

Fokker-Planck Simulations of X3 EC Wave Absorption Experiments in the TCV Tokamak 344
    P. Nikkola, S. Alberti, S. Coda, T. P. Goodman, R. W. Harvey,
    E. Nelson-Melby, and O. Sauter

Complete Suppression of the m=2/n=1 NTM Using ECCD on DIII–D 348
    C. C. Petty, R. J. La Haye, T. C. Luce, M. E. Austin, R. W. Harvey,
    D. A. Humphreys, J. Lohr, R. Prater, and M. R. Wade

Assessment of Electron Banana Width Effect in ECCD Experiments 352
    K. L. Wong, V. S. Chan, and C. C. Petty

ELECTRON BERNSTEIN WAVE HEATING AND CURRENT DRIVE

Electron Bernstein Wave Studies on COMPASS-D and MAST *(Invited)* 359
    V. Shevchenko, E. Arends, Y. Baranov, M. O'Brien,
    G. Cunningham, M. Gryaznevich, B. Lloyd, A. Sykes,
    F. Volpe, A. D. Piliya, A. N. Saveliev, and E. N. Tregubova

Electron Bernstein Wave Experiments in the MST Reversed Field Pinch 367
    J. K. Anderson, M. Cengher, P. K. Chattopadhyay, C. B. Forest,
    M. Carter, R. W. Harvey, R. I. Pinsker, and A. P. Smirnov

Electron Bernstein Wave Applications for Stellarators 372
    M. D. Carter, D. A. Rasmussen, G. L. Bell, T. S. Bigelow, and J. B. Wilgen

General Properties of Scattering Matrix for Mode Conversion Process between B Waves and External EM Waves and Their Consequence to Experiments 376
    T. Maekawa, H. Tanaka, M. Uchida, and H. Igami

Study of Electron Bernstein Wave Absorption Using VORPAL 380
    C. Nieter, J. R. Cary, R. W. Harvey, R. Dominguez, and A. P. Smirnov

Modeling of EBW Coupling with Waveguide Launchers for NSTX 384
    R. I. Pinsker, G. Taylor, and P. C. Efthimion

ECE in MAST: Theory and Experiment 388
    J. Preinhaelter, V. Shevchenko, M. Valovic, P. Pavlo, L. Vahala, G. Vahala,
    and the MAST Team

Relativistic Effects in Heating and Current Drive by Electron Bernstein Waves 392
    A. K. Ram, J. Decker, A. Bers, R. A. Cairns, and C. N. Lashmore-Davies

Electron Bernstein Wave Research on NSTX and CDX-U 396
    G. Taylor, P. C. Efthimion, B. Jones, G. L. Bell, A. Bers, T. S. Bigelow,
    M. D. Carter, R. W. Harvey, A. K. Ram, D. A. Rasmussen, A. P. Smirnov,
    J. B. Wilgen, and J. R. Wilson

RF SOURCES AND SYSTEMS

Archimedes Plasma Mass Filter *(Invited)* 403
    R. Freeman, S. Agnew, F. Anderegg, B. Cluggish, J. Gilleland, R. Isler,
    A. Litvak, R. Miller, R. O'Neill, T. Ohkawa, S. Pronko, S. Putvinski,
    L. Sevier, A. Sibley, K. Umstadter, T. Wade, and D. Winslow

Radio-Frequency Sustainment of Laser Initiated, High-Pressure Air Constituent Plasmas 411
    K. Akhtar, J. E. Scharer, S. M. Tysk, and M. Denning

RF Frequency Oscillations in the Early Stages of Vacuum Arc Collapse 415
    S. T. Griffin and Y. C. F. Thio

PSpice Model of Lightning Strike to a Steel Reinforced Structure 419
    N. Koone and B. Condren

Investigation of a Light Gas Helicon Plasma Source for the VASIMR Space Propulsion System 423
    J. P. Squire, F. R. Chang-Diaz, V. T. Jacobson, T. W. Glover, F. W. Baity,
    M. D. Carter, R. H. Goulding, R. D. Bengtson, and E. A. Bering, III

Experimental Measurements and Modeling of a Helicon Plasma Source with Large Axial Density Gradients 427
    S. M. Tysk, C. M. Denning, J. E. Scharer, B. O. White, and M. K. Akhtar

RF-Plasma Coupling Schemes for the SNS Ion Source 431
    R. F. Welton, M. P. Stockli, S. Shukla, and Y. Kang

THEORY AND MODELING

Effects of Non-Maxwellian Plasma Species on ICRF Propagation and Absorption in Toroidal Magnetic Confinement Devices *(Invited)* 439
    R. J. Dumont, C. K. Phillips, and D. N. Smithe

ECCD for Advanced Tokamak Operations Fisch-Boozer versus Ohkawa Methods *(Invited)* 447
    J. Decker

Towards Predictive Scenario Simulations Combining LH, ICRH, and ECRH Heating 455
    V. Basiuk, J. F. Artaud, A. Bécoulet, L. G. Eriksson, G. T. Hoang,
    G. Huysmans, F. Imbeaux, X. Litaudon, D. Mazon, C. Passeron, and
    Y. Peysson

Nonlinear Interaction between RF-Heated High-Energy Ions and MHD-Modes 459
    T. Bergkvist, T. Hellsten, T. Johnson, and M. Laxåback

Local Mode Analysis of 2D ICRF Wave Solutions 463
    D. A. D'Ippolito, J. R. Myra, E. F. Jaeger, L. A. Berry, and D. B. Batchelor

The Effects of Combined Parallel Gradient and Collisional Decorrelation in the Absorption of RF Waves 467
    B. Goode, J. R. Cary, and L. A. Berry

Calculation of Fokker-Planck Ion Distributions Resulting from ICRF
Full-Wave Code Fields and Collisions 471
    R. W. Harvey, N. Ershov, A. P. Smirnov, P. Bonoli, J. C. Wright,
    F. Jaeger, D. B. Batchelor, L. A. Berry, E. D'Azevedo,
    M. D. Carter, and D. N. Smithe

Mode Conversion Flow Drive in Tokamaks 475
    E. F. Jaeger, L. A. Berry, J. R. Myra, D. B. Batchelor, and E. D'Azevedo

Modelling of ICRH Induced Current and Rotation 479
    T. Johnson, T. Hellsten, L.-G. Eriksson, M. Laxåback, and T. Bergkvist

Local Solutions for Generic Multidimensional Resonant
Wave Conversion 483
    E. R. Tracy and A. N. Kaufman

Momentum Conservation and Nonlinear RF-Induced Flows 487
    J. R. Myra, D. A. D'Ippolito, L. A. Berry, E. F. Jaeger, and D. B. Batchelor

Hybrid Method for RF Heating Problem of Mirror
Confinement Machine 491
    B. H. Park, S. S. Kim, J. H. Yeom, K. I. You, J. Y. Kim, M. Kwon,
    and the HANBIT Team

Advanced 3-D Electron Fokker-Planck Transport Calculations 495
    Y. Peysson, J. Decker, and R. W. Harvey

Plasma Dielectric Tensor for Non-Maxwellian Distributions in the
FLR Limit 499
    C. K. Phillips, A. Pletzer, R. J. Dumont, and D. N. Smithe

Gabor Wave Packet Method to Solve Plasma Wave Equations 503
    A. Pletzer, C. K. Phillips, and D. N. Smithe

O-X Mode Conversion in an Axisymmetric Plasma at Electron
Cyclotron Frequencies 507
    K. Imre and H. Weitzner

Ultrahigh Resolution Simulations of Mode Converted ICRF and LH
Waves with a Spectral Full Wave Code 511
    J. C. Wright, P. T. Bonoli, E. D'Azevedo, and M. Brambilla

Author Index 515

PREFACE

The Fifteenth Topical Conference on Radio Frequency Power in Plasmas was held on May 19-21, 2003 in the Grand Teton National Park in the Jackson Lake Lodge Resort in Moran, Wyoming under the sponsorship of the University of Wisconsin, Madison, and the American Physical Society. The meeting was also endorsed by the Division of Plasma Physics of the American Physical Society. The Conference was chaired by Prof. Cary Forest of the University of Wisconsin. The Program Committee included Drs. Stefano Bernabei, Mark Carter, John DeGrassie, Fritz Leuterer, Yves Peysson, Bob Pinsker, Yuichi Takase and Steve Wukitch. Drs. Jean-Marie Noterdaeme and Cynthia Phillips also contributed to the meeting program.

A total of 110 papers were presented at the Conference, of which three were topical oral reviews, eighteen were invited oral papers, and the rest were poster presentations. As in past years, the presentations were predominantly on heating and current drive in fusion plasmas. In addition to the fusion talks, presentations were made in the areas of: waves in dusty plasmas, processing plasmas, plasma sources for accelerators, plasma thrusters, and nuclear waste remediation. A future goal for this conference might be to be more inclusive of the non-fusion areas; the scientific overlap between rf systems developed for fusion and those in other areas is strong. Of the 114 registered attendees, 69 were from the United States, 37 were from the European Union, 6 were from Japan, and 1 was from South Korea.

Papers were presented covering the full range of frequencies available for heating fusion plasmas: approximately 40 papers were presented on heating, flow drive, and current drive in the ion cyclotron range of frequencies; 14 papers in the lower hybrid range of frequencies; 25 papers in the electron cyclotron range of frequencies, 11 of which were in the emerging area of electron Bernstein waves. The major tokamak fusion experiments were well represented at the Conference, including JET, DIII-D, C-Mod, ASDEX Upgrade, JT-60U, FTU, and Tore Supra, covering the ICRF, LH, and ECRF ranges of frequencies.

We would like to acknowledge the efforts by Ms. Linda Jones in making this a very successful conference. Special thanks are due to Dr. John Goetz for assistance in assembling the Abstract Book.

The Sixteenth Topical Conference will be held in Utah in 2005, and chaired by Drs. Steve Wukitch and Paul Bonoli of MIT.

Cary Forest
Chairman

REVIEWS

Waves and Instabilities in Dusty Plasmas

Robert L. Merlino

*Department of Physics and Astronomy, The University of Iowa,
Iowa City, Iowa, 52242 USA*

Abstract. This review article is divided into five parts. The Introduction contains some examples of dusty plasmas in the Universe and in the laboratory. This is followed by a brief section describing the mechanisms by which dust particles acquire an electrical charge in plasmas. The next section provides theoretical background for understanding the effect of dust on waves and instabilities in dusty plasmas. A description of some of the experimental studies of waves in dusty plasmas is then given. A few concluding remarks are then made.

INTRODUCTION

Dust represents much of the solid matter in the Universe and often coexists with the ionized gas forming a dusty plasma. The need to understand dusty plasmas stems from the fact that the dust acquires an electrical charge, and is thus subject to electromagnetic as well as gravitational forces. Perhaps the most striking example of this was found in the Voyager 2 images of Saturn's rings. Fig. 1 is an image of Saturn's B ring showing radially elongated structures (spokes) [1]. The spokes, which

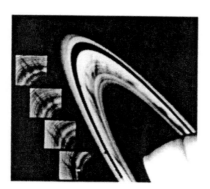

FIGURE 1. Voyager 2 image of spokes in Saturn's B ring. In insets show a sequence of images showing spoke formation. The time separations between the images are 10, 15, and 10 minutes, respectively.

are believed to contain micron size grains, appear quickly, extend radially for thousands of kilometers and disappear, all in a period of about half of a ring revolution around the planet. They are believed to be due to the levitation of the particles just above the rings by their interaction with Saturn's magnetic field or by electrostatic forces arising from ring-particle collisions [2]. The dusty plasma community is anxiously awaiting the arrival of higher resolution images of the rings when the Cassini-Huygens satellite encounters Saturn in July 2004.

Dust particles are often present in laboratory plasmas as well, either by choice or by circumstance. An example of a dusty plasma formed by circumstance is provided by the semiconductor manufacturing industry. Semiconductor processing facilities often use a parallel plate RF discharge as a plasma source for etching and deposition of silicon wafers. Typically, the plasma is formed by a mixture of the reactive gas silane (SiH_4) with argon and oxygen. In the resulting chemical reaction, SiO_2 particles (dust) are formed which grow in size over time. These dust particles become negatively charged in the plasma and become trapped in positive potential regions of the discharge. Eventually, some of the particles fall onto the wafer, contaminating it and rendering it useless. A photograph of dust clouds surrounding wafers in a typical processing device is shown in Fig. 2 [3]. As improvements to the cleanroom environments dramatically reduced external sources of wafer contamination, it became obvious that the largest source of contamination was due to the plasma process itself.

FIGURE 2. Photograph of rastered laser light scattered from dust particles formed during operation of a 200 mTorr Ar plasma processing tool. The dust particles are trapped in rings surrounding the Si wafers. The wafers in the foreground are 5.7 cm diameter and the one in the background is 8.3 cm.

Dust has also been found in fusion devices [4]. After periods of operation, particles ranging in size from fractions of millimeters to sub microns are found at the bottom of the vacuum vessel. The source of these particles is believed to be evaporation and sublimation of wall material which has been exposed to high thermal loads. Sputtering from particle impacts is also an important particulate source. Dust particles may also be generated by growth and agglomeration in the edge plasma regions. This process may be of particular concern for future long pulse devices. Some of the dust is ferromagnetic, so that it can easily be sucked into the main volume of the discharge vessel and accumulate on the inner wall where the magnetic field is strongest. From the plasma physics point of view, the smaller particles are of particular concern since they would be expected to remain at the edge of the plasma where they could alter the balance between the electrons and protons.

These are just a few examples of dusty plasmas. Other examples of dusty plasmas are discussed in a number of review articles that have appeared [2, 5, 6] as well as a recent monograph [7].

CHARGING OF DUST PARTICLES

The first consideration in dealing with dusty plasmas is to understand how the particles get charged. There are basically three processes that must be considered: (1) absorption of electrons and ions from the plasma, (2) secondary emission, and (3) UV induced photoelectron emission. Each process can be represented by a flux or current to the particle, and since the dust is electrically floating, the net current due to all processes must be zero, or

$$\sum I = I_e + I_i + I_{sec} + I_{pe} = 0 \qquad (1)$$

where the various terms represent the contributions from electrons, ions secondaries, and photoelectrons, respectively. Photoelectron emission is an important contribution for dust particles in a strong UV environment, but is typically not important in laboratory plasmas. Similarly, secondary emission is usually not important in low temperature plasmas. Under these conditions an isolated particle in a typical low temperature laboratory plasma will acquire a negative surface potential V_s relative to the plasma because of the higher mobility of the electrons. V_s is determined by the balance of electron and ion currents, $I_e + I_i = 0$. The electron current is given by

$$I_e = -4\pi a^2 e n_e \left(\frac{kT_e}{m_e}\right)^{1/2} \exp\left(\frac{eV_s}{kT_e}\right), \qquad (2)$$

where a is the radius of the particle, T_e is the electron temperature, n_e is the electron density and m_e is the electron mass. For the ion current to the particle, the orbital motion limited (OML) model is usually used:

$$I_i = 4\pi a^2 e n_i \left(\frac{kT_i}{m_i}\right)^{1/2} \left(1 - \frac{eV_s}{kT_i}\right) \qquad (3)$$

where T_i is the ion temperature, n_i is the ion density and m_i is the ion mass. The OLM model is valid in the limit where the particle radius is small compared to the Debye length. For the case of an isothermal hydrogen plasma, equations (1) – (3) give the particle's surface potential $V_s = -2.5kT_e/e$. The particle charge Q_d is determined from the capacitance model $Q_d = 4\pi\varepsilon_0 a V_s$. For a 1 micron diameter particle in a 2 eV hydrogen plasma the charge is then $Q_d = -1736$ e. This charging model applies strictly to the case of an isolated dust grain in a plasma has been verified experimentally [8]. In the case in which the dust grains are 'closely packed' so that the Debye sheaths of individual grains overlap, the particle charge is reduced [9].

LOW FREQUENCY WAVES IN DUSTY PLASMAS -THEORY

The presence of charged dust in a plasma influences the collective behavior in two ways. First, the excitation and propagation characteristics of the usual wave modes are modified even at frequencies at which the particles do not participate directly in the wave motion. This comes about through the effect on the plasma quasineutrality condition, which for negatively charged particles is

$$n_i = n_e + Z_d n_d \qquad (4)$$

where, $Z_d = Q_d/e$ and n_d is the dust particle density. Second, new very low frequency (few Hz) modes appear in which the dust particles participate in the wave dynamics. Since the dust particles can be imaged, the characteristics of these "dust modes" can be observed visually and recorded on tape.

The characteristics of waves in dusty plasmas have some similarities to waves in plasmas containing negative ions with important differences. The charge on a dusty plasma is not necessarily constant. Since the charge depends on the difference between the particle and plasma potential, the charge will fluctuate in response to fluctuations in the plasma potential. Another unique property of a dusty plasma is the fact that there may be a large distribution of dust sizes which results in a distribution of particle masses and charge. Theoretical work on dusty plasmas with broad size distributions is only in the preliminary stage.

A review of theoretical work on waves in dusty plasmas with particular emphasis on space plasmas was presented by Verheest [10].

The simplest approach to understanding wave phenomena in dusty plasmas is to apply the fluid equations, treating the dust as a third fluid component. This approach was used by D'Angelo in 1990 [11] to obtain a dispersion relation for low-frequency electrostatic waves in a magnetized dusty plasma (see also Merlino *et al.* [12]). Two types of low-frequency electrostatic modes are possible: acoustic modes which propagate along the magnetic field (perpendicular wavenumber $K_\perp = 0$) and cyclotron modes which propagate at a large angle to the magnetic field with a perpendicular wavenumber $K_\perp \gg K_\parallel$. Each type of mode is further subdivided into modes in which either the dust only provides a static background of negative charge or modes in which the dust particles participate directly in the wave dynamics (dust modes).

Dust Ion-Acoustic (DIA) and Dust Acoustic (DA) Waves

The dust ion acoustic mode is the usual ion acoustic mode that is modified by the presence of negative dust that is considered as a static background. The frequency of this mode is too high for the dust particles to respond. The DIA dispersion relation is given by

$$\frac{\omega}{K_\parallel} = \left[\frac{kT_i}{m_i} + \frac{kT_e}{m_i(1-\varepsilon Z_d)}\right]^{1/2} \equiv C_{DIA} \qquad (5)$$

where the quantity $\varepsilon Z_d = (n_{do}/n_{io})Z_d$ is the fraction of negative charge in the plasma on the dust particles. For the case in which no dust is present $\varepsilon = 0$, and equation (5) reverts to the usual ion acoustic velocity. In the presence of dust the phase velocity increases with increasing εZ_d.

The dust acoustic mode [13] is a very low frequency acoustic mode involving longitudinal dust density fluctuations. For this mode the electron and ion inertia can be neglected. The dispersion relation is

$$\frac{\omega}{K_\parallel} = \left[\frac{kT_d}{m_d} + \varepsilon Z_d^2 \frac{kT_i}{m_d} \frac{1}{1+(T_i/T_e)(1-\varepsilon Z_d)}\right]^{1/2} \equiv C_{DA} \qquad (6)$$

where T_d is the temperature of the dust fluid. Since the DA velocity, C_{DA} depends on the dust mass, m_d, the frequency of these waves can be quite low.

Electrostatic Dust Ion-Cyclotron (EDIC) Waves and Electrostatic Dust-Cyclotron (EDC) Waves

The electrostatic dust ion-cyclotron mode is the ion-cyclotron mode with a frequency $\omega \sim \Omega_i$ (the ion-cyclotron frequency), that is modified by the presence of negatively charged dust. At this relatively high frequency, the dust is considered to be immobile and the dispersion relation is

$$\omega^2 = \Omega_i^2 + K_\perp^2 C_{DIA}^2 \qquad (7)$$

where C_{DIA} is the dust ion acoustic speed defined in equation (5). The frequency of this mode increases with increasing εZ_d.

The electrostatic dust-cyclotron mode occurs in a dusty plasma in which the dust grains are magnetized. It is a low frequency "dust mode" in which the dust particles participate in the wave motion. The EDC dispersion relation is

$$\omega^2 = \Omega_d^2 + K_\perp^2 C_{DA}^2 \qquad (8)$$

where Ω_d is the dust cyclotron frequency and C_{DA} is the dust acoustic velocity defined in equation (6).

Wave Excitation and Damping

Ion acoustic waves are subject to ion Landau damping which is particularly strong in plasmas with $T_e = T_i$. This occurs because the wave phase velocity is close to the ion thermal velocity resulting in strong wave-particle resonance. In the presence of negatively charged dust however, the wave phase velocity increases as the fraction of negative charge on the dust increases, as seen from equation 5. With increasing phase velocity, the importance of Landau damping due to wave-particle interactions is reduced, so that propagation of ion acoustic waves becomes possible even in a plasma with equal ion and electron temperatures. This conclusion based on considerations of the dispersion relation derived from fluid theory, has also been confirmed from kinetic theory calculations which showed that the damping rate of the waves was drastically reduced when a sufficient amount of negatively charged dust was present [14].

The DA mode is one in which the wave inertia is provided by the heavy dust particles whereas the electrons and ion pressures provide the wave restoring force. This mode can be driven unstable by electrons and ions drifting with respect to the dust particles. This scenario may occur, for example, in planetary ring systems in which the plasma co-rotates with the planet, while the dust particles move in Keplerian orbits at a much slower speed. Calculations, based on the Vlasov equation show that relatively weak drifts, on the order of the dust acoustic speed (much less than the ion or electron thermal speed) are needed to excite the DA mode.

Electrostatic ion-cyclotron (EIC) waves can be excited in a plasma by an electron drift (relative to the ions) along the magnetic field. The critical electron drift velocity, v_{ec}, required to excite the EIC instability has been computed from kinetic theory. The results show that v_{ec} decreases as more negative charge is carried by the dust particles, thus the instability is easier to excite in a dusty plasma [15].

The kinetic instability of the EDC mode in the presence of streaming ions has also been investigated [7]. The calculations show that, as one might expect, the EDC waves would grow if the ion streaming velocity exceeded the parallel phase velocity, $\omega/K_\parallel$, where the mode frequency $\omega \cong \Omega_d$, the dust cyclotron frequency.

LOW FREQUENCY WAVES IN DUSTY PLASMAS – EXPERIMENTS

In this section we discuss some of the experimental work on waves in dusty plasmas. Two devices used to produce dusty plasmas, a Q machine and a discharge plasma are described. Experimental results on the dust ion acoustic (DIA) wave, electrostatic dust ion cyclotron (EDIC) wave and dust acoustic (DA) wave will be presented. At this point there have been no experimental observations of the EDC mode due to the difficulty in producing a dusty plasma in which the dust particles are magnetized. We conclude this section with a brief introduction to waves in strongly coupled dusty plasmas.

Dust Ion Acoustic Wave Experiment

The effect of negatively charged dust on the propagation of ion acoustic waves was studied in the dusty plasma device (DPD) shown schematically in Fig. 3. The DPD utilizes a single-ended Q machine as the plasma source and a rotating dust dispenser. The plasma is formed by contact ionization of potassium or cesium atoms from an atomic oven on a 6 cm diameter tantalum hot plate (~2200K) which also provides thermionic electrons. The plasma is confined by a uniform axial magnetic field with a strength up to 0.35 T. Typically, the electron and ion temperatures are $T_e = T_i \approx 0.2$ eV and the plasma density is in the range of $10^8 - 10^{10}$ cm^{-3}. Aluminum silicate powder in the micron size range, initially loaded into the bottom of the dust cylinder, is dispersed into the plasma by rotating the dispenser around the plasma column. As the dust falls through the plasma it acquires a negative charge. Further details of the operation of the DPD can be found elsewhere [16]. The fraction of negative charge on the dust, εZ_d, is determined from Langmuir probe measurements of the reduction in the electron saturation current that occurs when the dust is present as compared to the case with no dust [9]. The dust density can be controlled to some extent by varying the rotation speed of the cylinder.

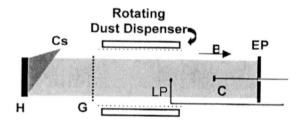

FIGURE 3. Schematic diagram of the dusty plasma device (DPD). The rotating dust dispenser is used to disperse dust into the plasma.

Ion acoustic waves were launched into the plasma by means of a grid (G) that was located ~ 3 cm in front of the dust dispenser and oriented perpendicular to the magnetic field. The grid was biased at several volts negative with respect to the plasma potential and a sinusoidal tone burst of frequency ~ 20 – 80 kHz and 4 – 5 V peak-to-peak amplitude was applied to it. This produced a density perturbation that traveled down the plasma column as an ion acoustic wave. Using an axially movable Langmuir probe the phase velocity ($v = \omega/K_r$), wavelength (λ), and spatial attenuation length ($\delta =$

$2\pi/K_i$) were measured as a function of εZ_d (K_r and K_i are the real and imaginary parts of the wavenumber). The results are shown in Fig. 4. The phase velocity increases with increasing εZ_d while the spatial damping length decreases with increasing εZ_d. In the absence of dust ($\varepsilon Z_d = 0$) ion acoustic waves are heavily damped and do not propagate more than one wavelength. However, when the negative dust is present, the phase velocity is increased so that Landau damping is significantly reduced.

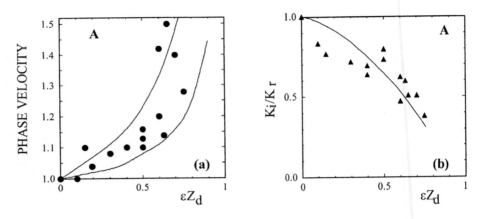

FIGURE 4. Properties of grid-launched ion acoustic waves in a dusty plasma. (a) Phase velocity versus εZ_d, normalized to the value with no dust present. The solid lines are theoretical predictions from fluid theory for two assumed values of the plasma drift speed. (b) Ratio of the inverse wave damping length to wavenumber versus εZ_d. All values are normalized by the value with no dust present. The solid curve is the prediction of kinetic theory.

Electrostatic Dust Ion-Cyclotron Wave Experiment

Unlike the IA waves which were launched into the plasma, the EIC waves were excited by driving an electron current along the magnetic field of the Q machine. The experimental setup was identical to that shown in Fig. 3, except that the grid was removed. A 5 mm diameter collector (C) was located near the end of the dust dispenser and was biased at ~ 0.5 – 1 V above the space potential to draw electron current. In a Q machine, this will produce an electron drift that is sufficient to excite EIC waves with a frequency slightly above the ion-cyclotron frequency which propagate nearly perpendicular to the magnetic field. The effect of the negative dust was studied by measuring the wave amplitude A_{nd} with no dust present and then turning the dust dispenser on an measuring the wave amplitude A_d with the dust present. The ratio A_d/A_{nd} is then an indication of the effect of the dust. This measurement was repeated for several values of εZ_d and the results are shown in Fig. 5. These results show that it becomes increasingly easier to excite EIC waves as more of the negative charge is carried by the dust. This result is in line with the theoretical calculation of Chow and

Rosenberg [15] that the presence of negative dust reduces the critical electron drift for excitation of the EDIC mode.

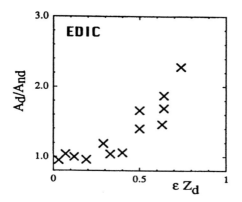

FIGURE 5. The electrostatic dust ion cyclotron wave amplitude as a function of εZ_d. The amplitudes are normalized to the EIC amplitude when no dust is present.

Dust Acoustic Wave Experiment

Observation of the dust acoustic (DA) wave requires that the dust particles be trapped within the plasma for a sufficient time. The dusty plasma device shown in Fig. 3 does not meet this requirement since the dust is continuously falling through the plasma column. We found that a simple DC glow discharge was ideal for this purpose, since the space potential within a DC glow discharge is positive with respect to the walls of the device. As a result, the plasma contains regions in which there are electric fields of appropriate strength and direction to levitate negative particles. A schematic of the DC glow discharge device used to observe DA waves is shown in Fig. 6. The

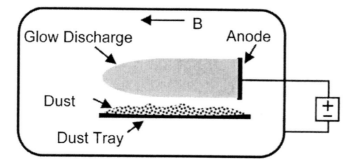

FIGURE 6. Schematic of the DC glow discharge device used to trap negative dust particles. Dust initially located on a tray below the anode is attracted into the plasma and levitated by the electric field associated with the glow.

glow discharge is formed in nitrogen at a pressure of ~ 100 mTorr by applying a positive potential of 200 – 300 V to a 3 cm diameter anode disk located in a large vacuum chamber. A longitudinal magnetic field of 0.01T is applied to provide radial confinement, which results in the formation of a cylindrical glow discharge. Aluminum silicate powder on a tray below the anode is attracted into the discharge and trapped in the positive potential region. The dust particles in the plasma are illuminated from behind by a high intensity lamp and imaged with a video camera.

Dust acoustic waves are excited spontaneously in the plasma, probably due to an ion-dust streaming instability. The waves appear as vertically elongated regions of enhanced intensity scattered light which propagate in the horizontal direction away from the anode. A single frame image of a DA wave captured on video tape is shown in Fig. 7. From analysis of the single frame images of this type, the wavelength and

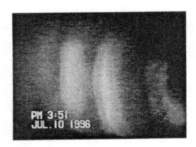

FIGURE 7. Single frame video image of a dust acoustic wave excited spontaneously in a DC glow discharge. The brighter regions are the wave crests which correspond to dust density enhancements.

phase velocity can be measured. The DA wave dispersion relation was obtained by applying (in addition to the DC bias) a sinusoidal modulation signal to the anode in the frequency range of 5 – 30 Hz to fix the DA wave frequency. For each applied modulation frequency, a video recording of the waves was obtained from which the wavelength was measured. The resulting dispersion relation, K versus ω is shown in Fig. 8.

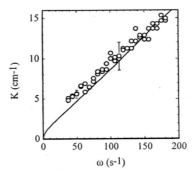

FIGURE 8. Measured dust acoustic wave dispersion relation. The wave phase velocity is 12 cm/s. The solid line is the prediction of fluid theory.

The measured dispersion relation was compared with one obtained from fluid theory (Eq. 6) modified to include the effect of collisions between the dust particles and the neutral atoms. Over this range of frequencies K ∝ ω, as expected for longitudinal compressional waves.

Waves in Strongly Coupled Dusty Plasmas

In ordinary electron/ion plasmas the ratio of the interparticle potential energy to thermal kinetic energy,

$$\Gamma = \frac{(eZ)^2}{4\pi\varepsilon_o \, d \, kT} \qquad (9)$$

where d is the interparticle spacing, is typically << 1. However, in a dusty plasma, Z can be on the order of thousands, and the so-called coupling parameter Γ can be much larger than one. In this case the dust particles actually arrange themselves in a regular lattice array [17]. The presence of short scale correlations gives rise to novel modifications of the collective behavior in these plasmas. Unlike ordinary plasmas which exist in the fluid state, these liquid and solid-like plasmas can accommodate both compressional waves and shear waves [18,19].

CONCLUDING REMARKS

Dusty plasmas are ubiquitous in the Universe, and it is now realized that they are not that uncommon in the lab as well. Methods have now been devised to create dusty plasmas for experimental study. This review has concentrated on how the presence of charged dust modifies the excitation and propagation characteristics of a number of well-known electrostatic waves in plasmas. In the presence of negatively charged dust, both ion-acoustic and ion-cyclotron waves are more easily excited. This result is well understood on the basis of fluid and kinetic theories, and has been verified by laboratory experiments. New waves, which involve in an essential manner the dynamics of the dust particles themselves, have also been investigated. Visual observations of the so-called dust acoustic wave have been presented here. Finally, we note that the presence of collective and coherent fluctuations in the density of a dust cloud may provide a mechanism for structuring on small scales.

ACKNOWLEDGMENTS

The work on dusty plasmas at the University of Iowa was supported by the Office of Naval Research and the National Science Foundation.

REFERENCES

1. Smith, B. A., Soderblom, L., Beebe, R., Boyce, J., Briggs, G., Bunker, A., Collins, S. A., Hansen, C. J., Johnson, T. V., Mitchell, J. L., Terrile, R J., Carr, M., Cook, A. J., Cuzzi, J., Pollack, J. B., Danielson, G. E. , Ingersoll, A., Davies, M. E., Hunt, G. E., Masursky, H., Shoemaker, E., Morrison, D., Owen, T., Sagan, C., Veverka, J., Strom, R., Suomi, V. E., *Science* **212**, 163-191 (1981).
2. Goertz, C. K., *Rev. Geophys.* **27**, 271-292 (1989).
3. Selwyn, G., Heidenreich, J. E., and Haller, K. L., *J. Vac. Aci. Technol.* **A 9,** 2817-2824 (1991).
4. Winter, J., *Plasma Phys. Control. Fusion,* **40**, 1201-1210 (1998).
5. Mendis, D. A., "Physics of Dusty Plasmas: Historical Overview," in *Advances in Dusty Plasmas*, edited by P. K. Shukla, D. A. Mendis and T. Desai, World Scientific, Singapore, 1997, pp. 3-19.
6. Mendis, D. A., *Plasma Sources Sci. Technol.* **11**, A219-A228 (2002).
7. Shukla, P. K., and Mamun A. A., *Introduction to Dusty Plasma Physics*, Institute of Physics, Bristol, 2002.
8. Walch, B., Horáni, M., and Robertson, S., *Phys. Rev. Lett.* **75**, 838-841 (1995).
9. Barkan, A., D'Angelo, N., and Merlino, R. L., *Phys. Rev. Lett.*, **73**, 3093-3096 (1994).
10. Verheest, F., *Space Science Rev.* **77**, 267-302 (1996).
11. D'Angelo, N., *Planet. Space Sci.,* **38**, 1143-1146 (1990).
12. Merlino, R. L., Barkan, A., Thompson, C., and D'Angelo, N., *Phys. Plasmas* **5**, 1607-1614 (1998).
13. Rao, N. N., Shukla, P. K., and Yu, M. Y., *Planet. Space. Sci.* **38**, 543-546 (1990).
14. Rosenberg, M., *Planet. Space. Sci.* **41**, 229-233 (1993).
15. Chow, V. W., and Rosenberg, M., *Planet. Space Sci.* **43**, 613-618 (1995).
16. Xu, W. Song, B., Merlino, R. L., and D'Angelo, N., *Rev. Sci. Instrum.* **63**, 5266-5269 (1992).
17. Ikezi, H., *Phys. Fluids* **29**, 1764-1766 (1986).
18. Nunomura, N., Samsonov, D., and Goree, J., *Phys. Rev. Lett.* 84, 5141-5144 (2000).
19. Piel, A., and Melzer, A., *Plasma Phys. Control. Fusion* **44**, R1-R26 (2002).

Electron Bernstein wave heating in fusion plasmas

H.P. Laqua, W7-AS Team, ECRH-Group*

*Max-Planck-Institut für Plasmaphysik, EURATOM Ass.
D-174898 Greifswald, Germany*
*Institut für Plasmaforschung, Univ. Stuttgart, D-70569 Stuttgart, Germany

Introduction

In the famous article of Ira. B. Bernstein [1] electrostatic waves propagating perpendicular to magnetic field and without Landau damping were postulated. The experimental verification of the so-called electron Bernstein waves (EBW) had been demonstrated by Crawford et al. [2] in 1964. Later on many Experiments had been performed to measure the transmission and emission of EBW in linear low temperature plasmas devices [3, 4]. EBW´s have no density limit for propagation and they are strongly damped by cyclotron absorption even for higher harmonic resonances. This makes them an attractive candidate for high-density fusion plasma heating.

Generation of EBW´s

EBW´s are eigenmodes of the hot magnetised plasma, therefor they cannot be launched from directly from the vacuum, but have to be either generated by mode conversion from the electromagnetic waves or exited by in-plasma antenna structures. The last is not applicable for fusion devices.

In the **high field launch scheme**, which is only applicable for the first harmonic heating, an X-wave is launched from the high field side of a toroidal plasma, passing the electron cyclotron (EC-) resonance. At the upper hybrid resonance (UHR) the wave is converted in EBW´s, which propagate back to the EC-resonance, where they are absorbed (see Fig. 1). This scheme was demonstrated in Tokamaks [5, 6, 7] and Stellarators [8, 9]. Unfortunately, this scheme is not applicable for high-density plasma heating, since the propagation of the X-waves is limited by the L-cut-off density.

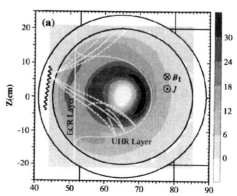

Figure 1. Tomographic reconstruction of Soft x-ray emission during ECRH. The white lines represent the calculated ray-trajectories.

For overdense plasma heating the **FX-SX-B scheme** was proposed. Here a so-called fast X-wave is launched from the low field side as shown in Fig. 2. If the density scale length is smaller than the wavelength, the X-mode can be transmitted through the evanescent region between the low-density cut-off (R-cut-off) and the UHR. Here a slow X-wave is excited, which is reflected by the high-density cut-off (L-cut-off) towards the UHR. If the distance between the cut-off and the UHR is well adjusted EBW´s can be generated with a high efficiency as demonstrated in [10]. Unfortunately this scheme is very sensitive on density fluctuation.

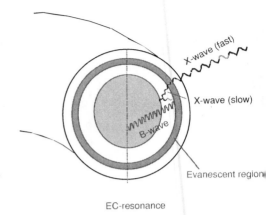

Figure 2. Schematic picture of FX-SX-B conversion.

Another possibility to excite EBW´s in overdense plasmas is the **OXB-scheme**, which was proposed in [11]. For an optimal parallel component of the refractive index $N_{z,opt} = \sqrt{Y/(Y+1)}$ the O-mode and the slow X-mode coincide at the O-mode cut-off as shown in Fig. 3.

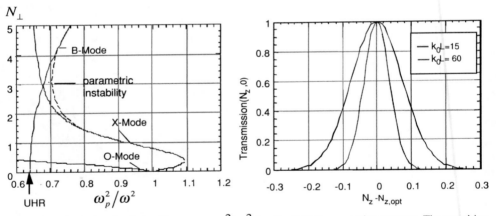

Figure 3. Left: Refractive index N_x versus ω_p^2/ω^2 for the O-X-B conversion process. The transition represents the connection of the X-mode and B-mode due to the hot dielectric tensor.
Right: Transmission through the O-cut-off as a function of the N_z component.

Here Y is the ratio of ω_{ce}/ω. This means that for an optimal launch angle, similar to the well-known Brewster angle in optics, a wave can be transmitted through cut-off surface without reflections. Even for non-optimal launch some power can be transferred through the cut-off, if the density scale length is short enough. The transmission for the angular window is given in [12]

$$T(N_y, N_z) = \exp\left\{-\pi k_0 L \sqrt{\frac{Y}{2}} \left[2(1+Y)(N_{zopt} - N_z)^2 + N_y^2\right]\right\},$$

where N_y and N_z are the poloidal and longitudinal components of the vacuum refractive index and k_0 the wave number and is calculated for two values of $k_0 L$ in Fig. 3. It is clear that for a steep density profile with small-scale length L the angular or N_z-window is broad, while for a flat profile it is narrow. Once the slow X-wave is generated it can only propagate up to the L-cut-off, where it reflected back to the UHR. There the X-waves couple to the EBW´s, which propagate into the overdense plasma, where they are absorbed by cyclotron damping.

Propagation of EBW´s

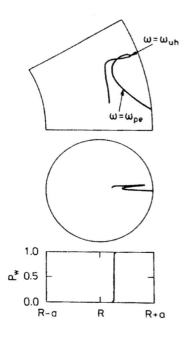

FIGURE 4. Ray trajectories in the toroidal (top) and poloidal (center) cross-section. At the bottom the power is plotted over the large plasma radius

The propagation of EBW´s has been calculated by ray-tracing methods. Maekawa et al. [13] showed for a slab geometry that for fusion plasma parameters the propagation and absorption of EBW´s is favourable. In further publications the propagation was calculated for the axial symmetric geometry of tokamaks [14][15] and finally for the full 3D-geometry of stellarators [9][16]. As an example in Fig. 4 ray calculation for the OXB-heating at the Princeton PLT tokamak is shown. Here already the EBW current drive (EBCD) was calculated.

Experiments with EBW´s

The first EBW experiments had been performed in small laboratory plasma devices. Here the dispersion relation could be experimentally verified [2,3,4].

In fusion plasmas, the generation of EBW by high field launch was demonstrated first. In experiments in tokamaks [5, 6] and in Stellarators [8] the cyclotron absorption and the parametric decay was measured.

The OXB- conversion was experimentally investigated by Sugai et al [18] in a low temperature linear plasma first. Although here the angular window could not be demonstrated, EBW-generation above the O-mode cut-off density could be shown.

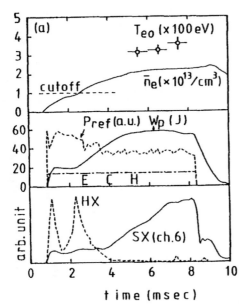

Figure 5. From the top: density, diamagnetic energy and x-ray emission during the discharge.

At the Heliotron DR device overdense plasmas could be achieved with 28 GHz [19], even for higher harmonic resonances. Although the wave was launched with x-polarisation and perpendicular to the magnetic field, there was evidence that the EBW's were generated by the OXB-process as shown in Fig (5). Especially the density threshold at the O-mode cut-off was an hint for OX-conversion. At Heliotron DR EBW heating up to the 5th harmonic resonance was demonstrated.

At the Wendelstein7 AS stellarator the first OXB heating experiments have been performed with 70 GHz at the first harmonic resonance [20]. Here with a movable antenna and a pure O-mode polarisation the angular dependence of the OXB mode conversion could be demonstrated (see Fig 6). In addition, the parametric decay at the XB-conversion could be measured (Fig. 6) and finally the propagation and resonant absorption of the EBW's was found (Fig 7).

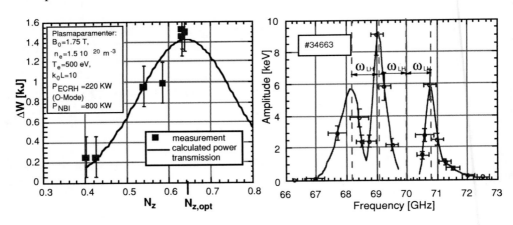

Figure 6. left: Increase of the plasma energy content by O-X-B-heating versus the longitudinal vacuum refractive index N_z of the incident O-wave.

Right: High frequency spectrum of the parametric decay waves generated in the O-X-B-process. The incident wave frequency is 70 GHz and the LH frequency is about 900 MHz.

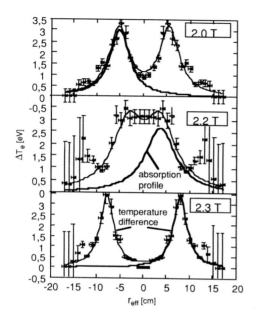

Figure 7. Changes of temperature 3 ms after O-X-B heating switch-off and the related ECRH absorption profiles different central magnetic fields.

Further, because of the symmetry of the mode conversion process, the measurement of the EBW emission (EBE) gives important information on the conversion efficiency. In addition EBE can also been used as a fast local temperature diagnostic like electron cyclotron emission measurement (ECE) below the critical density [21]. At Wendelstein7-AS up to 90 % conversion efficiency was found by this method.

The so-called high density H-mode [22] at the Wendelstein7-AS has promising properties for OXB heating with 140 GHz. The central density exceeds $3.5 \ 10^{20} \ m^{-3}$ and the density scale length at the ECRH launch position is 0.5 cm. Three 140 GHz beams with a total power of 1.5 MW were launched into a NBI sustained (up to 4 MW) HDH-plasma. The ECRH launch was optimised in respect to the angular window necessary for efficient EBW generation. Here the optimisation criterion was the increase of plasma energy shown in Fig. 8 and the reduction of the non-absorbed ECRH stray radiation [23].

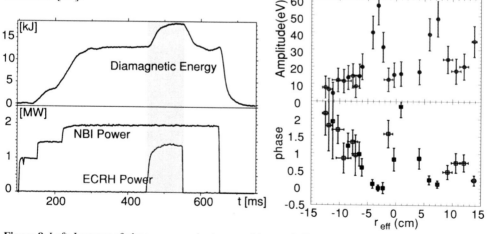

Figure 8. Left: Increase of plasma energy due to second harmonic Bernstein wave heating (B2) with 140 GHz. Right: Heat wave amplitude and phase generated by off-axis 140 GHz B2-heating and reconstructed from first harmonic EBE. The reason for the apparent asymmetry of the power deposition is due the strong Shafranov shift of the plasma, which could no be taken into account completely in the EBE temperature profile reconstruction.

Further on, a magnetic field scan was performed to achieve central power deposition. The deposition profile was estimated by power modulation and coherent heat wave detection of EBE [16]. For the 2nd harmonic we have got a power deposition at an effective radius of about 4 cm as shown in Fig. 8. The EBW-heating increased the power flow across the separatrix, which initiated a transition from detachment to attachment. Therefore the increase of plasma energy is not an adequate gauge for heating efficiency. It is more realistic to compare EBW heating with NBI. For this two NBI beams with 0.5 MW power each were replaced by about 1.1 MW ECRH. This did not degrade the plasma performance, thus demonstrating that the EBW heating efficiency was comparable with NBI. Even more, EBW heating becomes more effective with increasing density.

One key issue for commercial fusion reactors is the stability at high beta values. High beta experiments are preferably performed at low the magnetic field. For a fixed ECRH-frequency (140 GHz) this requires heating at a higher harmonic resonance. The plasma is usually optically thick for even higher harmonic EBW´s. Nevertheless the density has to expire the threshold for OXB-conversion. The accessibility of the plasma core can further be restricted by the appearance of the next resonance at the plasma edge. Therefore the experiments have mainly concentrated on the third and fourth harmonic heating at a magnetic field of 1.5-1.6T and 1.0-1.2T respectively. At least 3 MW of NBI heating power was necessary to sustain the OXB-threshold density at that low magnetic field. Due to the confinement degradation with heating power ($\sim P^{-0.6}$) only small effects on the plasma parameters could be expected by additional EBW-heating 1.1 MW power. Nevertheless in a magnetic field clear resonance effects could be found. The largest increase of the average plasma beta was found at 1.1 T. The power was mainly deposited at half the plasma radius. Central power deposition was not possible due to the appearance the next harmonic resonance at the plasma edge. This was confirmed by the tomographic reconstruction of the change of the SX-emission at the ECRH switch-off shown in Fig.9.

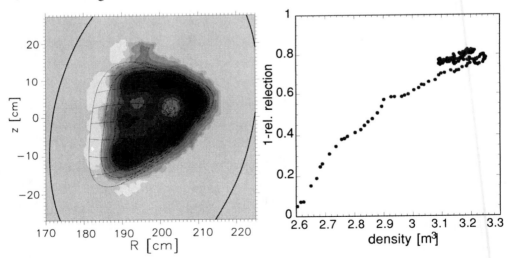

Figure 9. Left: Change of the SX-emission (tomographic reconstruction) after ECH switches off. The grey scale is linear and in arbitrary units. Right: Total coupling efficiency for the 140 GHz waves, deduced from the stray radiation level in the torus, versus the central density.

With the ECRH stray radiation diagnostic the total efficiency of OX- coupling could be estimated. An example is shown for the third harmonic EBW-heating in Fig 9. The maximum stray radiation was found near the cut-off density. Assuming that no power is coupled to the plasma in that case, a reduction of the stray radiation down to 20% with increasing density indicates that 80% of the ECRH radiation has been absorbed in the plasma. Of course some part of the ECRH power may be absorbed at the plasma edge and did not contribute much to the increase of plasma energy

The FX-SX-B scheme was investigated at the high beta spherical Tokamak devices Mast, NSTX and CDX-U and on the reversed field pinch MST. Reviews of the results are given in [10, 24]. In a sophisticated experiment on CDX-U [10], it was demonstrated that X-B conversion efficiency of up to 100% are attainable by a careful adjustment of the density gradient in front of the antenna as shown in Fig. 10. This was performed by movable local limiters. On the other hand the EBE measurements showed that the conversion can be significantly degraded by density fluctuation.

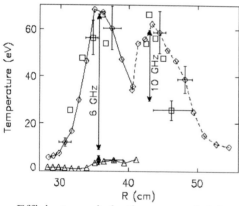

FIGURE 10. Fundamental EBW T_{rad} with the local limiter near the LCFS (diamonds, solid line) and with no local limiter near the LCFS (triangle, solid line). Second harmonic EBW T_{rad} with the local limiter near the LCFS is also shown (diamonds, dashed line). EBW T_{rad} with limiter near the LCFS is similar to Te from Thomson scattering (shown by squares).

Efficient non-inductive **current drive** (CD) may allow steady state operation of Tokamaks as a fusion reactor. Moreover, for high-density operation and in spherical Tokamaks like NSTX the plasma density can exceed the accessible plasma density for electron cyclotron current drive (ECCD) with electromagnetic waves. For the electrostatic EBW no upper limit exists. Even more, due to their electrostatic character EBW's can achieve parallel refractive indices ($N_\parallel$) larger than 1. This makes the EBW's an attractive candidate for efficient current drive as postulated by in [25].

At the COMPASS D Tokamak EBCD experiments have been performed with 600 kW ECRH at 60 GHz [7]. The waves were launched from the high field side in a low-density plasma of $1.8 \; 10^{19}$ m^{-3} at a temperature of 3.5 keV. Up to 100 kA EBCD current was estimated from the loop voltage change as shown in Fig 11. Remarkably, the direction of the current was mainly determined by the change of $N_\parallel$ during EBW propagation and not as usual for ECCD by the launch angle.

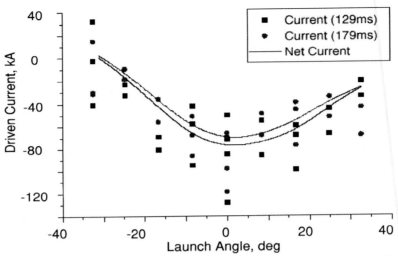

Figure 11. Non-inductive current driven in the plasma estimated from experimental data and the net current rom ray-tracing simulations.

At Wendelstein7-AS EBCD experiment have been performed with 400 kW ECRH at 70 GHz [24]. The EBW were generated by the OXB-process with a fixed optimal launch angle.

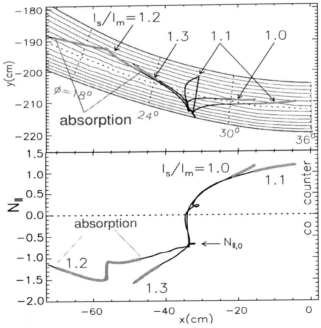

Figure 12. Top: Ray trajectories (black lines) in the equatorial plane for different "mirror ratios" I_s/I_m. Φ is the toroidal angle, x and y are the spatial co-ordinates. The dashed line represents the plasma center, the grey lines are the cuts through the flux surfaces. The thick grey parts of the ray traces indicate the region where 99% of the power is lost by cyclotron absorption. Bottom: Evolution of $N_\parallel$ for different ray trajectories according to the left figure.

The over-dense plasma was sustained by up to 1 MW NBI at a density $1.04 \cdot 10^{19}$ m^{-3} and a temperature of up to 0.8 keV. The co- and counter-EBCD could be achieved by changing the magnetic configuration at the launch position as shown in Fig 12. In both direction up to 2 kA of EBCD could be achieved. This is in consistency with the COMPASS-D results taking into account that CD efficiency scales like T_e/n_e and that in a Stellarator the trapped particle fraction is larger than in Tokamaks.

Conclusion

A large progress has been done in the application of EBW´s for fusion plasma heating, current drive and diagnostic. All applications for EC-waves below the cut-off density have been demonstrated with EBW´s in over-dense plasmas. Even more with higher harmonic heating high beta values are achievable now. This is essential for the new generation of high beta Tokamaks and Stellarators.

References

1. Bernstein, Ira B., Phys. Rev.,Vol. 109, p. 10 (1958)
2. Crawford, F. W., et al., PRL Vol.13, No. 7, 229 (1964)
3. Landauer G., internal report of the Max-Planck-Institut für Plasmaphysik, IPP-report 2/53 (1966).
4. Leuterer, F., internal report of the Max-Planck-Institut für Plasmaphysik, IPP-report 3/102 (1969).
5. McDermott, F. S., et al. Phys.Fluids 25, 1488, 1982
6. Maekawa, T., et al., PRL Vol.86, No. 17, 3783 (2001)
7. Shevchenko, V., et al., Phys. Rev. Lett., Vol. 89, 265005 (2003)
8. Wilhelm, R., et al. PPCF, Vol.26, No. 12A, pp. 1433-1444, 1984.
9. Nagasaki, K., et al., in Proc. of the 12th joint Workshop on ECE and ECH 2002, p. 317
10. Taylor, G., et al. in Proc. of the 12th joint Workshop on ECE and ECH 2002, p. 151
11. Preinhaelter, J. and Kopecky,´, V., J. Plasma Phys. 10 (1973) 1;
12. Mjølhus, E., J. Plasma Phys. 31 (1984) 7;
13. Meakawa, T., Tanaka, S., Terumichi, Y. and Hamada, Y., Phys. Rev. Lett. 40 (1978) 1379;
14. Hansen, F. R., et al., Plasma Phys.Controll. Fusion 27 (1985) 1077;
16. Volpe, F., and Laqua, H. P., Rev. Sci. Instrum. 74. Issue 1 (Jan. 2003)
17. Forest, C. B., et al., Phys. Plasmas 7, p 1352 (200)
18. Sugai H., et al. J. Phys. Soc. Jpn. Vol. 58, No.11, pp.3779-3782 (1987)
19. Morimoto, S., Yanagi, N., Sato, M., et al.,Nucl.Fusion 29 (1989) 1697
20. Laqua, H. P., et al., PRL 78, 18 (1997);
21. Laqua, H. P., et al., PRL 81, 10 (1998);
22. McCormick, K., et al. PRL 89, 015001 (2002)
23. Laqua H. P.,et al. in Proceedings of the 28th EPS Conf. Control. Fusion and Plasma Phys., Funchal (Madeira) 2001, edited by C. Silva, C. Varandas and D. Campbell (European Physical Society, Lisbon, 2001), P3.099
24. Forest, C. B., et al., in Proc. of the 12th joint Workshop on ECE and ECH 2002, p. 17
25. Litvak, A.G., et al. Phys.Letters A. 188, 64 (1994)

Lower Hybrid Current Drive: An Overview of Simulation Models, Benchmarking with Experiment, and Predictions for Future Devices

P.T. Bonoli[a], E. Barbato[b], R.W. Harvey[c], and F. Imbeaux[d]

[a]*PSFC, MIT, Cambridge, MA 02139, USA*
[b]*C.R.E. ENEA,CP 65 0004, Frascati, Italy*
[c]*CompX, Del Mar, CA 92014, USA*
[d]*CEA, Cadarache 13108, France*

Abstract. This paper reviews the status of lower hybrid current drive (LHCD) simulation and modeling. We first discuss modules used for wave propagation, absorption, and current drive with particular emphasis placed on comparing exact numerical solutions of the Fokker Planck equation in 2-D with solution methods that employ 1-D and adjoint approaches. We also survey model predictions for LHCD in past and present experiments showing detailed comparisons between simulated and observed current drive efficiencies and hard X-ray profiles. Finally we discuss several model predictions for LH current profile control in proposed next step reactor options.

I. INTRODUCTION

Lower hybrid current drive (LHCD) has been the subject of intense theoretical [1], numerical [2], and experimental [3] investigations over the past twenty years. LHCD has been shown to be an effective technique for maintaining steady state plasmas where the current is sustained solely by non-inductive means and also for localized current generation in applications involving sawtooth stabilization and shear reversal control. Furthermore, LHCD is a promising technique for off – axis (r/a ≥ 0.65) current profile control in reactor grade plasmas. Lower hybrid waves have the property of damping efficiently at high parallel (to **B**) phase velocities ($v_{//}$) relative to the electron thermal speed, where $v_{//} \geq 2.5\, v_{te}$ and $v_{te} = (T_e / m_e)^{1/2}$. Consequently these waves are well-suited to driving current in the plasma periphery where the electron temperature is lower. Also owing to their relatively high $v_{//}$, deleterious effects of particle trapping and parasitic absorption on alpha particles are minimized. Finally the higher phase velocity also leads to a higher current drive efficiency [1].

We are thus motivated to review the status of simulation models used for LHCD, assess their reliability by comparing model predictions with experiment, and finally to present the predictions of these models for current profile control in future devices (reactors). The first part of this paper will review different model treatments of LH wave propagation, absorption, and current drive. Emphasis will be placed on three different approaches to solving the Fokker Planck equation that vary considerably: exact 2-D ($p_\perp$, $p_{//}$) and 3-D ($p_\perp$, $p_{//}$, r) numerical solutions, modified 1-D ($p_{//}$) treatments, and adjoint methods. In the second part of this review we compare model predictions with experimental observations of the LHCD figure of merit, with measurements of hard X-ray profiles under conditions of varying density and plasma current, and with current profile modification experiments. The conundrum of the spectral – gap in LHCD experiments [4] will be reviewed and possible insights into the gap problem offered by present day models will be discussed. In the third part of the review we will present model predictions for LH current profile control in next step option (NSO) devices such as the proposed ITER – FEAT [5]. The predictions of exact 2-D, modified 1-D, and adjoint models will be discussed and reasons for differences in the model predictions will be given. Finally, recommendations for future work will be given in terms of model improvements and comparisons with experiment.

II. OVERVIEW OF SIMULATION MODELS FOR LOWER HYBRID CURRENT DRIVE

Simulation models for LHCD typically consist of combined wave propagation and Fokker Planck packages that iterate between each other to obtain a self-consistent quasi-linear electron distribution function [4,6]. This iteration process is necessary to reproduce the significant distortion or "quasi-linear flattening" that occurs in LHCD experiments because of the presence of a broad spectrum of LH waves. The converged wave absorption and rf current density are then incorporated into time dependent transport codes which evolve the electron and ion temperature, poloidal field, and MHD equilibrium based on these LH source terms [7-10].

Wave Propagation, Coupling, and Absorption

The preferred method for treating wave propagation in the LH range of frequencies is to integrate the ray equations of geometrical optics in toroidal geometry [11-15]:

$$\frac{d\mathbf{x}}{dt} = -\frac{\partial \varepsilon / \partial \mathbf{k}}{\partial \varepsilon / \partial \omega} \qquad \frac{d\mathbf{k}}{dt} = +\frac{\partial \varepsilon / \partial \mathbf{x}}{\partial \varepsilon / \partial \omega} \quad (1a,b)$$

where $(\mathbf{x}, \mathbf{k})$ are respectively the position and wave-number of the LH wave and $\varepsilon (\mathbf{x},\mathbf{k},\omega)$ is the local dispersion relation. This method is particularly useful because 2-D equilibrium effects [$\mathbf{B}$ (r,θ)] on the parallel wave-number ($k_{//}$) evolution can be accounted for in a straightforward manner. Note that in toroidal geometry the poloidal mode number of the LH wave is not a conserved quantity and can vary considerably since $k_{//} = \mathbf{k} \cdot \mathbf{b} = [(m/r)B_\theta + (n/R)B_\phi] / |\mathbf{B}|$. The ray tracing approach does neglect full-wave coherence effects such as focusing and diffraction that have been shown to cause significant spectral broadening of the injected wave spectrum [16]. Electromagnetic field solvers in the LH frequency regime are currently being adapted to massively parallel computing platforms [17] which should make the computation of full-wave effects practical in the near future.

Ray tracing codes are used to reconstruct the quasi-linear diffusion coefficient (D_{rf}) by tabulating the local wave polarization and group velocity of a "pencil" of rays of finite cross-section, emitted from a waveguide at the plasma edge [4,6]. The group velocity relation $\Gamma_k = V_g U_k$ is then used to relate the flux of power through the ray tube (Γ_k) to the electric field strength ($|E_{rf}|^2$) of the LH wave which is then used to evaluate D_{rf} from the Kennel – Engelmann form[18] of the diffusion coefficient.

Fokker Planck Analysis

Formation of a LH tail is described most generally by a Fokker Planck equation of the form:

$$\frac{\partial}{\partial t}f_e = \frac{\partial}{\partial p_{//}}D_{rf}(p_{//})\frac{\partial f_e}{\partial p_{//}} + C(f_e, p_{//}, p_\perp) + eE_{//}\frac{\partial f_e}{\partial p_{//}} + \Gamma_s \delta(p_{//}) + \frac{1}{r}\frac{\partial}{\partial r}r\chi_F\frac{\partial f_e}{\partial r}, \quad (2)$$

where $p_\perp$ ($p_{//}$) are respectively the perpendicular (parallel) components of momentum, Γ_s is a source of particles at low velocity, $E_{//}$ is the DC electric field, and χ_F is the fast electron diffusivity. Although numerical solution of Eq.(2) is computationally onerous, a number of widely used codes[19-23] have been developed to accomplish this task including the CQL3D[20], BANDIT[21], and DELPHINE[22] models. All these codes can exercise the simplifying option of solving Eq. (2) in the absence of the radial diffusion operator. Once $f_e(p_\perp, p_{//}, r)$ is known, the driven LH current density is straightforward to compute from:

$$J_{rf}(r) = \int d^3 p (n_e v_{//}) f_e. \quad (3)$$

Numerical 2-D ($p_\perp$, $p_{//}$) solutions of Eq. (2) are necessary to capture the complicated perpendicular velocity space dynamics of LH current drive.[19,24,25] In particular these simulations reveal that a distribution of LH waves at parallel velocities of $v_1 \leq v_{//} \leq v_2$ can cause significant pitch angle scattering of electrons into the perpendicular direction, both above and below the region of rf waves. This results in an electron distribution function characterized by a large perpendicular temperature where $T_\perp / T_e \gg 1$ within, above, and below the region of waves. Two-dimensional treatments of Eq. (2) are also necessary to include other important effects such as momentum conserving corrections to the background collision operator $C(f_e, p_\perp, p_{//})$[26] and toroidal or particle trapping effects.[27]

In order to incorporate combined ray tracing and Fokker Planck modules into closed transport and MHD equilibrium loops where the modules are called many times, some LHCD packages[4,6,28,29] perform numerically fast 1-D ($p_{//}$) solutions of Eq. (2). This is done by either neglecting the spatial diffusion operator or replacing it with a loss term. Also f_e is assumed to be of the form:

$$f_e(p_\perp, p_{//}) = \frac{F_e(p_{//})}{2\pi m_e T_\perp}\exp\left(-\frac{p_\perp^2}{2m_e T_\perp}\right), \quad (4)$$

where $T_\perp = T_\perp(p_{//1}, p_{//2}, Z_{eff})$ is prescribed analytically following the work of Fuchs et al.[25] One Fokker Planck package described in Ref (28) (the "Baranov code") refines this approach by retaining the spatial diffusion operator and solving for $f_e(p_{//}, r)$. This approach is still computationally fast enough to allow the code to be run in time dependent transport simulations. Both the Baranov code[28] and the FRTC code[29,30] account for 2-D velocity space effects due to pitch angle scattering by changing the leading coefficient in $C(f_e, p_\perp, p_{//})$ from its 1-D value of $(2 + Z_{eff})/2$ to the 2-D result of $(5+Z_{eff})/10$.[24,25,28,30] These types of treatments however do not account for effects due momentum conserving corrections in the collision operator[26] and particle trapping.[27]

A more recent treatment of the Fokker Planck equation that is both computationally fast yet retains important 2-D effects involves reformulating Eq. 2 as an adjoint problem[26] and solving for the Spitzer – Harm function (χ). The rf current density is then computed by convolving χ with the wave-induced rf flux:

$$J_{rf} = \int d^3p \frac{\partial \chi}{\partial p} \cdot \Gamma_{rf} , \quad \Gamma_{rf} = -\mathbf{D}_{QL} \frac{\partial f_e}{\partial p_{||}} . \qquad 5(a,b)$$

This approach is rather powerful because effects due to toroidicity (trapping), DC electric field, pitch angle scattering, and momentum conserving corrections can all be included in χ. Furthermore, the Spitzer – Harm function is fast to evaluate computationally[26,27] and has even been tabulated.[31] Typically a 1-D ($p_{//}$) approximation is used to evaluate the wave-induced flux[10,32] which again is computationally fast. The primary drawback with this approach is that 2-D effects in the wave damping that appear through Γ_{rf} in Eq. 5(b) are not accounted for in the rf current density computation [Eq. (5a)]. These effects are especially important when there is significant distortion of the quasi-linear distribution function from a Maxwellian, as is typically the case for a broad spectrum of LH waves. Despite this limitation, the adjoint – ray tracing method is widely used, most notably in LHCD modules that are run within the ACCOME MHD equilibrium and current drive code,[10] the Tokamak Simulation Code (TSC),[32] and the transport analysis code TRANSP[9].

III. BENCHMARKING OF LHCD SIMULATION MODELS WITH EXPERIMENT

Toroidal ray tracing codes have been invaluable in providing an understanding of the so-called "spectral-gap" phenomenon[4] in early LHCD experiments[3] where LH waves injected at $v_{//} / v_{te} \approx 5 - 10$ (where the Maxwellian electron population is effectively zero) are able to efficiently damp and drive current. In these studies it was found that multi-pass ray trajectories resulted in significant increases in the parallel wave-number due to toroidal variations in the poloidal mode number. The concomitant decrease in $v_{//} = \omega / k_{//}$ was found to be sufficient to cause wave absorption near the quasi-linear damping limit of $v_{//} / v_{te} \approx 2.3 - 2.5$. A simulation[4] of this physical upshift mechanism is shown in Figs. 1(a)-1(e) for an LHCD experiment in the PLT device[33] where $n_{//}(0) =$

1.33, $T_e(0) = 1.5$ keV, and $f_0 = 800$ MHz. From Figs. 1(c,d,e) it is clear that as the ray propagates, $n_{//}$ (and m) undergo large increases, resulting in power absorption which in turn causes a quasi-linear plateau to be populated. Once this elevated plateau is set up, strong damping of the LH wave also occurs at lower $n_{//}$ (<2). A variety of other $k_{//}$ upshift mechanisms can also be found in the literature. These include spectral broadening effects due to wave scattering from density fluctuations,[34,35] wave diffraction and focusing,[16] and magnetic ripple.[36]

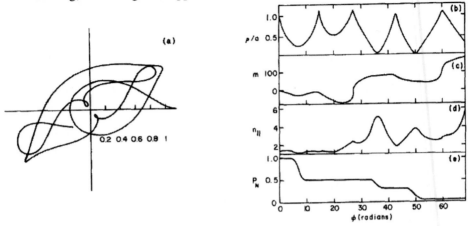

FIGURE 1. Typical ray trajectory for an LHCD experiment in PLT. (a) Projection of ray trajectory in the poloidal cross-section. (b) Normalized radius (ρ/a) versus toroidal angle along the ray path (ϕ). (c) Poloidal mode number (m) vs. ϕ. (d) Parallel refractive index ($n_{//}$) vs. ϕ. (e) Normalized wave amplitude (P_N) due to quasi-linear damping vs. ϕ.

Combined ray tracing and Fokker packages have also been successful in reproducing the macroscopic features of LHCD experiments such as the experimentally observed current drive figure of merit defined as $\eta_{CD} = <n_e(10^{20}\text{m}^{-3})>I(A)R_0(m)/P_{LH}(W)$. For the example shown in Fig. 1, a 1-D ($p_{//}$) Fokker Planck package[4] employing the $T_\perp$ model described in Ref.(25) predicted 196 kA of LH current for 100 kW of injected rf power, at $n_{eav} = 3.75 \times 10^{18}$ m^{-3}, in good agreement with the experiment.[33] Figure 2 is a comparison of the simulated[4] and measured values of sustained current during steady state LHCD experiments in the Alcator C[37] tokamak. The driven currents were simulated using the combined ray tracing-Fokker Planck package[4] discussed above. The agreement between simulation and experiment is clearly quite good. In these experiments the

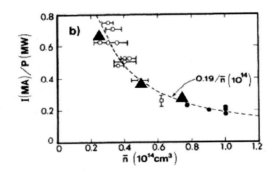

FIGURE 2. Comparison of experimental and simulated LH current drive (I/P) versus density for steady state LHCD experiments in Alcator C. Solid triangles are simulated points and open (solid) circles are the experimental data.

plasma temperature was typically $T_e(0) \approx 1.5$ keV and the injected $n_{//}(0)$ was $1.3 - 1.4$. Thus a sizeable spectral gap also existed in these experiments. It is also worth noting that these experiments were performed at high density [$n_{eav} = 0.3 - 1.0 \times 10^{20}$ m^{-3}] and magnetic field [$B_0 = 8 - 10$ T], thus requiring the use of a higher LH source frequency ($f_0 = 4.6$ GHz) in order to avoid parametric decay of the LH pump wave.[38] Nonetheless the combined ray tracing – Fokker Planck models work well in duplicating macroscopic features of the experiments over a disparate parameter range in density and field.

In the results shown above, the driven LH current was generated either on-axis or close to the plasma center. More recently, LHCD experiments were carried out in the PBX-M device[39] in which the injected $n_{//}(0)$ was systematically lowered, rendering the LH waves inaccessible to the plasma center. The LH source frequency in these experiments was 4.6 GHz and the density range was $n_{eav} = 0.1 - 0.3 \times 10^{20}$ m^{-3}. The lower magnetic field ($B_0 = 1.53$ T) and higher densities in PBX-M resulted in a higher limit for wave accessibility ($n_{//acc} \approx 2.0$) than in earlier experiments performed on PLT (low density) and on Alcator C (high field). The experimentally deduced figure of merit versus $n_{//injected}$ for these studies is shown in Fig. 3. Also plotted are the corresponding predictions for η_{CD} from a combined ray tracing and 2-D Fokker Planck code.[23] The agreement between experiment and simulation is good, especially as $n_{//}$ is increased from about 2.1 to 4.0. The figure of merit decreases as would be expected with increasing $n_{//}$ (decreasing wave phase velocity $v_{//}$). As $n_{//}$ is lowered below 2.0 however, the observed and simulated η_{CD} decrease even though the incident phase velocity is higher. Presumably this occurs because the injected waves become inaccessible to the plasma core at $n_{//} < 2.0$. Consequently, the wave absorption is reduced as waves are forced to damp at lower electron temperature.

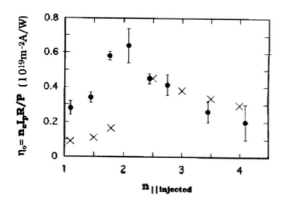

FIGURE 3. Comparison of experimental and simulated LHCD figure of merit in PBX-M versus $n_{//injected}$. Solid circles are the experiment data and crosses are the simulations.

LHCD simulation models have also been benchmarked against experiment at a more microscopic level. Fast electrons generated in LHCD experiments are characterized by energies that are typically in the range of $100 - 300$ keV. Thus the spatial profile of driven current density also corresponds at some level to the profile of hard X-ray (HXR) emission from the fast current – carrying electrons. The measured profile of HXR emission can be influenced by various physical effects such as plasma current[40] and the spatial diffusion of fast electrons.[7,20,21,28,41,42,43] An interesting set of HXR emission measurements were made in the TORE SUPRA device during LHCD experiments at different plasma currents. The measured emissivity profiles from these experiments and the simulated LH power deposition profiles from a combined ray tracing and 2-D Fokker Planck calculation[22,40] are shown in Figs. 4(a,b). The source frequency in these experiments was 3.7 GHz with $B_0 = 3.9$ T, $n_{eav} = 1.3 - 4.5 \times 10^{19}$ m^{-3}, and $n_{//}(0) = 1.8$.

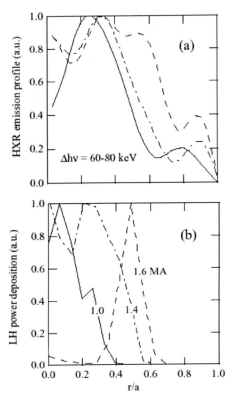

FIGURE 4. (a) Normalized HXR emission profiles in TORE SUPRA (60 - 80 keV) at different values of plasma current [1.0 MA (solid curve), 1.4 MA (dot-dash curve), 1.6 MA (dashed curve)]. (b) Simulated LH power density profiles.

Thus wave accessibility was quite good in these experiments. Figure 4(a) clearly shows the HXR profile broadening as the current increases from 1.0 to 1.6 MA. The simulated LH power deposition profiles are roughly in agreement with this experimental trend as the profile peak moves radially outward as the current increases. A closer examination of the LH ray trajectories in these cases using $(n_{//}, r)$ phase space plots [see Fig. 11 of Ref. 40] reveals that the envelope of wave absorption moves both outward in radius and upward in $n_{//}$ as I_p is increased. Recalling that for a LH wave, the toroidal variations in m are converted to changes in $k_{//}$ through the poloidal field, it can be seen that higher values of current (and therefore B_θ) will result in larger variations in $k_{//}$. The likely consequence of the higher $k_{//}$ is a radial outward shift in the deposition profile as seen in Fig. 4(b). One important point of disagreement between simulation and experiment is that while the simulated deposition profiles become hollow, the profiles of HXR emission only become broader, remaining flat in the plasma core ($r/a \leq 0.6$). This discrepancy could likely be due to the absence of fast electron diffusion effects in the computation of the rf power deposition and current density profiles, which have been shown to cause an inward diffusion of fast electrons, keeping the driven current profile monotonic.[7,28,41,42]

Attempts have been made to quantify fast electron diffusion effects in a number of numerical Fokker Planck model treatments.[7,20,21,28,41] One such study utilized the ray tracing – Fokker Planck package described in Ref. (28) to simulate LHCD experiments in the JET tokamak. A 2-D $(p_{//}, r)$ solution of Eq. (2) was performed using a collision operator with the empirical correction for 2-D effects [$(5+Z_{eff})/10$] described in Sec. II, a model for $T_\perp(p_{//})$ from Ref. (25), and a radial diffusion operator. The fast electron diffusivity was taken to have the form $\chi_F = \chi_0 (p_{//} / \gamma v_{te})$ with $\chi_0 = 0.5$ m^2/s. This model for χ_F assumed that the diffusion of fast electrons was determined by the stochasticity of the confining magnetic field.[44] The simulated profiles of rf power deposition and 'diffused" LH current density are shown in Fig. 5. Some of the relevant parameters used in the simulation were $T_e(0) = 1.5$ keV, $n_e(0) = 1.7 \times 10^{19}$ m^{-3}, $B_0 = 2.4$ T, $I_p = 0.37$ MA, $f_0 = 3.7$ GHz, and $n_\parallel(0) = 1.85$. Although the rf deposition profile is clearly hollow, the diffused LH current density profile is monotonic and broad. The simulated electron distribution function was also used to compute profiles of fast electron *bremsstrahlung* emission (FEB) that were measured in the experiment. The measured and simulated profiles are shown in Figs. 6(a,b). The best fit of the calculated to measured FEB signals was found for $\chi_0 = 0.5$ m^2/s, providing evidence that fast electron diffusion physics was playing an important role in the JET LHCD experiments.

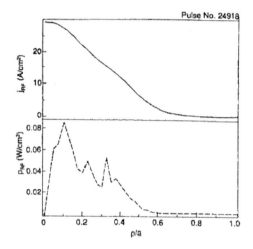
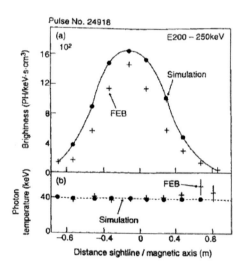

FIGURE 5. Simulated LHRF power dissipation and current density profiles, including fast electron diffusion effects for LHCD in JET.

FIGURE 6. Profiles of calculated and measured hard X-ray emission during LHCD in JET. (a) Brightness profile. (b) Photon temperature profile.

It is expected that fast electron diffusion will become less of a concern as one moves to the reactor regime for several reasons. First, the slowing down time (τ_S) of fast electrons is shorter as the density increases. Second, reduced wave accessibility at higher density requires the use of lower phase velocity waves, resulting in the production of lower energy electrons, which in turn take less time to thermalize. Finally, fast electron confinement (τ_F) should be better in a larger device since the bulk energy confinement time is longer. Thus in moving from the regime of present day experiments to reactor grade plasmas, one moves from LHCD experiments where $\tau_S \approx \tau_F$ to the limit where $\tau_S \ll \tau_F$.

IV. MODEL PREDICTIONS FOR FUTURE DEVICES

As pointed out in the Introduction, the primary role of LHRF power envisioned for next generation devices (fusion reactors) is that of off-axis current profile control. The ability to generate current efficiently in the plasma periphery provides a means to maintain and control profiles of reversed magnetic shear, thus accessing regimes of improved tokamak performance.[45,46] Recently 3 MW of LHRF power was injected into the ELMy H-mode phase of a discharge in the JET tokamak[47], generating about 0.5 MA of off-axis current and thus preventing q(min) from crossing two. This resulted in sustainment of the longest internal transport barrier (ITB) ever in JET (approximately 37 energy confinement times).[47] Off-axis LHCD power (2.5 MW) was also used in JT60-U in conjunction with 2.5 MW of NBI power to achieve quasi-steady state sustainment of an ITB for 4.7 sec., with good alignment between bootstrap and non-inductive currents.[48]

Given the great promise this current profile control technique holds, it is useful to examine the predictions for LHCD in a proposed reactor design such as the ITER – FEAT,[5] using several of the models discussed in the previous section. Parameters typical of the ITER advanced operating regime ("ITER – FEAT") are $B_0 = 5.18$ T, $I_p = 9$ MA, $R_0 = 6.35$ m, $a = 1.85$ m, $n_e(0) = 0.7 \times 10^{20}$ m^{-3}, $T_e(0) = 30$ keV, and $T_i(0) = 32$ keV. Central current profile control is provided by 30 MW of 1.0 MeV neutral beam injection (NBI) power and off-axis current profile control is provided by 30 MW of LHRF power at 5.0 GHz. The corresponding density and temperature profiles used in the predictive simulations are shown in Figs. 7(a,b) and correspond to volume average density and temperature values of $<n_e> = = 0.68 \times 10^{20}$ m^{-3} and $<T_e> = 11.3$ keV.

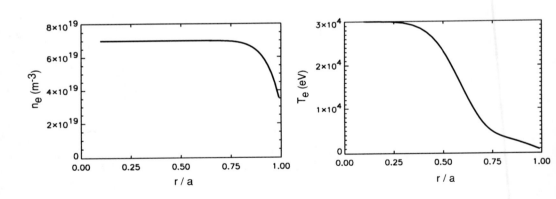

FIGURE 7. Density and temperature profiles used in predictive simulations of LHCD in ITER – FEAT.[5]

The ACCOME current drive and MHD equilibrium code[10] is first used to compute an MHD equilibrium that is consistent with the local current density source terms due to NBCD, LHCD, and bootstrap effects. The resulting current density profiles and profile of safety factor are plotted in Figs. 8(a,b). The total non-inductive current is 7.4 MA with 1.96 MA from NBI power, 1.16 MA from off-axis LHCD, and 4.37 MA from bootstrap effects (a bootstrap fraction of $f_{bs} = 0.59$). The profile of safety factor exhibits clear shear reversal due to a combination of off-axis LHCD and bootstrap current with $q(0) = 3.95$, $q(min) = 2.36$, and $q(95) = 5.19$. The current drive figure of merit for the NBCD is $\gamma_{NB} = 0.29$ (10^{20} A/W/m^2) and the figure of merit for the LHCD is $\gamma_{LH} = 0.17$ (10^{20} A /W/m^2).

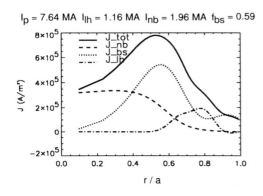
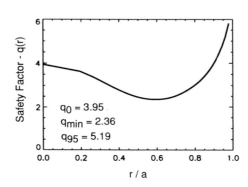

FIGURE 8. Current drive predictions for ITER – FEAT using the ACCOME code. (a) Current densities versus r/a. Curves are labeled as follows: Total current (solid line), neutral beam current (dashed line), bootstrap current (dotted line), and LH current (dot – dash line). (b) Safety factor versus r/a.

Recall from Sec. II that the LHCD model in ACCOME employs an adjoint solution of the Fokker Planck equation and a ray tracing module. Also the wave-induced flux is evaluated using a 1-D ($p_{//}$) approximation [see Eqs. 5(a) and 5(b)]. It can be shown that even for LHCD, the effects of particle trapping become more important as the deposition layer moves beyond r/a > 0.5. In fact for the case shown in Fig. 8(a), if particle trapping effects are ignored in the adjoint solution of the Fokker Planck equation then the LH current increases from 1.16 to 1.59 MA with a corresponding increase in γ_{LH} from 0.17 to 0.24 (10^{20} A/W/m^2).

It is interesting to compare the adjoint – ray tracing prediction for LHCD from ACCOME to the LH current predicted by the 2-D ($p_{//}$, $p_\perp$) code CQL3D.[20] The comparison is done by passing the final MHD equilibrium from the ACCOME code to CQL3D, where a ray tracing – 2-D Fokker Planck iteration is carried out. The computed LH current density profile is shown in Fig. 9. The integrated current is 1.6 MA and the corresponding figure of merit is $\gamma_{LH} = 0.24$ (10^{20} A/W/m^2). The discrepancy between the adjoint method (ACCOME prediction) and CQL3D is significant and deserves discussion, especially in view of the fact that both calculations include particle trapping effects. A closer examination of the two simulations reveals that the reason for the lower current predicted by the adjoint – ray tracing method is related to the use of a 1-D ($p_{//}$) damping model[4] for the wave-induced rf flux in Eqs. 5(a) and 5(b). Other aspects of the calculations are in fact identical. For example, LH ray trajectories computed by the ray tracing module in CQL3D are found to be identical to the ray paths in an $(\mathbf{x}, \mathbf{k})$ phase space computed by the ACCOME ray tracing module.[4]

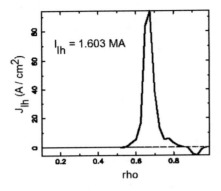

FIGURE 9. LHCD prediction for ITER – FEAT using the CLQ3D Fokker Planck code.

It also instructive to examine the predictions of two other LH simulation models for the ITER – FEAT equilibrium computed by ACCOME [see Figs. 10(a,b)]. These code predictions are from the 2-D $(p_\perp, p_{//})$ Fokker Planck code DELPHINE[22] and from the 1-D $(p_{//})$ Fokker Planck ray tracing package FRTC.[29,30] Recall that although FRTC is 1-D, the collision operator has been corrected to include 2-D effects as discussed in Section II. Both DELPHINE and FRTC were run in modes where particle trapping was ignored for these simulations. The RF power deposition profiles for all three codes (ACCOME, FRTC, and DELPHINE) are shown in Fig. 10(a). The deposition profiles from FRTC and DELPHINE tend to be more centrally peaked and narrower than the computed RF

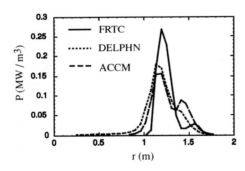
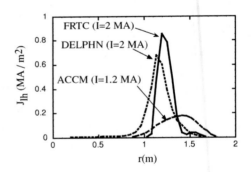

FIGURE 10. LHCD predictions for ITER – FEAT from three different simulation models. (a) Profiles of LHRF deposition versus minor radius. (b) Profiles of LH current density versus minor radius. ACCOME predictions are shown as the long dashed curves and are denoted by "ACCM", DELPHINE predictions are shown as the short dashed curves and are denoted by "DLPHN", FRTC predictions are shown as solid curves and are denoted by "FTRC". Code comparisons were done using only the forward injected LH power (26 MW).

absorption from ACCOME. In fact the deposition profile from ACCOME has a secondary peak at about r = 1.45 m. The corresponding current density profiles are shown in Fig. 10(b). The current density profile from ACCOME tends to be peaked farther out in radius because of the secondary peak in the RF deposition profile. While the FRTC and DELPHINE predictions for driven LH current agree with each other (2MA), they are both higher than the LH current predicted by CQL3D (1.6 MA). This discrepancy is presumably due to the fact that particle trapping was not included in the FRTC and DELPHINE simulations, which would reduce the driven LH current. Thus a careful comparison of four model predictions for LH current profile control in ITER – FEAT leads to the conclusion that the CQL3D result is the most realistic answer since this computation was done in 2-D ($p_\perp$, $p_{//}$) and included the effects of particle trapping.

IV. CONCLUSIONS

In conclusion, the status of LHCD simulation and modeling is such that detailed comparisons between simulation and experiment are possible at both the macroscopic and microscopic levels. Several different simulation approaches are available with the 2-D velocity space Fokker Planck models providing the most complete description of the current drive physics including effects due to particle trapping, momentum conserving corrections to the collision operator, and pitch angle scattering of fast electrons. It was argued that adjoint solution methods are powerful so long as an accurate estimate of the wave-induced rf flux is available. Also 1-D velocity space methods were shown to be useful owing to their minimal computational requirements and so long as the collision operators are corrected for 2-D effects. With the advent of massively parallel computing platforms, 2-D and 3-D ($p_\perp$, $p_{//}$, r) Fokker Planck solutions are fast becoming routine and should now be possible to carry out within closed computation loops with transport codes and MHD solvers. The most detailed benchmarking between LHCD simulation models and experiment has typically been done at the level of comparing computed and measured profiles of hard X-ray emission and fast electron *bremsstrahlung*. Lower hybrid current drive is envisioned as an off-axis current profile control tool in fusion reactor applications. In view of this fact, it will be necessary both in present day and future LHCD experiments to make more direct comparisons between computed LH current density profiles and the experimentally measured profiles from motional Stark effect (MSE), for example. Finally, the predictions of driven LH current for reactor devices such as ITER – FEAT using the most advanced ray tracing – Fokker Planck simulation models available are extremely encouraging. About 1.6 MA of current is predicted to be generated at r/a ≥ 0.6 using 30 MW of injected LH source power, for a current drive figure of merit of γ_{LH} = 0.24 (10^{20} A/W/m^2). Furthermore, the local LH current densities [see Figs. 8 and 9] should be sufficient to control the point of shear reversal and maintain profiles of negative magnetic shear in advanced tokamak operating regimes.

ACKNOWLEDGEMENTS

It is a pleasure to acknowledge many useful conversations with Dr. Cynthia Phillips, Dr. Yves Peysson, Professor Miklos Porkolab, and Professor Cary Forest. This work was supported by the US Department of Energy under Grant. No. DE-FC02-99ER54512 and Grant No.DE-FC02-01ER54648.

REFERENCES

1) Fisch, N.J., Reviews of Modern Physics **59**, 175 (1987).
2) Bonoli, P.T., in Radio Frequency Power in Plasmas, AIP Conf. Proc. 159 (AIP, NY, 1987) p. 85.
3) Peysson, Y., in Radio Frequency Power in Plasmas, AIP Conf. Proc. 485 (AIP, NY, 1999) p. 183.
4) Bonoli, P.T. and Englade, R.C., Physics of Fluids **29**, 2937 (1986).
5) ITER Technical Basis Document (IAEA, Vienna, 2001) Doc. No. GAO FDR 1 00-07-13 R1.0, Section 4.3.3.
6) Valeo, E.J.. and Eder, D.C., Journal of Computational Physics **69**, 341 (1987).
7) Bonoli, P.T. et al., Nuclear Fusion **28**, 991 (1988).
8) Pereverzev, G., et al., IPP Report 5/42, Max Planck Institute fur Plasmaphysik (1991).
9) Jobes, F.C., Ignat, D.W., and McCune, D.C., Bull. Am. Physical Society **41**, 1515 (1996).
10) Devoto, R.S. et al., Nuclear Fusion **32**, 773 (1992).
11) Baranov, Y.F., and Fedorov, V.I., Nuclear Fusion **20**, 1111 (1980).
12) Colestock, P., Nuclear Fusion **18**, 740 (1978).
13) Ignat, D.W., Physics of Fluids **24**, 1110 (1980).
14) Bonoli, P.T. and Ott, E., Physics of Fluids **25**, 359 (1982).
15) Brambilla, M., Computer Physics Reports **4** (3-4), 71 (1986).
16) Pereverzev, G., Nucl. Fusion **32**, 1091 (1992).
17) Wright, J.C. et al., 15th Topical Conf. On RF Power in Plasmas, Moran, Wyoming (May, 2003) paper C89.
18) Kennel, C.F. and Engelmann, F., Phys. Fluids **9**, 2377 (1966).
19) Karney, C.F.F. and Fisch, N.J., Phys. Fluids **22**, 1817 (1979).
20) Harvey, R.W. and McCoy, M.C., in Advances in Simulation and Modeling of Thermonuclear Plasmas (Proc. IAEA Tech. Comm. Mtg., Montreal, 1992), IAEA, Vienna (1993) p. 498.
21) McKenzie, J.S. et al., Computer Physics Communications **66**, 194 (1991).
22) Imbeaux, F., EURATOM – CEA Report EUR-CEA-FC 1679 (1999); also Peysson, Y. and Shoucri, M., Computer Physics Communications **109**, 55 (1998).
23) Shoucri, M. and Shkarofsky, I., Computer Physics Communications **82**, 287 (1994).
24) Fisch, N.J. and Boozer, A.H., Phys. Rev. Lett. **45**, 720 (1980).
25) Fuchs, V. et al., Phys. Fluids **28**, 3619 (1985).
26) Karney, C.F.F. and Fisch, N.J., Phys. Fluids **28**, 116 (1985).
27) Karney, C.F.F. et al., in Radio Frequency Power in Plasmas, AIP Conf. Proc. 190 (AIP, NY, 1989) p. 430.
28) Baranov, Yu. F. et al., Nucl. Fusion **36**, 1031 (1996).
29) Esterkin, A.R. and Piliya, A.D., Nucl. Fusion **36**, 1501 (1996).
30) Tala, T.J.J. et al., Nucl. Fusion **40**, 1635 (2000).
31) Ehst, D.A. and Karney, C.F.F., Nucl. Fusion **31**, 1933 (1991).
32) Ignat, D.W. et al., Nucl. Fusion **34**, 837 (1994).
33) Bernabei, S. et al., Phys. Rev. Lett. **49**, 1255 (1982).
34) Bonoli, P.T. and Ott, E., Phys. Rev. Lett. **46**, 424 (1981).
35) Andrews, P.L. and Perkins, F.W., Phys. Fluids **26**, 2537 (1983).
36) Bizarro, J.P. et al., Phys. Rev. Lett. **75**, 1308 (1995).
37) Porkolab, M. et al., Phys. Rev. Lett. **53**, 450 (1984).
38) Porkolab, M., Phys. Fluids **20**, 2058 (1977).
39) Bernabei, S. et al., Phys. Plasmas **4**, 125 (1997).

40) Peysson, Y. *et al.*, Plasma Phys. and Cont. Fusion **42**, B87 (2000).
41) Fuchs, V. *et al.*, Nucl. Fusion **29**, 1479 (1989).
42) Cairns, R.A. *et al.*, Nucl. Fusion **35**, 1413 (1995).
43) Barbato, E. *et al.*, Transport Task Force Meeting, Cordoba, Spain (2002).
44) Mynick, H.E. and Strachan, J.D., Phys. Fluids **24**, 695 (1981).
45) Kessel, C. *et al.*, Phys. Rev. Lett. **72**, 1212 (1994).
46) Levinton, F.M. *et al.*, Phys. Rev. Lett. **74**, 718 (1995).
47) Litaudon, X. *et al.*, Plasma Physics and Cont. Fusion **44**, 1057 (2002).
48) Ide, S. *et al.*. Proc. 17th Int. Conf. On Fusion Energy, (IAEA, Vienna, 1999), Vol. 2, p. 567.

ION CYCLOTRON RANGE OF FREQUENCIES

Bulk Plasma Rotation in the Presence of Waves in the Ion Cyclotron Range of Frequencies

L.-G. Eriksson, J.-M. Noterdaeme[1,2], S. Assas, C. Giroud[3], J. DeGrassie[4], T. Hellsten[5], T. Johnson[5], V.G. Kiptily[6], K. Kirov[1], M. Mantsinen[7], K.-D. Zastrow[6], M. DeBaar[3], J. Brzozowski[5], R. Budny[8], R. Cesario[9], V. Chan[4], C. Fenzi-Bonizec, A. Gondhalekar[6], N. Hawkes[6], G.T. Hoang, Ph. Lamalle[10], A. Meigs[6], F. Meo[1], F. Nguyen, E. Righi[11], A. Staebler[1], D. Testa[12], A. Tuccillo[9], H. Weisen[12], and JET-EFDA contributors[*]

Association EURATOM-CEA, CEA/DSM/DRFC, CEA-Cadarache, St. Paul lez Durance, France; [1]*Max-Planck IPP-EURATOM Assoziation, Garching;* [2]*Gent University, Department EESA, Belgium;* [3]*The Stichting voor Fundamenteel Onderzoek der Materie FOM, The Netherlands;* [4]*General Atomics USA;* [5]*Euratom-VR Association, Stockholm, Sweden;* [6]*Association Euratom-UKAEA, Culham Science Centre, Abingdon, United Kingdom;* [7]*Helsinki University of Technology, Association Euratom-Tekes, Finland;* [8]*PPPL, Princeton USA;* [9]*ENEA Euratom Association, Frascati, Italy,* [10] *EFDA-CSU JET, UK,* [11] *European Commission, Brussels, Belgium;* [12]*CRPP, Assoc. EURATOM-Confederation Suisse, EPFL, Lausanne, Switzerland;* *see J. Paméla, 19th Fusion Energy Conf., Lyon, France.*

Abstract. Experiments with directed ICRF waves have for the first time in JET demonstrated the influence of absorbed wave momentum on bulk plasma rotation. Resonating fast ions acted as an intermediary in this process and the experiments therefore provided evidence for the effect of fast ions on the plasma rotation. Results from these experiments are reviewed together with results from ICRF heated plasmas with symmetric spectra in JET and Tore Supra. The relevance of different theoretical models is briefly considered.

INTRODUCTION

Plasma rotation can have beneficial effects on the performance of a tokamak plasma. For example, it is widely believed that the shear in the toroidal velocity component associated with the radial electric field is an important factor for the formation of transport barriers [1, 2]. Furthermore, plasma rotation can influence MHD activity, and enhance the stabilizing effect of a resistive wall [3]. It is therefore important to understand the mechanisms behind plasma rotation.

Strong toroidal plasma rotation is normally induced by Neutral Beam Injection (NBI) heating in present day tokamaks. In a burning reactor plasma, however, there might not be any NBI or it will be used for current drive purposes. In the latter case, the injection energy must be very high for the injected neutrals to penetrate towards the center of the plasma. Since the injected momentum scales like $P_{NBI}/E_{inj}^{1/2}$, the toroidal torque on the plasma could be relatively modest. As a result, it is relevant to investigate other mechanisms that can give rise to plasma rotation.

Interesting observations of rotation in plasmas heated by waves in the Ion Cyclotron Range of Frequencies (ICRF) have been made in several tokamaks since the early nineties [4-9]. The presence of often very fast resonating ions is such plasmas means that there are significant similarities to alpha particle heated plasmas. It is therefore of particular interest to investigate the origin of the observed rotation. Especially intriguing is the fact that little or no external momentum injection has been involved in most of the reported cases of rotation during ICRF heating. Furthermore, rotation predominantly in the same toroidal direction as the plasma current has been observed in these conditions, excluding losses of fast ions as a source for the rotation. Nevertheless, fast ion effects have been proposed as a possible source for the rotation [10]. On the other hand, scalings of the experimentally measured rotation indicate that a bulk plasma effect could be involved [4-9], especially since rotation in ohmic H modes have been observed to generally follow the scaling of the rotation in ICRF heated plasmas in Alcator C-Mod [6, 7]. In fact, there is probably a combination of effects involved, with fast ions being one of them. Consequently, it is useful to identify and quantify their effect on plasma rotation experimentally.

In recent experiments on the JET tokamak with directed (or travelling) ICRF waves, the effect of fast ions on the rotation has been clearly identified for the first time. Furthermore, the use of directed waves offer a possibility to control the rotation. The principal part of this paper is therefore devoted to JET experiments with directed waves.

MHD activity has been observed to have a strong influence on ICRF heated JET plasmas. Such experiments are briefly discussed in the second section of this paper. In the last section, rotation in plasmas with little or no external momentum injection are discussed and new results from Tore Supra are reported together with a brief look at theoretical models.

PLASMA ROTATION IN THE PRESENCE OF DIRECTED ICRF WAVES

It would of course be advantageous to control plasma rotation. In principle this could be achieved by directed waves in the Ion Cyclotron Range of Frequencies (ICRF). Although such waves would globally provide much less torque than NBI injected ions, the maximum torque density would not necessarily be small since the power deposition profile for ICRF is much narrower than for NBI. Furthermore, the location and direction of the provided torque could be controlled by moving the cyclotron resonance and changing the spectrum of the waves. Directed waves are discussed in for instance [11]. The first clear evidence of an influence of directed ICRF waves on rotation has been observed in recent experiments on JET. Since the momentum carried by the waves is initially absorbed by resonating fast ions and subsequently transferred to the background plasma, the JET results provide evidence for fast ion influence on rotation. By monitoring the fast ions with a gamma-ray diagnostic, it has been possible to identify the relative importance of ions on topologically different orbits in the discharges with directed waves.

The equation of motion for a resonating ion combined with Maxwell's equations shows that a particle receiving a change in its energy ΔE during a resonant interaction with a wave also experiences a change in its toroidal angular momentum $\Delta P_\varphi = (N/\omega)\Delta E$ (c.f. quantum mechanics, absorption of wave quantum $\Rightarrow \Delta E = \hbar\omega$

and $\Delta P_\varphi = \hbar k_\varphi R = \hbar N$). The amount of toroidal momentum imparted to the plasma per unit time is therefore given by:

$$\sum_N (N/\omega) P_{ICRF}(N),$$

where $P_{ICRF}(N)$ is the power into the toroidal mode number N. Thus, by exciting directed waves, i.e. waves with an asymmetrical toroidal mode number spectrum, toroidal angular momentum can be imparted to the plasma.

In order to investigate the possibility of influencing toroidal rotation with directed ICRF waves, a set of experiments was carried out in JET. According to the formula above, a scenario with a low frequency should be used to maximize the imparted momentum for a given power. In the JET experiments discussed here, ^{3}He minority heating in deuterium plasmas, (^{3}He)D, was the chosen scenario. The ICRF frequency was 37 MHz and the central toroidal magnetic field 3.4 T, placing the cyclotron resonance slightly on the high field side of the magnetic axis (~0.15 m). The JET four strap antennas had a phasing between the currents in two neighboring straps of either +90° or –90°, producing waves propagating predominately in the toroidal direction parallel or anti-parallel to the plasma current, respectively. Characteristic toroidal mode number spectra for the two phasings, typically peaking at $N\pm15$ for $\pm$ 90° phasing, can be found in Ref. [13]. An overview of two discharges with +90° and –90° phasing respectively is given in Fig. 1. Neutral Beam Injection (NBI) blips and charge exchange spectroscopy, described comprehensively in [20], were used to measure the rotation in these discharges. The resulting profiles measured at t=51 sec are shown in Fig. 2.

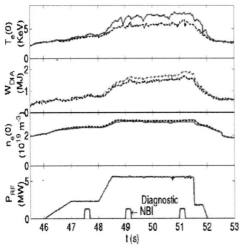

FIGURE 1. Overview of two discharges with +90° (solid line) and –90° phasing (dashed line).

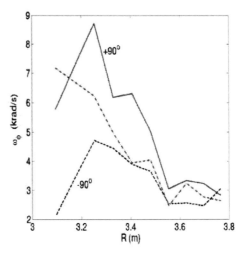

FIGURE 2. Rotation profiles for the discharges with +90° and –90° phasing. The rotation profile for a discharge where 2MW of +90° ICRF power was replaced by 2MW of LH power is also shown (dot dashed line)

As can be seen, the two discharges are very well matched in terms of the applied ICRF power and plasma density, whereas there is a small difference in the stored energy and a somewhat larger difference in the central electron temperature. These differences are consistent with the absorption of wave momentum and the concomitant inward/outward pinch effect on the resonating ions [14, 15]. While both discharges rotate in the co-current direction, there is a significant difference: the +90° discharge rotates much more strongly in the center.

The stronger co-current rotation of the +90° discharge is consistent with theoretical expectations since the waves in this case carry momentum in the co-current direction. In order to increase the confidence in that the absorbed wave momentum was the critical factor and not more efficient central heating in the +90° case, a discharge where 2 MW of +90° ICRF power was replaced by 2MW of LH power was carried out in an otherwise similar discharge (the LH wave also carries momentum, but is much smaller than for the ICRF waves, around a tenth). The stored energy for this discharge was somewhat lower (~10%) than for the –90° case, the rotation profile is shown in Fig.2. In spite of having a lower stored energy, the discharge with LH power rotates more strongly in the co-current direction than the –90° discharge. This is a strong indication that it is the wave momentum and not the power deposition or the efficiency of the heating that is the cause of the difference in rotation velocity between the +90° and –90° phasing discharges.

Important information on the fast ion behavior in these discharges has been provided by the gamma-ray measurements [12]. Figure 3 shows the line integrated emission, normalized to its maximum value, from the vertical lines of sight as a function of the major radius where the sight line crosses the mid-plane. As can be seen, the –90° discharge has an asymmetrical emission with respect to the magnetic axis whereas it is almost symmetric for the +90° case. Thus, the fast ion characteristics are clearly different in the two discharges. The asymmetrical emission for –90° is consistent with a strong population of trapped fast ions and the more symmetric for +90° with a significant presence of passing ions in the potato regime (i.e. they have orbits not covered by the small banana width limit [16]).

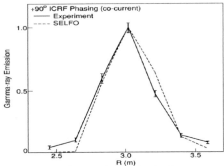

FIGURE 3a. Measured and simulated gamma ray emission as a function of the major radius for +90°

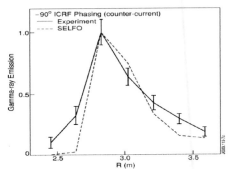

FIGURE 3b. Measured and simulated gamma ray emission as a function of the major radius for –90°

The presence of non-standard passing ions in the center should have an important effect on the torque provided by the fast ions to the bulk plasma in the central region. This is particularly interesting since the presence of such ions has been found to be an essential factor for co-current rotation driven by fast ions in the absence of external momentum injection [17, 18].

To gain further insight and to check in more detail if theory can explain the experimental observations it is necessary to carry out rather comprehensive simulations. For this purpose we have used the SELFO code [19], which self consistently calculates the ICRF power deposition and the distribution function of the resonating ions, including finite orbit width effects and wave induced transport in real space. From the simulated distribution function we have calculated line integrated gamma-ray emissions corresponding to the measured vertical channels. The results of these simulations have been added to Figs. 3a and b. As can be seen, the spatial distribution of the measured and calculated gamma-rays have very similar features. Only in the relative level is there a difference, the ratio between of the emissions at the normalization points for +90° and –90° phasing is roughly 4 experimentally and 2 in the SELFO simulations. This is probably due to a too low concentration of ^{3}He ions in the –90° simulation (there is a pump out of resonating ions due to the outward drift of ions in the -90° case). The simulations confirm that the asymmetrical shape of the emission seen in Fig. 3b is caused by a dominating presence of fast trapped ions whereas the more symmetric emission in Fig. 3a is the consequence of a large fraction of co-passing orbits in the potato regime.

The SELFO code has also been used to calculate the torque density absorbed from the ICRF waves by the resonating ions and subsequently transferred to the thermal background plasma. This torque has been inserted in a simple momentum diffusion equation of the type:

$$n_i m_i \frac{\partial V_\varphi}{\partial t} = \frac{1}{r}\frac{\partial}{\partial r}\left[r n_i m_i D \frac{\partial V_\varphi}{\partial r}\right]$$

where V_φ is the rotation velocity, m_i and n_i are the mass and density of the plasma ion species (summation over repeated index is assumed), D is a momentum diffusion coefficient. The momentum transport is likely to be anomalous, and in our estimates we simply take $D = a^2/(\alpha_M \tau_M)$, where a is the minor radius, τ_M the momentum confinement time assumed to be related to the energy confinement time ($\tau_M \sim \tau_E$), and α_M is a parameter adjusting the transport so that the global confinement, τ_M, time is obtained (in reality α_M depends on the rotation profile, but it should be of the order 5). Inserting the torque profile calculated by SELFO in the diffusion equation above using $\tau_M = \tau_E = 0.3s$ and $\alpha_M = 2$ results in the rotation profiles shown in Fig. 4. As can be seen, the difference in the simulated toroidal rotation is of the order 4 krad/s, in good agreement with the experimental finding. However, the underlying co-current rotation cannot be explained by the fast ion effects included in SELFO.

From the combination of experimental observations and numerical simulations with the SELFO code, a rather clear picture emerges. In the case of ICRF waves propagating in the direction of the toroidal current, the absorbed wave momentum is transferred from the resonating ions largely by an inward wave-induced drift of trapped ions and via collisions with co-passing ions in the potato regime. The presence of the latter is a consequence of the inward drift, as the turning point of a trapped ion reaches the equatorial plane it de-traps and is transformed into a co-passing orbit. The presence of co-passing orbits is important because they provide a co-current torque in the central part of the plasma. On the other hand, for waves propagating in the toroidal direction opposite the plasma current, the momentum transfer is mainly due to the outward drift of the resonating ions.

In earlier experiments reported in [20], very small differences were seen between +90° and -90° phasing. There could have been several reasons for this. Two important factors were the lower power levels used and the fact that the cyclotron resonance was placed much further off-axis, leading to a significantly lower predicted change of the rotation (~5) and greater uncertainties in the central rotation due to dependence on the momentum transport.

ON THE INFLUENCE OF MHD ACTIVITY ON PLASMA ROTATION

A number of JET experiments have been carried out to investigate if the position of the cyclotron resonance has an influence on the plasma rotation during ICRF heating. In a particular series of experiments, the magnetic field was ramped during the application of the ICRF power so that the cyclotron resonance was moved from the low field side (LFS) to the high field side (HFS) in one discharge and vice versa in another. During the course of the magnetic field ramp the cyclotron resonance passed through the centre, and the resulting central fast ion pressure led to the creation of monster sawteeth. At a monster sawtooth crash MHD activity was triggered, and locked modes appeared. The rotation profiles before and after the appearance of locked modes for the two discharges with increasing and decreasing magnetic field ramps are shown in Fig. 5. The MHD activity is found to wipe out the rotation over almost the whole plasma radius. This observation is consistent with similar observations in NBI heated plasmas [21].

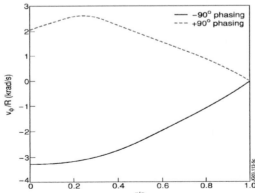

FIGURE 4. Rotation profiles obtained with the torque from SELFO simulations inserted in to a momentum diffusion equation.

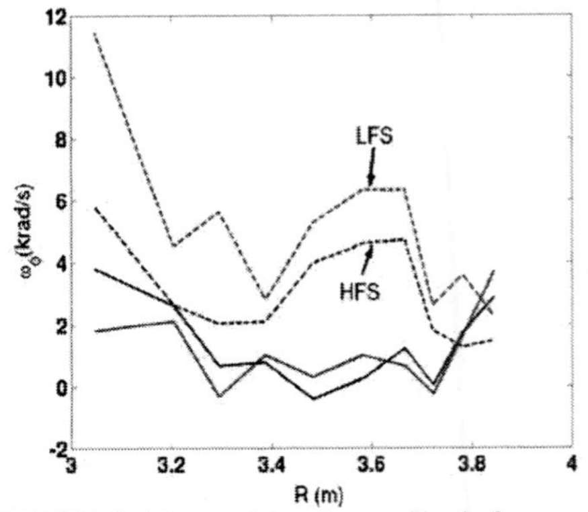

FIGURE 5. Measured rotation profiles before onset of MHD activity, dashed lines, and after, solid lines.

PLASMA ROTATION FOR SYMMETRIC ICRF ANTENNA SPECTRA

We have shown above that the change in toroidal rotation induced by directed waves can be understood in terms of the absorbed wave momentum and its subsequent transfer from fast resonating ions to the background plasma. However, we have not explained the underlying co-current rotation shown in Fig. 3 or the observed rotation in the discharges in Fig. 5, where symmetric antenna spectra were used. These are all in L mode and are all rotating predominantly in the co-current direction, but with rather modest rotation velocities. This observation is confirmed by the larger study presented in [20]. An important aspect to note in Fig. 5, and which has also been found to be a general feature in [20], is that there is little difference between HFS and LFS resonance. In particular, there is no change in the direction as is predicted by the most basic theory for fast ion induced rotation [10]. A key point of the theory in Ref. [10] is that the differential torque produced by the fast ions (i.e. separated regions of positive and negative torque) is sufficient to drive rotation even if the total torque from them on the background is zero. However, preliminary results from SELFO simulations for discharges similar to those in Fig. 3 but with symmetric spectra suggest that the differential torque created by the fast ions is not enough to explain the observed rotation when inserted into the momentum diffusion equation with the same momentum confinement used for Fig. 4.

It is especially interesting that rotation in the same toroidal direction as the plasma current has been observed in the three machines JET, Alcator C-Mod and Tore Supra [4-9]. Furthermore, a scaling of the co-current rotation with bulk plasma parameters (stored energy and/or bulk ion pressure) has been reported from all three machines. The JET and Alcator C-mode results showing this feature are all from H-mode plasmas. The recent rotation experiments carried out in JET investigated the effect of the resonance location mostly in L-mode at constant power [20]. Therefore they did not provide additional information on the relation between bulk plasma parameters and plasma rotation. New results from Tore Supra on this aspect are reported below. Comparing results from the different machines is especially interesting in view of their significant differences. One should first note that JET and Alcator C-Mod are equipped with divertors whereas limiter discharges are operated in Tore Supra. Furthermore, the collisional regimes in Alcator C-Mod and JET can be quite different due to the high densities with which Alcator C-Mod is normally operated. Tore Supra is in this respect in between the two other machines but closer to JET.

As was reported in [8, 9], Tore Supra plasmas have been found to be accelerated in the co-current direction during hydrogen minority ICRF heating with high concentrations of hydrogen. On the other hand, plasmas with low H concentrations are normally found to be accelerated in the counter current direction. A theory put forward in [8] is that the counter current rotation could be due to ripple losses of fast ions, which can be quite significant in Tore Supra due to the relatively large ripple, 7% at the plasma boundary. At lower concentrations the resonating ions become on average more energetic and thus more subjected to ripple transport. Thus, the counter torque produced by ripple lost ions could explain the counter current rotation at low H concentrations.

This picture has been reinforced by results from recent ³He minority heating experiments carried out in Tore Supra. The ³He minority scenario creates few fast ions in Tore Supra, and the ripple losses are therefore very low. As is shown in Fig. 5, the Tore Supra discharges with ³He minority heating are accelerated in the co-current direction, and the change in rotation velocity scales well with the diamagnetic energy content divided by the plasma current, c.f. similarities with Alcator C-Mod H-mode discharges. Another factor is the stronger ion heating expected during ³He minority heating. Its role has not yet been clarified, but it is possible that it could have an influence on the rotation as well.

With in many respects fairly similar results for the observed toroidal rotation during ICRF heating with symmetric antenna spectra in three rather different tokamaks, it is quite a challenge for theory to explain the observations. There are currently mainly three theoretical explanations put forward for the observed co-current rotation during ICRF heating, fast particle effects [10], neo-classical effects [22], and the so-called accretion theory [23]. The influence of fast resonating ions on the JET experiments has already been discussed above, and this influence does not seem sufficient to explain the bulk of the observed rotation. While it is not explicitly stated in the paper, the neo-classical theory of rotation developed in [22] for ohmic H-mode plasmas should be an important component also for ICRF heated H-modes. There are two key factors in this theory. Firstly the plasmas, and particularly the outer part, are assumed to be in the Pfirsch-Schlüter regime; secondly, the main contribution to the rotation comes from the strong ion pressure gradient in the pedestal region of the H-mode. The theory is therefore particularly suited to Alcator C-Mod H-mode plasmas, where plasmas with very high densities are operated, and fits the experimental measurements in this machine quite well. However, it is not clear that this theory can explain the bulk of the plasma rotation observed in JET and Tore Supra. For example, Tore Supra is a machine without divertor and consequently does not have an H-mode (at least not in the normal sense of a machine with a divertor). Furthermore, the collisional regimes in JET and Tore Supra are quite different from Alcator C-Mod. In fact both JET and Tore Supra are closer to the banana regime, where neo-classical theory predicts rotation in the counter current direction [24]. The accretion theory for toroidal rotation [23] predicts a reversal of the central toroidal rotation direction during a transition to H mode, with the plasma rotating in the co-current direction while in H mode. As can be seen from the JET rotation profiles presented here, the plasma often rotates in the co-current direction also in the L mode phase. This does not necessarily mean that the theory in [23] is wrong. It

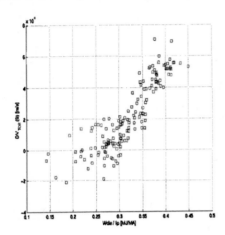

FIGURE 6. Change in toroidal rotation velocity as function of W_{DIA}/I_p during ³He minority ICRF heating.

could well describe the co-current acceleration of the plasma in H modes. However, one then needs another theory to explain the co-current rotation in L mode plasmas seen in JET.

We can conclude this section by saying that no single theory appears to explain all the observations of toroidal rotation during ICRF heating in JET, Alcator C-Mod and Tore Supra. It is possible that several effects are involved which conspire to give results which appear similar on the three machines. Alternatively, a more general explanation, yet to be clarified, exists.

References

1. T.S. Hahm and K.H. Burrell, Physics of Plasmas **2**, 1648 (1995).
2. E.J. Synakowski, Plasma Physics and Controlled Fusion **40**, 581 (1998).
3. A.M. Garofalo, et al., Nuclear Fusion **41**, 1171 (2001).
4. L.-G. Eriksson, R. Giannella, T. Hellsten, E. Källne and G. Sundström, Plasma Physics and Controlled Fusion **34**, 863 (1992).
5. L.-G. Eriksson, E. Righi and K.D. Zastrow, Plasma Physics and Controlled Fusion **39**, 27 (1997).
6. R. Rice, et al., Nuclear Fusion **38**, 75 (1998).
7. R. Rice, et al., Nuclear Fusion **39**, 1175 (1999).
8. L.-G. Eriksson, G.T. Hoang and V. Bergeaud, Nuclear Fusion **41**, 91 (2001).
9. G.T. Hoang et al, 14th Topical Conference on Radio Frequency Power in Plasmas (2001).
10. F.W. Perkins, R.B. White, P.T. Bonoli and V.S. Chan, Physics of Plasmas **8**, 2181 (2001).
11. F.W. Perkins, R.B. White and V.S. Chan, Physics of Plasmas **9**, 511 (2002).
12. O.N. Jarvis et al., Nucl. Fus. **36**, 1513 (1996); V.Kiptily, et al., Nucl. Fus. **42**, 999 (2002).
13. A. Kaye, et al., Fusion Engineering and Design **24**, 1 (1994).
14. L.-G. Eriksson, et al., Phys. Rev. Lett. **81**, 1231 (1998).
15. M.Mantsinen, at al., Phys. Rev. Lett. **89**, 115004-1 (2002).
16. L.-G. Eriksson and F. Porcelli, Plasma Physics and Controlled Fusion **43**, R145 (2001).
17. V. S. Chan, S. C. Chiu and Y. A. Omelchenko, Physics of Plasmas **9**, 501 (2002).
18. L.-G. Eriksson and F. Porcelli, Nuclear Fusion **42**, 959 (2002).
19. J. Hedin, T. Hellsten and L.-G. Eriksson, Nuclear Fusion **42**, 527 (2002).
20. J.-M. Noterdaeme, et al., Nuclear Fusion **43**, 274 (2003).
21. E. Lazzaro et al., proceedings of the 29th EPS conference Plasmas Physics and Controlled Fusion, Montreux, 2002, EPS ECA vol. 26B (2002) P-5.079.
22. A.L. Rogister, et al., Nuclear Fusion **42**, 1144 (2002).
23. B. Coppi, Nuclear Fusion **42**, 1 (2002).
24. M.N. Rosenbluth et al., Plasma Physics and Nuclear Fusion Research (IAEA Vienna, 1971) vol. 1, p. 495.

Study of Ion Cyclotron Range of Frequencies Mode Conversion in the Alcator C-Mod Tokamak

Y. Lin, S.J. Wukitch, P.T. Bonoli, A. Mazurenko, E. Nelson-Melby, M. Porkolab, J.C. Wright, I.H. Hutchinson, E.S. Marmar, D. Mossessian, S. Wolfe

MIT, Plasma Science and Fusion Center, Cambridge, MA 02139, USA

C.K. Phillips, G. Schilling

Princeton Plasma Physics Laboratory, Princeton, NJ 08543, USA

P. Phillips

Fusion Research Center, University of Texas, Austin, TX 78712,USA

Abstract. ICRF mode conversion (MC) in H(^{3}He, D) and D(H) plasmas have been studied in detail in Alcator C-Mod. In H(^{3}He, D) plasma, the mode converted ion cyclotron wave (MC ICW) was observed in tokamak plasmas for the first time using a phase contrast imaging system. The MC ICW was observed in the low field side of the ion-ion hybrid layer, and generally had a wavelength in-between the fast wave and the MC ion Bernstein wave (IBW). Localized mode conversion electron heating (MCEH) has been clearly observed for the first time in D(H) plasmas with moderate hydrogen concentration in Alcator C-Mod. Both on- and off-axis (high field side) MCEH have been studied. The MCEH profile was obtained from a break in slope analysis of T_e signals in the presence of rf shut-off. The experimental profiles were qualitatively in agreement with the predictions of the two-dimensional full-wave poloidal mode code TORIC. The electron heating contributions from MC ICW and MC IBW are examined from the TORIC simulations.

INTRODUCTION

In ion cyclotron range of frequency (ICRF) experiments in tokamaks, as demonstrated in Refs [1–4], mode converted (MC) waves can locally heat electrons, drive current and potentially drive plasma poloidal flow. Therefore, ICRF mode conversion is important to the advanced tokamak operational scenario. The fast magnetosonic waves (fast wave (FW)) launched from the rf (radio frequency) antenna on the low field side can be mode converted to ion Bernstein wave (IBW) and ion cyclotron wave (ICW) near the ion-ion hybrid layer [5], $n_\parallel^2 = S$, where S is the dielectric tensor component as in the Stix notation [6]. The ICW in tokamak plasmas has recently been observed experimentally for the first time in tokamak plasmas through a phase contrast imaging (PCI) system in H(^{3}He, D) plasmas in Alcator C-Mod [7]. The MC ICW exists in the low field side (LFS) of the ion-ion hybrid layer, and usually has a longer wavelength than the MC IBW, but shorter wavelength than the fast wave. The excitation of the MC ICW is the competition result of the poloidal B field (B_{pol}) and the temperature [5, 8]. The MC ICW favors higher B_{pol} and lower temperature. The experimental observation of MC ICW suggests that ICRF mode conversion is more complicated than previously thought.

In this paper, we present the experimental detection of the MC ICW in Alcator C-Mod, followed with the method to identify this wave. We also present a detailed measurement of the mode conversion electron heating (MCEH) in Alcator C-Mod in D(H) plasmas [9]. The study concludes that the MC IBW is the primary mode converted wave for on-axis MC, while for off-axis MC, the MC ICW can have comparable contribution to the MCEH as the MC IBW.

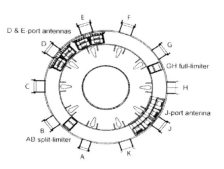

FIGURE 1. ICRF antennas in Alcator C-Mod.

EXPERIMENTAL OBSERVATION OF MC ICW

PCI Observation

Alcator C-Mod (R ~ 0.67 m, a ~ 0.22 m, $B_t \le 8.1$ T) [10] has three fast wave antennas (Fig. 1): Two 2-strap antennas (strap width 10 cm and separation of 25.75 cm center to center) at D port and E port [11], and a 4-strap antenna (strap width 8 cm and separated at 18.6 cm center to center) at J port [12]. The D and E port antennas are operated at 80.5 MHz and 80 MHz respectively. The J-port antenna is tunable from 40 MHz to 80 MHz.

The PCI system [13] in Alcator C-Mod is shown in Fig. 2. The PCI viewing CO_2 laser ($\lambda = 10.6$ μm, expanded to a width of 15 cm) is in front of the E port antenna. The PCI technique [14] relies on the interference of scattered and appropriately phase-shifted un-scattered radiation passing through a phase object such as plasma. The laser light is imaged onto a 12-element HgCdTe photoconductive linear array after passing through the plasma and reflecting from a 90-degree phase plate. The diagnostic is most

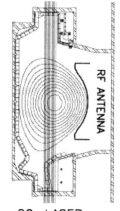

FIGURE 2. PCI system in Alcator C-Mod.

sensitive to density perturbations whose surfaces of constant phase are aligned vertically with the laser beam.

Experiments were performed to study the ICRF mode conversion in H(He3, D) plasmas [7]. In these experiments, the PCI laser intensity was modulated at a frequency offset typical 0.3 MHz from the rf frequency so that we can measure the rf signals at the beat oscillation signal frequency. Figure 3 shows the k-spectrum contour from the PCI signal of one of the plasma discharges in these experiments. The rf beat signal is at 350.9 kHz corresponding to the rf frequency at 80.5 MHz. A wave at k_R of +7 cm^{-1} is clearly seen. The wave is usually seen with a k_R at +4 to +10 cm^{-1} in these experiments. Based on the estimated plasma mixture concentration of different species, the PCI observation window is on the LFS of the ^{3}He-H hybrid layer (Fig. 4). The positive k indicates that the phase velocity of the wave is toward the LFS. The wave also has a longer wavelength than that of expected MC IBW on the high field side (HFS) of the ^{3}He-H hybrid layer. The observed short wavelength wave is identified as the slow electromagnetic MC ICW of the hydrogen species as shown in the next section.

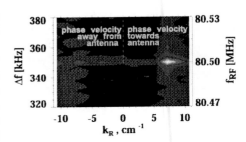

FIGURE 3. ICW appears in the k space contour of PCI signals. Plasma parameters: B_ϕ = 5.84 T, I_p = 800 kA, 59% H, 4% ^{3}He, 33% D, n_{e0} = 2×10^{20} m^{-3}, T_{e0} = 1.3 keV.

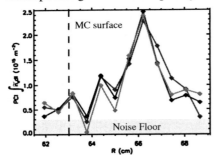

FIGURE 4. PCI observation window is on the lower field side of the MC surface.

Identification of MC ICW

To identify the physical origin of the wave, we compared the experimental observation with TORIC simulation [15, 16] and the solution of the full-wave hot plasma dispersion equation [17]. In Fig. 5(a), the E_z field contour from a TORIC simulation is shown for a H(^{3}He, D) plasma. A short wavelength wave structure appears clearly in the LFS of the ion-ion

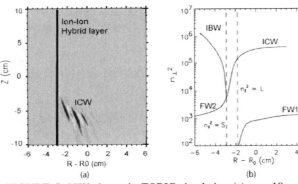

FIGURE 5. ICW shown in TORIC simulation (a), n_ϕ =10, and solutions of the dispersion equation (b). Plasma parameters: B_ϕ = 5.8 T, I_p = 800 kA, 33% H, 23% ^{3}He, 21% D, n_{e0} = 2.4×10^{20} m^{-3}, T_{e0} = 1.5 keV, f_{rf} = 80 MHz

(^{3}He-H) hybrid layer. A similar result is also obtained by simulations using AORSA [18]. This wave structure agrees with the experimental observation by PCI in all aspects such as wave location and wavelength. The origin of the wave is further clarified by solving the full electromagnetic hot plasma dispersion equation. We include the B_{pol} effect, and solve the equation around a magnetic surface. With the assumption that $k_\perp \sim k_{pol}$, and using $k_\parallel \approx k_{pol} B_{pol}/B + n_\phi B_{tor}/RB$, where n_ϕ is toroidal mode number, and B_{tor} is the toroidal B field, we find the root MC ICW. In Fig. 5(b), we plot the three waves involved near the mode conversion region. The ICW wavelength is similar to that shown in TORIC simulation. The MC IBW and FW roots are obtained on the mid-plane, while the MC ICW root is calculated on the magnetic surface tangential to the ion-ion hybrid layer. In Fig. 6, we compare the power deposition profile from TORIC simulation and that calculated based on the imaginary $k_\perp$ in the dispersion equation solution. The good agreement of these two calculations suggests that the wave appears on the LFS of the ion-ion hybrid layer in TORIC simulation, and detected by PCI, is the MC ICW. This is the first observation of MC ICW in tokamak plasmas.

The up-down asymmetry of the ICW wave front (Fig. 5(a)) is a consequence of the fact that the ICW only propagates to the LFS of the mode conversion layer. This propagation direction corresponds to positive m-numbers ($\sim k_{pol}$ r) below the mid-plane and negative m-numbers above. For a positive B_{pol}, the positive m-numbers below the mid-plane result in larger values of $k_\parallel$ which the local dispersion relation admits as a propagating ICW. In contrast, the negative m-numbers above the mid-plane yield reduced values of $k_\parallel$ which are evanescent modes of the local dispersion relation.

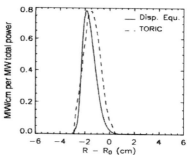

FIGURE 6. MC ICW power deposition profiles.

MODE CONVERSION DIRECT ELECTRON HEATING (MCEH)

In previous experiments in Alcator C-Mod, the MCEH in D(^{3}He) and H(^{3}He) plasmas have been studied in detail [19–22]. MCEH has also been studied in other tokamaks, e.g., D(^{3}He, ^{4}He) plasmas [1] and D(T) plasmas [23] in TFTR, ^{3}He(H) plasmas in ASDEX Upgrade [24] and Tore Supra [25, 26], and ^{4}He(^{3}He) in JET [27]. A numerical study based on METS95 1D integral wave code showed that mode conversion heating may be significant in D(H) plasmas with moderate hydrogen concentration in Alcator C-Mod [28]. However, a previous experimental study on D(H) plasmas in Alcator C-Mod as reported in Ref. [29] lacked the evidence of localized electron absorption, unlike that predicted by TORIC. The measured power deposition profile in the experiment was possibly limited by the accuracy of the species concentration measurement and the spatial/temporal resolution of the temperature measurement.

Detailed MCEH has recently been studied in D(H) plasmas in Alcator C-Mod [9]. Figure 7 shows one of the typical plasma discharges with a moderate H concentration in these experiments. The hydrogen concentration is obtained by a spectroscopic measurement of hydrogen and deuterium Balmer α-line level near the plasma edge [30]. We assume a constant concentration for the entire plasma. For this plasma, the rf power is applied consecutively by J (70 MHz), D (80.5 MHz) and E (80 MHz) antennas at a level about 1.5 MW. The D-H hybrid layer is nearly on axis when J port is on, while the layer is off-axis on the high field side when D or E is on. The direct electron heating can be measured by the break in slope of T_e signals during rf shut-off [31]. T_e is measured by a 2nd harmonic heterodyne ECE system with high spatial resolution (≤ 7 mm) and temporal response (≤ 5 μs) [32]. We can obtain the power deposition profile from the following equation:

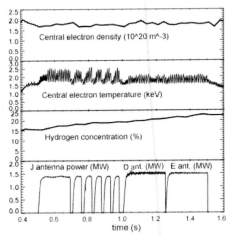

FIGURE 7. Plasma parameters for the MCEH study.

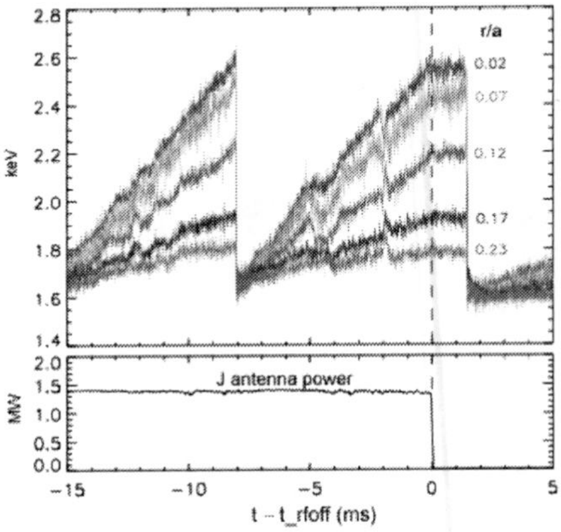

$$S(r) \approx \frac{3}{2} n_e \Delta \frac{\partial T_e(r)}{\partial t},$$ where $\Delta \frac{\partial T_e}{\partial t}$ is the difference of the slopes before and after the rf power transition. The time window used to calculate the break in slope is in the range of 0.5 to 1 ms, which is substantially shorter than the time scales of changes in radiated power, ohmic heating, and heat-exchange between ions and electrons. The fraction of rf power to electron heating is simply $\eta_e^{rf} = \int S(r)dV / P_{total}^{rf}$, where the volume integration is performed based on the magnetic surfaces reconstructed by EFIT.

On-axis Mode Conversion Heating

Figure 8 shows some temperature signals from the high resolution ECE system near the rf transition. The sampling rate is at 200 kHz. At t = 0.8744 sec, the rf is shut-off, and clear

FIGURE 8. Break in slope seen in ECE signals. t_{rfoff} = 0.8744 of the discharge in Fig. 7.

break in slope is seen in the temperature signals in the channels near the magnetic axis, while the break in slope is insignificant away from the axis. Figure 9 shows the experimentally obtained MCEH profile in comparison with the TORIC simulation result. The shaded area is the noise level indicating the level one would get if calculating the break in slope at time points other than the rf shutoff. The TORIC simulation result is done with toroidal modes $n_\phi = \pm[4, 9, 10, 11, 12, 13, 14, 15, 16, 17]$, and sum over all results by considering the antenna toroidal spectrum, which is peaked at $n_\phi = \pm 13$. A good agreement is shown between the experimentally measured MCEH profile and TORIC result in the expected mode conversion region $0 < r/a < 0.25$. In this region, the volume integrated MC power fraction is $\eta_e^{rf} \sim \mathbf{0.16}$ from the experimentally measured profile, and $\eta_e^{rf} \sim \mathbf{0.14}$ from the TORIC simulation. The apparent peak in the experimental result in the region of $0.25 < r/a < 0.4$ may be due to a spurious effect on the break in slope calculation

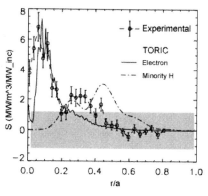

FIGURE 9. MCEH deposition profiles for on-axis mode conversion

from a partial reconnection in this region before the rf shut-off. The minority heating profile from TORIC is also shown in this figure. The TORIC simulation shows that the direct electron heating power is primarily from the MC IBW. The result is consistent with the fact that B_{pol}, which is crucial for the excitation of the ICW, is small near the magnetic axis.

Off-axis Mode Conversion Heating

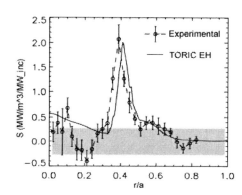

FIGURE 10. MCEH profile for off-axis mode conversion.

Figure 10 shows the MCEH profile obtained at t = 1.5024 sec in the plasma discharge shown in Fig. 7. The E port antenna is on at 80 MHz with power level at about 1.5 MW. The D-H hybrid layer is off axis on the HFS at about r/a = 0.36 for the dominant toroidal mode number $n_\phi = \pm 10$ of this antenna. The deposition profile from TORIC is also plotted in the figure. The TORIC simulation is done on $n_\phi = \pm [2, 4, 6, 8, 10, 12, 14, 16]$. In the expected mode conversion region $0.35 < r/a < 0.7$, the volume integrated total MCEH power from experiment is $\eta_e^{rf} \sim \mathbf{0.20}$, and $\eta_e^{rf} \sim \mathbf{0.18}$ from the TORIC simulation. The experimental result and TORIC result are similar in both location and shape. In Fig. 11, we plot the electron heating profile from TORIC simulation and the power partition of the MC ICW, MC IBW, and fast wave electron heating. The result suggests a comparable MC heating for the ICW ($\eta_e^{ICW} \sim \mathbf{0.087}$) and IBW ($\eta_e^{IBW} \sim \mathbf{0.093}$), while there is a small part of electron heating from the fast wave ($\eta_e^{FW} \sim \mathbf{0.03}$) near the hydrogen cyclotron resonance on axis. The ICW and IBW peak at approximately the same r/a. In Fig. 12, the 2-D contour of S_{ELD}, the power deposition through electron Landau damping (ELD), from the TORIC simulation (n_ϕ = 10) is plotted. Both MC IBW and ICW are clearly shown in the figure at a similar level, with the IBW on the HFS of the ion-ion hybrid layer and the ICW on the LFS. Generally, the total MC efficiency and power partition between the ICW and IBW are very complicated, which depend on a number of plasma parameters, such as plasma current, species mixture, density, and temperature. Higher B_{pol} and lower temperature are favored by the MC ICW.

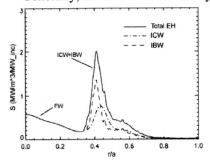

FIGURE 11. MC ICW, MC IBW and FW power partition for the off-axis mode MC.

It is difficult to experimentally distinguish the MC ICW and IBW contributions in heating because they generally peak at a similar r/a (near the hybrid layer location). The mode conversion power partition in this study is based on the agreement between the TORIC simulation and the experimental measurement. The good agreement suggests that the MC ICW plays an important role in the ICRF mode conversion heating. The ICW may also play a substantial role in other ICRF MC applications, such as plasma flow drive and current drive.

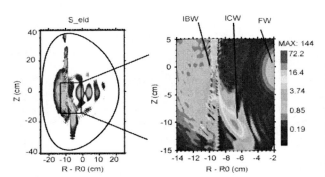

FIGURE 12. MCEH profile for off-axis mode conversion. S_{ELD} has units of MW/m^3 per m^2 area per MW antenna input power, and is plotted on a logarithmic scale.

CONCLUSION

The mode converted ICW was observed for the first time in tokamak plasmas in Alcator C-Mod using a PCI system. The wave is on the LFS of the ion-ion hybrid layer, and it has wavelength generally in-between the MC IBW and fast wave. Detailed measurement of the MCEH in D(H) plasmas with moderate H concentration and comparison with TORIC modeling show that the MC IBW is the primary MC wave for the case of on-axis mode conversion, while the MC ICW can have comparable contribution as that of MC IBW in electron heating when the mode conversion is off-axis.

ACKNOWLEDGMENTS

This work is supported at MIT by U.S. Department of Energy Cooperative agreement No. DE-FC02-99-ER54512.

REFERENCES

1. R. Majeski, J. H. Rogers, and S. H. Batha *et al.*, Phys. Rev. Lett. **76 (5)**, 764 (1996).
2. J. R. Wilson, R. E. Bell, S. Bernabei *et al.*, Phys. Plasma **5**, 1721 (1998).
3. L. A. Berry, E. F. Jaeger, and D. B. Batchelor, Phys. Rev. Lett. **82 (9)**, 1871 (1999).
4. C. K. Phillips, M. G. Bell, R. E. Bell, S. Bernabei *et al.*, Nucl. Fusion **40**, No. 3Y, (463) 2000.
5. F. W. Perkins, Nucl. Fusion **17**, 1197 (1977).
6. T. H. Stix, in *Waves in Plasmas* (American Institute of Physics, New York, 1992).
7. E. Nelson-Melby, M. Porkolab, P. T. Bonoli, Y. Lin, A. Mazurenko, and S. J. Wukitch, Phys. Rev. Lett. **90**(15), 155004(2003).
8. D. W. Faulconer, D. Van Eester, and R. R. Weynants, *Proc. 14th EPS Conference on Contr. Fusion and Plasma Phys.*, ECA Vol. **11D**, p932 (1987).
9. Y. Lin, S. J. Wukitch *et al*, Plasma Phys. and Control. Fusion **45** (6) 1013 (2003).
10. I. H. Hutchinson, R. L. Boivin, F. Bombarda *et al.*, Phys. Plasmas **1**, 1511 (1994).
11. Y. Takase *et al.*, *Proc. 14th Symp. Fusion Engineering, San Diego (Piscataway, NJ):IEEE (1992)*, p118.
12. S. J. Wukitch, R. L. Boivin, P.T. Bonoli *et al*, *Proc. 19th IAEA Fusion Energy Conference (Lyon) FT/P1-14 (2002)*.
13. A. Mazurenko, PhD thesis, Massachusetts Institute of Technology (2001).
14. H. Weisen, Rev. Sci. Instrum. **59**, 1544 (1988).
15. M. Brambilla, Nucl. Fusion **38**, 1805 (1998).
16. M. Brambilla, Plasma Phys. Control. Fusion **41** (1), 1(1999).
17. F. F. Chen, in *Introduction to Plasma Physics and Controlled Fusion*, Plenum Press, New York, 1984.
18. E.F. Jaeger, L.A Berry, J. Myra, D.B. Batchelor *et al*, Phys. Rev. Lett. (2003), accepted for publication.
19. P. T. Bonoli, M. Brambilla, E. Nelson-Melby *et al.*, Phys. Plasmas **7**, 1886 (2000).
20. P. T. Bonoli, P. O'Shea, M. Brambilla *et al.*, Phys. Plasmas **4**, 1774 (1997).
21. P. J. O'Shea, PhD thesis, Massachusetts Institute of Technology (1997).
22. E. Nelson-Melby, PhD thesis, Massachusetts Institute of Technology (2001).
23. R. Majeski, C. K. Phillips, and J. R. Wilson, Phys. Rev. Lett. **73**(16), 2207 (1994).
24. J-M. Noterdaeme, S. Wukitch, D. A. Hartmann *et al.*, *Proc. 16th IAEA Fusion Energy Conference (Montreal) IAEA-CN-64/F1-EP-4 (1996)*.
25. B. Saoutic, A. Becoulet, T. Hutter *et al*, Phys. Rev. Lett. **76** (10), 1647 (1996).
26. I. Manakhov, Yu. Petrov, V. Basiuk, A. Becoulet, and F. Nguyen, *13th Conference on Applications of Radio Frequency Power to Plasmas* (American Institute of Physics Conference Proceedings 485, New York, 1999, p136).
27. M. J. Mantsinen, M.-L. Mayoral, E. Righi *et al*, *14th Conference on Applications of Radio Frequency Power to Plasmas* (American Institute of Physics Conference Proceedings 595, New York, 2001, p59).
28. C. K. Phillips, P. T. Bonoli, J. C. Hosea *et al*, *12th Conference on Applications of Radio Frequency Power to Plasmas* (American Institute of Physics Conference Proceedings 403, New York, 1997, p265).
29. G. Taylor, B. LeBlanc, C. K. Phillips *et al*, *13th Conference on Applications of Radio Frequency Power to Plasmas* (American Institute of Physics Conference Proceedings 485, New York, 1999, p490).
30. T. E. Tutt, Department of Nuclear Engineering Master's thesis, Massachusetts Institute of Technology (1999). Also MIT PSFC/RR-99-11.
31. D. J. Gambier, M. P. Evrard, J. Adam *et al*, Nuclear Fusion **30**(1), 23 (1990).
32. J. W. Heard, C. Watts, R. F. Gandy, P. E. Phillips *et al*, Rev. Sci. Instrum. **70** (1), 1011 (1999).

The Analysis of ICRF Heating Experiment in View of the Confinement of High-energy Particles in LHD

K. Saito, R. Kumazawa, T. Mutoh, T. Seki, T. Watari, F. Shimpo,
G. Nomura, A. Kato, M. Yokota, M. Isobe, T. Ozaki, M. Osakabe,
M. Sasao[1], T. Saida[1], T. Yamamoto[2], Y. Torii[2], N. Takeuchi[2],
and LHD Experimental Group

National Institute for Fusion Science, Toki 509-5292 Japan
[1]*Tohoku University, Faculty of Engineering, Sendai 980-8579, Japan*
[2]*Nagoya University, Faculty of Engineering, Nagoya 464-8601, Japan*

Abstract. The confinement of high-energy particles produced by ICRF heating was studied in LHD experimentally setting magnetic axis at 3.6m and 3.75m in two methods. It was concluded that the confinement of high-energy particles with R_{ax}=3.6 m was better than that with R_{ax}=3.75m. Based on this result the long pulse experiments were conducted in the magnetic configuration of R_{ax}=3.6m. The plasma was sustained for 150sec. The total input power reached 71MJ though the pulse length was limited by the collapse of plasma due to density increase. The second harmonic experiment was done also with R_{ax}=3.6m. The stored energy increased and the particles with the high-energy were observed.

INTRODUCTION

In Large Helical Device (LHD) by ion cyclotron range of frequency (ICRF) heating high performance of plasma was attained with the configuration of R_{ax}=3.6 m. e.g. the plasma stored energy over 200 kJ and long pulse discharge sustained by ICRF heating [1-6]. In this configuration plasma had the same performance of confinement with tangentially injected NBI plasma [7].

In ICRF minority ion heating H^+ is used for minority ion and He^{2+} for majority ion. Figure 1 is the calculated position of resonance layers and cutoff layers in the cross section in front of ICRF antennas. In this case magnetic field strength is 2.75 T, the magnetic axis position is R_{ax}=3.6 m, and frequency is 38.47 MHz which is the same with that of experiment. The

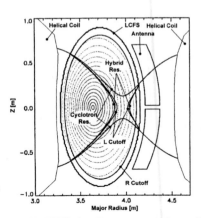

FIGURE 1. Resonance and cutoff layers in front of ICRF antennas.

electron density is 1×10^{19} m^{-3} and the minority concentration ratio, $n_H^+/(n_H^+ + n_{He}^{2+})$ is 5%. The ion cyclotron resonance layer locates near the saddle point of the mod-B surface. The fast wave which transferred through the evanescent region is damped mainly around saddle shape point because of large absorption area where the minority ion (H$^+$) is accelerated perpendicularly [5, 6]. High-energy particles are created by this mode of ICRF heating. The energetic particles can be observed with detectors installed in LHD.

Figure 2 shows orbits of the trapped particles in the case of (a) R_{ax}=3.75 m and (b) R_{ax}=3.6 m. As shown in this figure, the orbit is largely deviated from magnetic surface in the R_{ax}=3.75 m configuration. Therefore the confinement of the high-energy particles is thought to be better in the R_{ax}=3.6 m configuration than that in the R_{ax}=3.75 m configuration. With the good confinement of high-energy ions, bulk plasma will be enough heated by the high-energy particles, besides impurity due to high-energy ions decreases.

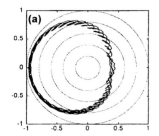
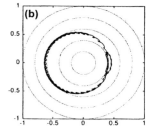

FIGURE 2. Orbits of trapped particles.
(a) R_{ax}=3.75 m (b) R_{ax}=3.6 m

TAIL TEMPERATURE ANALYSIS

In this section the loss of high-energy ions will be analyzed with the experimentally obtained tail temperature and Monte Carlo simulation in the cases of R_{ax}=3.6 m and 3.75 m configuration. Figure 3 was the discharge of ICRF heating with NBI. ICRF maximum power was 1.0 MW, then the power of NBI was 1MW too. The radius of magnetic axis was 3.6 m. Electron density was 0.3×10^{19} m^{-3} and electron temperature was 2 keV and plasma stored energy was 130 kJ. In this discharge high-energy

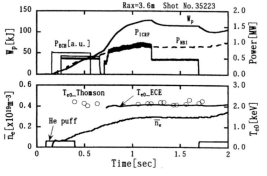
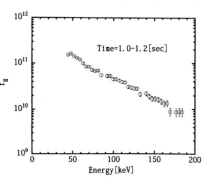

FIGURE 3. Time evolution of plasma parameters.

FIGURE 4. Energy spectrum of energetic ions.

particles were observed with Si-FNA (silicon-diode-based fast neutral analyzer), which detects the high-energy particles with almost perpendicular pitch angle [8]. As shown in fig.4 the particle energy was up to 200 keV and the tail temperature T_{tail} determined by following equation was 50 keV in this discharge.

$$T_{tail} = -[d(\ln f_H)/dE]^{-1} \tag{1}$$

On the other hand effective temperature T_{eff}, which is the tail temperature without loss of energetic ions, can be calculated with power density and plasma parameters,

$$T_{eff} = \frac{p_{abs} t_s / 2}{n_H} \tag{2}$$

where, p_{abs} is power density and t_s is slowing down time and $t_s/2$ is confinement time of high-energy ions. Figure 5 was the relation between T_{tail} and T_{eff}. It was seen that though experimental tail temperature increases almost linearly with T_{eff} in the case of R_{ax}=3.6 m, T_{tail} was saturated at around 35 keV in the case of R_{ax}=3.75 m. The difference between two cases in fig. 5 was due to the difference of loss time of high-energy particles. The real confinement time of high-energy particles τ_{tail} was written as,

$$\frac{1}{\tau_{tail}} = \frac{1}{\tau_s/2} + \frac{1}{\tau_{loss}} \tag{3}$$

where, τ_{loss} was the loss time of high-energy particles. Therefore T_{tail} was smaller than T_{eff}. The transfer efficiency η_{trns}, which was transferred power from tail to bulk divided by absorbed RF power by resonant particles, was written as [9],

$$\eta_{trns} = \frac{\tau_{tail}}{t_s/2} = \frac{T_{tail}}{T_{eff}} \tag{4}$$

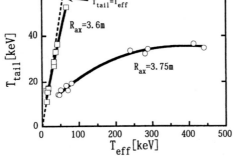

FIGURE 5. Relation between T_{eff} and T_{tail} with two radii of magnetic axis.

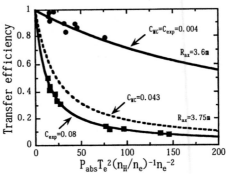

FIGURE 6. Transfer efficiencies and plasma parameters.

It was found that the transfer efficiency of R_{ax}=3.75 m configuration was smaller than that of R_{ax}=3.6 m configuration due to the loss of high-energy particles. The transfer efficiency η_{trns} was numerically calculated by the Monte Carlo simulation. η_{trns} was written with the following formula [9, 10],

$$\eta_{trns} = \frac{1}{1+CP_{abs}T_e^2(n_H/n_e)^{-1}n_e^{-2}} \tag{5}$$

where, units of P_{abs}, T_e, and n_e were MW, keV, and 10^{19}m^{-3}, respectively. C was 0.004 for R_{ax}=3.6 m and 0.043 for R_{ax}=3.75 m. In fig. 6 the transfer efficiencies in two case of R_{ax}=3.6 m and R_{ax}=3.75 m were plotted. The rounds and squares were the experimental results for the R_{ax}=3.6 m and R_{ax}=3.75 m configurations, respectively. Lines were the fitting curve of the experimental data and the results of Monte Carlo simulation. It was able to be noted that the transfer efficiency of R_{ax}=3.6 m was better than that of R_{ax}=3.75 m. The loss time calculated by eq. 3, 4 was order of 1 sec in R_{ax}=3.6 m configuration and 0.1 sec in R_{ax}=3.75 m configuration.

PHASE DELAY ANALYSIS

In the former section we could say that the confinement of high-energy particles were better in the configuration of R_{ax}=3.6 m than that of R_{ax}=3.75 m by the tail temperature measurement. The good confinement performance of R_{ax}=3.6 m can also be shown by the method of power modulation. Figure 7 was the discharge by ICRF heating and NBI heating. The magnetic axis was R_{ax}=3.6 m. ICRF power was modulated in the frequency of 4 Hz. The average power of ICRF heating and NBI were about 400 kW. Electron temperature was 2 keV and line averaged electron density was 0.3×10^{19} m^{-3}. In this discharge we detected high-energy particles with Si-FNA. The counts were plotted in the bottom row of fig. 7. The counts were modulated with ICRF heating power. But it was found that the phase was different in every

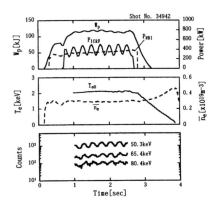

FIGURE 7. Time evolution of power-modulated plasma.

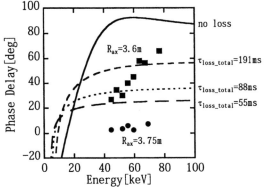

FIGURE 8. Energy dependence of phase delay.

particle energy. The phase delay (the difference between the phase of ICRF heating power and that of high-energy particle counts) was determined by Fourier analysis, and the relation with energy was plotted in fig. 8. In the case of R_{ax}=3.75 m the phase delay was small.

To calculate the phase delay Fokker-Planck equation was used [11,12]. The distribution function $f(v,t,\theta)$ was expanded to the second order of Legendre function.

$$f(v,t,\theta) = A(v,t) + \frac{1}{2}B(v,t)(3\cos^2\theta - 1) \tag{6}$$

Here θ was a pitch angle. Function A and B were calculate by using two differential equations,

$$\frac{\partial A}{\partial t} = \frac{1}{v^2}\frac{\partial}{\partial v}[-\alpha v^2 A + \frac{1}{2}\frac{\partial}{\partial v}(\beta v^2 A) + Kv\frac{\partial}{\partial v}v(A - \frac{B}{5}) - K(A + \frac{2}{5}B)v]$$
$$- A/\tau_{loss} + \tilde{N}(\frac{m_H}{2\pi T_{H\_bulk}})^{1.5}\exp(-\frac{m_H v^2}{2T_{H\_bulk}}) \tag{7}$$

$$\frac{\partial B}{\partial t} = -\frac{1}{v^2}\frac{\partial}{\partial v}(\alpha v^2 B) + \frac{1}{2v^2}\frac{\partial^2}{\partial v^2}(\beta v^2 B) - \frac{3}{2}\frac{\gamma}{v^2}B$$
$$+ \frac{K}{v^2}\frac{\partial}{\partial v}v\frac{\partial}{\partial v}v[-A + \frac{5}{7}B] - \frac{K}{v^2}[3A + \frac{30}{7}B] + \frac{K}{v^2}\frac{\partial}{\partial v}v[4A - \frac{5}{7}B] - B/\tau_{loss} \tag{8}$$

Coefficients α, β, γ were functions of velocity and K was the indication of the power density. The second term of eq. 7 and the last term of eq. 8 were the loss term, where τ_{loss} of the second term of eq. 7 is particle loss time. The third term of eq. 7 was source term. $\tilde{N}$ was selected to compensate the particle loss as,

$$\tilde{N} = 4\pi \int_0^\infty v^2 A/\tau_{loss} dv \tag{9}$$

The lines of fig. 8 were calculated phase delay depending on energy. In this calculation τ_{loss} was assumed inversely proportional to energy. Bulk ion and electron temperatures were 1.5 keV, electron density was 0.3×10^{19} m^{-3}, K=0.42$\times10^{13}$ W/kg, minority concentration ratio $n_H^+/(n_H^+ + n_{He}^{2+})$ was 0.2, and pitch angle was 90 deg. Then $t_s/2$ was 0.3 sec. τ_{loss_total} was total loss time $N/\tilde{N}$. The phase delay increased with energy and loss time. The loss times in both case of R_{ax} were too short compared with former results. The difference may be due to the integration effect along the line-of-sight of Si-FNA. But It can be said qualitatively that confinement of high-energy particles is better in the case of R_{ax}=3.6 m than that of R_{ax}=3.75 m.

LONG PULSE ICRF HEATING

So far the magnetic configuration of R_{ax}=3.6 m has the best confinement performance. Therefore the long pulse ICRF heating was conducted with this configuration. Figure 9 is the result of long pulse discharge by ICRF minority heating. Line averaged electron density was 0.6×10^{19} m^{-3}, net radiated power was 500 kW, electron and ion temperature were 2 keV. This plasma was sustained during 150 sec and total input power reached 71 MJ. High-energy particles were created during this discharge. Figure 10 shows an energy spectrum detected by natural diamond detector (NDD) [13]. This detects high-energy particles with pitch angle of about 90 deg. The particles with the energy up to 200 keV were well confined and observed by NDD. But in the last of the discharge the radiation power was increased and plasma was collapsed. This may be due to the density increase by the temperature increase in carbon divertor plate. The wall temperature increased only 3 °C but the local temperature of divertor plate, near the inner port of third toroidal section, increased up to 400°C and the intensity of H$_\alpha$ was most increased in this toroidal section. A hot spot on another divertor plate was also observed. This hot spot was observed only a few second after the start of ICRF heating but not observed in NBI plasma. Therefore it may be due to high-energy particles with large pitch angle produced by ICRF heating. These problems may be alleviated by increase of electron density. Another possible reason for density increase is out gas from carbon side limiter of ICRF antenna due to eddy current. By the increase of electron density may be effective because high loading resistance due to high electron density means small antenna current.

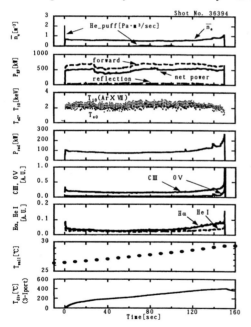

FIGURE 9. Time evolution of plasma parameters in the long pulse discharge.

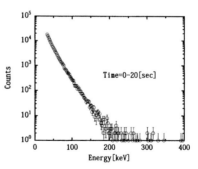

FIGURE 10. Energy spectrum of high-energy particles in the long pulse discharge.

SECOND HARMONIC HEATING

In the magnetic field configuration of R_{ax}=3.6 m a second harmonic ICRF heating experiment was conducted by lowering the magnetic field strength and clear evidence of heating was obtained. Figure 11 shows a discharge with second harmonic heating superposed on NBI target plasma. The magnetic field strength on the axis was 1.363 T. Two pulses with frequencies f=38.47MHz and 40.47MHz were applied from t=1.0 sec. The former had a pulse length of 0.6 sec and the latter one of 0.3 sec, and the total injected power was 2 MW in the early 0.3 sec. The plasma stored energy, W_p increased from 110 kJ to 130 kJ after RF was turned on. During ICRF heating the line averaged electron density was 0.75-1.0×10^{19} m^{-3} and electron temperature on axis was 1.0-1.4 keV. Figure 12 shows the energy spectrum of the high-energy particles measured with the NDD. The labels A, B, and C in Fig. 12 correspond to the times in Fig. 11. A high-energy tail was created on the application of RF (B) and it grows as time passes (C). The effective temperature reached 19 keV at time C. Figure 13 shows the beta dependence of the heating efficiency, which was defined as the absorbed power divided by injected ICRF power. It was found that the heating efficiency increased with beta value. This suggested that second harmonic heating was enhanced by the increased beta value due to the finite Larmor radius effect. Therefore, it was expected that

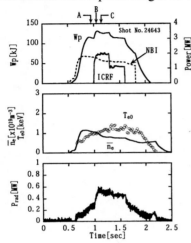

FIGURE 11. Discharge with second harmonic ICRF heating superposed on NBI target plasma.

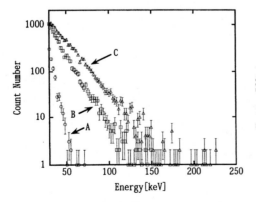

FIGURE 12. Energy spectrum of energetic particles in the second harmonic heating.

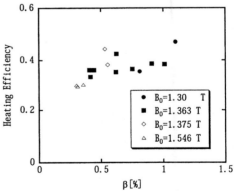

FIGURE 13. Relation between heating efficiency and beta.

the performance would be further improved in future experiments with increased RF power.

CONCLUSION

The confinement of the high-energy particles was studied in LHD experimentally for the plasma major radii of 3.6m and 3.75m by the measurement of tail temperatures with various power densities and by the measurement of phase delay of high-energy particle flux during power modulation. These results were analyzed with Monte Carlo simulation and Fokker-Planck equation, respectively. It was concluded that the confinement of high-energy particles with R_{ax}=3.6 m was better than that with R_{ax}=3.75 m.

Based on these results, the long pulse experiments were conducted in the configuration of the major radius of 3.6m. The plasma was sustained for 150 sec. Then electron density was 0.6×10^{19} m^{-3} and electron and ion temperatures were 2 keV on the magnetic axis. The total input power was 71 MJ. The high-energy particles were detected up to 200 keV. Pulse length was limited by the collapse of plasma due to density increase, which may be associated with the temperature increase in divertor plate or side limiter of antenna.

The second harmonic experiment was done also with R_{ax}=3.6 m configuration. The stored energy increased and the particles with the energy of up to 200 keV were observed.

These high performance were owing to the magnetic field configuration of R_{ax}=3.6 m, since high-energy particles were well confined with this configuration.

ACKNOWLEDGEMENTS

The authors wish to thank the technical staff of the LHD group in the National Institute for Fusion Science for their helpful support during this work.

REFERENCES

1. MOTOJIMA, O., et. al., IAEA Conference, 2002, Lyon, IAEA-CN-94/OV/1-6.
2. SEKI, T., et al., Journal of Plasma and Fusion Research SERIES **3** (2000) 359.
3. WATARI, T., et. al., Nucl. Fusion **41** (2001) 325.
4. MUTOH, T., et. al., IAEA Conference, 2002, Lyon, IAEA-CN-94/EX/P2-19.
5. MUTOH, T., et al., Phys. Rev. Let. **85** (2000) 4530.
6. SAITO, K., et al., Nucl. Fusion **41** (2001) 1021.
7. KUMAZAWA, R., et al., Physics of Plasmas **8** (2001) 2139.
8. OSAKABE, M., Review of Scientific Instruments **72** (2001) 788.
9. KUMAZAWA, R., et al., to be published in Plasma Physics and Controlled Fusion.
10. MURAKAMI, S., et al., Nucl. Fusion **39** (1999) 1165.
11. SAITO, K., et al., Plasma Physics and Controlled Fusion **44** (2002) 103.
12. STIX, T.H., Nucl. Fusion **15** (1975) 737.
13. ISOBE, M., et al., Review of Scientific Instruments **72** (2001) 611.

Recent ^{3}He Radio Frequency Heating Experiments On JET

D. Van Eester[1], F. Imbeaux[2], P. Mantica[3], M. Mantsinen[4], M. de Baar[5], P. de Vries[5], L.-G. Eriksson[2], R. Felton[6], A. Figueiredo[7], J. Harling[6], E. Joffrin[2], K. Lawson[6], H. Leggate[6], X. Litaudon[2], V. Kiptily[6], J.-M. Noterdaeme[8,9], V. Pericoli[10], E. Rachlew[11], A. Tuccillo[10], K.-D. Zastrow[6] & JET-EFDA contributors

[1] LPP-ERM/KMS, Association "EURATOM – Belgian State", TEC, Brussels, Belgium
[2] Association EURATOM-CEA, CEA-Cadarache, Saint-Paul-Lez Durance, France
[3] Instituto di Fisica del Plasma, EURATOM-ENEA-CNR Association, Milan, Italy
[4] Helsinki University of Technology, Association EURATOM-Tekes, Finland
[5] FOM-Rijnhuizen, Associatie EURATOM-FOM, TEC, Nieuwegein, The Netherlands
[6] EURATOM/UKAEA Fusion Association, Culham Science Centre, Abingdon, United Kingdom
[7] Associação EURATOM-IST, Centro de Fusão Nuclear, Lisboa, Portugal
[8] Max-Planck IPP-EURATOM Assoziation, Garching, Germany
[9] Gent University, EESA Department, B-9000 Gent, Belgium
[10] Associazione EURATOM-ENEA sulla Fusione, CR Frascati, Rome, Italy
[11] KTH Association EURATOM/VR, SE-10044 Stockholm, Sweden

Abstract Various ITER relevant experiments using ^{3}He in a majority D plasma were performed in the recent JET campaigns. Two types can be distinguished: dedicated studies of the various RF heating scenarios which rely on the presence of ^{3}He, and physics studies using RF heating as a working tool to provide a tunable heat source. As the success of a number of these experiments depended on the capability to keep the ^{3}He concentration fixed, real time control of the ^{3}He concentration was developed and used. This paper presents a brief overview of the results obtained, zooms in on some of the more interesting recent findings and discusses some of the theoretical background.

INTRODUCTION

One of the advantages of radio frequency (RF) heating in tokamaks is linked to the fact it is based on resonant and thus localized wave-particle interaction at the position where $\omega = k_{//}v_{//} + N\Omega(\bar{x})$ is satisfied (see e.g. [1,2]). Here ω is the antenna frequency, $k_{//}$ is the projection of the wave vector of the electromagnetic wave on the confining magnetic field $\bar{B}_o(\bar{x})$, $v_{//}$ is the parallel velocity and Ω is the cyclotron frequency. Via a proper choice of the ratio of ω and B_o one can therefore deposit power at specific desired radial positions. At small concentrations of the minority gas, efficient ion heating can be achieved at $\omega \approx \Omega$ while at harmonics of the majority gas ($\omega \approx N\Omega$; $N > 1$) finite Larmor radius effects allow direct heating of a subpopulation of the bulk ions. Whereas electron Landau or TTMP damping at $\omega = k_{//}v_{//}$ on the externally via RF antennae excited fast wave often gives rise to rather broad electron power deposition profiles, electron absorption on the short wavelength Bernstein wave branch to which the fast wave converts at the ion-ion hybrid layer is much more localized.

In ³He-D plasmas, efficient single pass direct ³He minority heating requires a small ³He concentration of 3-5 percent while efficient mode conversion, on the contrary, necessitates a large concentration of 15-20%. As the mode conversion position is a known function of the concentration, the mode conversion position can be fixed at a desired location by controlling the concentration. A proper choice of the concentration and ω/B_o therefore not only decides on which direct heating (i or e) is preferred but also at which radial position it takes place.

There are various reasons for examining shots involving ³He. Firstly, performing discharges using ³He as a minority allows studying basic RF physics. As the presence of a small RF heated ³He minority enhances the fusion reactivity in ITER [3], making ³He one of the standard working gases for this future machine, examining scenarii involving this gas is important. The ³He ion being more massive than the H nucleus, RF heated ³He ions have a lower velocity than H tails for a given RF power density and favor indirect heating of bulk ions rather than electrons. Moreover, JET is equipped with gamma ray diagnostics allowing the study of the spatial and energy distribution of fast ³He populations [4]. Finally, mode conversion in ³He plasmas gives rise to narrow electron power deposition profiles and thus constitutes a tool for perturbative transport studies relying on a well localized heat source.

The present paper is structured as follows: First, a few dedicated minority heating experiments are discussed. Before proceeding to a survey of the mode conversion experiments in L-mode, H-mode and ITB plasmas, the methods used to estimate the experimental power deposition profile as well as the real time control scheme set up to guarantee keeping the concentration constant are subsequently commented on. After discussing some of the results obtained at large concentration, conclusions are drawn.

MINORITY HEATING EXPERIMENTS

A number of ICRF physics experiments have been carried out using ³He minority heating. Gamma ray spectrometry and tomography were key diagnostics in these experiments.

RF heating of a small minority gives rise to a subclass of highly energetic ions that slow down on electrons rather than ions. This scheme therefore does not optimally use auxiliary power to boost the ion temperature and enhance fusion reactivity. Reducing the local power density reduces the tail energy and alleviates this deleterious effect. One way of doing so is using multiple frequencies simultaneously. JET polychromatic excitation experiments, using 37MHz and 33MHz while fixing the magnetic field at 3.7T, were compared to monochromatic excitation experiments at 37MHz in (³He)-D plasmas [5]. RF heated energetic ions have perpendicular energies significantly exceeding their parallel energy and thus tend to have trapped orbits. On account of their energy, these trajectories can have an appreciable radial extent that gives rise to exotic bean- or potato-shaped orbits. Gamma emission tomography allowed visualizing the different orbit topology when wave power was changed and/or mono- or polychromatic excitation was used. For high power and minority absorption in the center or on the low field side, a larger fraction of trapped particles and lower tail temperatures were observed for polychromatic than for monochromatic excitation. Nonstandard passing potato orbits were observed in the latter case. The SELFO code was adopted to confront theoretical and experimental findings, and allowed to demonstrate that various important aspects of minority heating (e.g. fast ions content) are within the reach of present-day theory [6].

As externally launched RF waves can reach the core of large, dense plasmas, RF heating is likely to play a role in the control of future machines of the ITER type. Since rotation impacts on plasma stability and confinement, the potential of RF induced momentum/torque transfer was looked into on JET. Toroidal rotation profiles, both when symmetric and asymmetric antenna spectra were used, were obtained via charge exchange spectroscopy [8]. The observed bulk ion rotation was found to be affected by the presence of energetic RF heated ^{3}He minority ions [7]. In particular, differences in toroidal rotation between discharges with $+90°$ and $-90°$ phasings were found to be consistent with absorption of wave momentum and its subsequent transfer from the fast ^{3}He minority ions to the background thermal plasma. Line integrated gamma ray emission revealed different radial profiles for the 2 phasings. The interpretation is that trapped ions dominate the fast ion population for $-90°$ phasing and that a significant fraction of the heated ions are on passing orbits in the potato regime for $+90°$.

As various nuclear reactions involving fast ^{3}He give rise to gamma ray emission, the gamma ray diagnostics available at JET provide important information on the presence of RF induced fast ^{3}He populations. Figure 1 shows the gamma ray yield of the 17MeV gamma rays produced in the reaction D(^{3}He,γ)^{5}Li as a function of time for 3 discharges with differing ^{3}He concentrations during which RF heating was applied. Only in the presence of RF power are fast ^{3}He ions observed. In agreement with theory, more very energetic RF heated ^{3}He particles are present (and more gamma rays are emitted) at low concentrations than at concentrations for which mode conversion is significant.

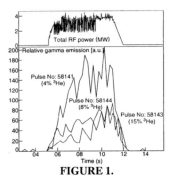

FIGURE 1.
RF power (top) and gamma ray emission (bottom) for 3 shots with different ^{3}He concentrations

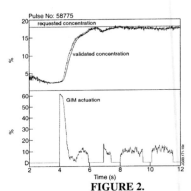

FIGURE 2.
Real time control of the ^{3}He concentration: requested & validated concentration (top) and valve opening (bottom)

FAST FOURIER TRANSFORM AND BREAK IN SLOPE ANALYSIS

Knowledge of the experimental power deposition profile is the key to understanding how a heating mechanism works. By studying the ion or electron temperature response to modulation of the RF power, the (direct or indirect) power deposition profile of RF heating scenarii can be checked. The Fast Fourier Transform

(FFT) technique is a powerful tool to study the response of a system to a periodic perturbation. The Break In Slope (BIS) method requires only a step change in the power level to estimate the temperature response. JET is equipped with a 96-channel electron cyclotron emission diagnostic with high temporal resolution providing detailed information on the electron temperature. FFT and BIS analysis rely on this detailed information to estimate the power deposition [9]. Both methods assume transport processes occur on a time scale much longer than the modulation period, a condition that is not fully fulfilled and can be dealt with in the frame of a full transport analysis, as the one presented later. As no detailed experimental data is available of the temporal evolution of the density, it was assumed that this quantity does not respond to the rapid changes of the modulated RF power. The momentary change of the RF power from one level to another then gives rise to a break in the slope of the temperature:

$$\frac{3}{2} k N \Delta \frac{\partial T}{\partial t} = \Delta P_{RF}.$$

Whereas the BIS analysis assumes the response of the temperature to the power level change is immediate at every location, the FFT allows to track the phase lag between the moment of the power level drop or rise and the associated break in the temperature slope. The advantage of the BIS is that it can be performed at any single moment the RF power level changes, but its drawback is that, due to the non-smoothness of the measured temperature signal, the estimate of the temperature slope has significant error bars. By subtracting a time-smoothed temperature from the raw temperature data, and by subsequently folding the data points from several periods on top of one another rather than only using the data points of a single period, the BIS error bars can be reduced.

REAL TIME CONTROL OF THE ^{3}HE CONCENTRATION

Mode conversion heating in (^{3}He)-D plasma gives rise to localized power deposition close to the ion-ion hybrid layer. As both the efficiency of the wave conversion and the location at which it takes place depend on the concentration, the ability to maintain the ^{3}He concentration (at 15-20%) is of uttermost importance. As ^{3}He gas is lost through transport, keeping the concentration at a specific level requires puffing extra ^{3}He gas into the machine during the discharge [10]. A robust way to do this is by setting up a real time control (RTC) scheme linking a measurement of the ^{3}He density to the opening of the gas injection valve. Such a scheme was successfully tested during the recent mode conversion experiments. Carbon is the main impurity in JET. Assuming it is the *only* one and making use of charge neutrality one finds that

$$\frac{N_{^3He}}{N_e} = \frac{6 - Z_{eff}}{5\frac{N_D}{N_{^3He}} + 8}$$

in which Z_{eff} is the effective charge. The relative density of the majority and minority ions is measured via the respective light in the divertor. The real-time control was developed in three stages. First, the observed open-loop response of the ^{3}He concentration to gas introduction was modeled using a 2nd order transfer function. Secondly, closed loop simulations of a PID control and the ^{3}He model were performed to tune the PID parameters. Finally, the PID controller was implemented in the Real-Time Central Controller (RTCC) system. Good control of the ^{3}He concentration was achieved in several pulses with different target concentrations. Figure 2 shows an example, the top figure depicting the target concentration as well as the measured concentration and the bottom figure showing the required gas puff. For diagnostic purposes, the gas puff was inhibited in various time intervals. Figure 3 depicts electron power deposition profiles at 3 different times obtained via FFT for a shot during which the magnetic field was ramped up. The RTC scheme keeps the concentration constant. Hence, the displacement of the location of the maximum of the absorption – close to the mode conversion layer – is solely due to the change of the magnetic field.

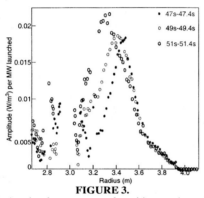 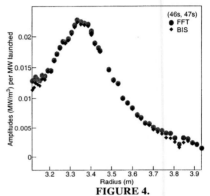

FIGURE 3. Moving the electron power deposition maximum by changing the magnetic field

FIGURE 4. L-mode power deposition profile from FFT/BIS

MODE CONVERSION HEATING

Building on earlier obtained experience [10], mode conversion experiments were performed in L-mode, H-mode and ITB plasmas. The electron temperature response to a 15Hz modulation of the RF power was studied via FFT and BIS analysis. Figure 4 shows a power deposition profile as a function of the major radius for an L-mode shot. The crosses give the BIS prediction, while the circles give the amplitude of the FFT response at the modulation frequency. The FFT and BIS analysis are in good agreement. The peak power density is of order 0.02MW/m^3 per MW launched. Integrating over the profile, one finds about 60% of the power is absorbed by the electrons.

Mode conversion experiments in H-mode were performed by adjusting the neutral beam power to ensure having grassy type III edge localized modes (ELMs), the characteristic time scale of which is very short compared to the period of the modulation. The adopted power levels typically were P_{NBI}=14.6MW and P_{RF}=3MW. On account of the ELMs and their impact on coupling, the temperature response shows fine structure. In spite of that, the quality of the FFT/BIS data is good on account of the different time scales.

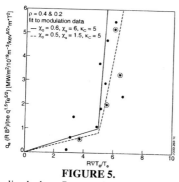

FIGURE 5.
Normalized e heat flux as a function of the inverse temperature gradient length

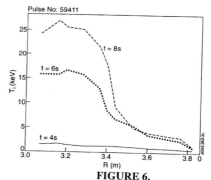

FIGURE 6.
Ion temperature profile of an ITB shot

L- and H-mode power deposition profiles obtained with FFT/BIS analysis have a very similar shape. They only represent the actual absorption profiles in a few cases, however: (i) in absence of transport and losses, or (ii) if the period of the modulation is much shorter than the time scale on which these processes take place. The drawback that the actual deposition cannot be observed becomes an advantage if one intends to perform a transport analysis: Mode conversion allows depositing power in a narrow region near the mode conversion layer and thus provides a well-localized heat source. Perturbative electron heat transport studies rely on such a heat source as they examine the heat waves propagating away from the source for a given, assumed, transport model. Adjusting the free parameters in such a model to minimize the differences between the predicted response of the harmonics for a localized source and the actual FFT response yields an estimate for the local diffusivity. It makes mode conversion heating a tool for transport analysis. Figure 5 shows the result of such an analysis for the L- and H-mode mode conversion shots using a stiffness transport model having 3 free parameters to model the diffusivity throughout the plasma [11,12]: (i) κ_{crit}, the critical inverse temperature gradient length above which turbulent transport is triggered, (ii) χ_o which, up to a gyro-Bohm-like multiplicative factor, is the background electron heat diffusivity, and (iii) χ_s which accounts for the enhanced transport when the threshold κ_{crit} has been exceeded:

$$\chi_T = q^{3/2}\frac{T}{eB}\frac{\rho_s}{R}[\chi_s(-R\frac{\partial T/\partial r}{T} - \kappa_{crit})H(-R\frac{\partial T/\partial r}{T} - \kappa_{crit}) + \chi_o].$$

The dots in the figure correspond to the time averaged temperature gradient and heat flux. The small dots correspond to shots in which ion heating dominates ($T_e/T_i \leq 1$) while the encircled dots represent shots in which electron (mode conversion) heating is dominant ($T_e/T_i \geq 1$). The lines indicate the fit to modulation data. Each line corresponds to a single set of ($\chi_o, \chi_s, \kappa_{crit}$) values for which the model's prediction of the T_e response (amplitude and phase) agrees satisfactorily well with the experimental findings. The plot shows that JET electron transport is rather stiff in ion heating dominated plasmas and is less stiff for weak ion heating. It is important to note that this result relies crucially on the use of modulation, because steady-state data alone do not allow to resolve the different degrees of stiffness.

The influence of the position of the mode conversion layer on global confinement in the L-mode mode conversion shots was studied by Lyssoivan [13].

Electron internal transport barriers (ITBs; see e.g. [14,15]) are formed using lower hybrid preheat in the low density startup phase of a discharge. Strong barriers are usually associated with reversed q-profiles favored by the slow penetration of the current in low density plasmas. In this phase, both the magnetic field and plasma current are being ramped up slowly. Adding NBI and RF power when the density is higher creates ion ITBs. The detailed physics of ITBs is still not fully understood but the safety factor, and via that parameter any mechanism that influences the evolution of the current, is recognized to be a key ingredient in their creation and sustainment. Like lower hybrid heating and current drive, mode conversion acts upon electrons and thus affects the electron temperature (slowing down the current diffusion) and influences the current. As in previous experiments [10], the ^{3}He puff was gradually increased from discharge to discharge to monitor the transition from minority to mode conversion heating. It was observed that strong ITBs can be formed and maintained when RF heating is located inside the ITB. These shots are characterized by reversed q-profiles and by high central temperatures. High ion and electron temperatures (up to, respectively, 25keV and 13keV) were recorded at ^{3}He concentrations well beyond those optimal for minority heating. Figure 6 shows an example.

Figure 7 summarizes the performance of the recent ITB plasma experiments (full circles) together with the other shots in the JET advanced scenario data base. The recent data points are underlining the efficiency of this heating scheme. At any auxiliary power level, the ion and electron temperatures attained in many of the recent ITB shots, for which $N_{^3He}/N_e > 10\%$, are among the highest ones achieved.

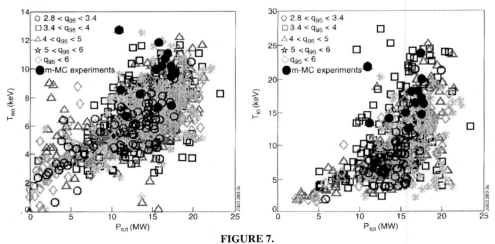

FIGURE 7.
Performance of recent m-MC heating ($N_{^3He}/N_e > 10\%$) and earlier ITB shots

CONCLUSIONS

Recent ^{3}He experiments in JET included studies of RF induced rotation, a performance comparison of mono- versus polychromatic excitation and an extensive mode conversion study. Mode conversion was adopted as a transport tool to demonstrate the stiffness of core electron transport both in JET L- and H-mode. Fine-tuning the mode conversion tool involved the successful development of a real time control scheme of the ^{3}He concentration.

REFERENCES

1. Stix, T.H., *Waves in Plasmas*, New York: AIP, 1992
2. Perkins, F.W., *Nucl. Fusion* **17**, 1197-1224 (1977)
3. Van Eester, D., et al., *Nucl. Fusion* **42**, 310-328 (2002)
4. Kiptily, V., et al., *Nucl. Fusion* **42**, 999-1007 (2002)
5. Mantsinen, M., et al., *"Comparison of monochromatic and polychromatic ICRH on JET"*, this conference
6. Laxaback, M., et al., *"Self-consistent modeling of polychromatic ICRH"*, this conference
7. Eriksson, L.-G., et al., *"Bulk Plasma Rotation in presence of waves in the ion cyclotron range of frequencies"*, this conference
8. J.-M. Noterdaeme, et al., *Nuclear Fusion* **43**, 274-289 (2003)
9. Gambier, D.J., et al., *Nucl. Fusion* **30**, 23-34 (1990)
10. M. Mantsinen, et al., *"Localized bulk electron heating with ICRF mode conversion in the JET tokamak"*, JET-EFDA paper EFD-P(03)21, submitted to *Nuclear Fusion*
11. Mantica, P., et al., "Transient heat transport studies in JET conventional and advanced tokamak plasmas", Fusion Energy 2002 (Proc. 19th Int. Conf. Lyon, 2002), EX/P1-04, IAEA, Vienna.
12. Mantica, P., *"Heat wave propagation experiments and modeling at JET: L-mode, H-mode and ITBs"*, to be presented at the 2003 EPS conference, St.Petersburg
13. Lyssoivan, A., et al., *"Effect of ICRF mode conversion at the ion-ion hybrid resonance on plasma confinement in JET"*, this conference
14. Joffrin, E., et al., *"Internal transport barrier triggering by rational magnetic flux surfaces in tokamaks"*, submitted to *Nuclear Fusion*
15. Mazon, D., et al., *"Real-time control of internal transport barriers in JET"*, submitted to *Plasma Phys. Contr. Fusion*

Local Improvement Confinement in the Ion Bernstein (IBW) Experiment on FTU (Frascati Tokamak Upgrade)

R. Cesario, A. Cardinali, C. Castaldo, M. Marinucci, G. Ravera, S. Bernabei[1], E. Giovannozzi, M. Leigheb, F. Paoletti[1], V. Pericoli-Ridolfini, F. Zonca, G. Apruzzese, R. De Angelis, E. Giovannozzi, L. Gabellieri, H.°Kroegler, M. Leigheb, G. Mazzitelli, P. Micozzi, L. Panaccione, P. Papitto, B. Angelini, M. L. Apicella, E. Barbato, L. Bertalot, A.°Bertocchi, G. Buceti, C. Centioli, V. Cocilovo, F. Crisanti, F. De Marco, B. Esposito, G. Gatti, C. Gormezano, M. Grolli, F.°Iannone, G. Maddaluno, G. Monari, P. Orsitto, D. Pacella, M. Panella, L. Pieroni, G.B. Righetti, F. Romanelli, E. Sternini, N. Tartoni, A.A. Tuccillo, O. Tudisco, V.°Vitale, G. Vlad, M. Zerbini

*Associazione EURATOM-ENEA sulla Fusione, Centro Ricerche Frascati,
C.P. 65, 00044 Frascati, Rome, Italy*
[1]*Princeton Plamsa Physics Laboratory, Princeton NJ, USA*

Abstract. New experiments with the ion Bernstein waves (IBW) has been performed on FTU both in Hydrogen (H) and Deuterium (D) majority plasmas, at higher radio-frequency power level, plasma density, current, lower effective ion charge than in previous campaign of 1999. Also in these conditions, no role is played by non-linear edge physics, which prevented, instead, the RF penetration in the plasma bulk of DIII-D. With resonant toroidal magnetic field ($\approx 8T$), improved confinement occurs inside a radial region of 1/3 of the minor radius operating in H-plasma, and 2/3 of the minor radius operating in D-majority plasma. Such behavior is consistent with the model of turbulence suppression by locally-induced-IBW-sheared flow expected to occur close the resonant layer. The FTU results provide support for active control of the pressure profile by IBW which is of relevance for advanced tokamaks.

INTRODUCTION

The model of IBW power coupling to tokamak plasma [1,2] expects that at sufficient RF power level, a local sheared flow is produced which, in turn, stabilizes the turbulence. Evidence of IBW-induced plasma flows was provided by the IBW experiment on TFTR [3], although the threshold for turbulence stabilization was not achieved. The first IBW experiment on FTU supported the IBW scheme finding a simultaneous increase of the electron density and temperature with profile peaking was found, for the first time in an IBW experiment [4]. Before presenting the IBW progress in FTU, we discuss the aspect of edge physics which made messy the IBW scenario in different machines. Indeed, the fact that the successful IBW experiments in PLT [5] and Alcator C [6] were not reproduced in Doublet III-D [7], built the idea that the test of the IBW scenario cannot be achieved. The present paper has two main focuses. i) Show that the edge physics relevant for the IBW experiment is well understood and that modeling and experiments show that a clear experimental scenario is available on FTU, which can complete the full assessment of the IBW concept. ii) Present the recent progress of the IBW-FTU experiment in improving local plasma confinement consistently with the model.

In the recent campaign, the IBW investigation continued on FTU operating at higher IBW power level (from 0.3 MW to 0.5 MW by the installation of a second launcher), higher plasma density (from $3\ 10^{19}\ m^{-3}$ to $6\ 10^{19}\ m^{-3}$), higher current (from 0.3 MA to 0.8 MA), and lower effective ion charge (Z_{eff} from 3.5 to 2.4, @ $6\ 10^{19}\ m^{-3}$, 0.8 MA), taking advantage from the boronization of the vessel. Operations both in Hydrogen and Deuterium majority plasma (only H-plasma operation in previous campaign) were performed at resonant (7.9T) and non-resonant (6T) toroidal magnetic field. The operating frequency is 433MHz. The modeling relevant for the IBW experiment on FTU is shown in Ref.[8].

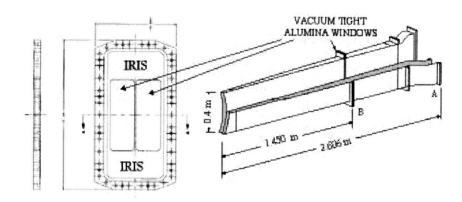

FIGURE 1a. Vacuum tight window   **FIGURE 1b.** IBW-FTU waveguide antenna

The higher IBW power level is allowed in FTU by the insertion of a second launcher. FTU utilizes waveguide antennas in place of standard loops, as in the other IBW experiments. The sketch of the antenna is shown in Fig. 1b. The antenna utilizes new alumina vacuum tight double windows in waveguide (Fig. 1a). The alumina (that has very big sizes) is brazed on the titanium frame that contains also the RF adapting iris. An anti-multipactoring rough gold coating allows achieving transmission of high RF power level (0.4 MW) [9]. At the present time, the coupled routinely power by the two IBW launchers is about 0.5 MW in total, limited by arcing at the antenna mouth. The coupled power is 80% more than in the previous FTU experiment. Further conditioning is necessary for coupling the maximum power supplied by the RF generators (two 0.5 MW klystrons).

Edge Physiscs detrimental for IBW Coupling

The IBW experiment on DIII-D was dominated by Parametric Instability (PI) producing full RF deposition at the edge [7,10]. Ponderomotive reduction of the plasma edge density occurred too, producing non-linear behavior of the antenna coupling with injected RF power [11]. Impurity influx made difficult the interpretation of some IBW experiments.

The existing models of edge physics for the IBW scheme are sufficient for understanding the edge phenomena: PI [12], ponderomotive effects [13-16], gas desorbtion by ion impact on metallic walls produced by RF sheaths [17]. A method supported by experiments for reducing the PI activity in IBW experiments was also proposed [10]. However such phenomena can be all avoided on FTU taking advantage by the high operating frequency and the waveguide antennas.

Following, we show that the FTU antenna works consistently with modeling [18] and properly launching the Lower Hybrid wave (LHW) which is necessary for efficient IBW coupling, as was not the case of DIII-D [11] and TFTR [19]. In FTU The antenna coupling behavior with plasma density during the experiment is in agreement with the LHW launch (see Figure 2).

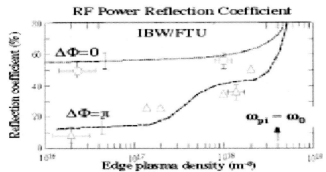

FIGURE 2. IBW-FTU waveguide antenna coupling.

The waveguide phasing $\Delta\Phi=0$ and $\Delta\Phi=\pi$ are considered corresponding in launching a power spectrum in parallel refractive index ($n_{//}$) peaked around $n_{//}\approx 1$ and $n_{//}\approx 5$, respectively. As expected, in the experiment the best antenna coupling is obtained for the spectrum with high $n_{//}$. At high plasma density, operating with antenna mouth close to the IBW mode conversion layer (close to the ω_{pi} layer) no direct IBW coupling is possible to be performed.

Ponderomotive force reduces the plasma density at the edge. Figure 3 shows modeling results [15] relevant for the IBW experiment on DIII-D (Figure 3a) and FTU (Figure 3b), respectively. The strong change of the edge density expected for DIII-D is in agreement with the experimental antenna loading non-linear behavior [11]. Conversely, in FTU, no significant change of the edge density and found in the experiment, as expected. The reflected power rate is independent of RF power and depends on plasma density at the antenna mouth, which is determined only by the antenna position in the plasma.

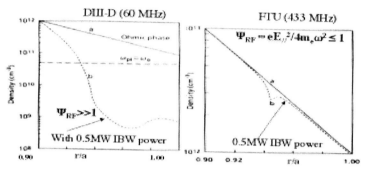

FIGURE 3. Modeling of ponderomotive plasma density reduction induced by IBW power in IBW experiments operating at different frequencies: DIII-D (a) and FTU (b)

The sidebands of PI occurring in the IBW scheme are LHWs produced by the launched (LH) pump and driven by ion cyclotron quasi-modes [12]. PI occurs in the LHW propagating region, before reaching the layer of IBW mode-conversion (located close to the ω_{pi} layer). The role of convective loss due to the plasma inhomogeneity is crucial mechanism for determining the PI RF power threshold in the IBW launch scheme [12]. Convective loss due to plasma inhomogeneity depends on the radial size of the propagation region of the LHW (Fig.4).

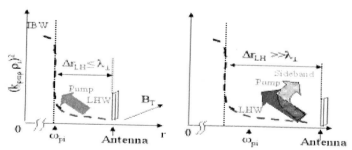

FIGURE 4. Operation inhibiting (a) and enhancing (b) the parametric instability

Operating with a wide size of the LHW region ($\Delta r_{LH} \gg \lambda_\perp$), as shown in Figure 4a, the RF power threshold is very low (≈ 10 kW), and is determined only by the finite extent of the antenna size. Such operating condition is very similar to the scheme of the original LH ion heating experiments (LHW propagating in presence of the IBW mode-conversion layer). Such experiments found strong PI and no penetration of RF power in the bulk [20]. In the IBW scheme operating with $\Delta r_{LH} \gg \lambda_\perp$, strong PI activity, as found in DIII-D. The condition $\Delta r_{LH} \gg \lambda_\perp$ was actually met in DIII-D is supported by the strong reduction of the plasma edge density expected by modeling [15] and measured during the experiment [11].

Conversely, the modeling [12] shows that operation with small radial size of the LHW propagation region (see Figure 4b), increases dramatically the PI RF power threshold. It was verified during the IBW experiments in PBX-M [12], and recently on HT-7 [21], by operation with antenna plasma distance shorter than in the standard operation designed for optimizing the antenna loading.

The IBW-FTU experiment naturally meets the condition suitable for avoiding strong PI activity [8]. Consistently, during the experiment, very low PI activity is found, as shown in Figure 5. Only small pump broadening (<1MHz @10 dB below the pump power level) and low sidebands (>30 dB below the pump power level).

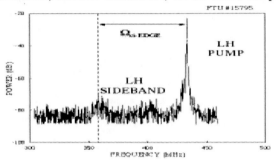

FIGURE 5. RF probe spectrum during the IBW phase in FTU

Such a low PI activity is fully compatible with almost full penetration of the launched RF power in the bulk. Indeed, similar spectra with such low PI activity were found in successful LH current drive experiments [22]. Such result indicates that in FTU the LHW is proper launched by the antenna, and the low PI level does not inhibit IBW mode conversion and RF power penetration in the bulk.

The impurity influx in not significant in FTU During injection of high IBW power the concentration of Oxygen, the main light impurity, (0.3%) does not change during the IBW phase. The heavy impurities show the following behavior: the Iron concentration increases a bit (of 10%), the Nickel does not change and the Molibdenum decreases (of 30%). Such results is consistent with the negligible effect expected in FTU of the RF sheaths in producing gas desorbtion by metallic walls [17].

Progress in Improved Confinement in the IBW-FTU Experiment

For testing the effects produced by different locations of the IBW resonant layer, operation both in pure Hydrogen (with about 1% of Deuterium) and D-majority plasmas (with 10% of H-minority) have been performed. The ion cyclotron harmonic layer $4\Omega_H$ is located at r_{abs}/a~0.3, and $9\Omega_D$ at r_{abs}/a~0.65 (a indicates the minor radius of plasma). The IBW power is fully absorbed at the first pass at about one third of the minor radius in H-plasma, and at two third of the minor radius in D-plasma [8]. At these layers, plasma flows are expected in FTU to be ponderomotively generated by about near the resonant layer [8]. The shearing rate produced by about 0.3MW of IBW power is expected to be sufficient for suppressing the electrostatic turbulence in FTU, operating at mid/low averaged plasma density ($\approx 5\ 10^{19}\ m^{-3}$). Higher densities require higher IBW power (roughly: $8\ 10^{19}\ m^{-3}/0.5MW$).

Regarding the experimental results, the electron pressure profiles (obtained by kinetic measures) of the ohmic and the IBW phases operating in H and D-majority plasmas are shown in Fig. 6.

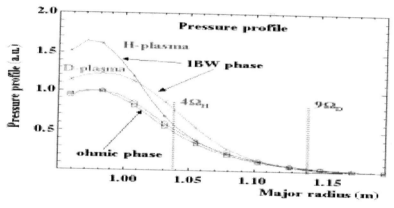

FIGURE 6. Pressure profile peaking in H and D plasmas with toroidal magnetic field of 7.9T.

The plasma current is 0.4 MA, toroidal magnetic field 7.9T (resonant), plasma density $4\ 10^{19}\ m^{-3}$, and about 0.35 MW of IBW power is coupled. In D-majority plasma (10%), the pressure profile has a wider radial foot (located in the outer half of plasma) than in pure H-plasma (in the inner half of plasma). Such result is consistent with the expected IBW deposition, respectively, in D-majority and pure H (with less than a few percent of D) [8]. No comparable peaking of the pressure profile is found operating in similar conditions, but with non-resonant field (B_T=6T), corresponding to IBW deposition at the very periphery.
Both in D and in H operations, the central ion temperature increases (of about 0.25 keV), and the effective ion charge decreases (from 3 to 2.4 during the IBW phase).

Only operating in H-plasma, an increase is found of both the line averaged plasma density (30%) and the central electron temperature ($T_{e0} \approx 2.5$ keV, $\Delta T_{e0} \approx 1$ keV), accompanied by profile peaking (see Figure 7). The density increase is never due to

FIGURE 7. Density and electron temperature profiles during the ohmic and IBW phases operating

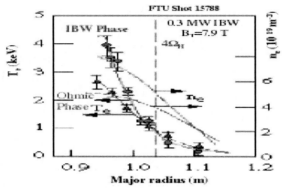

in hydrogen plasma with resonant toroidal magnetic field (7.9T)

increase of the recycling during the IBW phase, as documented by some drop of the Ha emission during the IBW phase. In D-majority, only the density peaking occurs, without significant change of the electron temperature. The transport analysis was performed by the JETTO code [23]. No significant improvement of confinement is found operating at plasma current of 0.4MA. Conversely, operating at higher plasma current (0.8 MA) and higher density (6 10^{19} m^{-3}) a reduction of 40 % of the electron thermal conductivity and a significant confinement improvement (of a factor 2) are found, in plasma region bounded by the resonant layer. It is accompanied
by a significant reduction (20%) of the ohmic power density, which provides by main power input of the experiments. The confinement improvement is well outside the error bar of the analysis, which is mainly due to the uncertainties of the inputted kinetic profiles. We attribute the peaking of the pressure profile to confinement improvement inside the layer bounded by the IBW resonant layer.
A larger plasma volume is involved operating in D-majority and, consequently, the improved confinement is more evident when operating with the higher ohmic input due to the high current.

Conclusions

The recent campaign performed in FTU at higher IBW power than in the previous campaign [17] confirms that the experimental scenario is free of non-linear edge plasma physics, as occurred instead in DIII-D. The IBW-FTU waveguides antennas works properly exhibiting the linear behavior with plasma density expected by the modeling. The low parametric instability activity is similar to the one observed in standard LHCD experiment, and is compatible with full RF power penetration in the bulk. Free of non-linear plasma edge phenomena and, FTU can achieve the full assessing the IBW scheme.

Operation in Deuterium plasma (at I_p=0.8 MA, $n_e \approx 0.5 \; 10^{20}$ m^{-3}) with resonant field of 7.9 T, shows a peaking of the pressure profile (20%). It is accompanied by a significant reduction (20%) of the ohmic power density, which provides the main power input of the experiments. Transport analysis shows a uniform decrease of the electron thermal conductivity by 40% over the region inside the absorption radius. The radial foot point of the ITB obtained in D plasma occurs in the outer half of plasma, than in the inner half as in H-plasma. These results are consistent with the models of IBW deposition and local generation of plasma flows, useful for turbulence stabilization.

The recent results of the IBW experiment on FTU give further support to the IBW scheme for producing ITB with controlled radial width, as requested by advanced tokamaks. The measure of turbulence by microwave reflectometry is in progress now in FTU for assessing further the IBW concept.

REFERENCES

1. Ono, M., Phys. Fluids **B2** 241 (1993)
2. Biglari, H., et al., Phys. Fluids **B 2**, 1 (1990)
3. LeBlanc, B., et al., Phys. Rev. Letters, 82 331 (1999)
4. Cesario, R., et al. Phys. of Plasmas 8 n. 11 4721 (2001)
5. Ono, M., et al., Phys. Rev. Lett., 60 294 (1988)
6. J.,D. Moody, et al., Phys. Rev. Lett., 60 298 (1988)
7. Pinsker, R., et al., in Proceedings of the 8th Top. Conf. on Radio Frequency Power in Plasmas, Irvine, CA 1989 (American Inst. of Physics, NY 1990) p. 314
8. Cardinali, A., Castaldo, C., Cesario, R., De Marco, F., Paoletti, F., Nucl. Fus., **42**,427,(2002)
9. Cesario, R., et al., in 2nd Europhysics Topical Conference on Radio Frequency Heating and Current Drive of Fusion Devices, Brussels (Belgium) 1998, Edited by J. Jaquinot, G. Van Oost, R.R. Weynants, European Physical Society, Brussels (Belgium) 1998, Vol. 22A, p. 65
10. Pinsker, R., et al., Nuclear Fusion Vol 33 n. 5 (1993) 777
11. Mayberry, et al. Nuclear Fusion Vol 33 n.4 (1993) 627
12. Cesario, R., et al., Nucl. Fusion, 11, 261 (1994)
13. Morales, G.J., Phys. Fluids 20 1164 (1977)
14. D'Ippolito, D., et al., Phys. Rev. Lett., 58, 2216 (1987)
15. Cardinali, A., et al., in Proceedings of the 21st EPS Conference on Controlled Fusion and Plasma Physics, Montpellier, 1994 (European Physical Society, Petit-Lancy, 1994) p. 968
16. Russel, D.A., et al., Phys. of Plasmas, **3** 743 (1998)
17. Cesario, R., et al., Nucl. Fusion **34** 1527 1994
18. Cesario, R., De Marco, F., Sauter, O., Nucl. Fusion **38** 31 (1998)
19. Rogers, J.C., et al., in Proceedings of the 12th Top. Conf. on RF Power in Plasmas (Savannah, 1997) (American Institute of Physics, New York 1997) **403**, p. 13
20. Cesario, R., and Cardinali, A., Nuclear Fusion **29**, 1709 (1989)
21. Jianganag Li, et al., Plasma Phys. and Contr. Fusion 2001
22. R. Cesario, et al., Nucl. Fusion, **32**, 2127 (1992)
23. Cenacchi, G., Taroni, A., in Proc. 8th Computational Physics, Computing in Plasma Physics, Eibsee 1986, (EPS 1986), Vol. 10D, 57

Properties of segmented Ion Cyclotron antennas

G. Bosia, S. Bremond

Association EURATOM-CEA, CE Cadarache, F-13108 ST-PAUL-LEZ-DURANCE

Abstract. A possible issue for Ion Cyclotron Heating and Current Drive systems in next step fusion devices is related to the high electric field at which these systems are planned to operate, which may limit the power transfer efficiency to the plasma core. This paper addresses the problem of maintaining a high power handling in an IC launcher at high power density, with some suggestion for a solution.

INTRODUCTION

The 16-straps ITER IC array is designed for a RF output power of 20 MW, or 1.25 MW/strap. A strap coupling impedance $Z_L \sim 1.2 + i\ 18\ \Omega$ is expected at 53 MHz (1). In these conditions the maximum RF voltage on the strap is $V_{L\ max} = 20.7\ kV$, leading to an on-strap peak electric field of ~ 2 kV/mm, just within the self imposed maximum voltage reference limit, which is based on best experimental values today obtained in Tokamaks.

However, there is no guarantee that the ITER IC system will operate at these reference conditions, as these are computed for a nominal antenna /plasma separatrix gap of $g = 120$ mm and $0,\pi, 0,\pi$ phasing. This is likely to be a compromise between several constraints in addition to IC H&CD system requirements, including wall and divertor thermal loading constraints, plasma radial control accuracy and plasma geometry parameters such as triangularity.

As plasma coupling could be several times lower than the reference value, a perhaps useful exercise is to assess if and how the antenna power handling capabilities could be maintained at such low coupling. This problem is addressed in this paper, with some suggestion for a solution, which however requires an antenna design different from the one presently adopted and restrictions in the frequency band.

If we assume, as someone does, that the electric field in the main (in general pressurized) transmission line can be kept below the breakdown value in any operating conditions, the limit to the amount of power coupled to the plasma by an IC array is set by the value of the electric field generated at the current strap and/or other components, operating in the in-vessel "vacuum".

Voltage breakdowns occur when the RF electric field temporarily exceeds the local dielectric strength of the in-vessel "vacuum". This appears to significantly vary with the torus operating conditions. Arcs, often showing marks in locations where the electric field is not the highest, are frequent in IC operation at high power level. They are usually extinguished by early detection and interruption of the power flow for a time interval sufficient for the dielectric strength to be restored. This process constitutes a main source of inefficiency in power handling.

REDUCING THE OCCURRENCE OF IN-VESSEL BREAKDOWN

The problem of avoiding in–vessel arcs can be addressed i) by reducing the RF electric field to a value lower of any possible breakdown threshold in the torus vacuum or ii) by using dielectric materials in critical points of the system, to substantially increase the local dielectric rigidity. The latter is applicable only to equipment not directly facing the plasma, as metal deposition during plasma operation would rapidly degrade the rigidity of the dielectric material used.

An example of application of ii) is the use of vacuum coaxial capacitors (sketched in Figure 2) in the tuning system of Tore Supra. This component, sketched in Figure 1 uses a private high vacuum, preserved by a ceramic enclosure as a dielectric. The breakdown electric field in the ceramic enclosure is in excess of 10kV/mm and the external electric field at full voltage is < 800 V/mm. The ceramic is only marginally involved in the main RF current conduction and dielectric losses are small. Capacitors of electrical specifications similar to the ones in Table 1 have been routinely operated in Tore Supra up to 50 kV peak.

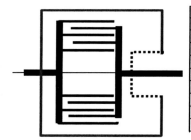

Maximum RF voltage (kV peak)	80.0
Maximum RF current (kA peak)	2.0
Capacitance range (pF)	50-350
Max resolution (pF)	1
Series Resistance	<10 mΩ
Stray inductance	<10 nH
Operating temperature range °C	20-150
Cooling	water
Actuation	hydraulic
Operation	CW

FIGURE 1. TS vacuum capacitor scheme

TABLE 1 Specifications for a similar capacitor applicable to ITER

In a system where antenna and tuning system are integrated, the maximum strap voltage (including also the voltage drop across the strap feeder) is proportional to the strap length ($V_{Lmax}= I_{res}X_L$ where $I_{res}=(2P/R_L)^{1/2}$ $Z_L=R_L+iX_L$ $R_L \sim R'l$, $X_L \sim \omega L l'$). If the tuning losses are sufficiently low, the strap voltage can be reduced by shortening the strap with no significant reduction in RF power transfer, as the reduction in plasma coupling, due to the shorter length, would be compensated by an equivalent increase in the resonant current. A single strap could be therefore replaced by a poloidal array of short, individually powered elements, so achieving a large power transfer at reduced strap voltage.

If the number of segments is high, this poloidal array may be difficult to construct. However, there is more than one way to reproduce the desired electrical properties of a poloidal array, using a single supply.

ELECTRICAL PROPERTIES OF A SEGMENTED ANTENNA

We consider a two-conductor periodic structure made of an arbitrary sequence of cascaded strap segments shown in Figure 2).

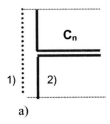 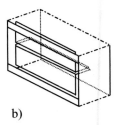

FIGURE 2 a) Periodic structure segment; b) practical example with FS partially removed

The lumped impedance, assumed capacitive, determines the conduction properties of the structure. We consider here the case of resonant segments, i.e. :

$$C_n = 1/\omega_0^2 L_n$$

where L_n is the segment inductance. If the overall periodic structure is short compared with the vacuum wavelength, the tuning capacitances are essentially equal, and increase in value with the number of segments,

The current along the structure is basically constant and independent on the number of segments. The reactive voltage is essentially periodic (a slight aperiodicity is due to the complex value of the strap characteristic impedance), and its maximum value is inversely proportional to the number of segments (Figure 3).

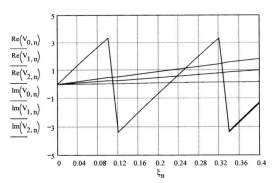

FIGURE 3) Resistive and reactive voltage pattern along a half section described in Table 2 and Fig 4.

The resistive voltage increases along the segment and the structure and reaches its maximum value. $V_{Lmax} = R'N\,l$ at the input of the strap.
The resistive voltage is also independent on number of segments and increases linearly from the short circuit to the input as in a single strap.

Strap length and number of segments need to be optimised together for best performance when frequency and available space are known. The practical limit to the total number of elements of the array is set by the mechanical accuracy needed to operate all segments tuned at the same frequency.

The segmented antenna can be matched against load variations in the same way as an usual single strap antenna. An ITER-like resonant double loop (RDL) scheme is applicable, to maintain tolerance to load variations. The reactive part of the antenna tuning system circuit is now the one of the feeders (rather than the one of the antenna itself), and can be rather freely chosen.

TABLE 2 Segmented ITER antenna electrical performance

Antenna parameter	Ref. design	Segm. design
Load range (Ω/m)	> 3	>0.7
Frequency range (MHz)	35 - 60	52-54
Strap length (m)	0.3	0.4
Number of segments	1	2
Strap $Z_c(\Omega)$	~ 35.0	~ 35.0
Feeder length (m)	0.3	0.2
Feeder $Z_c(\Omega)$	10	30
Match reactances (Ω)	-38.3 & -49.7	-4.45 & -9.5
Match C (pF)		667 & 312
RDL $Z_{in}(\Omega)$	6	4
Input power (MW)	2.5	5.0
Input voltage (kV)	5.5	6.3
Max. strap voltage (kV)	26 93	2.3
Max. strap current (kA)	1.30	2.58
MTL Voltage (kV)	12.25	12.5
V_{max} in tuner (kV)	32.1& 40.0	10.2 & 22.2 kV
E_{max} in strap (V/mm)	1.3	
E_{max} in VTL (V/mm)	0.53	
P-transfer efficiency (%)	~95	~95

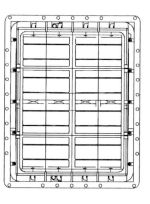

FIGURE 4. Segmented ITER antenna

In Table 2 a comparison between the electrical parameters of the ITER reference design and of a 2x2 RDL elements array of straps segmented in two sections (shown in Figure 5). The array would operate at fixed frequency (53 MHz), low loading and low electric field, and require no adjustable tuning component.

Conclusions

Operating ITER-like IC array at a power density of 10 MW/m², at plasma loading much lower than in present experiment (0.5 < R' < 5 Ω/m) and moderate electric field (<1 kV/mm) in the torus "vacuum" may be possible using a "segmented " ITER-like launching structure, with an integrated tuning system and using tuning components having voltage standoff improved by the use of ceramic materials.

The array would operate at a single frequency (f = 53 MHz), a limitation that does not appear prohibitive for ITER, as this frequency is used by both ICRF heating scenarios ($2\Omega_T$ and Ω_{He3}) currently envisaged .

References

1) ITER Detailed Design Description (DDD) No. 5.1 - Ion Cyclotron Heating and Current Drive System. IdoMS G AM MD 3 00-02-21 F1 (2000).

Interaction of Neutral Beam (NB) Injected Fast Ions With Ion Cyclotron Resonance Frequency (ICRF) Waves

M. Choi, V.S. Chan, S.C. Chiu,[a] Y.A. Omelchenko,[b] Y. Sentoku, and H.E. St. John

General Atomics, P.O. Box 85608, San Diego, California 92186-5608
[a]*Sunrise R&M.*
[b]*Dynamics Research Corporation.*

Abstract. Existing tokamaks such as DIII-D and future experiments like ITER employ both NB injection (NBI) and ion-cyclotron resonance heating (ICRH) for auxiliary heating and current drive. The presence of energetic particles produced by NBI can result in absorption of the Ion cyclotron radio frequency (ICRF) power. ICRF can also interact with the energetic beam ions to alter the characteristics of NBI momentum deposition and resultant impact on current drive and plasma rotation. To study the synergism between NBI and ICRF, a simple physical model for the slowing-down of NB injected fast ions is implemented in a Monte-Carlo rf orbit code. This paper presents the first results. The velocity space distributions of energetic ions generated by ICRF and NBI are calculated and compared. The change in mechanical momentum of the beam and an estimate of its impact on the NB-driven current are presented and compared with ONETWO simulation results.

MODELING OF STEADY-STATE SLOWING DOWN OF NB INJECTED FAST IONS

Assuming that the injected neutral particles are all ionized by the background plasma in the equatorial plane, Z=0 ($\vec{v} \cdot \vec{\theta}_p = 0$), an initial source rate of fast ions generated due to NB injection, $S(E_b, \xi, \psi, \theta)$, is calculated with the given NB injection power (P_{NB}), beam ion energy (E_b) and tangential radius (R_T) in phase space. Here the pitch of particle is given by $\xi = v_{//}/v = (B_T R_T)/(BR)$ [1]. The weightings of each test particle (fast ions), W_k (k=1,n: n is a total number of test particles), is calculated using the relation, $W_k = S \times t / n$, in order to represent the real number of fast ions generated from NB source. These injected energetic ions will lose their energy and momentum to the background plasma ions and electrons through pitch-angle scattering and drag [2], and eventually will be thermalized. To model a steady state slowing-down regime of NB injected fast ions, the thermalized fast ions (<1.5T_i) are re-injected into the plasma at the birth energy with a source rate compatible with P_{NB}. The weightings of re-injected fast ions are readjusted to be consistent with the constant input power.

SIMULATION OF ICRH INTERACTION WITH NBI

A D-H plasma is simulated with the plasma and heating parameters shown in Table 1. For the simulation of interaction of minority ions with ICRF wave, the wave zone is set to a wedge model in the poloidal cross section [2]. To relate the electric field magnitude (only left-hand component, E_+, is considered in this work) to the absorbed rf power, $|E_+(r)|^2 = (1 - r^2/a^2)$ is modeled [2]. The background minority H ions are modeled initially as Maxwellian distribution. Fast ions are produced by ICRF minority heating, and are subsequently slowed down by Coulomb collisions. They are then re-injected within their initial phase space to achieve steady state. Concurrently a neutral H beam is injected in tangential direction (on DIII-D, R_T =104 cm). The energy of resultant H ions is assumed to be mono-energetic with a full energy 80 keV (half and third energy ion beams are ignored in this work). Therefore, the ICRF wave will be simultaneously interacting with thermal minority ions from Maxwellian background as well as the energetic tail minority ions from NBI. To model the NB fast ion source properly, two kinds of ions produced by different heating methods are treated as two different groups of Monte Carlo test particles where they are separately coupled with background plasma through the slowing-down collisions and pitch angle scattering. The simulation was followed for a few slowing-down times to reach a steady-state solution. For the preliminary results presented here, typically 1000 test particles were used to represent the background ions (500) and NB injected fast ions (500).

Table 1 Input Parameters for Simulations.

Plasma Parameters		RF Heating	NB Heating
$T_D(0)$ =3.0 keV	$B_t(0)$ = 18.6 kg	P_{RF} = 3 MW	P_{NB} = 8.0 MW
$T_H(0)$ =2.0 keV	a = 50 cm	f_{RF} = 34.5 MHz	E_{NB} = 80 keV
$T_e(0)$ =3.0 keV	R_{maj} = 186 cm	n_ϕ = 7.0	R_T =104 cm
$n_e(0)$ =1.0×10^{14} cm^{-3}	n_H/n_D = 0.04		

Case With ICRF Only Fast Ions. The ICRH alone simulation is first done with the plasma and rf heating parameters in Table 1. With f_{RF} = 34.5 MHz, a fundamental cyclotron resonance surface is located at $(R - R_0)/a \approx -0.51$ on the high-field side. Figure 1 shows the energetic ion distribution [Fig. 1(b)] evolved from initial Maxwellian distribution [Fig. 1(a)] in velocity space. A strong energetic tails up to 80 keV are seen due to the interaction of resonant background H minority ions with the ICRF wave.

Case With ICRF + NBI. Almost all the injected fast ions have gone through one thermalization within two slowing down times. For the plasma parameters in Table 1, the slowing down time of NB injected H ions is about 70 ms. At this time, a steady state is reached by NB injected fast ions and an energetic tail is formed. The ICRF wave is turned on at the end of two slowing-down times and the simulation continues for one additional slowing down time. Figure 2(a) shows the initial distribution of test minority ions (Maxwellian + mono-energetic) in velocity space. Figure 2(b) shows the

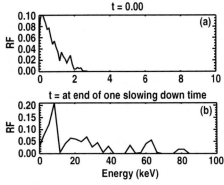

Figure 1. Distribution of energetic ions excited from thermal ions by ICRF.

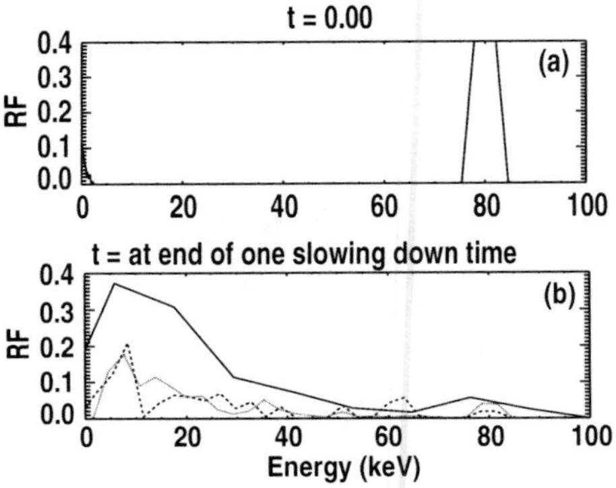

Figure 2. Distribution of energetic ions from NB energetic ions by ICRF.

energetic ions distribution generated due to the interaction with ICRF wave (from interactions with both the background and NB injected ions). Due to the resonant heating of NB injected energetic ions with ICRF wave, it is seen that an enhanced tail with energies up to 100 keV are generated (solid line). A slowing down distribution for NBI without ICRF is also plotted [Fig. 2(b) dotted line]. The dashed line is for ICRF only.

THE NET NB ION DRIVEN CURRENT

It is well known [3] that a directed NB with a net momentum can generate a current when the Z_{eff} of the bulk ions is different from the Z of beam ions. To study the impact of ICRF on NB current drive (NBCD), the net current generated by energetic beam ions can be calculated from momentum balance between NB injected fast ions and background electrons. Defining $\Delta M_{f \to e} \equiv \Sigma \Delta$ (momentum transfer from fast ions to electrons), v_f (steady state fast ion mean velocity) is obtained by [4].

$$v_f(\psi) = \frac{\Delta M_{f \to e}(\psi)}{n_f(\psi) m_f} \left[1 + \frac{Z_f^2}{Z_{eff}} \right] .\tag{1}$$

Including the opposite contribution due to electrons and accounting for toroidal trapping effect, the net NB ion driven current can be expressed as [5],

$$J_{net} = J_f + J_i + J_e = en_f v_f Z_f \times \left\{ 1 - \frac{Z_f}{Z_{eff}} + 1.46\varepsilon^{\frac{1}{2}} \left(\frac{Z_f}{Z_{eff}} - \frac{Z_{eff}}{Z_f} \frac{m_f}{m_i} \right) A(Z_{eff}) \right\} .\tag{2}$$

where $A(Z_{eff})$ is 1.36 for Z_{eff} = 2 [6] and ε is the inverse aspect ratio, $\varepsilon = r/R$. Figure 3(a) shows the net NB ion driven current (A/cm^2) calculated from ORBIT-RF at the end of two slowing down time (before the ICRF wave is turn on). Here the solid line is J_f (A/cm²) and the dotted line is J_{net} using (1) for v_f. This result is crosschecked using the definition on a flux surface, $v_f(\psi) = \int d^3 v_{//}(\psi) f_f(\bar{v})$ [the dashed line (J_f)]. The two calculations agree qualitatively, however, because of finite size orbits, a larger number of test particles to cover the phase space might be needed to achieve better agreement. In Fig. 3(b), a ONETWO transport code calculation using similar plasma and NB heating parameters is displayed. A reasonable agreement is obtained with the momentum balance approach. For this particular case, the change in NBCD due to the addition of ICRF is insignificant.

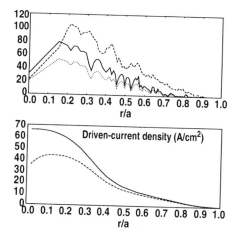

Figure 3. NB ion driven current density from (a) ORBIT-RF [solid and dashed lines, net (dotted)] and (b) ONETWO [solid, net (dashed)] in A/cm².

FUTURE WORK

Preliminary results in the modeling of ICRF interaction with NBI appear qualitatively as expected. Since the number of test particles used is relatively too small to achieve satisfactory statistics in these first simulations using a single processor, adaptation of the code to parallel processors will be required to increase the number of test particles for quantitative benchmarking. Further progress will also be made by coupling a full wave solver to ORBIT-RF to describe self-consistent ICRF wave propagation. In addition, the interaction of minority ions with ICRF at higher harmonics is planned in order to model the actual experiments on DIII-D.

ACKNOWLEDGMENTS

This work is supported by the U.S. Department of Energy under Grant No. DE-FG03-95ER54309.

REFERENCES

1. Chan, V.S., et al., *Phys. Plasmas* **9**, 501 (2002).
2. White, R.B., Chance, M.S., *Phys. Fluids* **27**, 2455 (1984).
3. Ohkawa, T., *Nucl. Fusion* **10**, 185 (1970).
4. Cordey. J.G., "A Review of Current Drive by Fast Ion Injection," Proc. of IAEA Current Drive Workshop, Abingdon, United Kingdom, 1983.
5. Cordey. J.G., *Nucl. Fusion* **16**, 3 (1976).
6. Connor, W., Cordey, J.G., *Nucl. Fusion* **14**, 185 (1974).

RF-Sheath Assessment of ICRF Antenna Geometry for Long Pulses

L. COLAS, S. HEURAUX[+], S. BREMOND

Association Euratom-CEA pour la Fusion Contrôlée, CEA Cadarache, 13108 St Paul lez Durance
[+]LPMIA, Unité du CNRS 7040, Université Nancy 1, BP 239, 54506 Vandœuvre Cedex, France

Abstract. Monitoring powered Ion Cyclotron Resonance Frequency (ICRF) antennas in magnetic fusion devices has revealed localized modifications of the plasma edge in the antenna shadow, most of them probably related to an enhanced polarization of the Scrape-Off Layer (SOL) through Radio-Frequency (RF) sheath rectification [1,2,3]. Although tolerable on present short RF pulses, sheaths should be minimized, as they may hinder proper operation of steady-state antennas and other subsystems connected magnetically to them, such as lower hybrid grills [3]. As a first step towards mitigating RF sheaths in the design of future antennas, the present paper analyses the spatial structure of sheath potential maps in their vicinity, in relation with the 3D topology of RF near fields and the geometry of antenna front faces. Various combinations of poloidal radiating straps are first considered, and results are confronted to those inferred from transmission line theory. The dependence of sheath potentials on RF voltages or RF currents is studied. The role of RF near-field symmetries along tilted field lines is stressed to interpret such effects as that of strap phasing. A generalization of the "dipole effect" is proposed. With similar arguments, the behavior of Faraday screen corners, where hot spots concentrate on Tore Supra (TS), is then studied. The merits of aligning the antenna structure with the tilted magnetic field are thus discussed. The effect of switching from TS (high RF voltage near corners) to ITER-like [6] electrical configurations of the straps (high voltage near equatorial plane) is also analyzed.

SIMULATION TOOLS

This study relies on self-consistent 3D RF field and current computations in realistic plasma and antenna geometry with ICANT [4]. Maxwell's equations are solved in a vacuum shell between the metallic chamber wall and a plasma frontier. At this boundary, $E_{//}$ is supposed to vanish, while propagation and damping of the fast wave in the target plasma are accounted for by a surface impedance matrix representative of a characteristic TS ICRF scenario. To resolve near field characteristic lengths, which are much smaller than the tokamak size, 1300 toroidal and 300 poloidal harmonics were introduced. ICANT was modified to accommodate a pitch angle α of the magnetic field with respect to the toroidal direction [7]. Parallelogram-shaped current elements were also created to build antennas structures aligned with tilted field lines. RF fields were mapped in 16 radial planes in the vacuum shell, normalized to 1MW coupled power and integrated along straight tilted field lines, consistently with the pitch angle in ICANT, to yield $V_{sheath} = \int E_{//} dz'$. It was verified with a remarkable precision that V_{sheath} vanished at the plasma boundary, and that self-consistent field calculations were essential to yield reliable results.

ARRAYS OF POLOIDAL RADIATING STRAPS

Figure 1 maps $Im(E_{//})$ in front of a finite size poloidal strap radiating 1MW into a TS plasma. The pattern shows qualitative similarity with the Transverse Electromagnetic Mode (TEM) in strip lines : the RF field topology is similar in each poloidal section y=const. From one section to the other the field amplitude scales with $V_{RF}(y)$, the RF voltage profile of strip line theory, which in the case of TS Resonant Double Loops (RDL) is zero at the strap center and maximal at the ends.

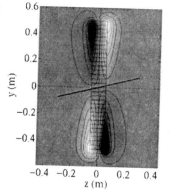
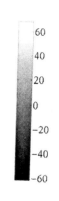

FIGURE 1. Mapping of $Im(E_{//})$ in a radial plane x=2cm in front of a TS like RDL strap. Plasma at x=0, strap at x=a=5.5cm, wall at x=$a+d$=11cm, α=14°. Unit is kV/m for 1MW coupled to plasma.

Integrating such a RF field only on one side of the strap yields huge V_{sheath} scaling as $V_{RF}(y)$ along the strip line. Yet, since $E_{//}$ changes signs from one side of the strap to the other, some compensation occurs when it is integrated from $-\infty$ to $+\infty$. Compensation is however partial : $E_{//}$ is not purely transverse to the strap, the integration path is tilted by α and V_{RF} depends on y. V_{sheath} then scales as $\partial V_{RF}/\partial y \propto I_{RF}(y)$, the RF current profile along the strap. In the framework of ICANT, and assuming that the radial dimensions in the vacuum shell are small compared to the poloidal characteristic lengths, V_{sheath} in the shell can be approximated by

$$\int_{-\infty}^{+\infty} E_{//}(x,y',z')dz' \approx i\mu_0 \omega_0 f(x)\sin\alpha \int_{-\infty}^{+\infty} j_y(y',z')dz'; \quad f(x) = \begin{cases} dx/(a+d) & x<a \\ \text{or else} \\ a(a+d-x)/(a+d) \end{cases} \quad (1)$$

where z' is the (tilted) parallel direction ; x, a and $a+d$ are respectively the radial distances between the plasma boundary and the field line, the straps and the metallic back wall ; y' is orthogonal to x and z' ; ω_0 is the wave pulsation, and j_y is the RF current density in the poloidal straps.

Formula (1) suggests simple design options for sheath mitigation. First the straps should be orthogonal to the local magnetic field, or as close as possible to this orientation. Moreover, if the field line crosses RF currents of opposite signs, a second compensation occurs upon integration. Particularly for two straps phased in dipole, it can be shown from equation (1) that $|V_{sheath}| \propto \tan\alpha^*|\tan\alpha-\tan\gamma|^*\partial I_{RF}/\partial y$, where γ is the pitch angle of a line containing the two strap geometric centers. This demonstrates that the "generalized dipole effect" is all the more effective as the strap geometric centers are aligned on the same field line. Incidentally V_{sheath} in dipole scales again as $V_{RF}(y)$.

Figures 2 and 3 show poloidal profiles of the rigorous V_{sheath} computed from ICANT, for the cases of two straps phased in monopole and dipole respectively. The main trends of formula (1) are recovered : poloidal and radial variations of V_{sheath}, the reduction of V_{sheath} in dipole phasing. Some discrepancy is however observed near the straps ends, where large poloidal gradients arise and simplifying assumptions fail.

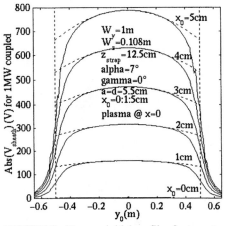
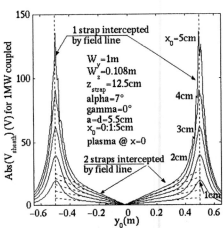

FIGURE 2. V_{sheath} poloidal profiles for two straps with monopole phasing, at various radial locations in front of the straps. Dashed : $I_{RF}(y)$ profiles.

FIGURE 3. V_{sheath} poloidal profiles for two straps with dipole phasing, at various radial locations in front of the straps. Dashed $V_{RF}(y)$ profiles.

SPECIFIC EFFECTS AT ANTENNA BOX CORNERS.

Realistic antennas cannot be completely reduced to sets of poloidal straps. This is illustrated in the present section, which investigates specific effects arising in antenna box corners, where hot spots have been observed on several machines [2,5].

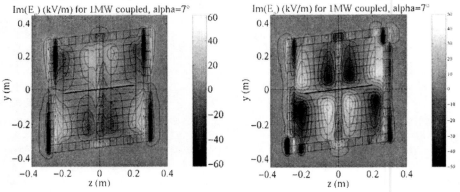

FIGURE 4. Mapping of $E_{//}$ in front of a two-strap antenna with present TS electrical circuit.

FIGURE 5. Mapping of $E_{//}$ in front of a four-strap antenna with ITER-like electrical circuit.

Figures 4 and 5 sketch the metallic structures simulated, together with the radiated $Im(E_{//})$. For simplicity, Faraday Shields (FS) were removed, results are qualitatively similar with FS. Figure 4 includes two TS RDL phased in dipole, put in a thick box with a thick septum. Figure 5 features a new TS antenna, designed with an ITER-like matching circuit [6]. Each RDL is replaced by two straps of half length and maximal RF voltages are displaced to the equatorial plane. Toroidal phasing is dipole. In front of the main straps, the lateral parts of the box and the septum behave as passive poloidal straps and a TEM-like RF field structure is preserved. Near corners, part of the RF current flows toroidally from the lateral parts to the top and bottom parts of the box, breaking the symmetries of $E_{//}$ on both sides of the (passive) straps in figure 1.

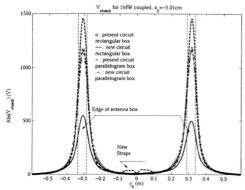

FIGURE 6. V_{sheath} poloidal profiles for 4 antenna geometries.

Figure 6 shows poloidal profiles of V_{sheath} with such antenna structures. Four cases were envisaged : TS or ITER-like matching circuit, rectangular box or parallelogram aligned with the field line direction ($\alpha=7°$). All cases have similar coupling resistance within 10%. In front of the straps the profiles are consistent with section 2, i.e. $V_{sheath} \propto V_{RF}(y)$. Near antenna corners peaks of sheath potential arise, that are attributed to the symmetry breaking in RF field maps. Simulations suggest that displacing the maximum of $V_{RF}(y)$ does not remove the peaks, while lining up the antenna on the local magnetic field seems beneficial.

SUMMARY AND PROSPECTS

In front of the radiating straps, an ICRF antenna can be approximated by a set of poloidal straps. The structure of the RF field transverse to the straps is close to a TEM mode. Sheath potentials could be evaluated by a simple formula, suggesting several options for sheath reduction. As far as possible, the straps should be perpendicular to the local magnetic field, and their geometric centers should be aligned on a field line. Dipole phasing is beneficial. Depending on the integration boundaries and the strap phasing, V_{sheath} varies poloidally as the RF voltages or as the RF currents.

Near antenna box corners this simple picture fails and large peaks of sheath potential arise. This is attributed to RF current flowing toroidally on the box frame, breaking some symmetries of $E_{//}$ mappings. Simulations with ICANT suggest that changing the poloidal profile $V_{RF}(y)$ does not remove the corner effects, while designing a parallelogram shaped antenna box aligned with the field lines is beneficial.

More insight is needed to cast the corner effects into a simple model and refine the antenna description. However reducing the toroidal RF currents seems desirable.

REFERENCES

1. Noterdaeme J.-M. and Van Oost G. *Plasma Phys.Control. Fusion* **35** (1993) p. 1481
2. L. Colas, L. Costanzo, C. Desgranges, S. Brémond, & al, *Nucl. Fusion* **43** (2003) 1–15
3. A. Ekedahl, L. Colas, M.-L. Mayoral & al., this conference.
4. Pécoul, S., et al., *Computer Physics Communications* **146**, 166-187 (2002).
5. Myra J., D'Ippolito D. and Ho Y. *Fusion Eng. Design* **31** (1996) p. 291
6. G. Bosia, *Fusion Science & technology*, **43** (2003) pp. 153-159
7. S. Lapuerta & L. Colas, PHY-NTT 2001.009 private communication

ELM Resilient External Matching System for the ICRF System of ITER: 2. Design of the Components and Implementation

P. Dumortier[1], F. Durodié[1], A. Messiaen[1], S. Brons[2]

[1]*LPP/ERM-KMS, Euratom-Belgian State Association, Brussels, Belgium, TEC Partner*
[2]*IPP, FZJ, Jülich, Germany, TEC Partner*

Abstract. This paper summarizes the design and implementation of an ICRF system for ITER capable of radiating 20 MW and matched to the power source outside the machine vacuum. The key points are the absence of in-vessel remotely operated parts, the ELM resilient *conjugate-T* matching through conventional line stretchers, the possibility to modify the tuning layout without breaking the machine vacuum and the reduction of the number of matching circuits to four.

INTRODUCTION

The electrical design of a 20 MW ICRH system with external matching has been summarized in a companion paper [1] (see Fig. 1 of [1]). This system avoids any in-vessel adjustable components and insures the ELM's load variation resilience by the *conjugate-T* matching method. The tuning of the matching is obtained by means of conventional line stretchers. The maximum voltage in the transmission line is limited by using 24 short radiating straps and by the choice of the characteristic impedance in the various line sections. The number of *conjugate-T* matching circuits (each fed by a 5 MW power source) is reduced to 4 by the use of 4-ports passive junctions. The design avoids exceeding an electric field of 20 kV/cm in any component for the reference distributed antenna load resistance of 4 Ω/m and a VSWR of 1.5 during the ELMs. It also avoids the use of remotely operated parts and of ceramic parts (except for the feedthroughs) inside the vessel. Fig. 1 displays the antenna plug and the matching system integrated in the ITER environment. In front appears the antenna plug, which includes the antenna array with its electrostatic screen, the transmission line links to the 4-ports junctions and the 4-ports junctions connected to the vacuum feedthroughs. Behind the RF antenna plug closure plate the set of coaxial lines are crossing the cryostat space and the bioshield up to the port cell where they are connected to the line stretchers. The output lines of the line stretchers go vertically upwards to the T-junctions located at the top of the port cell. Finally the central parts of the four Ts are connected through the Klopfenstein tapers to the four main transmission lines in the gallery and further to the four power sources of 5 MW each. Part of the surrounding walls of ITER is shown to illustrate the implantation.

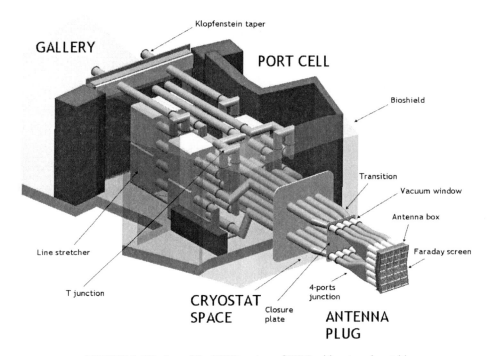

FIGURE 1. 3D view of the ICRH system of ITER with external matching.

DESIGN OF THE ANTENNA PLUG

Details of the RF plug are shown on Fig. 2A and 2B. One can distinguish the following parts: (i) The Faraday – or electrostatic - screen assembly made from cylindrical tubes. This assembly closes the set of antenna boxes to decrease as much as possible spurious coupling between the straps. (ii) The antenna boxes with the radiating straps, each fed by a stripline (section A-F). (iii) Coaxial line sections of $Z_0 = 15\ \Omega$ (F-VT) followed by $Z_0 = 60\ \Omega$ ones (VT-J), which connect the striplines to the input of the 4-ports junctions. (iv) The 4-ports junctions, which perform smooth tapered transitions at constant $Z_0 = 20\ \Omega$ from stripline sections to coaxial ones. The aim of these junctions is to feed as symmetrically as possible three straps in parallel. They are placed near voltage anti-nodes to minimize the effects of possible asymmetry between the 3 straps' circuits. (v) The connections of the junctions by means of 20 Ω coaxial line sections to double vacuum feedthroughs (of the type described in [2]) that make the links with the lines outside the plug.

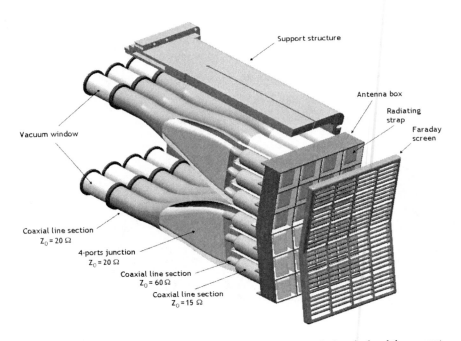

FIGURE 2A. 3D view of the ICRF antenna plug. The Faraday screen is detached and the support structure and front 4-ports junctions cut in order to show the interior.

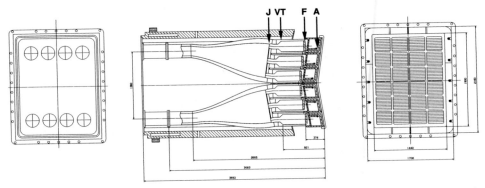

FIGURE 2B. Rear view, side cut and front view of the complete antenna plug. The labels correspond to those of Fig. 1 of [1].

IMPLEMENTATION OF THE EXTERNAL LINE SYSTEM

From the vacuum feedthroughs the lines are connected through the cryostat space and the bioshield to the line stretchers that are placed in the port cell (see Fig. 1).

There is first a transition from the 270 mm lines used inside the plug to commercially available 14,, pressurized lines (air, 3 bar, $Z_0 = 20\ \Omega$). The line stretchers have 2 trombone stages of 2 m stroke in series and are based on existing technology (similar ones have been delivered to JET). They have an RF current capability of 3.33 kA. This allows a good safety margin with respect to a maximum current in the lines of less than 1.8 kA for the reference antenna loading resistance of 4 Ω/m. As shown in Fig. 1 the 8 line stretchers are placed by pairs on top of each other, their trombones being horizontal. Each pair of line stretchers is provided with rollers for easy displacement in order to be removed when the cask has to come into operation.

Fig. 1 also shows at the output of the line stretchers the vertical connecting lines to the T-junctions placed at the top of the port cell. From the central connection of each T a $Z_0 = 6.67\ \Omega$ line followed by a Klopfenstein taper makes the link towards the 5 MW power source. The Klopfenstein taper (Fig. 3) of 5.16 m length (for a $|\Gamma| < 0.02$ on matched load in the frequency band 40-55 MHz) is implemented near the T in the straight section of the line going to the power source. Details on the T-junction connecting the two 20 Ω lines to the 6.67 Ω one are also seen on Fig. 3.

FIGURE 3. Klopfenstein taper and its connection to the T-junction.

By changing the connections between the vacuum feedthroughs and the line stretchers this design allows to easily modify the matching procedure without breaking the machine vacuum (e.g. use as *conjugate-T* other pairs of 3 straps to decrease the mutual coupling effect between them).

CONCLUSIONS

A flexible 20 MW ELM-resilient ICRF system with external matching is designed and implemented in the ITER environment. It uses only 4 *conjugate-Ts* that are operated by means of line stretchers outside the vacuum vessel. It uses present-day technology and an affordable electric field (≤ 20 kV/cm) is easily maintained in the lines and components. It also allows modifying the matching layout without breaking the vacuum of the machine. A detailed account of the design is given in [3].

REFERENCES

1. A. Messiaen et al., *ELM Resilient External Matching System for the ICRF System of ITER: 1. Principle and Performance*, this conference
2. R.C. Walton, *A continuous Wave RF Vacuum Window*, JET-R(99)03
3. Report LPP-ERM/KMS **121**, to be published

The ITER-like ICRH Launcher Project For JET

F.Durodié[1], G.Agarici[2], G. Amarante[1], F.W.Baity[3], B.Beaumont[2], S.Brémond[2], P. Chappuis[2], C.Damiani[4], J.Fanthome[5], R.H.Goulding[3], J.Hosea[6], G.H.Jones[6], A.Kaye[5], P.U.Lamalle[1], G.D.Loesser[6], A.Lorenz[4], M.Mead[5], M.A.Messineo[6], I.Monakhov[5], B.Nelson[3], M.Nightingale[5], J.Paméla[4], C. Portafaix[2], V.Riccardo[5], L.Semeraro[7], C.Talarico[7], E.Turker[5], K.Vulliez[2], A.Walden[5], R.Walton[5], J.R.Wilson[6], P.Wouters[1]

1. Laboratory for Plasma Physics, Association EURATOM-Belgian State, TEC, Royal Military Academy, B-1000 Brussels, Belgium. 2. Association EURATOM-CEA, CEA/Cadarache, 13108 Saint Paul-lez-Durance, France, 3. Oak Ridge National Laboratory, USA, 4. EFDA-JET Close Support Unit, Culham Science Centre, Abingdon OX14 3DB, UK, 5. UKAEA/EURATOM Fusion Association, Culham Science Centre, Abingdon OX14 3DB, UK, 6. Princeton Plasma Physics Laboratory, USA, 7. Associazione Euratom-ENEA sulla Fusione, CR Frascati,CP 65, 00044 Frascati, Italy

Abstract. The paper reports on the status of the JET ITER-Like ICRF Antenna project and highlights main challenges that have come up during its design phase.

INTRODUCTION

The design phase of the JET ITER-like ICRF Antenna[1] is complete and the procurement phase is now under way. It is expected that the launcher will be available for installation onto the JET torus by mid 2004 permitting an experimental campaign starting spring 2005, aimed at validating the proposed antenna concept in terms of ELM resilience at a power density of 8 MW/m^2 in conditions relevant to ITER. Key challenges of this design are its compatibility with worst case scenario disruptions and the high accuracy required for the in-vessel capacitor positioning systems. For predictive evaluation of the expected performance of the launcher as well as in support of the development of a suitable matching algorithm, a substantial modeling effort has been carried out[2]. In addition to testing the ELM tolerance of the ITER-like antenna, 3dB couplers will be used to provide resilience to load variations on two of the existing JET A2 antenna modules[3]. A third scheme based upon trombone-based conjugate-T matching is also under assessment by the UKAEA[4]. The High Power Prototype[5] (HPP) of one of the four loops has been completed by US teams of ORNL and PPPL and is under test at ORNL.

MAIN CHALLENGES

Disruption Loads : For the JET ITER-like antenna (see figure 1), the compatibility with worst case 6MA/4T plasmas disruptions on JET - taking into account assembly procedures and requirements for maintenance of the in-vessel matching capacitors - constitutes one of the main mechanical challenges. The induced mechanical loads are mainly due to the plain copper cylinders and flanges of the matching capacitors located

calculation of the induced perturbation on the in-vessel magnetic field. Consequently, the layout of Antenna Pressurized Transmission Lines (APTL), that interface the main transmission lines with the RF vacuum windows, has been made compatible with the routing of cable and piping for the antenna in-vessel capacitors and servo-hydraulics components and instrumentation, and for secondary containment and vacuum inter-space monitoring services. Prototype testing in order to determine: minimum hydraulic pipe diameters; hydraulic fluid properties; and assess the accuracy of this system, is nearing completion: at the time of writing the an accuracy in positioning well below 0.1mm and even below 50μm is demonstrated.

Project Management : Finally, a non-negligible challenge on the level of the management of the project is formed by the stringent procurement rules imposed by the European Commission and the nature of the constitution of the project team: distributed administratively and geographically over the different associations in Europe as well as the US, each taking care of a specific component (see table 1) or prototyping effort for the global project.

STATUS OF SUPPORTING AND PROTOTYPING ACTIVITIES

Supporting prototyping activities are coming to completion : the main one, the HPP allows benchmarking of the modeling efforts and designing and testing to some extent of the matching algorithm, to assess the final operation limits of the design and allow to tweak the design prior to starting the manufacturing of the antenna components. In the operational diagram shown in figure 3, the voltages obtained on the HPP for short subsecond pulses show that there are margins with respect to the voltage stand-off required to couple 8 MW generator power to a plasma with a base coupling of 2 Ohm/m. Further tests, in particular long high voltage pulses, are scheduled in the near term.

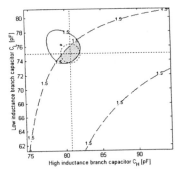
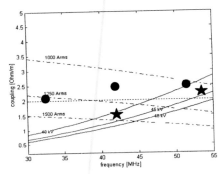

FIGURE 2 : For a single RDL at 55MHz and matching to 6Ω at the junction, without mutual coupling effects, the VSWR 1.5 contours in the domain of the values of the matching capacitors are shown for two different values of the resistive coupling with the plasma, 3 Ω/m (plain) and 9.5 Ω/m (dashed). In order to be resilient to the variation in resistive coupling the position of the capacitors the matching algorithm needs to position the capacitors in the shaded area indicated by the dotted crosshair.

FIGURE 3 : accessible operational domain expressed as iso-curves of capacitor voltage (≈ maximum feeder voltage) and current required to launch nominally 8MW into the plasma in the coupling vs. frequency domain. The stars correspond to the voltages achieved on the HPP for short pulses. The circles represent estimates of the coupling of the new launcher whereby the plasma was modeled as a dielectric slab of material using Microwave Wave Studio[8].

The testing of the finger contact at the front of the matching capacitors required for the assembly and maintenance of the matching capacitors, is, at the time of writing, on-going at CEA(F). The validation of the spring damper that is part of the ex-vessel support structure and limits the inertial forces appearing on the Main Horizontal Port during disruptions as well as the displacement of the launcher with respect to the JET Torus Hall floor, was completed successfully at ENEA(I). In order to validate and consolidate the manufacturing processes prior to the start of the main EU procurement contract, two first of a series capacitors (LPP-ERM/KMS(B)) will be produced.

PROJECT STATUS AND CONCLUSIONS

The procurement of the hardware has been divided into 12 packages shown in table 1 and figure 1. At the time of writing, all but one packages were successfully tendered for and approved by the European Fusion Development Agreement's (EFDA) Administrative and Financial Advisory Committee (AFAC). The manufacturing times confirmed by the selected manufacturers are in line with the project plan and the installations of the systems during the 2004 shutdown. One remaining package on the capacitors' actuator hydraulic system will be presented at the committee's next meeting.

If, in addition to the main ITER-like ICRF Antenna project, the other presently foreseen extensions to the JET ICRF plant all come to implementation, two strategies aiming at mitigating the effects of the ELMs on the matching of ICRF launchers will be available at restart of the experimental campaign after the 2004 shutdown: a high power density ITER-like launcher with in-vessel matching capacitors and the 3 dB hybrid couplers as on ASDEX-Upgrade implemented on two of the four A2 arrays. The comparison of the relative performance of these circuits will most certainly reveal further insight and valuable information guiding the design the ITER ICRF system. Furthermore the demonstration of high power density achieved by using short straps should validate ICRF as a strong candidate for plasma heating on ITER.

REFERENCES

1. Durodié, F., et al., in Radio Frequency Power in Plasmas, AIP Conference Proceedings **595**, New York: Melville, 2001, pp. 122-125.
2. Lamalle, P.U., et al., "Three-dimensional electromagnetic modelling of the JET ITER-Like ICRF antenna", these Proceedings.
3. Wesner, F., et al., in Fusion Technology (19th. SOFT, Lisbon, 1996), Amsterdam: North Holland, 1997, pp. 597-600
4. Monakhov, I., et al., "New techniques for the improvement of the ICRH system ELM tolerance on JET", these Proceedings.
5. Goulding R.H., et al., "Initial operation of the JET ITER-Like High Power Prototype ICRF Antenna", these Proceedings
6. Monakhov, I., et al., "Studies of JET ICRH antenna coupling during ELMs", these Proceedings.
7. Lamalle, P.U., et al., "Radio-frequency matching studies for the JET ITER-Like ICRF system", these Proceedings.
8. CST MICROWAVE STUDIO Version 4.0, CST GmbH, Darmstadt, Germany, 2002; http://www.cst-world.com.

Initial operation of the JET ITER-like High-Power Prototype ICRF Antenna*

R.H. Goulding[1], F.W. Baity[1], F. Durodié[3], A. Fadnek[1], J.C. Hosea[2],
G.H. Jones[1], G.D. Loesser[2], B.E. Nelson[1], D.A. Rasmussen[1],
P. M. Ryan[1], D. O. Sparks[1], D.W. Swain[1], R. Walton[4]

[1]*Oak Ridge National Laboratory, PO Box 2009, Oak Ridge, TN, 37831-8071, USA.*
[2]*Princeton Plasma Physics Laboratory, P.O. Box 451, Princeton, New Jersey, 08543-0451, USA.*
[3]*Laboratory for Plasma Physics, Association EURATOM-Belgian State, TEC, Royal Military Academy, B-1000 Brussels, Belgium.* [4]*UKAEA/EURATOM Fusion Association, Culham Science Centre, Abingdon OX14 3DB, U.K.*

Abstract. Fabrication and assembly of a High Power Prototype (HPP) of the JET ITER-like Ion Cyclotron Range of Frequencies (ICRF) launcher have been completed at Oak Ridge National Laboratory (ORNL), and high power tests have begun. The HPP consists of one quadrant of the full 7.5 MW antenna (1). The prototype is the product of a collaboration between ORNL, Princeton Plasma Physics Laboratory, and EFDA-JET/UKAEA. Internal matching capacitors are utilized in a circuit that maintains a voltage standing wave ratio (VSWR) at the input < 1.5 over a factor of ten range in resistive loading. Short (.05 s) pulses have achieved > 45 kV peak voltage at the internal matching capacitors, which is greater than the original design voltage. High power pulses up to 2s have been run. Diagnostics include thermocouples, voltage probes at the capacitors and along the integral $\lambda/4$ matching transformer, and an optical temperature sensor for in-situ measurements of capacitor temperatures. Low power measurements of electrical characteristics of the antenna have been made and compared with a 3-D electromagnetic model.

ANTENNA DESIGN FEATURES

The current strap configuration and internal matching network of the HPP are designed to minimize the magnitudes of both voltages and electric fields on launcher structures, and the physical extent over which high values of these quantities are present. The largest electric fields are confined to internal electrodes in the matching capacitors. Figures 1 and 2 show the principle components of the device. Two electrically short current straps, grounded at one end, are each connected in series to a matching capacitor. The capacitors are adjustable to provide a tuning range of 30-55 MHz. The other ends of the capacitors are connected in parallel to form a resonant double loop configuration (2). During operation, the capacitors are adjusted to produce complex-conjugate admittances having a small real part at the connection (feed) point. The vacuum transmission line (VTL) feeding these capacitors consists of a rectangular outer conductor and an inner conductor with a racetrack shaped cross-section. It serves as a $\lambda/4$ impedance transformer, transforming a low, real impedance at the feedpoint (3-6Ω) to the 30 Ω characteristic

---

* Oak Ridge National Laboratory, managed by UT-Battelle, LLC, for the U.S. Department of Energy under contract number DE-AC05-00OR22725.

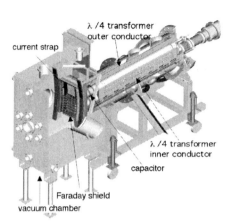
FIGURE 1. HPP installed in RFTF

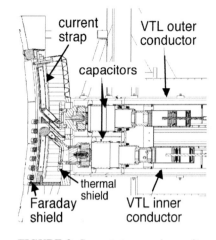
FIGURE 2. Current straps and capacitors

impedance of the main transmission line. In this configuration, the voltage standing wave ratio (VSWR) at the input to the impedance transformer is insensitive to increases in resistive loading over a large range(3).

Several novel features of the ITER-like antenna design are being tested on the HPP. Unlike previous RDL implementations, the variable ends of the tuning capacitors are not grounded, and are exposed to machine vacuum. This requires a double bellows arrangement. In addition, it is required that the capacitors be easily replaced in case of malfunction. The solution chosen was to make the entire VTL inner conductor removable, with the capacitors "plugged in" to the current strap feeds. This cannot be done while still actively cooling the fixed ends (nearest the current straps) of the capacitors, as is typically the case. Instead, the fixed end electrodes are cooled by conduction through the capacitor ceramics between pulses. The current is carried through silver plated copper-beryllium Multilam® contacts (type LA-CUT) between mating flanges, which also feature a ball and socket joint to prevent crushing of the contacts during installation, while allowing the capacitors to rotate relative to the straps to accommodate thermal expansion (Fig. 2).

Another unusual feature is the low characteristic impedance ($Z_0 = 9.4\ \Omega$) quarter-wave impedance transformer. The inner conductor is large enough to contain the capacitor drive motors, which are located in a separate vacuum interspace. The low Z_0 requires a minimum gap between inner and outer conductors of only 30 mm (Fig. 2). Elongated pumpout holes are present in both inner and outer conductors to maintain good pumping speed in this region of low vacuum conductance. The entire weight of the inner conductor assembly is supported through the insulating ceramics located in the bottom matching capacitor at one end, and the vacuum window at the other. These must also withstand disruption loads in the actual antenna (1).

LOW POWER MEASUREMENTS

Measurements of scattering parameters at the inputs to the feeds of the two current straps have been made using a network analyzer, with type N adaptors installed where the capacitors normally plug in to the strap feeds. These measurements compare well with modeling results obtained using the commercial 3-D electromagnetic solver, Microwave Studio (MWS) (4). An example is shown in figures 3 and 4. Figure 3 shows the model constructed in MWS, incorporating all important antenna structures including Faraday shield elements. The type N adaptors used in the measurements are also modeled, but not shown in the figure. Figure 4 shows a comparison of the magnitude of the measured and modeled mutual coupling between top and bottom straps, indicating very good agreement over a wide frequency range. This is important because it increases confidence in the accuracy of calculations now underway using MWS to model the full 8 strap ITER-Like antenna, where coupling between antenna elements may significantly affect the load-tolerant behavior (3). An extensive set of other low power measurements have been made, and will be reported at a later time.

MULTIPACTOR CONDITIONING

The large surface area, low electric field, and small gaps present in the $\lambda/4$ impedance matching transformer increase its susceptibility to multipactor breakdown. However, little indication of this behavior has been seen during operation to date at frequencies of 42.5 MHz and 50 MHz. On the first day of high power tests, the antenna was operated continuously at power levels between 10 and 1500 W. There was some evidence of multipacting in that the pressure was observed to rise quickly above 10^{-4} torr (compared to a base value of 10^{-6} torr) when power was applied, but the pressure increase lasted for only a short time (~15m) in comparison to other antennas we have operated for the first time in this so-called "multipactor conditioning" mode. In addition, an impedance mismatch caused by multipactor loading was only observed in the 10 W to 80 W power range, which is significantly smaller than the range seen for past antennas we have tested.

INITIAL HIGH POWER OPERATION

The first high power operation involved short pulses (0.05 s) repeated at 4-20 s intervals, in order to maintain an average power below 1500 W. This power level

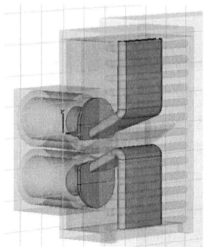

FIGURE 3. MWS model

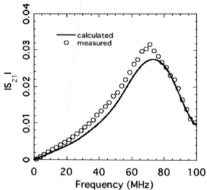

FIGURE 4. Comparison of modeled and measured cross coupling magnitude (S_{21})

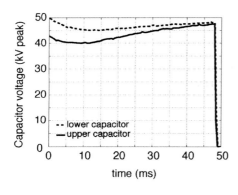
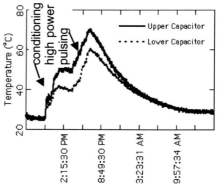

FIGURE 5. Highest measured voltages at capacitor fixed (closest to current strap) ends.

FIGURE 6. Example of the capacitor fixed end optical temperature measurement

allowed long periods of operation (several hours) with only moderate increases in temperature of the non-actively cooled regions of the matching capacitors. Voltage levels at the fixed ends of the capacitors > 45 kV were achieved, as measured with capacitive voltage probes that were absolutely calibrated in place using a high impedance Agilent model 41800 A active probe (fig 5). Optical measurements of the temperatures of the capacitor ceramics at a location 1 cm behind the fixed end electrodes have been made using a Luxtron Fluoroptic thermometer, in order to accurately determine the thermal behavior in the non-actively cooled regions. Figure 6 shows an example in which continuous power multipactor conditioning was followed by high voltage pulsing at average powers up to 1500 W, with maximum input powers of ~ 300 kW. The power was then switched off and the temperatures decayed, primarily due to heat conduction through the capacitor ceramics to the water jackets inside the capacitor fixed ends. Simple thermal calculations using the data shown in the figure predict that 10s pulses at the maximum capacitor operating current of 1250 A rms will produce a temperature rise between 15 – 21 °C, which is within the acceptable range. Long pulse testing has recently begun, with 2s pulses achieved at an input power level of ~ 100 kW.

REFERENCES

1. Durodié, F, et al., "The ITER-Like ICRH Launcher for JET", these proceedings.
2. T. L. Owens, F. W. Baity, and D. J. Hoffman, "ICRF Antenna and Feedthrough Development at Oak Ridge National Laboratory", AIP Conf. Proc. **129**, (Calloway Gardens, 1985), p.95
3. P. U. Lamalle, et al., "Radio-frequency matching studies for the JET ITER-Like ICRF System", these proceedings.
4. http://www.cst.de/

Measurements and Calculations of Electrical Properties of ICRF Antennas

D. A. Hartmann, R. Bilato, D. Birus, M. Brambilla, F. Braun, J.-M. Noterdaeme, F. Wesner

Max-Planck Institute for Plasma Physics, EURATOM-Association, 85478 Garching, Germany

Abstract. The S-matrix of mock-up ICRF antennas and the ASDEX Upgrade antenna were measured and compared with the commercial code HFSS ™, designed to calculate electromechanical properties of structures, and with FELICE, written to calculate the coupling of ICRH antennas to a warm plasma in slab geometry, but also usable for calculating antenna properties without plasma. Proper assessment of the boundary conditions led to excellent agreement between the measurements and the results of HFSS for a wide range of antennas. Good agreement between the measurements and the calculations with FELICE was found for those cases were the actual antenna geometries and those that could be implemented in FELICE were not too different.

INTRODUCTION

The need to design ICRF antennas for the stellarator W7-X and the planned modification of the present antennas of ASDEX Upgrade require computational tools that can reliably predict the electromagnetic properties of antennas with and without plasmas. A number of such tools already exist, albeit, with limited range of applicability. Commercial codes (HFSS™ [1], Microwave Studio™ [2]) are well suited to complicated geometries, but up to now they cannot treat a plasma dielectric, whereas codes written by the plasma physics community (FELICE [3, 4], TORIC [4], RANT3D [5]) retain most of the plasma properties, but need to greatly simplify the geometry.

At the Max-Planck Institute for Plasma Physics in Garching and Greifswald HFSS and FELICE are the most often used tools. HFSS is a well-established tool for investigating antennas as well as other RF components. FELICE has a long history at the IPP and has the advantage of facilitating fast calculations of the antenna properties and of the coupling to a plasma which retains all relevant plasma waves. In order to gain confidence in the prediction of the codes, we compared calculation results of the scattering matrix [6] of the codes with each other and with measurements done on simple mock-up antennas. Thus it was possible to assess the influence of some antenna elements such as side walls, thickness of the current strap, Faraday screen etc. on the antenna properties which cannot - or only in simplified form - be implemented in FELICE. In addition, one could confirm how to properly implement the boundary conditions in the calculations with HFSS.

In the following we report calculations and measurements of the scattering matrix of simple mock-up antennas and the present ASDEX Upgrade antennas without plasma. With plasma the present setup of the ASDEX Upgrade antennas does not allow full scattering matrix measurements but only measurements of the overall antenna loading where both antenna straps are connected to one transmitter.

S-PARAMETER OF ANTENNAS WITHOUT PLASMA

Four different mock-up antennas were investigated. They all had in common that their current straps were parallel to a large plain conducting back plane used as the return conductor and RF ground:
1) a feeder-to-short antenna consisting of a current strap fed on one side and short circuited to the back plane on the other.
2) a simple antenna consisting of a current strap fed on both ends
3) a push-pull antenna consisting of a current strap fed on both ends but short circuited to the back plane in the midsection of the strap.
4) a folded antenna consisting of a current strap folded in such a way that some parts of the antenna have more than one layer of RF current.

Schematics of antennas 3 and 4 are shown in Figure 1 a and b. The antenna feeders are indicated with "fdr". For the measurements these mockup antennas were simply manufactured out of 1 mm thick aluminum sheets mounted on a 5 mm thick aluminum sheet of about 1.5 m x 1.5 m. All antenna straps had a length of L = 1 m (poloidally[1]) and a width of W = 18 cm, the distance to the back plane was varied between H = 5 and 45 cm. The measurement device as well as all other bodies were at a distance of several meters.

In FELICE the current layers are assumed to be infinitely thin. FELICE evaluates the current distribution using the Ritz variational method first proposed by Theilhaber and Jacquinot [7]. The value of the variational functional corresponding to the "best" current distribution gives the complex input impedance of the antenna. Currents in the radial conductors are taken into account, but in the present implementation they are not part of the variational problem: they are taken to be constant (propagation constant equal to zero), which is equivalent to assume that the radial conductor segments have a negligible electrical length. The resonance frequency predicted in this way is systematically higher than the measured one, and the discrepancy increases linearly with the depth of the antenna, strongly suggesting that this assumption is not valid. Although the algebra to take into account a finite propagation constant in the radial conductors has been done, its numerical implementation is not trivial, and will require some time. For the moment, a partial solution to the problem has been found as follows:
1) The "test functions" used for the variation have been changed to take into account that the short boundary condition ($\partial J/\partial y = 0$ where y is the ordinate along the conductor) occurs not at the end of the tangential conductor, but farther away at a distance equal to the length of the radial segment which follows it;
2) The input impedance, evaluated by the code at the entry point of the tangential conductor, is transformed by adding a segment of transmission line equal to the distance from the true feeder to this point, assuming that this additional line has the same characteristics as the rest of the antenna. Although not exact, this procedure is quite plausible, and indeed gives results in very good agreement with the measurements and with the predictions of the commercial code as will be shown further on.

On a workstation a frequency scan typically took a few minutes. Presently mutual coupling - possibly important for two-port antennas - cannot be treated.

In HFSS the mock-up antennas were modeled with actual geometry and material properties. For the calculations the antenna was installed on one side of a box. Care had to be taken to make the box sufficiently large so that outward radiation conditions could be imposed on the five remaining walls. This required the box to be a cubicle with 5m long sides. On a PC a frequency scan typically took several minutes.

---

[1] The terms poloidal, radial and tangential are used with reference to the orientation of the conductors in a tokamak vessel.

Figure 2 shows the measurements and the calculations done with FELICE and with HFSS of the scattering matrix S at the feeder a folded antenna. The parameters of the antenna are L = 100 cm, H2 = 2 cm, L2 = 80 cm, H2 = 15 cm. Even for such a rather complicated antenna with internal coupling the agreement between the measurements and HFSS is excellent; the agreement with FELICE is good, in particular the resonance frequency (i.e. the frequency where the imaginary part of S_{11} is zero) approximately agrees.

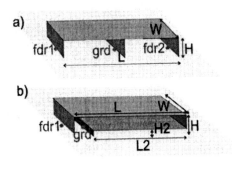

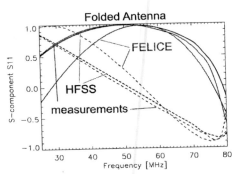

FIGURE 1 Schematic of the push-pull antenna (a) and the folded antenna (b).

FIGURE 2 S-values (real part-solid, imaginary part-dashed) of folded antenna.

The mock-up antennas were investigated in the given frequency range. In all cases the agreement between HFSS and the measurements was excellent. This also included 2-port antennas. For example, here the off-diagonal elements of the push-pull antenna had a magnitude of less than 0.2 and the deviation between the measurements and the calculations was less then 0.02. The agreement between FELICE and the measurements was still good, in particular the resonance frequency agreed within about 5 MHz in all cases. The off-diagonal elements of the scattering matrix cannot be calculated with FELICE.

Side walls were added to the mock-up antennas. They consisted of 1 mm thick aluminum sheets that extended by 5 cm over the length and by about 1 cm over the height of the antennas. They were mounted onto the back-plane at various toroidal distances from the antenna straps. At small toroidal distances the resonance frequency was increased, e.g. for the feeder-to-short antenna the resonance frequency increased by about 5 MHz at a distance of 5 cm. Increasing the thickness of the current straps at the edge to a radius of 5 mm (from an estimated radius of less than 0.5 mm) did not change this dependence, thus the effective propagation velocity along the strap was lowered by an increase of the inductance. Again the agreement between HFSS and the measurements was excellent.

An additional Faraday screen was added to the mock-up antennas. It consisted of 3mm diameter copper rods mounted on a 2 mm thick PVC sheet. The rods were electrically connected to the side walls. The influence of a Faraday screen can also be calculated with FELICE. If the side walls were sufficiently far (30 cm) from the antennas then the calculated decrease in the resonance frequency (due to the increased capacitance of the current strap) quantitatively agreed with the measurements. The effect of the Faraday screen on the resonance frequency was considerably reduced for antennas with current straps that had thick edges with 5 mm radius. Also these measurements were in quantitative agreement with HFSS.

Plasma loading was experimentally simulated using 1m x 1m sheets of water or steel wool that were placed at some distance parallel to the current straps. In the first case 16 mm PVC tubes were filled with water and mounted on a PVC sheet with 2 centimeters distance between each other. In the second case steel wool of a thickness of several centimeters was sealed in a PVC bag. It was found that the dominant damping mechanism of water was capacitative, therefore water is a poor absorber for testing ICRH antennas. Nevertheless, also here good agreement was found between HFSS and the measurements. Also the directional dependence caused by the capacitative coupling was well reproduced. The damping mechanism of steel wool was found to be predominantely inductive, thus steel wool is a suitable damping substance.

The scattering matrix of a replica of the ASDEX Upgrade antenna mounted in a teststand was also measured. One port was connected to a replica of the vacuum feedthroughs used in the experiment. In order to obtain the scattering matrix at the antenna inputs of the antenna the transformation done by the feedthroughs had to be evaluated. This was done with HFSS for each element and compared with measurements of the total assembly. Since the agreement was good, the scattering matrix at the antenna feeders was derived from the measurements of antennas and feedthroughs and the HFSS calculations of the feedthroughs. The diagonal elements of the inferred values of the scattering matrix at the feeders were in good agreement with the HFSS calculations; the off-diagonal elements were found to be sensitive to the details of the shorts in the antenna and need to be further investigated.

CONCLUSION

The tools easily accessible at IPP for designing ICRH antennas for a possible upgrade at ASDEX Upgrade and for W7-X were benchmarked for simple antenna structures. Using proper boundary conditions for HFSS and taking into account the radial currents with finite propagation velocity in FELICE, both codes have good predictive qualities for the S-matrix of antennas without loading. In a next step the RF E- and B-fields in the antenna vicinity will be investigated in more detail to gain insight into the limitations of the codes in predicting the coupling of the plasma and the launching of the fast wave.

ACKNOWLEDGMENTS

The assistance of Mr. G. Heilmaier und Mr. G. Siegl with the setup and the machine shop is gratefully acknowledged.

REFERENCES

1. *"High Frequency Structure Simulator (TM)"*, v.8.5, Ansoft Corp., Pittsburgh, PA, 2002.
2. *"CST Microwave Studio (TM)"*, Computer Simulation Technology, Wellesley Hills, 2002.
3. M. Brambilla, *Nuclear Fusion* **28**, 549-563 (1988).
4. M. Brambilla,*"Recent progress in IC wave codes at IPP Garching"* in Series *Radio Frequency Power in Plasmas*, edited by P. M. Ryan, et al., AIP Conference Proceedings, 403, Savannah: American Institute of Physics, 1997, pp. 257-263.
5. M. D. Carter, et al., *Nuclear Fusion* **36**, 209-223 (1996).
6. D. M. Pozar, *"Microwave Engineering"*, New York, John Wiley & Sons, 1998, pp. 194-204.
7. K. Theilhaber and J. Jacquinot, *Nuclear Fusion* **24**, 541-554 (1984).

The Effect of Weak Single Pass Damping on the Coupled ICRH Power Spectrum

T. Hellsten and M. Laxåback

Alfvén Laboratory, Association VR-Euratom, Sweden

Abstract. Consequences of coupling to eigenmodes during ICRH for weak single pass damping are discussed. A degradation of the central heating and an enhancement of parasitic absorption are found as the single pass absorption decreases.

INTRODUCTION

The often seen poor performance of weak single pass damping scenarii is common to different devices, but the performance depends also on the antennae. The heating efficiency of the JET A2 antenna was found to be strongly degraded for phasings coupling to modes with low toroidal mode numbers [1] whereas the heating efficiency of the A1 antenna was reported to be high, about 75% [2]. To get agreement between modelling and experiments for scenarii with weak single pass damping it has often been necessary to introduce an ad hoc parasitic damping [3]. Great efforts have been devoted to identifying possible parasitic damping mechanisms. In general, however, these mechanisms have been found to produce negligible damping. Large misalignments between the Faraday screen and the magnetic field was found to give rise to large parasitic absorption at rectified RF-sheaths [4,5], which had been explained to cause the observed impurity sputtering during RF-heating [6]. Recent experiments on JET showed that the degradation of the heating efficiency for scenarios with low single pass damping is caused by a large fraction of the power being not absorbed in the plasma, which is consistent with enhanced losses at RF-sheaths [7].

COUPLING TO A SPECTRUM OF WEAKLY DAMPED MODES

The power coupled by the generator, P_{GEN}, into the transmission line is distributed in power absorbed in the plasma, in rectified RF-sheaths, in the transmission lines and in the antennae. Separating the power absorbed in the plasma into its toroidal Fourier components one obtains

$$P_{GEN} = \left(\sum_n R_n I_n^2 / I^2 + R_{vac} + R_{RFC} \right) (U_M / Z_0)^2 \tag{1}$$

where U_M is peak to zero voltage on the transmission line, Z_0 the impedance of the transmission line, I_n the toroidal Fourier components of the antenna current, R_n the Fourier components of the Poynting flux normalised with the current, R_{vac} the resistance in the antenna and the transmission line and the losses related to RF-sheaths have been modelled as $(U_M/Z_0)^2 R_{RFS}$, which were observed to be proportional to the coupled power [5]. In general R_{RFS} is a non-linear function of U_M.

well inside the main toroidal magnetic field. These loads appear as torques about the axis of the capacitor's electrodes as well as perpendicular to it. Therefore, the design of the capacitors (CAP) – Inner Vacuum Transmission Line (VTL) – Vacuum Window (VCW) system, has been modified from the original concept, in order to minimise and resist the EM induced forces. Design validation tests on mock-ups have also been carried out. In particular: the capacitor's OHFC copper to ceramic brazes have been reinforced; its outer envelope redesigned and copper partially replaced by SS; and the shaft of the variable electrode has been designed to be able to react the torque of about 600 Nm induced on the electrode. After redesign, a torque of about 5 kNm remains, which is reacted through the inner-VTL by the RF vacuum window. In order to allow for this, the inner-VTL has been stiffened and the RF window increased in size.

Matching Algorithm : One of the outstanding challenges is the design of a matching algorithm that is able to take into account the complexity of the matching problem of several coupled Resonant Double Loops (RDL). Even for a simple RDL, and not taking into account the cross-coupling between straps, the matching capacitors must be adjusted within a few tenths of mm in order to achieve the desired resilience to resistive load variations (see figure 2). Cross-coupling and load variations that may not be purely resistive further compound the issue[6,7], and a global solution is to be found that keeps the VSWR as low as 1.5 for real ELMs on JET.

Actuators : The actuators positioning the matching capacitors will be based on servo hydraulics. Originally a solution based on vacuum-compatible electric-motors was identified (and used on the HPP) but this had to be abandoned after the detailed

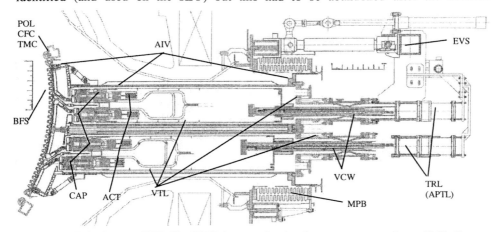

FIGURE 1. Cut view ITER-like ICRH Antenna showing the procurement packages (Table 1).

TABLE 1. Procurement Packages for the ITER-like ICRH Antenna

In-Vessel or on Tritium Boundary		Ex-Vessel
AIV- Antenna In-Vessel Compenents (UKAEA,UK)	VCW- Vacuum Window (UKAEA,UK)	EVS- Ex-Vessel Support Structure (ENEA,I)
MPB- Main Port Bellows (UKAEA,UK)	PLB- Poloidal Limiter Beams (CEA,F)	TSB- Testbed Components (UKAEA,UK)
VTL- Inner Vacuum Transm. Lines (ERM,B)	CFC- Material Blanks f. Limiter Tiles (UKAEA,UK)	TRL- Antenna Pressurized Transmission Lines (APTL), Second Stage Matching Components and Main Transmission Lines (UKAEA,UK)
BFS- Beryllium Faraday Screen (UKAEA,UK)	TMC- Pol. Limiter Tile Machining (CEA, F)	
CAP- Matching Capacitors (ERM,B)	ACT- Servo Hydraulic Actuators (ENEA, I)	

An ICRH antenna couples in general to a wide spectrum of toroidal and poloidal modes, which for strong single pass absorption gives rise to an evanescent wave field localised in the vicinity of the antenna. The presence of high-Q magnetosonic eigenmodes results in large oscillations in the coupling resistance, with large variations in the coupling to individual modes when the equilibrium parameters vary. When the plasma parameters are such that a high-Q eigenmode is in resonance most of the power is coupled to this mode. The weaker the single pass damping is the larger the fraction of the coupled power going to the resonating mode will be and the lower the wave field at the antenna will be for constant power. For weak single pass damping and short wavelengths compared to the dimension of the plasma the coupled mode spectrum becomes sensitive to small changes of the equilibrium parameters, in particular density and magnetic field, justifying a statistical treatment of the coupling. Because of the large number of eigenmodes the overlap of modes plays an important role. Noting that the spectrum of a plane slab described the coupling rather well [8], we propose as a generic model of the spectrum for a large aspect ratio torus with a circular cross section that of the equation

$$\frac{d^2 f}{dr^2} + \frac{1}{r}\frac{df}{dr} + \left(\frac{\omega^2}{V_A^2(r)} - k_z^2 - \frac{m^2}{r^2}\right) f = 0 \qquad (2)$$

For a constant $V_A(r)$, in the absence of a sink term at $r = 0$ and assuming $f(r_0) = 0$ the solutions are given by the Bessel functions, $J_m(pr)$, where $p^2 = (\omega/V_A)^2 - k_z^2$. The frequency of the eigenmode is determined by p so that $j_{m,s} = pr_0$, where $j_{m,s}$ denotes the s:th zero of the Bessel function J_m, where r_0 is the plasma boundary. Eigenmodes with large $|m|$ are localised at the edge. For $\omega/V_A \gg 1$ the eigenmodes with small values of $|m|$ have several radial nodes and can be approximated by $J_m(pr) \approx \sqrt{2/\pi pr}\{\cos(pr - m\pi/2 - \pi/4)\}$ giving rise to the resonance condition $r_0\sqrt{\omega^2/V_A^2 - n^2/R^2} = (j + m/2 + 3/4)\pi$. Between eigenmodes with consecutive radial mode numbers (j, m, n) and $(j+1, m, n)$ there is one eigenmode $(j, m+1, n)$ and a large number of eigenmodes $(j, m, n+k)$. For typical JET parameters k goes from 1 to about 40.

A model of the power coupled to the various modes can be obtained by using the facts that the averaged coupling resistance, $\langle R_{m,n} \rangle$, to a mode (m, n) is independent of the single pass damping [9], and that the maximum value of $R_{m,n}$ is inversely proportional to the single pass damping $a_{m,n}$ and that the width is proportional to it

$$R_{m,n} = \frac{\langle R_{m,n}\rangle}{a_{m,n}\sqrt{2\pi}} \exp-\left(\frac{\omega - \omega_{m,n}}{a_{m,n}\omega}\right)^2. \qquad (3)$$

The power coupled to the plasma is then given by $P_{PL} = \sum_{n,m} R_{m,n} I_n^2 / I^2 (U_M/Z_0)^2$. With the conditions that P_{GEN} is given and that $U_M < U_0$ this results in

$$P_{PL} = P_{GEN} \sum R_{m,n} \frac{I_{m,n}^2}{I^2} \left(\sum R_{m,n} \frac{I_{m,n}^2}{I^2} + R_{vac} + R_{RFS} \right)^{-1}. \quad (4)$$

The averaged coupled power for strong and medium strong single pass damping then becomes

$$\langle P_{PL} \rangle = P_{GEN} \left(\sum \langle R_{m,n} \rangle \frac{I_{m,n}^2}{I^2} \right) \left(\sum \langle R_{m,n} \rangle \frac{I_{m,n}^2}{I^2} + R_{vac} + R_{RFS} \right)^{-1} \quad (5)$$

which gives the same value as one obtains when assuming complete single pass damping [9]. In the limit of weak single pass damping, $a_{m,n} < nV_A^2 / \omega^2 R^2$, one obtains

$$\langle P_{PL} \rangle = \sum a_{m,n} P_{GEN}. \quad (6)$$

Best (worst) coupling is obtained when the toroidal mode number for the resonant eigenmode coincides with the maximum (minimum) of $\langle R_{m,n} \rangle$. When they do not the voltage has to be increased to couple the same amount of power, resulting in increased losses by RF-sheaths and a launched spectrum which will differ from the aimed. The narrower the maximum of the toroidal mode spectrum $\langle R_{m,n} \rangle$ is, the larger the reduction in coupling will be. The critical single pass damping for this degradation is defined by the overlap between m and $m + 1$ modes, resulting in the condition $a_{m,n} > \pi^2 V_A^2 / 4\omega^2 r_0^2$, for typical JET parameters $\pi^2 V_A^2 / 4\omega^2 r_0^2 \approx 0.6$. The expected power coupled to the plasma versus single pass damping is illustrated in Fig. 1, for $R_{vac} = 0.5\Omega$ [4]. R_{RFC} is assumed to be proportional $\sin\gamma$ and fitted to agree with 40% losses for the monopole phasing of the A1 antenna at an angle of 22° between the magnetic field and the screen, the width of the spectrum $\Delta n = 16$ and $\langle R \rangle = 4\Omega$ [5]. For monopole phasing of A2 $\langle R \rangle = 4\Omega$, the width $\Delta n = 8$. For 90° $\langle R \rangle = 2.6\Omega$ and $\gamma = 7°$. The effects of the two limits are clearly seen. Since the single pass damping decreases, in general, with $|n|$ the difference between the A1 and the A2 antennae should be caused by the lower effective single pass damping for lower $\langle |n| \rangle$ and by the narrower spectrum of the A2 antenna.

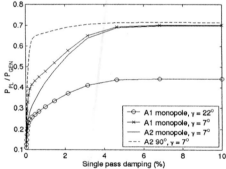

Fig. 1 The expected power coupled to the plasma versus single pass damping

CONCLUSIONS AND DISCUSSIONS

For weak single pass damping the radiated power spectrum will change with the change of the equilibrium parameters in particular density. Because of the infrequent coupling to modes with low single pass damping and on averaged high expected coupling the fraction of power coupled to modes with high single pass damping and/or low expected coupling resistance will increase. In general these modes have lower heating and current drive efficiencies because of less centrally peaked power deposition and reduced current drive caused by interactions with electrons at lower phase velocities for which trapped particle effects become important. Furthermore, the reduced coupling will increase the voltage and field at the antenna resulting in an enhanced absorption at RF-sheaths. Because of the large variation in the radiated wave spectrum there will be a large variation in the heating efficiency in agreement with observations [1]. The condition $a_{m,n} > \pi^2 V_A^2 / 4\omega^2 r_0^2$ defines the condition for coupling to a broad spectrum of toroidal eigenmodes, if this condition is not satisfied the superposition of the coupled eigenmodes can give rise to large localised fields producing rectified RF-sheaths at walls and limiters. This should give rise to losses caused by currents driven by the rectified voltage flowing along the field lines into the walls, consistent with a recent analysis of arcs produced when antennae with different phasings are used producing different rectified voltage at the antennae [10]. The condition on the single pass damping for good overlap, is of the order of 6% for JET, which magnitude is consistent with experiments at DIII-D [3] showing that in the limit of weak single pass damping the efficiency decreases with the damping when it is below a value of the order of 4% [3]. It is also consistent with that the lost power does neither appear in the form of radiation measured with bolometer camera nor in the divertor measured with thermocouplers in JET [7]. The degradation of the current drive with the ELM frequency observed by Petty et al [11] can also be explained by losses in RF-sheath enhanced by the increased density in the SOL after an ELM. For monopole phasing the losses are not caused by coupling to low $|n|$ modes, but by the increased voltage on the antenna caused by the reduced coupling when these modes are not resonant, which is consistent with the observation of a reduced impurity sputtering when the coupling resistance is high and increased sputtering when it is low [12]. The reduced heating efficiency of the A2 antennae in JET is consistent with lower single pass damping and the narrower spectrum of the A2 antennae for monopole phasing compared to that of the A1 antennae.

The coupling to edge localised modes with high poloidal mode numbers will on the averaged be nearly independent of the strength of the absorption at the plasma edge where the modes are localised because of the weak coupling. However, because of the low coupling the degradation of the heating efficiency will also be low.

REFERENCES

[1] J. Hedin, TRITA-ALF-1996-01 report ISSN 1102-2051, ISN KTH/ALF/R—96/1—SE
[2] J. Jacquinot, et al, Dubrovinik 1988 Plasma Physics and Contr. Fusion, **30**, (1988)1467
[3] C.C. Petty, et al, Nucl. Fusion **35** (1995) 773.
[4] M. Bures, J.J. Jacquinot, D.F.H. Start and M. Brambilla Nuclear Fusion **30** (1990) 251.
[5] M. Bures, et al, Nuclear Fusion **32** (1992) 1139.
[6] F. W. Perkins, Nuclear Fusion **29** (1989) 583.
[7] T. Hellsten and M. Laxåback, this conference.
[8] T. Hellsten et al, 13th EPS on Contr. Fusion and Plasma Heating, Schliersee 1986. Vol. II pp. 129.
[9] L-G. Eriksson and T.Hellsten, Physica Scripta **52** (1995) 70.
[10] D. A. D'Ippolito et al, Nuclear Fusion **42** (2002) 1357.
[11] C. C Petty, et al, Nucl. Fusion **39** (1999) 1421.
[12] M. Bures, et al, Plasma Physics and Controlled Fusion **33** (1991) 937.

The Radial Mode Transition of Excited Fast Alfvén Waves in the Mirror Plasmas

H.Higaki, M.Ichimura, Y.Yamaguchi, S.Kakimoto, K.Horinouchi, K.Ide, H.Utsumi, D.Inoue, H.Hojo, and K.Yatsu

Plasma Research Center, University of Tsukuba, Ibaraki, Japan 305-8577

Abstract. For producing a high density and high energy plasma in the GAMMA10 tandem mirror, the high harmonics ICRF waves were employed. The axial wave number of excited fast Alfvén waves with the frequency of 41.5 MHz was measured for the wide range of plasma parameters with magnetic probes. It was found that the plasma density increased clearly by applying the high harmonics ICRF (f = 41.5MHz) and the radial mode transition of excited fast Alfvén wave was observed at nl ~ 5 x 10^{13} cm^{-2}.

INTRODUCTION

The RF oscillator (RF3) for the high harmonics ICRF waves was introduced for producing a high density and high energy plasma in the central cell of the GAMMA10 tandem mirror. Its frequency is variable from 36 to 76 MHz. Although the effect of the RF3 (f = 63MHz) for the plasma production was reported[1], the physical mechanism for the higher density plasma production is still unknown. Recently, the axial wave number of excited fast Alfvén waves with the frequency of 41.5 MHz was measured for the wide range of plasma parameters with magnetic probes. The measured axial wave numbers were compared with simple numerical calculations.

EXPERIMENTAL SET UP

The GAMMA10 tandem mirror has the central cell and plug-barrier cells which have the axisymmetric mirror configuration. In addition, the anchor cells between the central cell and plug-barrier cells provide the MHD stability as a whole with the minimum-B mirror configuration having non-axisymmetric magnetic field. The strength of the magnetic field in the central cell and the anchor cell is shown in Fig.1. The central cell is a simple magnetic mirror with the mirror ratio of 5. The minimum field strength in the central cell is about 4kG.[2]

For the plasma production, short pulse (~1msec) gun-produced plasmas are injected from both ends with H$_2$ gas puffing. The initial plasma is then sustained by ICRF systems named RF1, RF2, and RF3 with H$_2$ gas puffing. RF1 supplies its RF power through Nagoya-type III antennas for the plasma production in the central cell and the plasma heating in the anchor cells with frequencies of 9.9 and 10.3 MHz. The plasma heating in the central cell is provided by RF2 with the fundamental ion cyclotron frequency Ω_{ci} of 6.36 MHz through the double half-turn antenna. Since the frequency of RF3 is much higher than the ion cyclotron frequency in the central cell, RF3 excites the fast Alfvén wave through the double half-turn antenna shown in figure 1 and creates plasma particles and high energy ions.[3]

The axial wave number of the excited waves was measured with 2 pairs of magnetic probes (Z1,2), (MP4,5) in the central cell, as shown in Fig.1. A pair of probes is separated 16 cm. Signals from those probes were recorded with analog to digital converter. Then, FFT power spectra, a phase difference and a coherence were analyzed every millisecond for the excited wave frequency.

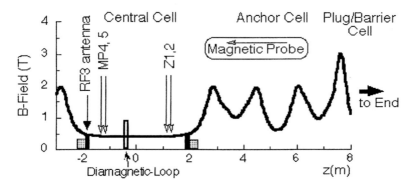

FIGURE 1. The strength of the magnetic field in the central cell and the anchor cell. The positions of the double half-turn antenna for RF3 and two pairs of magnetic probes are shown by arrows.

RESULTS

The high harmonics ICRF is thought to be effective for the plasma production since there are more radial eigenmodes.[1] To investigate the physical mechanism of the plasma production by RF3, the axial wave number of excited fast Alfvén waves was measured for f = 41.5MHz ~ 6.5Ωci.

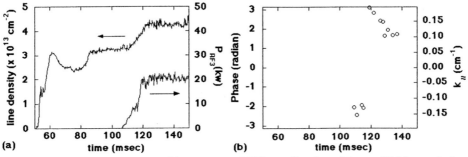

FIGURE 2. (a) The line density and input power of RF3 as a function of time. (b) Measured phase difference for the same plasma shot.

Examples of the line density at the central cell and the measured phase difference (and axial wave number) are shown in Fig.2. It is clearly seen in Fig.2(a) that the line density increased drastically by applying RF3. In Fig.2(b), the phase difference measured with a pair of magnetic probes (MP4,5) is plotted as a function of time for the same plasma shot.

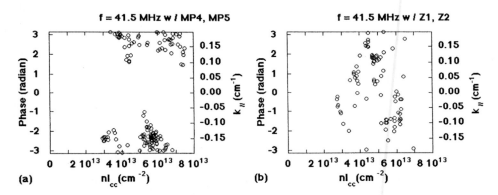

FIGURE 3. Phase difference of the excited fast Alfvén wave (f = 41.5MHz) measured with magnetic probes (a) (MP4,5) and (b) (Z1,2).

Shown in Fig.3 are the phase and axial wave number of the fast wave plotted as a function of the line density. The data shown in Fig.3(a) were obtained with a pair of probes (MP4,5) placed near the RF3 antenna. The data in Fig.3(b) were detected with a pair of probes (Z1,2) installed further away from the RF3 antenna. It is seen that there is a correlation between the phase difference and the line density in Fig.3(a). On the other hand, no correlation is observed in Fig.3(b). It is suggested that the excited fast Alfvén wave was absorbed when it propagated through the plasma.

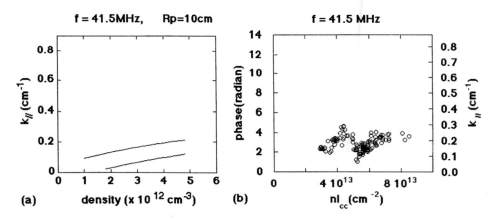

FIGURE 4. (a) The dispersion relation of the fast Alfvén wave in a uniform density cylindrical plasma. There are two radial modes. (b) The measured axial wave number reconsidered from Fig.3(a).

Measured axial wave numbers are compared with a simple numerical calculation in Fig.4. Here, a uniform density cylindrical plasma with the radius of 10 cm is assumed to approximate the plasma in the central cell. Although it might be a rough approximation, the obtained results are reasonable. The calculated dispersion relation of the fast Alfvén wave with the frequency of 41.5 MHz is shown in Fig.4(a). The axial wave length is plotted against the plasma density. It is seen that there are two radial modes. On the other hand, the experimental results shown in Fig.3(a) are reconsidered in Fig.4(b), so that it becomes consistent qualitatively with the calculation. The axial wave length and phase difference are plotted as a function of the line density in Fig.4(b). Considering the fact that the line density is about 20 times larger than the density in the present experiment, we can compare Fig.4(a) and (b). Figure 4(b) shows that two radial modes are observed experimentally with magnetic probes and that the radial mode transition is observed at $nl \sim 5 \times 10^{13}$ cm^{-2}. Currently, the reason for this mode transition is unknown. The antenna loading in the real configuration must be taken into account for the calculation to reveal the physical mechanism of the mode transition.

SUMMARY

It was found that the measured dispersion relation of the fast Alfvén wave for f = 41.5 MHz reproduced the calculated one qualitatively when the plasma density increased with RF3. Obtained results suggest that the radial mode transition of the excited fast Alfvén wave is an important factor for the higher density plasma production in GAMMA 10.

ACKNOWLEDGMENTS

The authors acknowledge the support of GAMMA10 group. This work is partly supported by University of Tsukuba Research Projects.

REFERENCES

1. M.Ichimura, et al., Phys. Plasmas **8** (2001) 2066.
2. M.Ichimura, et al., Plasma Phys. Rep. **28** (2002) 727-733.
3. M.Ichimura, H.Higaki, S.Saosaki, S.Kakimoto, Y.Yamaguchi, K.Horinouchi, H.Hojo, and K.Yatsu
   Transactions of Fusion Science and Technology **43**, 1T, (2003) pp.69-72
   4th International Conference on Open Magnetic Systems for Plasma Confinement
   Jeju, Korea, Jul.1-4, 2002

Radio-frequency matching studies for the JET ITER-Like ICRF system

P U Lamalle[1], F Durodié[1], R H Goulding[2], I Monakhov[3], M Nightingale[3], A Walden[3], P Wouters[1] and EFDA JET Workprogramme Contributors*

[1]Laboratory for Plasma Physics, Association EURATOM-Belgian State, TEC, Royal Military Academy, B-1000 Brussels, Belgium. [2]Oak Ridge National Laboratory, USA. [3]UKAEA/EURATOM Fusion Association, Culham Science Centre, Abingdon OX14 3DB, U.K.

Abstract. The transmisssion and matching system of the JET ITER-Like ICRF antenna includes specific design features, contributing to tolerance of the generators to large (dominantly) resistive increases in loading, and to be tested on JET in ITER-relevant ELMy H mode conditions for the first time. Beside the "conjugate-T" circuit, internally matching the launcher to a very low reference impedance, an original adjustable wideband transformer has been designed for compatibility with various ancillary functions. Optional 3dB splitters provide futher generator isolation. The paper discusses design choices leading to the final layout and briefly presents the simulated performance of the system.

TRANSMISSION AND MATCHING SYSTEM

Beside the characteristics of the antenna array [1], the transmission and matching system of the JET ITER-Like ICRF antenna includes a number of specific design features, Figure 1: (1) The conjugate-T circuit [2] provides internal matching of the launcher over its operating frequency range (30 to 55MHz) by feeding pairs of radiating straps in parallel through adjustable capacitors. (2) A very low reference characteristic impedance Z_0 is used for matching each of the four antenna inputs. These two ingredients make the system tolerant to large increases of the resistive loading of each pair of straps from a matched reference configuration (Figure 2). In the JET implementation a range of values of Z_0 from 3Ω to $\sim 9\Omega$ can be selected, allowing comprehensive investigation of the new scheme. In contrast with the ELM-tolerant scheme based on 3dB hybrids diverting reflected power to dummy loads [3], the internal conjugate-T matching circuit intrinsically keeps reflection low, once suitably configured. Conversely to the wideband matching system previously studied at JET [4], it does not require any adjustment on the ELM timescale. (3) The internal matching stage is followed by an original adjustable impedance transformer between Z_0 and the 30Ω of the JET ICRF transmission lines, described in the following section. (4) Two hybrid 3dB splitters (not shown) share RF power between the four antenna ports. Poloidal and toroidal splitting schemes are under consideration; we focus on the former, which allows full toroidal phasing capability. Line stretchers between splitters and each transformer allow two types of operation: either at

---

* See Annex of J Pamela et al, in Proc. 19th IAEA Conf. on Fusion Energy (Lyon), Vienna: IAEA, 2002.

maximum generator isolation from residual reflected power, which requires a 90° poloidal phasing between pairs of radiating straps and reduces antenna coupling; or at maximum array coupling (achieved by imposing 180° between input currents due to relative orientation of radiating loops), but reduced generator isolation.

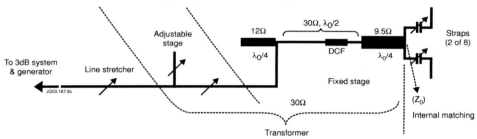

FIGURE 1. Schematic layout of transmission and matching circuit for one quarter of the JET ITER-like antenna. Midband quarter wavelength: $\lambda_0/4$=1.765m. DCF: vacuum window. Z_0: target reference impedance for internal matching.

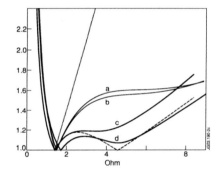

Normalized branch resistance $r = R/Z_0$	Input VSWR	$\arg(\frac{I_2}{I_1})$
0	∞	π
$S(1-\sqrt{1-1/S^4})$	S	
$r_1 = 1-\sqrt{1-1/S^2} = 1-\cos\alpha$	1	$\pi - \alpha$
$x_1 = 1/S = \sin\alpha$	S	$\pi/2$
$r_2 = 1+\sqrt{1-1/S^2} = 1+\cos\alpha$	1	α
2		0
$S(1+\sqrt{1-1/S^4})$	S	

FIGURE 2. Left: Input vswr vs. branch input resistance (55MHz, Z_0=3Ω). Dotted line: conventional matching circuit. Dashed line: ideal conjugate-T circuit (symmetrical branches, no mutual coupling between straps). More realistic models include nonsymmetric straps (curves c, d) and mutual coupling (a, b). Curves a, c correspond to a top and curves b, d to a bottom antenna feed. **Right:** Ideal conjugate-T characteristic, relationships between matched loads (r_1, r_2), the vswr at local maximum (S), two other loads producing vswr S, and the relative phase between branch input currents.

DESIGN OF THE IMPEDANCE TRANSFORMER

The transformer specifications are lowest possible return loss between 30 and 55MHz and large transformation ratio (up to 10, to obtain a highly load-tolerant internal matching circuit). Its layout is severely constrained: part of it is in-vessel; capacitor hydraulic actuating and cooling circuits must penetrate the vacuum boundary along a RF-free path, i.e. inside coaxial line internal conductors. The transformer must also accommodate the RF vacuum window, for which a validated 30Ω double conical feedthrough (DCF) design was available [5]. These boundary conditions made a standard maximally flat or equal ripple transformer impractical (such a design would also have required a very low first stage impedance and a fixed reference $Z_0 \geq 6\Omega$ for internal matching). The layout selected for the JET ITER-like system, shown on

Figure 1, consists of a fixed and an adjustable stage. No section of the circuit actually requires the lowest reference impedance Z_0, as the antenna matching capacitors are directly connected to an in-vessel 9.5Ω quarter-wave stage ($\lambda_0/4=1.765$m, the broad racetrack-shaped vacuum transmission line [1]). This section matches a 3Ω load at midband. It is followed by a fixed half-wave 30Ω section, partly in vessel and including the DCF window. At the end of this section is inserted a fixed $\lambda_0/4$ low-impedance (12Ω) stub, providing access for hydraulic and cooling circuits. These stub and line sections also play an active role in the RF transformer, and are designed for additional adaptation of a 3Ω load at the edges of the frequency band, Figure 3. As a bonus, the first stage also matches $Z_0=6Ω$ at 33 and 52 MHz (Fig.3, right). The design is completed by adjustable line stretcher and stub, to cancel the residual reflection of the fixed stage, and for which a set of reference settings will be obtained at the operating frequencies and Z_0 of interest. The combined stages therefore enable ideal impedance transformation over the operating domain.

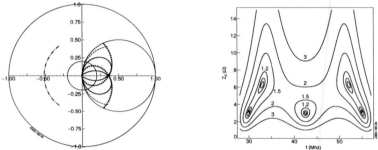

FIGURE 3. Design of the fixed transformer stage. Left: Smith impedance chart showing reflection at various points between 30 and 55MHz for $Z_0=3Ω$. Dash-dots: end of 9.5Ω section; dashes: onto 30Ω section; dots: before service stub; thick continuous line: after service stub, showing 3 matched frequencies. Right: VSWR induced by the fixed stage vs. frequency and Z_0.

SIMULATED BEHAVIOUR OF THE SYSTEM

From fast RF measurements on the JET A2 antennae, e.g. [6], the ELMs are known to induce very large increases of resistive antenna loading (factor of 4 commonly observed), and at the same time significant reactance decreases of up to ~ 10%. A variety of reactive variations are observed, and at present the precise effect of the ELMs on the input impedance matrix of a coupled ICRH array is open to various interpretations. Until more detailed information becomes available from ongoing analysis, the system response to ELMs is investigated using independent variations of the array resistance and reactance matrices obtained from 3D simulations [7], and assessed against the above orders of magnitude (Figure 4). Even with asymmetries and cross-talk between straps included in the model, the tolerance of the system to realistic load increases is still manifest (assuming the reference ELM-free matched configuration has been reached). A sensitivity analysis, Figure 5, shows that a sub-millimeter accuracy on the capacitor settings (i.e. better than 1pF at 55MHz) is mandatory to achieve optimum VSWR performance. This specification should be met by the selected hydraulic capacitor actuators. Cross-talk is found to affect voltages and power balance between straps significantly, and to increase the complexity of the matching procedure.

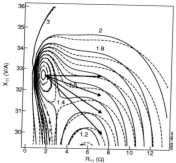 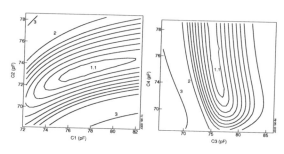

FIGURE 4. Input VSWR vs. scalings of 4 by 4 resistance and reactance matrices (coupled array, toroidal dipole phasing). Plain / dashed lines: respectively top / bottom feed. Arrows indicate typical trajectories from reference match during ELM rise (55MHz, $Z_0=3\Omega$).

FIGURE 5. Input VSWR in the top feed line vs. its two own capacitor settings (left, plot range 10pF), and vs. settings of the capacitors of the bottom line (right, plot range 20pF).

CONCLUSIONS

Within demanding constraints, a flexible implementation has been worked out for the matching circuit of the JET ITER-like ICRF antenna that will enable its comprehensive investigation on ITER-relevant ELMy H modes. Experimental information on the load perturbations due to ELMs and antenna impedance matrix from 3D simulations contribute to the realism of ongoing circuit modelling. The impact of cross-talk between straps on maximum performance and control is a serious concern under intensive assessment. Uncertainty remains on the array impedance matrix and its variations during ELMs, and only tests of the launcher in 2005 can provide a definitive answer on performance. Great care is required in meeting demanding accuracy goals, and in developing robust matching algorithms for the coupled array.

ACKNOWLEDGMENTS

This work has been performed under the European Fusion Development Agreement

REFERENCES

1. Durodié, F., et al., in Radio Frequency Power in Plasmas, AIP Conference Proceedings **595**, New York: Melville, 2001, pp. 122-125; Durodié, F., et al., these Proceedings.
2. Bosia, G., Fusion Science and Technology **43**, 2003, pp. 153-159.
3. Wesner, F., et al., in Fusion Technology (19th. SOFT, Lisbon, 1996), Amsterdam: North Holland, 1997, pp. 597-600.
4. Lamalle, P. U., et al., in Radio Frequency Power in Plasmas, AIP Conference Proceedings **595**, New York: Melville, 2001, pp. 118-121.
5. Walton, R., *A Continuous Wave RF Vacuum Window*, JET Report JET-R(99)03, 1999.
6. Monakhov, I., et al., two contributions to these Proceedings.
7. Lamalle, P.U., et al., these Proceedings.

Three-dimensional electromagnetic modelling of the JET ITER-Like ICRF antenna

P U Lamalle[1], F Durodié[1], A Whitehurst[2], R H Goulding[3], P M Ryan[3] and EFDA JET Workprogramme Contributors[*]

[1]Laboratory for Plasma Physics, Association EURATOM-Belgian State, TEC, Royal Military Academy, B-1000 Brussels, Belgium. [2]UKAEA/EURATOM Fusion Association, Culham Science Centre, Abingdon OX14 3DB, U.K. [3]Oak Ridge National Laboratory, USA.

Abstract. During its design phase, the JET ITER-Like ICRF antenna array has been modeled in great detail with the 3D electromagnetic software CST MICROWAVE STUDIO®. The resulting rf field and current density patterns have guided the optimization of the antenna feeder shapes, leading overall to a strong reduction (~25%) of the maximum electric field, and to a factor-of-three reduction of the inhomogeneity of rf current at the matching capacitors. The computed frequency response of the array is now used in matching studies and development of a control algorithm. Comparison with the experimental frequency response of the High Power antenna Prototype developed by ORNL and PPPL shows fair agreement.

INTRODUCTION

As of today, successful quantitative predictions of the coupling properties of ICRF antennae on plasma remain to be demonstrated. This requires an accurate computation of both active and reactive antenna properties. The reactance evaluation demands detailed calculation of near fields and self-consistent surface currents on complex metallic structures, and was probably the weakest point of earlier simulations. Accordingly, a key area of work during the design of the JET ITER-Like ICRF antenna [1] has been the selection, development and application of improved 3D electromagnetic simulation tools, which have become quite powerful and user-friendly since the last generation of antennae were designed. In the present paper, we summarize the results of simulations of the JET ITER-Like antenna array with the CST MICROWAVE STUDIO® (MWS) software [2], which was used to optimize the complex 3D geometry of the antenna feeders. Operation of the antenna will provide a crucial assessment of this effort, to benefit future designs, the ITER one in particular. A first benchmark of vacuum code predictions against recent measurements on the High Power Prototype antenna developed by ORNL and PPPL [3] has successfully been carried out. A comprehensive 3D study of the current JET A2 antenna, to be reported elsewhere, has also been initiated and will allow detailed comparison with ICRF coupling data.

---

[*] *See Annex of J Pamela et al, in Proc. 19th IAEA Conf. on Fusion Energy (Lyon), Vienna: IAEA, 2002.*

THE ANTENNA MODEL

Figure 1 shows the main features of the MWS RF antenna model, fully consistent with the drawing office blueprints: antenna box, coaxial feeds, tapered feeder sections through the back of the box, connected to oblique feeder sections by capacitor protection plates; radiating straps, Faraday shield, poloidal and toroidal septa. Note that the poloidal and toroidal curvatures of the system, the finite thickness of all conductors and the individual rods of the shield are taken into account. Reduced models of 1 or 2 straps allow investigating specific issues. MWS solves Maxwell's equations with the finite integration technique [4] and finite differencing in time.

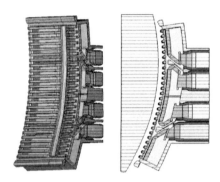

FIGURE 1. RF MWS model of the ITER-Like antenna (final design layout). Left: partial front view; right: side view with dielectric loading (both are cut vertically through the axes of 4 matching capacitors to show feeder profiles).

Typical outputs are the array scattering parameters as a function of frequency, and 3D field maps at selected frequencies. The software allows various time-dispersive media such as magnetized ferrites, but does not include a magnetized plasma dielectric tensor. The curved solid shown in the right view Figure 1 is a high permittivity dielectric mimicking plasma loading. Its steep edge is representative of worst-case coupling conditions to a high-power ELMy H mode discharge, and its dielectric constant is chosen to reproduce the magnetosonic wavelength at a typical frequency and toroidal wavenumber. The boundary conditions are a metal back plane (not shown) and radiation conditions on all other boundaries, which amounts to assuming single pass absorption in the plasma bulk and absence of global toroidal eigenmodes. There is clearly scope to improve this loading model for future comparisons with plasma heating experiments. Nevertheless, the design optimization has proceeded by evolution of an initial drawing, and comparison of the RF characteristics of the successive geometries has been performed in normalized conditions (unit currents at the strap short-circuits), which strongly reduces the influence of assumptions on resistive loading. The ad-hoc dielectric then mainly contributes through its influence on the antenna reactance.

OPTIMIZATION OF THE FEEDER REGION

The purpose of this activity was to improve the initial design and widen the operational domain of the launcher. This required compromise between several conflicting goals: (a) minimizing the maximum electric field, which requires maximum gaps between conductors and smooth transitions between feeder sections; (b) minimizing input voltages (at fixed radiating strap geometry), which requires the shortest feeders possible and small gaps for low inductance; (c) achieving voltage homogeneity between radiating loops for optimal operation close to limits; (d) achieving maximum azimuthal symmetry for the rf current density at the feeder-capacitor connections to make best use of capacitor specifications. The main constraints were mechanical strength, JET port size, vessel poloidal curvature, access for capacitor cooling through the feeders, requirement to shield the capacitors from direct heat flux, space requirements for RF gaskets at the back of the antenna box. The result of many progressive design modifications is summarized on Figures 2 and 3:

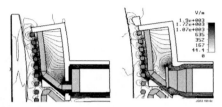
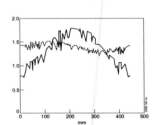

FIGURE 2. Top strap with its feeder and capacitor sliding contact flange. Left: initial drawing; right: final layout. Shading according to electric field magnitude (Normalization to 1A short circuit current, scale from 0 to 1.9kV/m).

FIGURE 3. Numerical current distribution along flange perimeter (for 1A at short-circuit). Plain line: initial drawing; discontinuous line: final layout (max. deviations from average: respectively 27% and 9%).

The initial design had tapered oblique feeder sections and racetrack-shaped horizontal sections directly connected to the capacitor sliding contact flange; the final layout has thinner septa, wider oblique sections, capacitor shielding plates, wide cylindrical sections through the antenna box connecting to the flange, and the capacitors moved outward by a few cm. Overall these improvements reduce the maximum electric field by ~25% (up to 40% at the capacitor flange). Moreover, Figure 3 shows a reduction of the surface current inhomogeneity at the flange by a factor of 3, which decreases peak and total dissipation on the capacitors.

FREQUENCY RESPONSE

The input scattering matrix of the loaded antenna array has been obtained (Figure 4) and is currently used in matching studies [5]. Maximum strap reactances are found smaller than the upper estimate initially used to define capacitor specifications. Final design refinements are under study to reduce poloidal cross-talk S_{23} between inner straps (Figure 5). Measurements on the antenna prototype [3] are well reproduced by its MWS model (Figure 6), which increases confidence in other predictions.

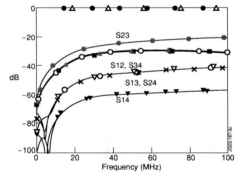

FIGURE 4. Input scattering parameter amplitudes vs frequency in toroidal dipole phasing (ports are numbered from top to bottom).

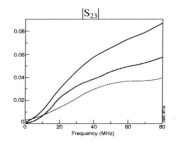

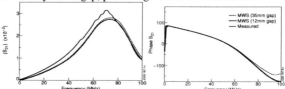

← **FIGURE 5.** Simulated poloidal cross-talk |S23| between inner straps. Plain, dotted, dashed lines: respectively, reference design; after extension of equatorial septum; and after closure of Faraday shield gap, predicting a factor-of-2 reduction.

FIGURE 6. Frequency response of scattering parameter S21 on 2-strap antenna prototype. Comparison between measurement and MWS model (the latter was run for 2 positions of the Farady shield). Left: amplitude; right: phase in degree.

CONCLUSIONS

The design of the JET ITER-Like antenna has triggered unprecedented RF modelling of ICRF antennae with the most advanced tools available today. This was instrumental to optimize the details of the strap and feeder geometry: since the early design, the maximum electric field has been reduced by ~25% (up to 40% near the capacitors), and the azimuthal inhomogeneity of currents near the capacitors has been reduced by a factor of three. The modelling is now used to adjust the final design of the equatorial septum, to reduce the cross-talk between straps, and to assess its influence on the matching performance using the computed input impedance matrix. Successful code benchmarking against measurements on the High Power Prototype of a quarter antenna increases confidence in the results. Based on this work, the use of the ITER-like antenna on JET will not only provide demonstration of ICRF coupling in ITER-relevant conditions, but also enable a validation of modelling useful to progress in the understanding of RF coupling in fusion devices, and useful to optimize the design of the ITER antenna itself.

ACKNOWLEDGMENTS

This work has been performed under the European Fusion Development Agreement. It has greatly benefited from numerous discussions within the JET ITER-Like antenna Design Team [1], with Dr R Ehmann and the CST GmbH Support Team. We gratefully thank Dr J Paméla for his stimulating support.

REFERENCES

1. Durodié, F., et al., these Proceedings.
2. CST MICROWAVE STUDIO User Manual, Version 4.0, CST GmbH, Darmstadt, Germany, 2002; http://www.cst-world.com.
3. Goulding, R.H., et al., these Proceedings.
4. Weiland, T., *International Journal of Numerical Modelling: Electronic Networks, Devices and Fields* **9**, New York: Wiley, 1996, pp.295-319.
5. Lamalle, P. U., et al., these Proceedings.

Self-Consistent Modelling of Polychromatic ICRH in Tokamaks

M. Laxåback*, T. Johnson*, T. Hellsten* and M. Mantsinen[†]

*Alfvén Laboratory, Association EURATOM-VR, Sweden
[†]Helsinki University of Technology, Association EURATOM-TEKES, Finland

Abstract. Polychromatic, multi-frequency, ion cyclotron resonance heating provide a useful tool for the optimization of plasma performance in fusion devices by tailoring the fast ion distribution function. Not only can the radial profile of the fast ion distributions be modified, but also the fast energy content, the power partition on resonant species and the bulk plasma ion- and electron heating rates. This work describes finite orbit effects of polychromatic ICRH which are demonstrated using the SELFO code.

INTRODUCTION

Polychromatic ICRH, where the RF power is divided over several frequencies, can be used to reduce the RF power density and fast energy content and increase the bulk ion heating. The trade off is an outward shift of the power absorption as some thin orbits in the center of the plasma will not be in resonance with all frequencies. Polychromatic operation can also affect the power partition on resonant species, which is often closely related to the RF power density through the shapes of the distribution functions.

During the JET 1997 DT-campaign polychromatic ICRH was used routinely to increase the bulk ion heating [1] and recently experiments were performed at JET comparing polychromatic and monochromatic ICRH of ^{3}He(D) and H(D) plasmas [2].

FINITE ORBIT EFFECTS ON POLYCHROMATIC ICRH

The distribution functions of the resonant ions can be described in terms of the invariants of motion of unperturbed drift orbits; $E = v^2/2$, $\Lambda = B_0 v_\perp^2/(Bv^2)$ and $P_\phi = Rv_\parallel + (q/m)\Psi$, where ϕ increases in the direction opposite to the current. The change in energy of an ion passing a Doppler shifted cyclotron resonance, $\omega = n\omega_c + k_\parallel v_\parallel$, can be calculated using the stationary phase method [3], yielding:

$$\Delta E = qv_\perp^{res}|E_+ J_{n-1}(k_\perp \rho) + E_- J_{n+1}(k_\perp \rho)|\sqrt{\frac{2\pi}{n|\dot{\omega}_c|}}\cos\varphi \quad (1)$$

where E_+ and E_- are the co- and counter ion rotating electric field components, n is the harmonic of the resonance, J_n are Bessel functions, $\dot{\omega}_c$ is the rate of change of the cyclotron frequency along the drift orbit and the difference φ between the gyro- and wave phase at the resonance can be taken to be a random number. The corresponding changes in Λ and P_ϕ can be shown to be [4, 5]:

$$\Delta \Lambda = \left(\frac{n\omega_{c0}}{\omega} - \Lambda\right) \frac{\Delta E}{E} \qquad (2)$$

$$\Delta P_\phi = \frac{n_\phi}{\omega} \Delta E \qquad (3)$$

showing that the increase in energy by RF-interactions will drive Λ towards $\Lambda_{res} = n\omega_{c0}/\omega$, which is to say that the turning points of trapped orbits will be driven towards the unshifted resonance, and P_ϕ will either increase or decrease depending on the sign of the toroidal mode number n_ϕ of the wave. Since P_ϕ determines the flux surface of the turning points of trapped orbits (where $v_\parallel = 0$), interactions with waves having positive n_ϕ will drive the turning points outwards and vice versa.

For monochromatic scenarios with a single toroidal mode number equations 1-3 describe one dimensional diffusion in phase space (E, Λ, P_ϕ), where the characteristics described by equations 2 and 3 have the asymptote $\Lambda = \Lambda_{res}$ as the energy of the ion increases. For a spectrum of toroidal modes the diffusion becomes two dimensional as a resonant ion may approach Λ_{res} along different characteristics since the evolution of P_ϕ depends on which toroidal modes the ion is in resonance with. Polychromatic scenarios add a third dimension to the diffusion process. Since ions may be in resonance with several frequencies they will end up with different $\Lambda = \Lambda_{res}$ depending on with which frequency the resonance is strongest.

The complex nature of polychromatic ICRH is illustrated when investigating possible evolutions of trapped orbits. All frequencies with which an orbit is in resonance will drive the turning points towards their unshifted resonances. As trapped orbits extend only to the low field side (LFS) of the turning points a significant Doppler shift is required to drive the turning points towards a resonance on their high field side (HFS). Depending on the amount of RF power applied at the different frequencies, their toroidal mode spectrums and spatial separation the dominating Λ drift is likely to be towards the lower frequency resonances. For lower ion energies pitch angle scattering will tend to restore the Λ spectrum and spread the location of the turning points. For ion energies high enough however, such that the pitch-angle scattering is reduced, orbits may completely detach from higher frequency resonances, leading to a drift of turning points in the direction of the major radius as is illustrated in figure 1.

THE SELFO CODE

The SELFO code [6, 7, 8] simulates ICRH by self-consistently coupling the Monte-Carlo code FIDO [9] to the global wave code LION [10, 11]. FIDO evolves resonant ion distribution functions for a given wave field and computes their dielectric tensor susceptibility contributions. LION iteratively updates the wave field and power partition on resonant species using the computed tensor.

SELFO is able to simulate polychromatic scenarios using an arbitrary number of frequencies. Since each frequency requires tensor and wave field calculations care must

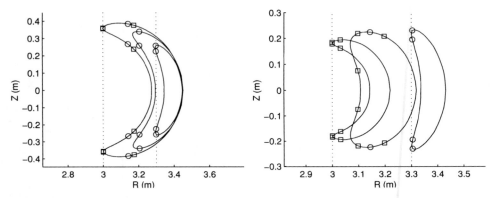

FIGURE 1. Left: Orbit in resonance with two frequencies is pulled to the LFS resonance, detaching from the central resonance. Right: Orbit only in resonance with central resonance gains enough energy and Doppler shift to start resonating with LFS resonance and is pulled over, detaching from central resonance. Doppler shifted resonance points are indicated by squares (central resonance) and circles (LFS resonance).

be taken to keep the computation time down. Instead of solving the orbit equations and engaging the standard RF-operator once per frequency, essentially slowing down the simulation by a factor of two to four for a polychromatic scenario, a cyclic-frequency RF-operator is implemented. Each time step the total RF power is applied to a single frequency chosen randomly from the spectrum with a probability proportional to the power injected at that frequency. Since the number of time steps in a simulation is large, ≈ 30000, the average power applied at each frequency will still be statistically correct.

SIMULATIONS

Monochromatic and polychromatic ICRH of JET-like ^{3}He(D) plasmas with parameters: $R_0 = 3m, a = 1m, B_0 = 3.45T, I_p = 2.6MA, n = n_0(1 - 0.98(r/a)^2)^{0.5}, n_{^3He} = 6.0 \times 10^{17} m^{-3}, n_D = 2.0 \times 10^{19} m^{-3}, Z_{eff} = 2.8, T = T_0(1 - 0.4(r/a)^2)^4, T_i = T_e = 6.0 keV$ is simulated for symmetrical toroidal mode spectra $n_\phi = \pm 25$ with frequencies around the fundamental ^3He cyclotron frequency. Figure 2 illustrates the fraction of ^{3}He ions with energies above 1MeV and turning points within three cm of the resonances. In the left plot a polychromatic scenario with two resonances at $R_{res} = 2.8m$ (3.2MW) and $R_{res} = 3.2m$ (3.2MW) is compared to a monochromatic scenario with $R_{res} = 3.0m$ (6.4MW). Interactions with the LFS resonance detaches turning points from the HFS resonance in the polychromatic scenario, leading to an accumulation of turning points on the LFS resonance and a significant reduction in the total number of high-energy trapped orbits compared to the monochromatic scenario. In the right plot a polychromatic scenario with three resonances at $R_{res} = 3.0m$ (3.2MW), $R_{res} = 3.15m$ (1.6MW) and $R_{res} = 3.3m$ (1.6MW) is compared to a monochromatic scenario with $R_{res} = 3.0m$ (3.2MW), i.e. the same parameters but without the LFS resonances of the polychromatic scenario. While the accumulation of turning points on the LFS resonances is not as clear in this simulation, note that the fraction of high energy trapped ions on the central resonance is lower in the polychromatic scenario than it is in the monochromatic scenario despite the same power being applied there. Thus *adding* RF power at LFS resonances *decreases* the number of high-energy trapped ions with turning points on the central resonance due to detachment.

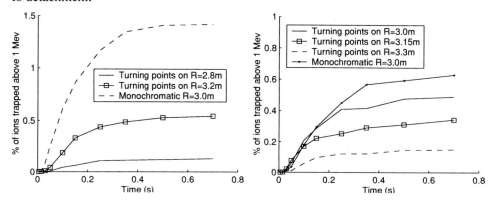

FIGURE 2. Left: Accumulation of high energy trapped particles on LFS resonance. Right: Reduction in number of high energy ions trapped on central resonance by adding LFS resonances.

CONCLUSIONS

Orbit effects in toroidal plasmas during polychromatic ICRH increases the complexity of the physics significantly beyond the simple reduction in RF power density by spreading the resonances. Due to the complexity of the problem and the many variables involved a careful treatment of the physics of polychromatic ICRH requires numerical modelling including finite orbit width effects of each individual scenario.

REFERENCES

1. F.G. Rimini, et al., *Nuclear Fusion*, **39** 1591 (1999)
2. M. Mantsinen, et al., In *Radio Frequency Power in Plasmas*. 15:th RF Topical Conference, (2003)
3. T.H. Stix, *Nuclear Fusion*, **15** 737–754 (1975)
4. J. Carlsson, T. Hellsten, and L.-G. Eriksson, Technical Report ALF-1996-104, Alfvén Laboratory, Royal Institute of Technology, SE-100 44 STOCKHOLM, SWEDEN, (1996)
5. L.-G. Eriksson, et al., *Physics of Plasmas*, **6** 513–18 (1999)
6. J. Hedin, T. Hellsten, and J. Carlsson, In J. W. Connor, E. Sindoni, and J. Vaclavik, editors, *Theory of Fusion Plasmas 1998*, 467–472, Varenna, (1999). Società Italiana di Fisica
7. J. Hedin, et al., *Nuclear Fusion*, **42** 527 (2002)
8. M. Laxåback, T. Hellsten, and T. Johnson, In O. Sauter J. W. Connor and E. Sindoni, editors, *Theory of Fusion Plasmas 2002*, 411–416, Varenna, (2002). Società Italiana di Fisica
9. J. Carlsson, L.-G. Eriksson, and T. Hellsten, In *Proceedings of the Joint Varenna-Lausanne Workshop "Theory of Fusion Plasmas"*, 351, (1994)
10. L. Villard, et al., *Computer Physics Reports*, **4** 95 (1986)
11. L. Villard, S. Brunner, and J. Vaclavik, *Nuclear Fusion*, **35** 1173 (1995)

Effect of ICRF Mode Conversion at the Ion-Ion Hybrid Resonance on Plasma Confinement in JET

A.Lyssoivan[a], M.J.Mantsinen[b], D.Van Eester[a], R.Koch[a], A.Salmi[b], J.-M.Noterdaeme[c,d], I.Monakhov[e], and JET EFDA Contributors

[a] *Laboratory for Plasma Physics Royal Military Academy ERM/KMS, Association EURATOM-BELGIAN STATE, 1000 Brussels, Belgium**
[b] *Helsinki Univ. of Technology, Association EURATOM-TEKES, FIN-02015 HUT, Finland*
[c] *Max-Planck Institut für Plasmaphysik, EURATOM Association, D-85748 Garching, Germany*
[d] *Gent University, EESA Department, B-9000 Gent, Belgium*
[e] *Association EURATOM-UKAEA, Culham Science Centre, Abingdon OX14 3DB, UK*
* Partner in the Trilateral Euregio Cluster (TEC)

Abstract. The objective of the present study is to find out the range of the IIHR layer radial position within which the plasma confinement in conventional *L-mode* regime may be improved. The recent ICRF-MC heating experiments on JET (RF power dominated $D(^3He)$-discharges, $P_{RF}/(P_{tot}-P_{RF})\approx 2.0$) were analyzed. The RF power coupled into plasma in the ICRF-MC scenario improved confinement properties of discharges with respect to the $(OH+NBI)$-phase at the central locations of IIHR ($|r_{ii}/a_{pl}| < 0.3$) with the best result (up to ~60% improved confinement) closer to axis and at dipole antenna phasing. A shift of IIHR towards the plasma edge ($|r_{ii}/a_{pl}| > 0.4$) resulted in degradation of confinement. An analysis of plasma confinement based on plasma diamagnetic and thermal energy content, and results of modeling of the absorbed power at ICRF-MC are discussed.

INTRODUCTION

The control of plasma heating efficiency and local transport are the principal goals for the achievement of enhanced tokamak operation. ICRF mode conversion (ICRF-MC) in plasma containing two ion species has a potential to solve the mentioned problems due to efficient local heating of electrons [1] and possibility to generate ponderomotively driven local toroidal/poloidal flow near the ion-ion hybrid resonance (IIHR) [2].

The objective of the present study is to find out the limits of the IIHR layer radial position within which the plasma confinement in conventional *L-mode* regime may be improved. The recent ICRF-MC heating experiments on JET ($D(^3He)$-discharges in the RF power dominated regime, $P_{RF}/(P_{tot}-P_{RF}) \approx 2.0$) were analyzed. Magnetic configurations with the single-null divertor and toroidal magnetic field / plasma current ratio of 3.44 T/1.35 MA, 3.34 T/2 MA and 3.7 T/2 MA were used in the experiments.

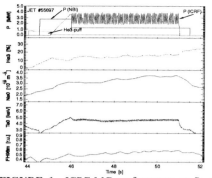
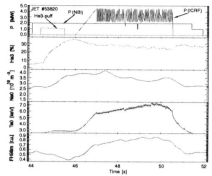

FIGURE 1. ICRF-MC performance at $B_T=3.44$ T, $I_p=1.35$ MA, $Z_{eff}\approx3.2$, $P_{rad}/P_{tot}\approx40\%$, A2 antennas powered at $+90°$ phasing, $f=33.8$ MHz.

FIGURE 2. ICRF-MC performance at $B_T=3.70$ T, $I_p=2.0$ MA, $Z_{eff}\approx2.5$, $P_{rad}/P_{tot}\approx36\%$, A2 antennas at dipole $(0\pi0\pi)$ phasing, $f=33.8$ MHz.

Up to 5 MW of the RF power was applied at frequencies of 33.8 or 37.2 MHz using four A2 ICRF antennas powered at $\pm 90°$, dipole $(0\pi0\pi)$ or $(00\pi\pi)$ phasing. By applying ^3He-puff of variable gas injection rate before the RF power, the concentration of ^3He ions was varied. As a result, the location of IIHR moved from the ω_{He3} layer towards the ω_{cD} resonance as the ^3He concentration was increased. For the analyzed shots, location of the IIHR layer across the plasma was varied in the range $-0.75 \leq r_{ii}/a_{pl} \leq 0$ by changing the ^3He concentration in deuterium plasma in the wide range $0 < n_{He3}/n_e <$ 0.35. The enhancement confinement factor FH98m defined as a total energy confinement time normalized to the ELMy H-mode confinement scaling IPB98(y,2) [3] was used to analyze plasma confinement as a function of the IIHR layer position. Since the concentration of the ^3He ions in deuterium plasma reached relatively high values (up to ~35%) in the experiments, the confinement scaling IPB98(y,2) was *corrected* following an established mass-dependence: $\tau_{E-IPB98(y,2)} \propto \left(\sum A_i n_i / \sum n_i\right)^{0.19}$. However, in terms of total coupled power, the scaling was *not corrected* for plasma radiation.

EXPERIMENTAL RESULTS

Figures 1 and 2 show time traces of the main plasma parameters in discharges performed in two different magnetic configurations with the similar RF power. In both discharges, the IIHR layer was located very close to axis, $r_{ii}/a_{pl} \approx -0.14$ at $n_{He3}/n_e \approx 14\%$ (Fig.1) and $r_{ii}/a_{pl} \approx -0.12$ at $n_{He3}/n_e \approx 25\%$ (Fig.2). Heating of electrons is clearly seen in both cases, which indicates established ICRF-MC conditions: FW launched by LFS antennas convert near IIHR into short-wavelength plasma waves absorbed by electrons. The change in the stored diamagnetic energy per MW of RF power coupled was about of 1.7 times higher in the discharge with antenna dipole phasing (Fig.2) suggesting more efficient absorption of RF power in the plasma. As a result, the enhancement confinement factor FH98m (here based on the diamagnetic energy content) was higher in this case (Fig.2). Note, however, that the presence of fast deuteron populations with energies above the maximum beam injection energy of 135 keV was registered by the gamma-ray emission spectroscopy in the latter case [1].

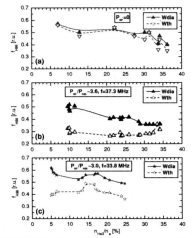
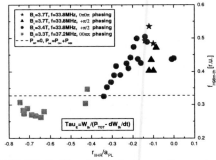

FIGURE 3. Comparison of confinement in discharges with ICRF-MC: (c) at $B_T=3.44$ T with +90° phasing, ##55394, 398, 403, 410, 695, 697; (b) at $B_T=3.34$ T, (00ππ) phasing, ##55708-711.

FIGURE 4. Trends in plasma confinement vs. IIHR radial location for the shots mentioned in Figs.1-3.

Further assessments of the plasma confinement properties vs. ^3He concentration and, consequently, vs. IIHR radial location were undertaken in terms of the total plasma diamagnetic energy content W_{dia} and thermal energy content W_{th} deduced from measured plasma densities and temperatures, respectively. The result of such analysis is shown in Figure 3. Several features may be stressed: (i) some decrease in the FH98m factor (here without mass-correction) on increasing the ^3He concentration during both, RF and non-RF phases of discharges, (ii) a minor improvement in the confinement within the concentration range $n_{He3}/n_e \approx 13$-17% corresponding to central locations of ICRF-MC (Fig.3c), (iii) a noticeable difference in confinement properties based on W_{dia} and W_{th} energy content during the RF+NBI power on phases of discharges. The main difference seems to be at low ^3He concentration (Fig.3b, 3c), at which He_3 minority damping may give rise to some non-thermal population. Trends in the plasma confinement properties in terms of the FH98m factor (based on W_{th} energy content and mass-corrected) vs. the IIHR radial position are shown in Figure 4. It is clearly seen that an improved confinement with respect to the (OH+NBI)-phase of discharges was only observed at the central ICRF-MC locations ($|r_{ii}/a_{pl}| < 0.3$). The best result (up to ~60% improved confinement at $r_{ii}/a_{pl} \approx -0.12$) was achieved so far in the 3.7T/2 MA magnetic configuration with antenna dipole phasing (star symbols in Fig.4). As it was mentioned before, the ICRF-MC heating efficiency was higher with dipole phasing than with +90° phasing. It resulted in larger T_{e0} increase (Fig.2) and better confinement. A shift of MC layer towards the HFS edge ($|r_{ii}/a_{pl}| > 0.4$) resulted in degradation confinement (Fig.4).

MODELING OF ICRF-MC ABSORBED POWER

To investigate the tendencies in the confinement/heating properties of the ICRF-MC heated plasmas, modeling of absorbed power per full (double) transit over the

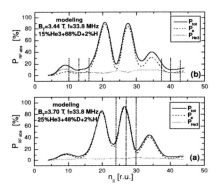
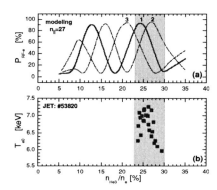

FIGURE 5. Calculated absorption in % of the incoming RF: (a) - for the shot #53820 (Fig.2); (b) - for the shot #55697 (Fig.1); $n_{e0}=3.2\times10^{19}$ m^{-3}. Vertical dashed/dot-dashed lines note the widths and positions of maxima of the antenna spectra.

FIGURE 6. Calculated absorption (a) and measured T_{e0} (b) vs. He_3 concentration: curve 1 – $n_{e0}=3.2\times10^{19}$ m^{-3}; curve 2 and 3 correspond to ±10% modeled variations in n_{e0}, respectively.

plasma with TOMCAT 1D RF code was undertaken [4]. Figure 5 shows how the absorption varies as a function of the toroidal *wave* mode numbers n_{tor} for the given plasma parameters and the A2 antenna $k_∥$-spectrum [5]. For the case of better ICRF-MC performance (Fig.2), the code predicts efficient central ($r/a_{pl} \approx -0.15$) absorption of the mode-converted power by electrons (>90% of total absorbed power) in the n_{tor}-range corresponding to the width of spectrum launched with dipole phasing (Fig.5a). A reasonable correlation of the calculated power (Fig.6a) with the measured T_{e0} (Fig.6b) was found. Note, however, that the absorbed power as a function of n_{tor} and the ^3He concentration reveals an oscillation nature, which makes difficult the detailed comparison with experiment. Small (±10%) variations in plasma density (curves 2, 3 with respect to curve 1, Fig.6b) resulted in the radial shift of *max* and *min* positions of the oscillations thus indicating sensitivity to the conditions of the computation.

The present study shows that the plasma confinement properties in discharges with ICRF-MC conditions are related to the electron heating efficiency which depends on the IIHR location and the antenna launched $k_∥$-spectrum.

ACKNOWLEDGMENTS

This work has been performed under the European Fusion Development Agreement.

REFERENCES

1. M.J.Mantsinen, et al., *28th EPS Conf. on CFPP*, ECA Vol.**25A** 1745 (2001).
2. L.A.Berry, et al., *Phys.Rev.Lett.* **82**, 1871 (1999).
3. ITER Physics Basis, *Nucl. Fusion* **39**, 2201-2209 (1999).
4. D.Van Eester and R.Koch, *Plasma Phys. Control. Fusion* **40**, 1949-1975 (1998).
5. P.U.Lamalle, et al., *22nd EPS on CFPP*, ECA Vol.**19C**, Part II, 329 (1995).

Efficient Self Consistent 3D/1D Analysis of ICRF Antennas

R. Maggiora, G. Vecchi, V. Lancellotti, V. Kyrytsya

Politecnico di Torino, Dipartimento di Elettronica, 10129 Torino, Italy

Abstract. An innovative tool has been realized for the 3D/1D simulation of Ion Cyclotron Radio Frequency (ICRF), i.e. accounting for antennas in a realistic 3D geometry and with an accurate 1D plasma model. The approach to the problem is based on an integral-equation formulation for the self-consistent evaluation of the current distribution on the conductors. The environment has been subdivided in two coupled region: the plasma region and the vacuum region. The two problems are linked by means of a magnetic current (electric field) distribution on the aperture between the two regions. In the vacuum region all the calculations are executed in the spatial domain while in the plasma region an extraction in the spectral domain and an analytical evaluation of some integrals are employed that permit to significantly reduce the integration support and to obtain a high numerical efficiency leading to the practical possibility of using a large number of sub-domain basis functions on each solid conductor of the system. The plasma enters the formalism of the plasma region via a surface impedance matrix; for this reason any plasma model can be used; at present the FELICE code has been adopted, that affords density and temperature profiles, and FLR effects. The source term directly models the TEM mode of the coax feeding the antenna and the current in the coax is determined self-consistently, giving the input impedance/admittance of the antenna itself. Calculation of field distributions (both magnetic and electric), useful for sheath considerations, is included. This tool has been implemented in a suite, called TOPICA, that is modular and applicable to ICRF antenna structures of arbitrary shape. This new simulation tool can assist during the detailed design phase and for this reason can be considered a "Virtual Prototyping Laboratory" (VPL). The TOPICA suite has been tested against assessed codes and against measurements and data of mock-ups and existing antennas. The VPL is being used in the design of various ICRF antennas and also for the performance prediction of the ALCATOR C-MOD D antenna.

THEORY AND CODE

The problem we are facing deals with the solution of a boundary-value problem for the Maxwell equations in a structure consisting of (lossless) conductors embedded in a background medium that is infinitely extended and invariant in the (y, z) plane, and exhibiting discontinuities along the x axis. The conductors are composed of the antenna, its housing structure, and the Faraday shield (FS), and can have a fairly general 3D geometry. The background medium is comprised of a vacuum region and a magnetized slab plasma half-space beyond the vacuum-plasma interface (plasma region). We do not need to specify the plasma parameter variation along x, provided we are able to solve for the appropriate evolution equations within it, and find a plasma impedance relation (presently the FELICE code [1] has been adopted).

The solution will be sought for via the Integral Equation formulation, in which the conductors are replaced by equivalent electric surface current distributions (as per the equivalence theorem). These currents generate a field, called "scattered" field, so that the total field is the sum of the driving term and of a function of the unknown currents. Enforcing the proper boundary conditions on the areas occupied by conductors ensures the self-consistency of the equivalent currents, and gives rise to an electric field integral equation (EFIE) for the latter. Note that the "scattered" field is the field produced by the currents (considered now as a source term) in the presence of the background medium, and is obtained via the Green's function itself. Note that the integral equation must be enforced only at the points on the metalizations, a fact that has to be kept in proper account when transferring to the spectral domain, since the Fourier transforms are always defined over infinite domains.

The EFIE is originally formulated in an infinite-dimensional function space, and the first step involved in its numerical solution is clearly the reduction into a finite-dimensional space, obtained applying to it the Method of Moments technique [2]. This technique is based on the approximation of the unknown current by a set of so-called basis functions (presently rooftop functions have been adopted). This approximated current is inserted into the EFIE, and the equation is enforced in a weak sense by setting to zero its projections (tests) over a set of testing functions (same as the basis functions). Taking into account the linearity of the operator, the EFIE turns into a linear system where the unknown current coefficients are grouped in a vector, the projections of the known driving term on the testing functions are collected in another vector, and the matrix is called "impedance" or MoM matrix, whose entries are the reaction integrals.

The global environment of the problem has been subdivided in two separate domains (a vacuum region and a plasma region) linked by means of a magnetic current (electric field) distribution on the aperture in a perfect conducting plane between the two domains (see Fig. 1).

In the vacuum region all the calculations are performed in the spatial domain while in the plasma region they must be carried out in the spectral domain because the plasma surface impedance matrix is naturally expressed in this domain. This approach permits to obtain a high numerical efficiency leading to the practical possibility of using a large number of sub-domain basis functions on each solid conductor of the system. The plasma enters the formalism of the plasma region via a surface impedance matrix; for this reason any plasma model can be used; at present the FELICE code has been adopted, that affords density and temperature profiles, and FLR effects.

The source term is given by the electric field distribution generated by the access coaxial lines (coax) that feed the antenna. In the coax the voltage and current are unambiguously related to fields (via the coaxial guide eigenmodes). This will be our starting point, in that we will take as an input term the voltage associated to the TEM mode of the coax; also we will eventually end our analysis by (self-consistently) determining the current that is established in the coax itself as a result of the plasma-facing antenna, whose ratio gives exactly the input impedance/admittance of the antenna itself.

The VPL, consisting of the simulation core, a pre-processing tool for meshing structures, and a post-processing tool to analyze results, has been named TOPICA.

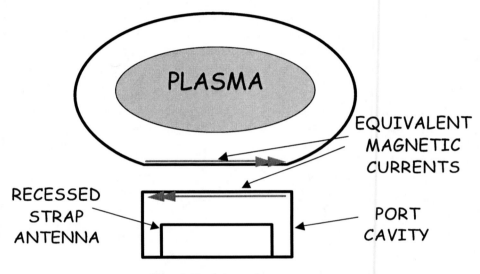

Figure 1. Simulation environment.

TESTING

In order to verify the computational reliability of the implemented simulation tool, extensive testing has been carried out [3]. Here some comparisons of antenna loading are presented performed employing a complete model of a compact loop antenna structure in DIII-D. The antenna, sketched in Fig. 2, is recessed in a single midplane port on DIII-D and it features frequency tunability and internal impedance matching by means of a variable capacitor (included in our simulation model)[3]. Low power ICRF coupling measurements have been carried out under a variety of plasma conditions including L-mode and H-mode discharges. Measurements of the edge plasma density profile, useful for simulation, using the Thomson scattering diagnostic and the movable Langmuir probe diagnostic are available.

Excellent quantitative agreement has been obtained between simulation and experiment, including the correct dependence of the loading resistance on antenna position and frequency (see Fig. 3).

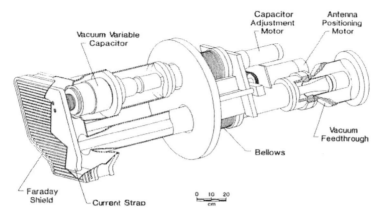

Figure 2. Antenna under test.

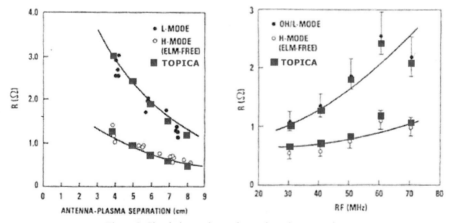

Figure 3. Simulation and experimental results comparison.

REFERENCES

[1] M. Brambilla, *Nuclear Fusion 35* (1995), p. 1265.
[2] R.F. Harrington, *Field Computation by Moment Methods*, Macmillan, New York (1969).
[3] R. Maggiora, et al., *AIP Conference Proceedings 595* (2001), p. 498.
[4] M.J. Mayberry, et al., *Nuclear Fusion 30* (1990), p. 579.

Comparison of monochromatic and polychromatic ICRH on JET

M.J. Mantsinen[1], M. Laxåback[2], A. Salmi[1], V. Kiptily[3], D. Testa[4],
Yu. Baranov[3], R. Barnsley[3], P. Beaumont[3], S. Conroy[2], P. de Vries[3],
C. Giroud[5], C. Gowers[3], T. Hellsten[2], L.C. Ingesson[5,6], T. Johnson[2],
H. Leggate[3], M.-L. Mayoral[3], I. Monakhov[3], J.-M. Noterdaeme[7,8],
S. Podda[9], S. Sharapov[3], A.A. Tuccillo[9], D. Van Eester[10]
and EFDA JET contributors[*]

[1]*Helsinki Univ. of Technology, Association Euratom-Tekes, P.O.Box 2200, FIN-02015 HUT, Finland*
[2]*Euratom-VR Association, Swedish Research Council, SE 10378 Stockholm, Sweden*
[3]*Association Euratom-UKAEA, Culham Science Centre, Abingdon OX14 3DB, UK*
[4]*CRPP, Association EURATOM-Switzerland, 1015 Lausanne, Switzerland*
[5]*FOM-Rijnhuizen, Ass. Euratom-FOM, TEC, PO Box 1207, 3430 BE Nieuwegein, NL*
[6]*present address: EFDA CSU-Garching, Boltzmann-Str.2, D-85748 Garching, Germany*
[7]*Max-Planck IPP-EURATOM Assoziation, Boltzmann-Str.2, D-85748 Garching, Germany*
[8]*Gent University, EESA Department, B-9000 Gent, Belgium*
[9]*Associazione EURATOM-ENEA sulla Fusione, CR Frascati, C.P. 65, 00044 Frascati, Rome, Italy*
[10]*LPP-ERM/KMS, Association Euratom-Belgian State, TEC, B-1000 Brussels, Belgium*

Abstract Experiments have been carried out on the JET tokamak to investigate differences between multiple and single frequency ICRH operation with ICRH power in the range of 3 to 8 MW using H and ^{3}He minority heating. High-energy neutral particle analysis and gamma-ray emission tomography are used to measure fast ions, including their radial localisation. For 3 MW of ^{3}He minority heating, the fast ^{3}He ion profile is broader according to the gamma emission data and the fast ion tail temperature T_{tail} and energy content W_{fast} are lower with polychromatic ICRH than monochromatic ICRH. Polychromatic ICRH has the advantage of producing smaller-amplitude and shorter-period sawteeth, consistent with a lower fast ion pressure inside q = 1, and higher T_i/T_e ratios (i.e. similar T_i at lower T_e). At high powers with resonances in the centre and/or on the low field side, the data indicates a larger fraction of trapped ions and a lower T_{tail} for polychromatic ICRH. Experimental results are compared with theoretical predictions.

INTRODUCTION

The frequency of waves launched by the four JET A2 ICRH antennas can be chosen independently. Hydrogen minority heating with multiple frequencies was used routinely in high-performance NBI-dominated plasmas before and during the 1997 deuterium-tritium campaign [1]. Using up to four frequencies, resonances were spread over a 30-40 cm wide region in the plasma centre to decrease the power density in order to improve bulk ion heating and plasma stability. However, one-to-one

---

[*]see annex of J Pamela *et al.*, Fusion Energy 2002 (Proc. 19th Int. Conf. Lyon, 2002) IAEA, Vienna (2002).

comparisons of monochromatic and polychromatic ICRH with identical coupled powers, including information on the fast ion radial localisation, were not obtained. Such comparisons are important in order to improve the understanding of ICRH physics and to benchmark numerical modelling codes against experiments. Recent simulations [2] suggest certain benefits for polychromatic ICRH at high powers available in the JET enhancement project, which require experimental verification.

OVERVIEW OF THE EXPERIMENTS

This paper reports recent experiments with ^{3}He and H minority heating on JET ($R_0 \approx 3$ m, a ≈ 1 m) to investigate in detail the differences between multiple and single ICRH frequency operation. The experiments were carried out in a single-null divertor configuration at a magnetic field of 3.3–3.7 T and a plasma current of 1.8–2 MA. High-energy neutral particle analysis [3] and gamma-ray emission tomography [4] were used to obtain information on the fast ions, including their radial localisation. The available gamma ray reactions and the average energies of ^{3}He ions and protons required for these reactions to take place [4], together with the available powers for each ICRH frequency, were the determining factors when the choice between H and ^{3}He minority heating was made.

Frequencies f_{ICRH} in the range of 33-37.5 MHz were used for ^{3}He minority heating, with total coupled power in the range of 3-4.5 MW (Fig. 1). Higher powers, in the range of 7-8 MW, were obtained with H minority heating using f_{ICRH} in the range of 46.5-51.5 MHz. For monochromatic ICRH, the resonance was located in the plasma centre and the spread in f_{ICRH} around the nominal f_{ICRH} delivered by the four antennas resulted in a natural spread ΔR_{res} of 5-10 cm in the major radius of the minority ion cyclotron resonance R_{res}. For polychromatic operation, the resonances with $\Delta R_{res} = 30\text{-}35$ cm were located in the plasma centre and on the low or high field side. The dipole ($0\pi0\pi$) phasing of the antennas was used, the ^{3}He concentration $n(^3\text{He})/n_e$ was about 1 %, and the H concentration n_H/n_e was in the range of 3-7 %.

EXPERIMENTAL RESULTS

Figure 2 summarises the experimental results in terms of the plasma diamagnetic energy W_{DIA}, fast ion energy $W_{fast} = 2(W_{DIA} - W_{th})/3$ and volume-averaged electron temperature $<T_e>$ as a function of ICRH power (W_{th} is the thermal plasma energy content deduced from measured plasma densities and temperatures). As we can see, W_{DIA}, W_{fast} and $<T_e>$ are somewhat smaller with polychromatic ICRH for $P_{ICRH} = 3\text{-}4.5$ MW. Here, the difference in W_{fast} is responsible for about 50 % of the difference in W_{DIA}, and is consistent with smaller-amplitude and shorter-period sawteeth produced with polychromatic ICRH (cf. Fig. 3a) suggesting a lower fast ion energy in the plasma centre inside q = 1. For discharges with higher-power H minority heating, not only W_{fast} but also $<T_e>$ and W_{DIA} are identical within the error bars for monochromatic and polychromatic ICRH with similar P_{ICRH}. Simulations with PION [5] and SELFO [6] codes suggest that this change in the overall behaviour is due to the fact that as P_{ICRH} increases, the fast ion orbit width increases and becomes comparable to or larger than the spread in the ICRH resonances. Throughout the scan in P_{ICRH}, ion temperatures and ion temperature profiles were similar with polychromatic and monochromatic ICRH (cf. Fig. 3b). Unfortunately, there was no time to optimise bulk ion heating in these discharges. Nevertheless, the observed higher T_i/T_e ratios and shorter-period sawteeth make polychromatic ICRH promising for further applications.

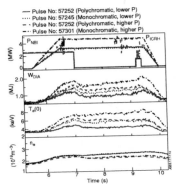

FIGURE 1. Overview of main plasma parameters for discharges with polychromatic and monochromatic ^{3}He minority heating.

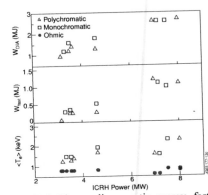

FIGURE 2. Plasma diamagnetic energy, fast ion energy and volume-averaged electron temperature as a function of ICRH power.

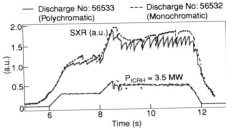

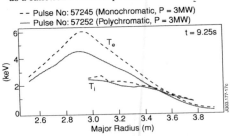

FIGURE 3. (a) Soft X-ray emission and ICRH power and (b) electron and ion temperature profiles for two discharges with monochromatic and polychromatic ^{3}He minority heating.

Important information on the fast ion orbits and radial localisation was obtained with gamma emission tomography. Figure 4 shows gamma emission profiles for three discharges with ^{3}He minority heating. When P_{ICRH} is increased from 3 to 4.5 MW using polychromatic ICRH (discharges 57252 and 57300), the gamma emissivity increases and the shape of the gamma emission changes, suggesting a larger fraction of trapped ions located on the low field side at high P_{ICRH}. This change is consistent with the expected increase in the average energy of the fast ions above the energy $E \approx 34\, T_e$ (keV) $\approx$ 150-200 keV at which pitch-angle scattering becomes weak [7], with a concomitant increase in the fraction of trapped ions. With monochromatic ICRH (discharge 57301), the maximum gamma emission is located on the low-field side of the resonance, but closer to it than in discharge 57300 with polychromatic ICRH, and the intensity of the emission increases with P_{ICRH}. This suggests a larger fraction of fast non-standard (co-passing and potato) orbits on the low-field of the resonance with monochromatic ICRH. PION and SELFO code modelling of the discharges show similar trends as the measurements (cf. Fig. 5).

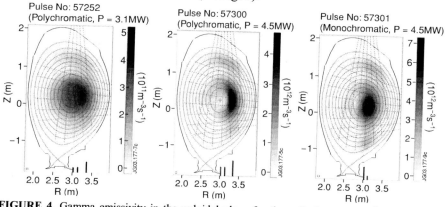

FIGURE 4. Gamma emissivity in the poloidal plane for three discharges with ^{3}He minority heating. Gamma emission is dominantly from reactions between fast ^{3}He and ^{9}Be which take place when $E(^3\text{He}) > 0.9$ MeV. The vertical bars indicate the relative fraction of total P_{ICRH} applied at a given R_{res}.

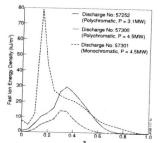

FIGURE 5. Fast ion energy density as given by PION for discharges in Fig. 4. Here, $s = \sqrt{\psi_p / \psi_{pa}}$, where ψ_{pa} is the poloidal flux at the edge.

Gamma-ray data for discharges with high-power H minority heating show similar trends as for discharges 57300 and 57301 with lower-power ^{3}He minority heating, but the peak gamma emissivity is located further to the low-field side and the profile of the gamma emission is broader, consistent with larger fast ions orbits. Measurements with high-energy NPA during H minority heating indicate about a 25% higher T_{tail} for fast protons with monochromatic ICRH than with polychromatic ICRH.

To conclude, experimental results show trends expected from theory. Detailed comparisons with simulations are underway and will be reported elsewhere.

REFERENCES

1. Rimini, F.G., et al., *Nuclear Fusion* **39**, 1591 (1999).
2. Mantsinen, M.J., Kihlman, M.D., and Eriksson, L.-G., 29th European Physical Society Conference on Plasma Physics and Controlled Fusion, 17-21 June 2002, Montreux, Switzerland, Europhysics Conference Abstracts Vol. 26 B, P-1.033 (2002).
3. Korotkov, A.A., Gondhalekar, A., and Stuart, A.J., *Nuclear Fusion* **33**, 35 (1997).
4. Kiptily, V.G., et al., *Nuclear Fusion* **42**, 999 (2002).
5. Eriksson, L.-G., Hellsten, T. and Willén, U., *Nuclear Fusion* **33**, 1037 (1993).
6. Carlsson, J., Hellsten, T., and Hedin, J., *Phys. Plasmas* **5**, 2885 (1998).
7. Stix, T.H., *Waves in Plasmas* (AIP, New York, 1992).

ELM Resilient External Matching System for the ICRF System of ITER: 1. Principle and Performances

A. Messiaen, P. Dumortier, F. Durodié

LPP/ERM-KMS, Euratom-Belgian State Association, Brussels, Belgium, TEC Partner

Abstract. The design of an ICRF antenna plug for ITER capable of radiating 20 MW and matched to the power source outside the machine vacuum is summarized in this paper and in a companion one [1]. The present paper describes the general layout of the system and how the ELM resilience is obtained through a *conjugate-T* external matching network that uses robust adjustable components only outside the vacuum vessel.

INTRODUCTION

For the heating of the International Thermonuclear Experimental Reactor (ITER) 20 MW of RF power in the frequency range 40-55 MHz are foreseen. This power is radiated by an antenna array placed along the wall of the reactor in a surface of approximately 1.45 x 1.9 m^2. As the load seen by the antenna can rapidly vary with large fluctuations of the ionized gas (plasma) facing the antenna it is requested to have a matching system which can cope with large variations of the antenna loading in order to maintain the VSWR seen by the generators lower than 1.5. Furthermore all components must have CW capability.

Two matching methods are presently considered. The first uses adjustable components inside the vessel of the reactor, which must be remotely controlled in high vacuum and under a large neutron flux [2]. The second uses adjustable components only outside the reactor and requires lines with high VSWR between the antenna array and the matching components. This last system is called *external matching* and is considered in this paper and in [1]. A detailed account of the design is given in [3].

LAYOUT OF THE ICRH SYSTEM WITH EXTERNAL MATCHING

On Fig. 1 is shown the antenna array consisting of 24 radiating straps grouped in 4 columns of 6 straps. Each column has its own matching system and is fed by a 5 MW RF power source (obtained by grouping generators in parallel by means of broadband hybrid junctions). The number of straps has been increased from 16 to 24 with respect to [2] to decrease the antenna voltage. Each strap is characterized by its distributed loading resistance R_A (reference: $R_{A0} = 4$ Ω/m), inductance L_A and capacitance C_A. Fig. 1 shows the layout of the system. In order to reduce to 2 the number of lines per column coming out of the vessel and of the tuning units (which could interact with each other) the 3 upper straps and the 3 lower straps are grouped in a symmetrical way by passive 4-ports junctions. Load resilience is achieved by the *conjugate-T* method [4]. This is performed by joining the two lines in a T through adjustable line stretchers (LS_U and LS_D). The ouput of the T is connected through a broadband impedance transformer of the *Klopfenstein* taper type [5] to the power source.

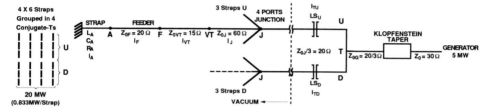

FIGURE 1. Proposed solution for the external matching; l_{TU} and l_{TD} are respectively the distances J-T along the U and the D lines and are adjustable by the line stretchers.

DESIGN OF THE CONJUGATE-T SYSTEM

If the normalized impedances seen in T towards the U and D branches $z_{TU}=1/y_{TU}$ and $z_{TD}=1/y_{TD}$ are given by $z_{TU}=r+i|x|$ and $z_{TD}=r-i|x|$ with x independent of r, the admittance $y_T=y_{TU}+y_{TD}=g+ib$ seen on the generator side is given by:

$$y_T = g = (1/|x|) \, 2\zeta/(1+\zeta^2) \text{ with } \zeta = r/|x| \tag{1}$$

The stationary character of the function $y_T(\zeta)$ and of the resulting VSWR seen by the generator, S, in the vicinity of $\zeta=1$ allows to obtain the load resilience of the *conjugate-T* matching. Its optimization is largely facilitated by the use of the Smith's chart (Fig. 2). On this chart the loci of the conjugate impedances z_{TU} and z_{TD} for which $S<S_{max}=1.5$ (value of S_{max} that normally ICRH generators can tolerate for any phase of the reflection coefficient Γ) are situated on the two dashed arcs of circle. They correspond to $x=\pm 1/1.5$ and $S<1.5$ in the range $0.156<r<2.84$. The ideal matching $S=1$ is obtained for $r_U=r_D=0.255$ and 1.745. The corresponding loci of y_{TU} and y_{TD} are also shown (using the Smith's chart for the admittances). These are displaced of 180° in the polar diagram. The matching $S=1$ corresponds to $g_U=g_D=0.5$ and $S=1.5$ to $g_U=g_D=0.75$ and 0.333. The transmission line system which links the T-junction to the corresponding antenna pair makes impedance transformations such that z_{TU} and z_{TD} are in good approximation conjugate and on the above-defined path in the Smith's chart for the range of plasma loading resistance R_A considered. On Fig. 2 are indicated the normalized impedance transformations performed at each line discontinuity from the antenna (A) to the T along the line stretcher U or D. At the 4-ports junction J there is no impedance jump because Z_0 is decreased by a factor 3 between J and T. The circles on each path correspond to $R_A=1,2,\ldots,20\Omega/m$. The last paths T_U and T_D normalized with respect to Z_{0G} are within a good approximation complex conjugate and lie on the dashed ideal path. The *conjugate-T* matching requires positioning the T near a voltage node (on both sides of the node) by means of LS_U and LS_D. The resilience of the active power to the resistive part r of the load has a drawback: the phase difference $\Delta\phi$ between the input currents in the two lines U and D is depending on r through the relation $\Delta\phi=2\arctg(|x|/r)$.

PERFORMANCES AND ELM RESILIENCE

Fig. 3 shows that a large resilience to R_A can be achieved for the complete frequency band. The best results are for the mid-band parameters (around 47.5 MHz) for which the optimization is performed. The ELMs not only increase the antenna loading resistance R_A but also slightly reduce its inductance L_A. Measurements performed on JET [6] show that the mean path of L_A versus R_A during the ELMs is roughly linear. We assume that the same relative effect will occur on the antenna impedance of the ITER antenna, i.e. a factor 3 of increase of R_A with respect to the reference value of 4 Ω/m with a decrease of L_A with a slope $\Delta L_A/R_A=-2.1nH/(\Omega/m)$. This path is shown in Fig. 5 on a contour plot $S=f(R_A,\Delta L_A)$ for conditions of *conjugate-T* matching. The VSWR remains lower than 1.5 all along the ELMs' path. This ELM resilience can be obtained in the complete frequency band.

Fig. 4a and b show the variation of the phase difference $\Delta\phi$ between the current flowing in the 3 upper straps and the 3 lower straps of each poloidal row and the fraction of the 2.5 MW radiated by these upper and lower branches of the T. This slight unbalance is due to asymmetry between g_U and g_D. Fig. 6 shows as a function of the frequency the voltage and current expected at resp. voltage and current anti-node between the locations A and J of Fig. 1 for the reference loading resistance (4 Ω/m). Between J and T the maximum voltage is the same and the maximum current is three times higher. The voltage and current vary with R_A as $R_A^{-0.5}$.

The performances are similar to those obtained with an internal matching system (tuning stubs or capacitors near the antenna straps inside the vacuum vessel). The maximum tuning stub or capacitor voltages are comparable to the maximum voltage in the line of the external matching case for the same number of radiating straps [3].

REFERENCES

1. P. Dumortier et al., *ELM Resilient External Matching System for the ICRF System of ITER: 2. Design of the Components and Implementation*, this conference
2. Detailed Design Description Ion Cyclotron Heating and Current Drive System WBS 5.1 (DDD)
3. Report LPP-ERM/KMS **121**, to be published
4. R.H. Goulding et al., Bull. Am. Phys. Soc., 46(2001), no. 8, p.154
5. R. Klopfenstein, Proc. IRE 44(1956)31
6. I. Monakhov et al., *New Techniques for the Improvement of the ICRH System ELM Tolerance on JET*, this conference

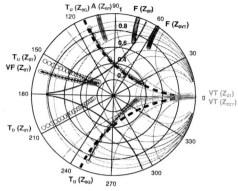
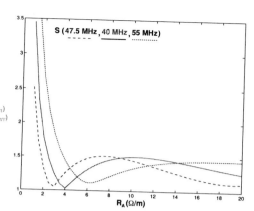

FIGURE 2. Impedance transitions to achieve the *conjugate-T* matching conditions. The positions and values of the characteristic impedance used for the impedance normalization correspond to the indications of Fig. 1.

FIGURE 3. Evolution of S in function of R_A for the limit and mean frequencies.

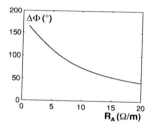

FIGURE 4. Evolution of (a) $\Delta\Phi$ and (b) $P_{U,D}/P_{tot}$ as a function of R_A (f = 47.5 MHz).

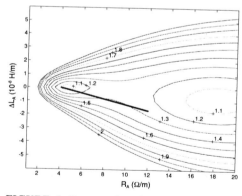
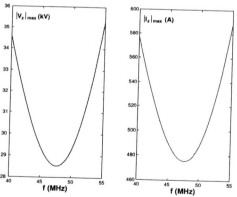

FIGURE 5. Contour plot of S in $R_A, \Delta L_A$ plane (f = 47.5 MHz, l_{TU} = 3.28 m + nλ/2, l_{TD} = 2.72 m + nλ/2, L_{A0} = 197 nH/m). The thick straight line indicates the ELM's path.

FIGURE 6. Amplitude of the voltage and current anti-nodes in the line (section VT-J) as a function of f for R_A = 4 Ω/m.

Studies of JET ICRH Antenna Coupling During ELMs

I. Monakhov[1], T. Blackman[1], M.-L. Mayoral[1], M. Nightingale[1], A. Walden[1], P. U. Lamalle[2], P. Wouters[2] and JET EFDA contributors*

[1] *Euratom/UKAEA Fusion Association, Culham Science Centre, Abingdon, OX14 3DB, UK*
[2] *LPP-ERM/KMS, Association Euratom-Belgian State, Brussels, B-1000, Belgium*
Appendix of J. Pamela, et al., "Overview of JET Results", Fusion Energy 2002, IAEA, Vienna, (2002).

Abstract. Details are provided of a new fast data acquisition system that records RF data for a complete four-strap JET ICRH antenna array with sampling rates up to 250 kHz, triggered by the rapid increase in plasma D_α emission intensity during an ELM. The coupling properties are deduced from the transmission line voltage amplitude and phase measured using 80dB directional couplers installed close to the antenna. Resistive and reactive components of antenna loading perturbation are reported during different types of ELMs and discharge magnetic configurations.

INTRODUCTION

Detailed knowledge of ICRH antenna array coupling characteristics during ELMs is valuable both for the design of ITER-relevant ELM-resilient matching systems and for understanding ELM phenomena [1]. JET plasmas offer an excellent opportunity for such studies given their abundance of various types of ELMs [2,3] and the availability of the JET multi-strap phased ICRH antenna arrays [4].

Previous measurements [5], using a PC-based fast data acquisition system connected to directional couplers located ~84m from the antenna, focused on the behaviour of individual antenna straps during conventional "low-triangularity" JET discharges. The coupling resistance measured during an ELM comprised a fast (~150 μs) rise up to 8 Ω followed by slow (~2-3 ms) decay, accompanied by an alternating and relatively small (±10 cm) strap electrical length change. The fast data acquisition system has now been upgraded and used for ICRH antenna coupling studies in wider range of JET discharge scenarios.

NEW FAST DATA ACQUISITION SYSTEM

A new fast data acquisition system has been installed to provide RF data for a complete four-strap antenna array. It comprises 2×16 fixed channels, each providing up to 250kHz sampling rate and ~65000 points, triggered by the increase of plasma D_α emission intensity during an ELM.

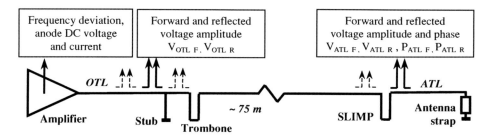

FIGURE 1. Schematic representation of fast data acquisition pickup points location and signal names.

The ELM-related coupling perturbations are deduced from the antenna transmission line (ATL) voltage amplitude and phase measurements produced by accurate 80dB directional couplers installed close (~13m) to the antenna array (Fig.1). The resistive and reactive components of antenna strap loading are characterised respectively in terms of the coupling resistance $R_{IN\_ELM}=Z_0 (V_{ATL\_F} - V_{ATL\_R})/(V_{ATL\_F} + V_{ATL\_R})$ and the electrical length $L=(P_{ATL\_F} - P_{ATL\_R})/(2\beta)$, where Z_0 is transmission line characteristic impedance (30 Ohm) and β is the propagation factor. Because of the high VSWR in a mismatched ATL, the accuracy of coupling resistance measurement defined above quickly deteriorates at low loading between ELMs. For this reason, signals from the matched amplifier output transmission line (OTL) are used additionally to calculate the coupling resistance during a low quasi-stationary loading $R_{BETWEEN\_ELM}=Z_0 (V^2_{OTL\_F} - V^2_{OTL\_R})/(V_{ATL\_F} + V_{ATL\_R})^2$ and the generalised coupling resistance in an ELMy plasma is estimated to be $R=R_{BETWEEN\_ELM}+\Delta R_{IN\_ELM}$.

It is also worth noting that the JET A2 antenna is a complicated system of coupled resonant loops and a comprehensive description requires an advanced formalism (e.g. [6]). The approach described above is a traditional simplification aimed mainly at facilitating the matching analysis. Care needs to be taken in extrapolating the results to other antennas and in deductions of coupling physics.

STRAP COUPLING DURING DIFFERENT TYPES OF ELMS

The general features of coupling resistance changes (Fig.2) are consistent with previous observations [5]: a fast increase up to 8-9Ω followed by a slow and highly variable decay. The electrical length behaviour, however, was found to be both quantitatively and qualitatively different: higher (up to ~40cm) perturbation magnitudes were detected with the length predominantly decreasing.

The response to ELMs within the array depends on the array phasing and is not particularly sensitive to frequency in the 28-51MHz band. The pair of straps which are close to the antenna central septum (septum straps) typically demonstrate smaller coupling perturbations compared with the outer pair (limiter straps). This can be explained by the different strap and feedthrough design.

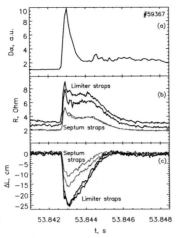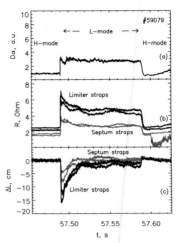

FIGURE 2. Temporal behaviour of strap coupling perturbations during a big Type I ELM (left) and a small ELM, accompanied by an H-L mode transition (right): (a) - D_α signal, (b) - coupling resistance and (c) - strap electrical length change. The different traces on plots (b) and (c) correspond to four straps of JET antenna array with grey shade used to denote the inner straps located close to the antenna central septum. Both discharges correspond to operations at f=42.1 MHz and [0 π 0 π] array phasing.

No differences between the coupling perturbations induced by similar ELMs were found during the full array operation and the single strap operation, which indicates that the strap cross-talk is not likely to play a significant role in ELM coupling.

Although "big solitary" Type I ELMs produce the largest registered coupling perturbations, "small frequent" ELMs, including Type III ELMs can also cause substantial coupling resistance changes, which has strong implications for the RF plant matching. In general, the relationship between the level of D_α activity and coupling perturbations' magnitude and temporal behaviour is not straightforward. Short time-scale (≤100 ms) H-L-H mode transitions, occasionally accompanying ELMs, were found to produce noticeable coupling resistance changes, which can contribute to and partly explain the coupling behaviour following an ELM (Fig.2).

COUPLING PERTURBATION DEPENDENCE ON PLASMA MAGNETIC CONFIGURATION

Fig.3 shows that the plasma magnetic configuration affects the magnitude of ELM-induced strap coupling resistance and electrical length perturbations. The analysis of *max(ΔR)* and *max(ΔL)* values in a series of shots with "similar" Type I ELMs shows that the perturbation magnitude increases with increasing plasma triangularity and decreasing antenna-plasma separation. The first dependence partially explains the discrepancies with earlier results [5] and implies an unfavourable scaling for ITER for the Type I ELM regime examined. The second should be scaled with care, as the larger outer gap also strongly reduces the coupling during the period between ELMs.

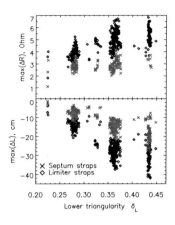 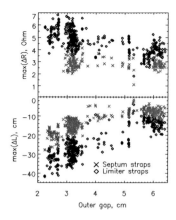

FIGURE 3. Dependence of strap coupling resistance and electrical length perturbation magnitudes on plasma lower triangularity (left) and outer gap (right). Data from ~55 pulses with the same RF frequency (f=42.1 MHz) and array phasing [0 π 0 π]. "Similar in magnitude" Type I ELMs are selected by choosing ELMs with mid-plane D_α signal changes in the range of 0.8-1.5 V.

CONCLUSIONS

A new fast data acquisition system measures JET ICRH antenna array loading perturbations and fast RF plant responses during ELMs. The results have contributed to better understanding of ELM and ICRH physics and have helped to identify problems relevant to the design of ITER-oriented antennas and matching systems.

ACKNOWLEDGEMENTS

The authors are grateful to the JET Task Forces, Session Leaders and CODAS personnel for their co-operation. This work was performed under the European Fusion Development Agreement, funded jointly by the United Kingdom Engineering and Physical Sciences Research Council and by EURATOM.

REFERENCES

1. Connor, J.W., *Plasma Phys. Control. Fusion* **40**, 531-542 (1998).
2. Saibene, G., et al., *Plasma Phys. Control. Fusion* **44**, 1769-1799 (2002).
3. Loarte, A., et al., *Plasma Phys. Control. Fusion* **44**, 1815-1844 (2002).
4. Kaye, A., et al., *Fusion Engineering and Design* **24**, 1-21, (1994).
5. Simon, M., et al., "ICRH coupling studies and wideband matching in ELMy plasmas on JET" in *Proc. 2nd Europhysics Topical Conference on RF Heating and Current Drive of Fusion Devices*, edited by J. Jacquinot et al., Europhysics Conference Abstracts 22A, Mulhouse, 1998, pp. 105-108.
6. Lamalle P. U., et al., "Physical aspects of coupling the ICRF A2 antennae to JET discharges" in *Proc. 22nd EPS Conference on Controlled Fusion and Plasma Physics*, edited by B. E. Keen et al., Europhysics Conference Abstracts 19C, Geneva, 1995, pp. 329-332.

New Techniques For The Improvement Of The ICRH System ELM Tolerance On JET

I. Monakhov[1], T. Blackman[1], A. Walden[1], M. Nightingale[1],
A. Whitehurst[1], F. Durodie[2], P. U. Lamalle[2] and JET EFDA contributors*

[1] *Euratom/UKAEA Fusion Association, Culham Science Centre, Abingdon, OX14 3DB, UK*
[2] *LPP-ERM/KMS, Association Euratom-Belgian State, Brussels, B-1000, Belgium*
*Appendix of J. Pamela, et al., "Overview of JET Results", Fusion Energy 2002, IAEA, Vienna, (2002).

Abstract. Two complementary improvements to the ELM tolerance of the existing A2 antennas on JET are being assessed. The use of external conjugate-T matching of straps of adjacent antenna arrays could reduce the VSWR levels at RF amplifier output during fast load perturbations. The scheme under consideration uses coaxial line-stretchers (trombones) for tuning the conjugate-T to low resistive impedance (3-6 Ohm) with subsequent stub/trombone circuit impedance transformation to 30 Ohms. Another technique is to modify the RF plant protection system logic to reduce the high VSWR trip duration to an absolute minimum corresponding to a typical ELM response (~1-2ms) without compromising the plant safety. Both projects are presently being tested and could increase the average power delivered by RF plant into ELMy plasmas at JET.

INTRODUCTION

The capability of RF plant to inject high power levels into ELMy H-mode plasmas is essential both for the JET research program and successful ITER operations. The main research in this field involves the new ITER-like antenna being developed on JET [1,2]. Two complementary approaches to improve the ELM tolerance of the existing JET A2 antennas [3] are underway and discussed below.

The present-day JET ICRH system [4] comprises 16 "similar" circuits for independent energising of four straps of four existing antenna arrays, each using a stub/trombone matching and real-time frequency control to track slow evolutions of plasma loading. Because of insufficient speed and load tolerance of the present matching system, the amplifier output VSWR reaches high values during ELMs. The plant VSWR protection system, being unable to distinguish between arcs and ELMs, then produces relatively long RF power trips with slow post-trip recovery, resulting in impaired RF plant performance and significantly reduced average power delivered into ELMy plasmas. Two ICRH system modifications are under consideration to overcome this loss in average coupled power: an external conjugate-T matching system has been modelled and is now proceeding to prototype test; modifcations to the existing trip management system have been implemented and are ready for testing.

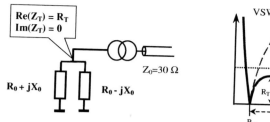 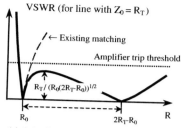

FIGURE 1. Generalised scheme (left) and loading response (right) of conjugate-T matching circuit

EXTERNAL CONJUGATE-T MATCHING AT JET

Principles

Conjugate matching comprises the parallel connection of two complex conjugate impedances, tuned to make the resulting impedance purely real (Fig.1). The main advantage of such a system is that the simultaneous resistive load increase in both legs of the circuit does not cause a severe transmission line mismatch. Simple analytical treatment shows that the load tolerance quickly improves with increasing the value of unperturbed loading R_0, which is favourable for the A2 antenna case as compared with short-strap antenna design [1,2]. Load tolerance optimisation also dictates the necessity to tune the T-junction to low impedances ($R_T << Z_0$) and, hence to use an additional impedance transformer $R_T \rightarrow Z_0$. The conjugate-T matching reactance $X_0^2 = R_0(2R_T - R_0)$ dependence on R_0 and R_T implies adjustments for variable loading R_0 and enables load-tolerance fine-tuning via R_T control.

Features

The proposed conjugate-T system involves four amplifiers and two antenna arrays. The power from each amplifier (Fig.2) is split between two straps of *different* arrays, to retain array-phasing flexibility. With the existing ~1 MW per strap power limit of A2 antennas the maximum values of delivered power remain unaffected.

An external (out of vessel) coaxial T-junction will be tuned to a low resistive impedance ($R_T = 3 - 6 \, Ohm$) by coaxial line-stretchers (trombones). The use of the *external* T-junction allows the whole system to be built from standard coaxial components and facilitates conventional strap loading diagnostics. Trombone-based conjugate-T tuning also offers a number of advantages for the system design: the elements readily provide the required reactances of both signs; they have no constraints on location along the line; the tuning accuracy is manageable and the line dispersion compensation is straightforward.

A prototype test of the proposed configuration has now been installed onto one pair of straps and is ready for testing.

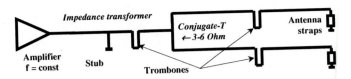

FIGURE 2. Schematic diagram of conjugate-T matching for JET A2 antenna straps (4 similar circuits)

Simulations

Computer simulations have been undertaken to analyse the circuit load tolerance, tuning accuracy and electrical strength under different loading conditions. Available A2 antenna loading data were used to provide the following input parameters [5]: the static (between ELMs) strap coupling resistance R_0= 1-3 Ω; the perturbed (during ELM) coupling resistance $R \to 10$ Ω and the ELM-related electrical length change $\Delta L=+2 \to -20$ cm. The transmission lines lengths between the T-junction and the strap (31 m) and the T-junction and the stub (2.9 m) were chosen to optimise the frequency coverage in the 31-57MHz band, assuming 0-1.5 m trombone length variation and 3 m maximum stub length.

The simulations demonstrate an excellent circuit tolerance to the resistive loading perturbations and a possibility to fine-tune the loading characteristics depending on strap loading between ELMs (Fig.3a). The circuit shows reasonable robustness with respect to asymmetry of resistive loading perturbation between straps (Fig.3b) and to the accompanying strap electrical length changes up to -20 cm (Fig.3c).

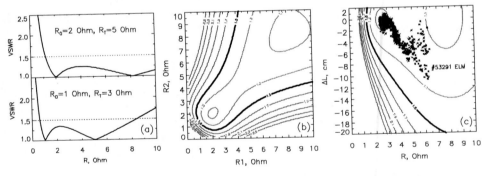

FIGURE 3. Conjugate-T circuit load tolerance at f=42.5 MHz. (a) - VSWR dependence on strap coupling resistance in case of symmetric resistive loading perturbation; two examples of the circuit tuning are shown corresponding to relatively high (top) and low (bottom) strap coupling resistance between ELMs; (b) - VSWR contours in case of asymmetric resistive loading perturbation (R_0=2 Ohm, R_T=5 Ohm), (c) - VSWR contours in case of symmetric resistive loading, accompanied by strap electrical length change (R_0=2.5 Ohm, R_T=5 Ohm). Load tolerance domain VSWR=1.5 is shown by the bold line. The set of dots represents a <u>measured</u> "footprint" of a moderate ELM.

IMPROVED TRIP MANAGEMENT LOGIC

In parallel with the above developments, a trip management system is being updated on JET to improve the RF plant performance in situations where conjugate-T matching may not cope, in particular for ELMs involving a large strap electrical length perturbation [5]. At present, this system trips the RF power if a high VSWR is detected, with a trip threshold depending on RF power and ELM activity. This introduces a long trip (5→20 ms) and recovery (10→80 ms) duration, with both durations increasing with trip frequency. The result is that the average power delivered in ELMy plasmas is reduced, and in constant-power control setting the power feedback loop increase forward voltage demand to maintain the requested average power level, causing even more trips.

Given these problems, the present system is being upgraded to include a check as to whether a high measured VSWR is accompanied by an ELM (D_α signal increase rate is above the set threshold). If *no* ELM is detected, then the long trip & recovery time described above is applied to quench any arc. If an ELM *is* detected, the trip & recovery time is reduced to ~2 ms only, to protect the amplifier during the ELM. If a high VSWR is re-detected immediately after the shortened (ELM-triggered) trip then the long trip & recovery time is applied, regardless of ELM detection, as ELM-related arcing is likely. This procedure should increase the average power delivered in ELMy plasmas without detriment to the RF plant protection.

CONCLUSIONS

A combination of external conjugate-T matching and modifications to the RF plant protection system logic should improve the performance of the JET ICRH system during ELMy plasma discharges.

ACKNOWLEDGEMENTS

This work was performed under the European Fusion Development Agreement. The work carried out by the UKAEA personnel was funded jointly by the United Kingdom Engineering and Physical Sciences Research Council and by EURATOM.

REFERENCES

1. Durodie, F., et al., "The ITER-like ICRF launcher project for JET", this conference
2. Lamalle, P.U., et al., "Radio-frequency matching studies for the JET ITER-like ICRF system", this conference
3. Kaye, A., et al., *Fusion Engineering and Design* **24**, 1-21, (1994).
4. Wade, T., et al., *Fusion Engineering and Design* **24**, 23-46, (1994).
5. Monakhov, I., et al., "Studies of JET ICRH antenna coupling during ELMs', this conference

Status and development of the ICRF antennas on ASDEX Upgrade

J.-M. Noterdaeme [1,2], W. Becker [1], V. Bobkov [1],
F. Braun [1], D. Hartmann [1], F. Wesner [1]

[1] *Max-Planck-Institut für Plasmaphysik, Euratom Association, D-85748 Garching, Germany*
[2] *Gent University, EESA Department, B-9000 Gent, Belgium*

Abstract. The 3 dB couplers on the ICRF system of ASDEX Upgrade allow operation of the ICRF system, even in the presence of strong ELMs. The refurbishing of the power tetrodes has restored the nominal power of the generators. Under those conditions, the limiting factor on the power to the plasma can become the voltage stand-off of the antennas. The paper reports the status, progress and plans in this area. By doubling the radius of the sides of the antenna straps (from 5 mm to 10 mm) the edge electric field of the antenna straps was reduced. Experiments in a vacuum test set-up confirmed an increase of the voltage stand-off of the antenna, from typically 27 kV to 41 kV for 10 s pulses and from 42 kV to 58 kV for 1 s pulses. The effect of plasma scrape-off layer variations on the voltage stand-off was investigated with a high voltage RF probe of a simple geometry inserted in the plasma edge. The voltage stand-off of the probe is reduced during the period of the edge density increase correlated with an ELM. The reduction in voltage stand-off can be understood as an effect due to plasma getting into the antenna. This implies that in addition to the need to make a system ELM resilient from the point of view of matching, and optimised from the point of view of vacuum electric fields, one still needs to consider the possible influence of the edge plasma on the voltage stand-off. To prevent plasma from penetrating into the antenna during the ELM phase an optically closed Faraday screen could be used.

INTRODUCTION

The ICRF system of ASDEX Upgrade has 4 antennas [1] with 2 straps each. The two straps are connected in parallel to a double stub matching system. The antennas are then connected in pairs, through two 3-dB couplers to 2 high power generators with a frequency range of 30 to 120 MHz. Each of the 4 generators has a nominal power of 2 MW between 30 and 80 MHz dropping to 1 MW at higher frequency.

The system has been in operation since 1992, at first without 3 dB couplers. The strong reflections due to the change in coupling during ELMs limited the operation of each generator to about 1 MW.

Since installation of 3 dB couplers, the ICRF system can operate even in the presence of strong ELMs. The recent refurbishing of the power tetrodes has restored the nominal power of the generators to 2MW. Under those conditions, a limiting factor on the power to the plasma can become the voltage stand-off of the antennas.

Over the years minor changes have been made to the antenna. Ceramic insulation originally foreseen to avoid induced currents was removed. The thermal capability of the limiters was improved. New antenna straps were installed when the antenna was adapted to plasmas with higher triangularity.

The voltage stand-off of the antenna required to couple 2 MW is 25 kV peak for an antenna resistance of 2 Ohm, a typical value for H-mode plasmas at a plasma-antenna distance of 3 cm. This voltage stand-off of 25 kV could often be achieved, but it was noticed that the limit strongly depends on plasma conditions. A two pronged approach was taken. Antenna straps with reduced edge electric fields were developed and a study programme to identify and understand the influence of plasma on the antenna voltage limitation was under taken.

STRAPS WITH REDUCED EDGE ELECTRIC FIELD

The radius of the sides of the antenna straps was doubled from 5 mm to 10 mm. The toroidal distance between the edge of the strap and the septum is 9 cm, so that the average electric field is 2.77 kV/ cm for 25 kV on the antenna.

The actual value of the electric field of course varies and depends on the geometry of the edge of the strap. For the radius of curvature of 5 mm, the maximum electric field at the strap is (based on numerical calculations) 11 kV / cm. This maximum electric field is reduced to 8.9 kV/cm for a radius of curvature of 10 mm (1/1.23 times lower). Experiments in a vacuum test set-up lead

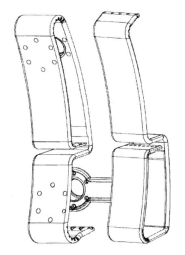

FIGURE 1. New straps with increased radius of curvature of the edges, leading to a lower edge electric field.

to an increase of the voltage stand-off of the antenna, from typically 27 kV to 41 kV (1.51 times higher) for 10 s pulses and from 42 kV to 58 kV (1.38 times higher) for 1 s pulses. The new straps have replaced the old ones in two of the four antennas in ASDEX Upgrade. The systems with the new straps were conditioned to 30 kV, 1 sec, whereas for the ones with the old straps it was difficult to achieve in vacuum at 30 kV a pulse length longer than 200ms. For short pulses (5ms) all systems were conditioned to 40 kV. A possible benefit with plasma could not be investigated so far. Without reaching the voltage limit, voltages of up to 30 kV were achieved in ICRH pulses of some seconds, for a power of 1.3 MW launched per antenna, under low coupling conditions. Differences in the voltage strength behavior will be investigated in special test series.

VOLTAGE LIMITATION STUDIES

The effect of plasma scrape-off layer variations on the voltage stand-off was investigated with a high voltage RF probe of a simple geometry inserted in the plasma edge [2]. The probe is the open end of a high-Q transmission line resonator. The matching of the resonator is realized by a change in frequency and a variable short. The inner conductor of the probe can be DC biased and the rectified current to the inner conductor can be measured. The probe was exposed to the plasma boundary of ASDEX Upgrade at same radial position as the ICRF antenna straps.

The voltage stand-off of the probe was measured using two power ramps providing an increase of voltage up to 80 kV in 50 ms and 300 ms respectively. Fig. 2 shows the breakdown voltages achieved in plasma compared to the voltages in vacuum. The vacuum voltage stand-off was sustained at a level of 75 kV by conditioning between the ASDEX Upgrade discharges while in plasma the voltage was limited by appearance of strong (type I) ELMs. ELMs lead to a breakdown, despite the simultaneous reduction of the voltage due to the better coupling. Note that the voltage presented in Fig. 1 is the maximal voltage, i.e. voltage between ELMs.

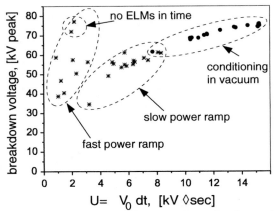

FIGURE 2. Voltage achieved on the probe for various condition (vacuum = circles, plasma = stars) and different power ramps

The mechanism leading to a breakdown during an ELM is the following. The appearance of an ELM is accompanied by a series of bursts leading to an intermittent increase the plasma density in the scrape-off layer and in the coaxial electrode gap of the probe. The plasma density during these bursts is sufficiently high to decrease the RF sheath thickness to dimensions smaller than the gap distance. Thus a high electric field appears on the electrode surface. Together with bombardment of surface by ions with energy defined by the applied voltage, the electric field can lead to a spark evolving into an arc discharge after several RF periods. This mechanism of arc initiation is dependent on the electrode surface conditions. Indeed, a conditioning effect is observed for each new breakdown, i.e. the voltage limited by the arrival of an ELM burst is increased from shot to shot (the higher voltages from Fig.2 are obtained later than the lower voltages). The bursts of a single ELM were detected by the measurement of the rectified current, which is a measure of the ion flux collected by the inner conductor of the probe. As shown in Fig. 3, the DC biasing of the antenna affects the rectified current and the voltage stand-off that can be obtained. A reduction of rectified current leads to a higher voltage stand-off.

Initiation of arcing by strong ELMs was confirmed by measurements on the real ICRF antennas. The breakdown on the antenna initiated by ELMs can also be attributed to a second mechanism. Since the relevant distances in the antenna are often comparable with the ionization length, ignition of a semi-self-sustained RF glow discharge can be followed by a glow-to-arc transition as observed using the RF probe in a test facility.

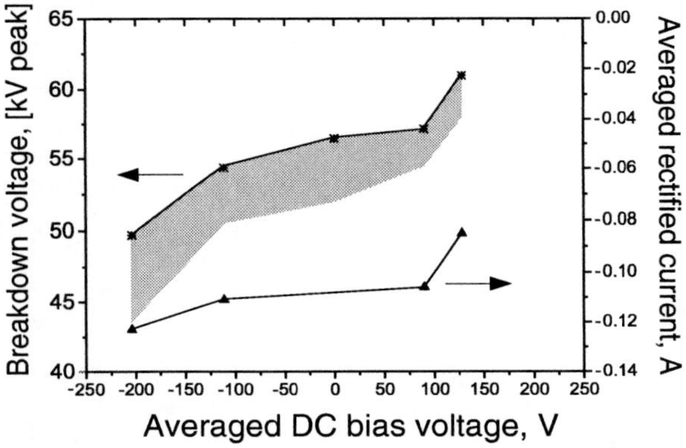

Thus the voltage stand-off in the plasma discharges is strongly reduced during the period of the edge density increase correlated with an ELM. Strong interaction between the incoming plasma and the RF electric fields often leads to local power dissipation and to arcing on the antenna. Therefore the increase of the plasma density at the antenna electrodes should be avoided by a proper antenna design, for example, by introducing an optically closed Faraday screen.

FIGURE 3. Dependence of the breakdown voltage on the DC boundary conditions of the probe. A DC condition that reduces the rectified current to zero leads to the highest value of the breakdown voltage.

OPTICALLY CLOSED FARADAY SCREEN

Since 1986, when experiments on ASDEX have shown that an optically closed and an optically open Faraday screen are in many areas equivalent [3], most ICRF antenna have been equipped with optically open Faraday screen. This has strong advantages from the manufacturing point of view. Experiments were even performed on TEXTOR and ASDEX Upgrade [4] without Faraday screen. The experiments on ASDEX however could not test a difference in voltage limitation for an open or a closed Faraday screen for two reasons. The antennas did not operate under strong type I ELMs conditions and the same voltage that could be reached in both cases was not limited by the voltage stand-off of the antenna. Since there are strong indications that the plasma effect on the voltage limits in the antenna may be due to plasma getting into the antenna, it may be worth reconsidering the use of a Faraday screen designed specifically to avoid this. A compromise must be found between having the screen as open as possible for good coupling, and as closed as necessary to avoid plasma getting into the antenna.

PLANS FOR NEW ANTENNAS

Plans have also been made for new antennas to replace the present 2 strap antennas. The design goals were threefold: reduced the sensitivity of the antenna to plasma shape variations, more clearly defined spectrum, and higher voltage stand-off capabilities. The reduced sensitivity to plasma shape variations can be obtained by making reducing the poloidal extend of the antenna. The spectrum can be improved by a larger number of straps in the toroidal direction. Since the number of feedpoints to an antenna is limited by the ports in ASDEX Upgrade to two, an internal distribution over a larger number of straps was chosen. By combining a shorter poloidal extend with a distribution network in a separated box with good vacuum, a larger voltage stand-off of the antenna is expected.

CONCLUSIONS

The antennas of ASDEX Upgrade have performed well with some evolutionary changes over the years. The capabilities to operate under strong type I condition made the influence of the plasma on the antenna voltage stand-off an issue. It becomes clear that in addition to the need to make a system ELM resilient from the point of view of matching, and optimised from the point of view of vacuum electric fields, two further points have to be addressed. The DC boundary condition of the antenna must be critically assessed and plasma should be prevented from entering the antenna. This may require the reconsideration of the use of optically closed Faraday screens.

REFERENCES

1. Noterdaeme J.-M., Wesner F., Brambilla M., Fritsch R., Kutsch H.-J., Söll M., and the ICRH team, "The ASDEX Upgrade ICRH antenna", Fus. Eng. Des. **24**.1-2 (1994) 65-74.
2. Bobkov V.V., Noterdaeme J.-M., Wesner F., Wilhelm R., and Team A.U., "Influence of the Plasma on ICRF Antenna Voltage Limits", in *Plasma Surface Interactions in Controlled Fusion Devices*, (15th Int. Conf. (PSI-15), Gifu (Japan), 2002), Journal of Nuclear Materials
3. Noterdaeme J.-M., Ryter R., Söll M., and ICRH Group, "The role of the Faraday screen in ICRF antennas : Comparison of an optically open and an optically closed screen on ASDEX", in *Controlled Fusion and Plasma Heating*, (13th EPS Conf., Schliersee, 1986), Europhysics Conference Abstracts Vol. 10C(I), (G. Briffod and M. Kaufmann ed.), EPS 137-140.
4. Noterdaeme J.-M., Becker W., Braun F., Faugel H., Fuchs J.C., Herrmann W., Hoffmann C., F. Hofmeister, Neu R., Suttrop W. et al., "Achievement of the H-Mode with a Screenless ICRF Antenna in ASDEX Upgrade", in *Radio Frequency Power in Plasmas*, (11th Topical Conf., Palm Springs, CA, 1995), Vol. 355, (R. Prater and V.C. Chan ed.), AIP Press (1996) 47-50.

Investigation of ICRF coupling resistance in Alcator C-Mod Tokamak

A. PARISOT [1], S.J. WUKITCH, A. RAM, R. PARKER, Y. LIN, J.W. HUGHES, B. LABOMBARD, P. BONOLI and M. PORKOLAB*

Plasma Science and Fusion Center, MIT, Cambridge MA 02139, USA

Abstract.
The ICRF coupling resistance for Alcator C-Mod antennas has been studied for low H minority fractions discharges. We have observed a strong correlation between the coupling resistance and the electron radial density profile at the plasma edge, which is routinely monitored on C-Mod with millimeter resolution Thomson scattering. The fast wave cutoff is in the limiter shadow for typical operations, and the strongest influence is found with the density at the top of the edge pedestal. The coupling also increases in EDA H-mode with the neutral pressure at the wall, which is related to changes in the density in the scrape-off layer. This dependence on density profiles is found to be in agreement with a 1D wave propagation model developed by Ram and Bers in 1984. With better knowledge of the density profile in the scrape-off layer, this work can allow relevant comparison of antenna performance, especially between the two straps and four straps antennas in C-Mod.

INTRODUCTION

The coupling of an ICRF antenna can vary significantly during plasma discharges in tokamaks, and the corresponding potential increase of the standing-wave ratio is a major issue for high power operations. Several effects which can explain these variations in different regimes have been identified [1]. The distance between the fast-wave cutoff and the straps is often one of the dominant factors ; in addition, the density profile at the edge of the plasma can also influence the antenna loading.

Alcator C-Mod tokamak presents interesting characteristics regarding ICRF coupling studies. It is a compact, high density (up to $10^{21} m^{-3}$) and high toroidal field (3-8 T) device, equipped with three ICRF antennas [2]. The two straps center-grounded antennas at the D and E horizontal ports are operated at 80 and 80.5 MHz respectively, and up to 1.5 MW, while the antenna at J-port has a compact four-strap antenna structure, which can couple up to 3 MW at a given frequency in the 40-80 MHz range. The straps for all three antennas have the same length, 44 cm, and they are separated on center by 25.75 cm for D and E ($n_\parallel \approx 10$ for $0-\pi$ phasing), by 18.6 cm for J ($n_\parallel \approx 13$ for $0-\pi-0-\pi$ phasing). The antenna coupling resistance is defined as the ratio $\frac{V_{max}^2}{P_{net}}$ of the maximum voltage to the net power in the transmission line.

---

[1] Email contact : alexandre.parisot@polytechnique.org

DEPENDANCE ON DENSITY PROFILES

We have studied the dependance of the measured coupling resistance for C-Mod antennas with various plasma conditions, in an effort to identify the most important parameters and understand their effect.

1. Density at the top of the pedestal The electron density profile at the plasma edge is measured routinely in C-Mod with millimeter resolution Thomson Scattering. For all three antennas, the coupling resistance decreases with the formation of the transport barrier in the L-H transition, and the experimental data shows a strong correlation with the density at the the top of the pedestal.

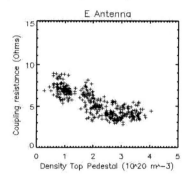 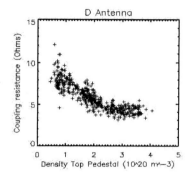

FIGURE 1. Dependance of the coupling resistance (Ohms) with the density at the top of the edge transport barrier (run 1021010)

2. Density in the scrape-off layer The experimental data deviates significantly from this behaviour especially during long EDA H-mode, and we observed a correlation between the increase of the loading for all three antennas and the increasing neutral pressure at the wall. This can be related to an increase of the density in the scrape-off layer.

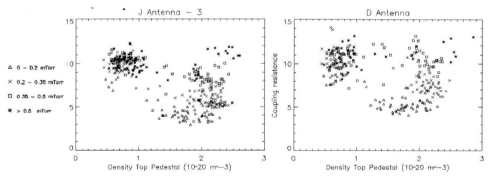

FIGURE 2. Significant increases of coupling resistance (Ohms) in EDA H-mode are correlated with the increase in the neutral pressure at the limiter wall. (run 1021124)

3. Outer gap and evanescent decay length No clear dependance is found between the loading resistance and the outer gap in C-Mod, and gap control is relatively ineffective. The typical density in the scrape-off layer for C-Mod is above $10^{19} m^{-3}$ and the cutoff density is in the upper $10^{18} m^{-3}$ range ; the fast wave cutoff is therefore typically located between the limiter and the Faraday screen (1 cm apart), and not in the near scrape-off layer like in many other tokamaks, where its radial excursions have a strong influence on the coupling through changes of the evanescent decay length.

WAVE PROPAGATION MODEL

The measured dependence finds a simple interpretation in wave propagation theory for 1D inhomogeneous cold plasma. Following the approach used by A. Ram and A. Bers in [3], we write the ICRF equation [4] in the limit $n_{poloidal} = 0$:

$$\frac{d^2 E_\perp}{d\xi^2} + K(\xi) E_\perp = 0$$

where $K(\xi) = S - \frac{k_\parallel^2 c^2}{\omega^2} - D^2/(S - n_\parallel)$ in the Stix notation, and $\xi = \frac{x\omega}{c}$; S and D are functions of the density and the magnetic profile. Assuming perfect heating in the core, we take a plane wave solution towards the absorption layer as boundary condition somewhere near the core and integrate backwards to the edge using a fourth order Runge-Kunta method.

The power coupled to the plasma is : $P_{abs} = \frac{1}{2} Re(\int_S E \times H^* dS) = \frac{1}{2} \int_S Re(\frac{E_\perp}{H_\parallel}) |H_\parallel|^2 dS$ in this simple geometry, with S denoting the cutoff surface. The term $Re(\frac{E_\perp}{H_\parallel})$ corresponds to the plasma surface impedance. As $|H_\parallel|^2$ is proportionnal to the square of the input current, the surface impedance is expected to scale with the loading resistance ; the proportionnaly factor accounts for structural properties of the antenna and plasma shape considerations.

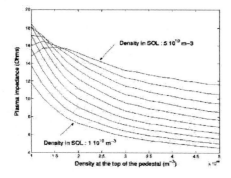 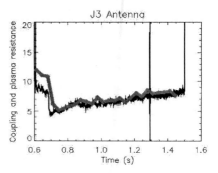

FIGURE 3. Calculated dependance of the coupling resistance (Ohms) with the density profile and comparison of the model (black) with the experimental data (grey) with early L-mode in a long EDA H-mode. (plasma discharge 1021105008)

As seen in the Fig 3, the model reproduces qualitatively the observed dependance with the density at the top of the pedestal and shows an increase related to the scrape-off layer (SOL) density. We have attempted to use more acurrate input density profiles, using the diagnostics presently available in C-Mod. The edge and core Thomson diagnostics provide a good knowledge of the electron density profile at the edge of the plasma, but a better resolution or other diagnostics would be required for the profile in the SOL during ICRF. We had to assume there a flat profile and appropriate linear variations with the neutral pressure at the limiter. By doing so, a good and reproductible agreement between the model and the observed variations can be obtained for L-mode, ELM-free and EDA H-mode separately, but L-H transitions could not be rendered without changing the SOL input relation. As probe measurements in the Ohmic regime show significant changes in the SOL profile during mode transitions, this suggests that a better experimental knowledge of this profile could lead to an good understanding of the dominant loading variations in C-Mod.

CONCLUSION AND FURTHER WORK

The ICRF coupling in Alcator C-Mod is strongly correlated with the electron density profile at the edge and in the scrape-off layer. The fast wave cutoff, which is typically located in the limiter shadow, does not play a dominant role, and the coupling is rather determined by the transport barrier height and the scrape-off profile. These variations stem mainly from wave propagation effects in an inhomogeneous plasma, and with a cold plasma 1D slab model, a good agreement with the data can be obtained for all antennas from sufficient knowledge of the density profiles.

The remaining factor of proportionality between the measured coupling resistance and the calculated surface impedance could be considered as a measure of structural antenna performance. While this factor is typically the same for D,E and J operating as two straps antennas, it is twice as high per strap for J operating as a four strap antenna. A possible explanation lies in the high coupling between the straps, which might affect the current pattern and therefore the radiation and coupling properties. This point will be further investigated.

ACKNOWLEDGMENTS

This work is supported by USDOE Coop. Agree. No. DE-FC02-99-ER54512.

REFERENCES

1. Mayberry, M., et al., *Nuclear Fusion*, **30**, 579 (1990).
2. Wukitch, S., et al., , in *Proc. 19th of IAEA Fusion Energy Conference (Lyon) FT/P1-14*, 2002.
3. A. Ram, A. Bers., *Nuclear Fusion*, **24**, 679 (1984).
4. Stix, T., *Theory of plasma waves*, AIP, 1992.

Optimized AT Target Plasma Studies in Alcator C-Mod with I_p-Ramp and Intense ICRH

M. Porkolab,[1] P.T. Bonoli,[1] Y. Lin,[1] S.J. Wukitch,[1] C. Fiore,[1] A. Hubbard,[1] J. Irby,[1] L. Lin,[1] E. Marmar,[1] R. Parker,[1] J. Rice,[1] G. Schilling,[2] J.R. Wilson,[2] S. Wolfe,[1] H. Yuh,[1] K. Zhurovich,[1] and the C-Mod Team

[1] *MIT, Plasma Science and Fusion Center Cambridge MA USA*
[2] *Princeton Plasma Physics Laboratory, Princeton NJ USA*

Abstract. Exploratory experiments have been carried out in the Alcator C-Mod tokamak to develop target plasmas for the upcoming lower-hybrid current drive experiments (LHCD). The goal was to establish reversed shear plasmas with intense ICRF heating early in the current ramp-up phase so as to delay the current diffusion as long as possible while establishing multi-keV electron temperatures, suitable for single pass absorption of lower hybrid waves. The ICRF power was injected from 3 antennas in several different configurations, namely: (1) simultaneous central and low field side off-axis heating; (2) central and high field side off-axis heating; and (3) maximum RF power in the central minority heating mode. Suitable target plasmas for future LHCD have been established and are described below.

INTRODUCTION

The goal of the future C-Mod AT program is to establish and maintain reversed shear plasmas with q_{min} above 2.0, with off-axis lower hybrid current drive localized near r/a=0.7, with beta-normal, β_N, near 3 and bootstrap currents of the order of 70%.[1] It is desirable to form a multi-keV (T_e=3-5 keV) target plasma with ICRH injected as early into the current ramp-up phase as possible before injecting the LHCD. Such a scenario would lead to maximum duration of the reversed shear plasma, allowing long pulse (compared to the L/R time) studies. Discharges with high electron temperatures, reversed shear q-profile and relatively low densities at r/a $\geq$ 0.7 ($n_e \leq 1.5 \times 10^{20}$ m^{-3}) where the LH current is driven, are highly desirable for good accessibility, good single pass absorption and optimized LHCD efficiency. The purpose of the present experiments was to explore the feasibility of forming such target plasmas and study their stability and confinement properties. As described below, several ICRF scenarios have been investigated, including multi-frequency heating techniques so as to explore the possibility of ITB formation with off axis heating, while providing strong central heating. Several advanced diagnostics have been used to characterize the target plasmas: q-profile (MSE diagnostic), plasma rotation (HIREX) and fluctuations (PCI) measurements, but analysis of these measurements is incomplete.

The ICRF power was injected from 3 antennas in several different configurations: (A) simultaneous central and low field side off-axis D(H) heating at 70 and 80 MHz at 5.4 T; (B) central and high field side off-axis D(H) heating at 70 and 80 MHz at 4.5 T; (3) maximum RF power at 78 and 80 MHz in the central minority D(H) heating mode at 5.4T. Initial results from cases A and B were previously reported.[2]

MULTI-FREQUENCY ICRF HEATED TARGET PLASMAS

A. Central and Low-field Side Minority D(H) ICRH Resonance

Operating at 5.4 T central magnetic field, we applied up to 2.5 MW power in a D majority plasma at 80 MHz, with the H minority resonance at 1.7 cm on the low field side from the plasma center, using the E and D-port dipole antennas. Simultaneously, up to 1 MW power at 70 MHz was injected via the J-port 4-strap antenna, with the H minority resonance being at 11.5 cm to the low field side.[3] Typical E-FIT equilibrium is shown at 3 different times in Fig. 1, indicating that early into the ramp the plasma was limited on the inner nose of the divertor plate (up to 0.22 sec) and after 0.24 sec

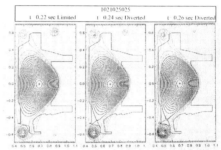

FIGURE 1. E-FIT equilibrium reconstruction at three different times of shot number 1021025005. This time sequence is typical of many of the shots studied.

the plasma became diverted. Typical time evolution of the discharge is shown in Fig. 2, indicating that the plasma entered an ELMy H mode phase at about 0.20 sec, followed by an ELM free H mode after 0.24 sec, with only one or two ELMs. Impurity accumulation followed from 0.24-0.32 sec, when the H mode collapsed. The sawteeth begin at 0.30 sec, and therefore we deduce that q_{min} remained above unity up to this time, with possibly reversed shear at earlier times, and flattening out just before sawteeth appear.[4] Unfortunately, the MSE measurements during these experiments were not working reliably and hence a time evolution of the q profile is unavailable. During these experiments, at ramp-up times of 0.17-0.24 sec, the edge electron temperatures on the top of the pedestal achieved values of 0.8-1.0 keV, and the edge densities remained at or below 1.5×10^{20} m^{-3} while central electron (ion) temperatures rose to 4.0 (3.0) keV and central densities rose to 3×10^{20} m^{-3}. These parameters would, in fact, form suitable target plasmas for off-axis LHCD. It is interesting that with this combination of heating powers, strong ELM phenomena, possibly type-I, was observed on C-Mod, perhaps for the first time.[5] An expanded version of the D-alpha light is shown in Fig. 3. It is believed that the low collisionality in these high edge temperature and low edge density plasmas are suitable to support large ELMs, in contrast to the usual EDA H mode that are collisional.[6] We have also studied the fluctuation spectra, and it is often characterized by high frequencies (250kHz) and relatively low k_R (=1.5 cm^{-1}) in contrast to the usual quasi-coherent mode of the EDA H mode (f=100 kHz, k_R =5-6 cm^{-1}, Ref. 6). Finally, it should be noted that at these low densities no ITBs were observed in the initial 0.3 seconds of the plasma evolution.

B. Central and High-field Side Minority D(H) ICRH Resonance

Operating at 4.5 T central magnetic field, we applied up to 2.0 MW power in a D majority plasma at 80 MHz, with the H minority resonance at –9.6 cm on the high field side from the plasma center, using the E and D- port dipole antennas. Simultaneously, up to 1 MW power at 70 MHz was injected via the J-port 4-strap antenna, with the H minority resonance being at –1.5 cm to the high field side. This scenario was limited in performance by the low central heating power that could be coupled to the plasma (1 MW only). While weak H mode was observed, including ELMs, the edge pedestal temperatures attained only 0.5 keV at edge pedestal densities of 1.0×10^{20}m^{-3}, and the central stored energies were only about 2/3 that of the 5.4T case. This case deserves a repetition with the improved J-port antenna that is able to couple 2.5 MW of power (see below). Again, no ITBs were obtained at the low densities used in these experiments.

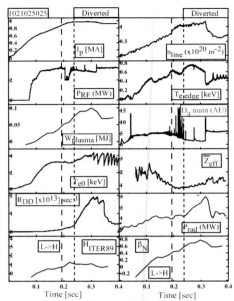

FIGURE 2. Discharge evolution of shot number 1021025025.

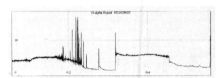

FIGURE 3. Expanded view of the D_α signal, showing large ELMs during the current ramp.

C. Central Minority D(H) ICRH Resonance Case at High Power

This case was studied very recently when the J port antenna performance was improved significantly and up to 2.5 MW of power at 78 MHz could be coupled using the [0,π,π,0] phasing. Up to 2.5 MW power at 80 MHz was coupled simultaneously using the D and E port dipole antennas. However, it is not clear that the overall plasma performance was significantly better than case A, at the same total injected power level. The ELM phenomena was much less obvious in the present case than in case A. It is interesting that the plasma equilibrium was very similar to case A, namely the plasma being limited on the inner divertor nose-plate until 0.24 sec, at which time the plasma became diverted. A typical shot sequence is shown in Fig. 4. It can be seen that H-factor up to $1.7 \times \tau_{ITER89P}$ was obtained, with central electron and ion temperatures of 4.0(3.0) keV at total RF power levels of about 4.0 MW. The edge electron temperature was of the order of 0.6 keV, slightly less than case A with off-axis heating. Note that a weak H-mode is obtained at 0.17 sec as the plasma is still limited by the divertor plate, and a robust H mode is obtained at 0.24 sec, as the plasma becomes purely diverted. Sawteeth did not arise until 0.33 sec, therefore q should be above unity everywhere in the plasma up to this time. Again, no ITB was obtained, however, we did not expect any owing to the purely central heating scenario. This scenario would again form a suitable AT target plasma for LHCD injection.

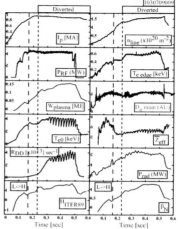

FIGURE 4. Discharge evolution with central heating only at 5.4T.

CONCLUSIONS

We examined three different scenarios for AT target plasma formation with intense ICRH using the D(H) minority heating scenario, before LHCD is injected. We find that at least two cases formed excellent target plasmas, with high edge and central electron and ion temperatures and suitably low edge densities for good lower hybrid wave penetration and absorption. Suitable target plasmas were obtained by initiating intense ICRF plasma heating early in the current ramp, and keeping the density low, by either applying only central, or central plus off-axis heating. ELMs were obtained for the first time in C-Mod by operating in the collisionless regime such that the quasi-coherent mode was not present. In the case of central heating only, we observed the transition from the ELMy H mode regime to the EDA regime dominated by the QC mode with excellent confinement times.

ACKNOWLEDGMENTS

Work supported by US DoE Cooperative Agreement DE-FC02-99ER54512 and Contract DE-AC02-76-CH0-3073. We thank the Alcator technical support staff for their contributions.

REFERENCES

1. P.T. Bonoli et al., Plasma Physics and Controlled Fusion, **31**, 223 (1997).
2. M. Porkolab, et al, Bulletin of APS **47**(9), Orlando, FL, Nov. 2002, pp. 232.
3. S. J. Wukitch et al., presented at the 19th IAEA Fusion Energy Conference, Lyon, France, 14 – 19 October 2002, FT/P1-14.
4. M. Porkolab et al., Proc. 24th EPS, Vol. **21A**, 569, (1997).
5. D. Mossessian et al, Physics of Plasmas **10**, 1720 (2003).
6. A. Mazurenko et al., Phys. Rev. Lett. **89**, 225004 (2002).

Analysis of 4-strap ICRF Antenna Performance in Alcator C-Mod

G. Schilling[1], S. J. Wukitch[2], R. L. Boivin[3], J. A. Goetz[4], J. C. Hosea[1], J. H. Irby[2], Y. Lin[2], A. Parisot[2], M. Porkolab[2], J. R. Wilson[1], and the Alcator C-Mod Team

[1]*Princeton University Plasma Physics Laboratory, Princeton NJ 08543,* [2]*MIT Plasma Science and Fusion Center, Cambridge, MA 02139,* [3]*General Atomics, San Diego CA 92186,* [4]*University of Wisconsin, Madison WI 53706*

Abstract. A 4-strap ICRF antenna was designed and fabricated for plasma heating and current drive in the Alcator C-Mod tokamak. Initial upgrades were carried out in 2000 and 2001, which eliminated surface arcing between the metallic protection tiles and reduced plasma–wall interactions at the antenna front surface. A boron nitride septum was added at the antenna midplane to intersect electric fields resulting from rf sheath rectification, which eliminated antenna corner heating at high power levels. The current feeds to the radiating straps were reoriented from an $E \| B$ to $E \perp B$ geometry, avoiding the empirically observed ~15 kV/cm field limit and raising antenna voltage holding capability. Further modifications were carried out in 2002 and 2003. These included changes to the antenna current strap, the boron nitride tile mounting geometry, and shielding the BN-metal interface from the plasma. The antenna heating efficiency, power and voltage characteristics under these various configurations will be presented.

INTRODUCTION

The antenna design provides four vertical current straps in a configuration that allows efficient heating as well as providing a directed launched wave spectrum for current drive by changes in current strap phasing.[1] An antenna's ability to deliver useful power to the plasma may be limited by the injection of impurities into the plasma or by arcing at high voltage limits. The 4-strap antenna power capability has increased from an initial value of 5 MW/m^2 to ~11 MW/m^2 by eliminating impurity generation and improving high voltage handling.[2,3,4]

IMPURITY GENERATION BY PLASMA-FACING SURFACE INTERACTIONS

Initial antenna operation in 1999 resulted in high levels of metallic impurity influx at heating power levels above ~1.3 MW. The impurity source was identified from the melt damage found upon inspection after the initial commissioning campaign. The molybdenum tiles on separate ground elements had melt damage, while those on the same ground element did not. This suggests a voltage was being developed across tiles on separate ground elements. Induced RF currents of ~25 A resulted in a tile-tile potential of ~100 V at 78 MHz, sufficient to arc across the gaps under the local edge plasma conditions. The gaps were short-circuited in 2000 with stainless steel straps installed underneath the plasma protection tiles, eliminating this problem.

Operation with the metal plasma-facing components was satisfactory, but the level of Mo impurity at the plasma core was found to scale with the rf power. Although the source rate was low, plasma screening was poor.[5] The antenna's plasma protection tiles were therefore changed from the original molybdenum to boron nitride. No deleterious effects have been observed on plasma operation resulting from the boron nitride.

A new front surface interaction limit appeared later in 2000 above 2.5 MW. Camera images of ICRF operation revealed antenna side and corner hot spots that were aligned along the edge magnetic field lines and resulted in impurity injection and disruption (Figure 1). An analysis of the hot spot mechanism suggests that the tokamak's field line pitch in front of the antenna results in nonzero rf magnetic flux linkage to tokamak field lines connecting antenna surfaces. The resulting rf electric field expels electrons, and plasma neutrality results in ion acceleration leading to an enhanced sheath potential.[6] All front protection tiles were realigned with side tiles, all remaining exposed metal surfaces were covered with boron nitride or removed, and a central boron nitride septum was installed to reduce the tokamak field line connection length.

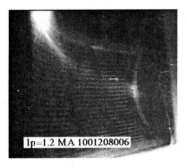

FIGURE 1. Antenna front surface hotspots

Several of the top and bottom tiles fractured during the 2002 run period, with the fragments falling through the plasma into the divertor chamber. The fragments appeared not to have a major impact on the plasma, but the newly exposed metal surfaces reduced the antenna power level before metallic impurity injection set in once more. Disruption forces induced in the metal mounting structure were transmitted to the boron nitride, which yielded under tensile stress. The tiles and their fasteners were redesigned, and no losses have been observed in 2003 so far.

RF-induced arcing was detected in the metal spine supporting the central septum tiles. This was originally designed with slots to reduce induced currents, but sufficient rf voltage developed across the slots in (0,0,0,0) phased operation to arc across the gaps. A new spine without slots was fabricated, and operation in 2003 so far has been successful.

ARCING IN ANTENNA INTERNAL STRUCTURE

During the 1999 operation arcing was observed along the direction of the tokamak magnetic field between the high voltage portion of the antenna current straps and adjacent resistive terminations of the Faraday shields. Grounded stainless steel cups were placed around the base of the Faraday shield rods to protect the resistive terminations in 2000.[7] Subsequent inspections showed no damage.

Extensive arc damage was observed in 2000 between the striplines feeding rf current to the antenna straps, in a direction along the tokamak edge magnetic field. An effective stripline voltage limit of ~15-20 kV in plasma (45 kV in vacuum) limited the antenna heating power to ~2.5 MW. This corresponded to an empirical electric field limit of ~15 kV/cm under the local conditions, i.e. **E∥B**, and plasma edge neutral gas pressure up to ~0.5 mTorr. The mechanism for this breakdown is not clear. Field emission initiation requires local field strengths considerably higher than those present. For gas breakdown, the Paschen curve minimum is ~Torr-cm, while at the antenna we have ~mTorr-cm, with mean free paths much greater than the electrode spacing. Multipactoring initiation would require lower electric fields or greater path lengths.

The striplines had been designed with **E∥B** in order to achieve maximum compactness, but a redesign was performed in 2001 to reorient the striplines to an **E⊥B** configuration (Figure 2). High voltage gaps were increased to reduce electric fields, and in the case of arcing at the current strap crossover, electrodes were reshaped to reorient the region of highest field.

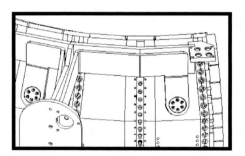 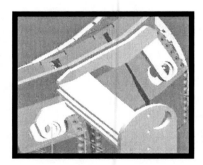

FIGURE 2. Original **E∥B** current feed design (left) and modified **E⊥B** design (right). The tokamak magnetic field is roughly horizontal on left, and rises at ~30° on right.

Series arcing was observed in 2002 in bolted contacts both in the current feeds and the antenna mounting plate. These have been redesigned with more bolts, improved mating surfaces, and copper plating where needed to improve electrical contact.

SUMMARY

C-Mod has presented a challenge to install a high power (~4 MW) 4-strap ICRF antenna in a tight space. Modifications have been made to the antenna plasma-facing surfaces and the internal current-carrying structure. At the present time the antenna has performed up to 3 MW into plasma with heating phasing, with good efficiency and no deleterious effects (Figure 3).

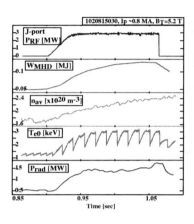
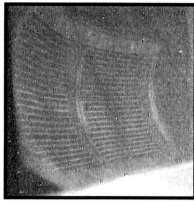

FIGURE 3. 3 MW pulse into C-Mod.

ACKNOWLEDGEMENTS

Work supported by US DoE Contract DE-AC02-76-CH0-3073 and Cooperative Agreement DE-FC02-99ER54512. Engineering support by W. Beck and R. Vieira, MIT PSFC, is gratefully appreciated.

REFERENCES

[1] G. Schilling et al., "Extension of Alcator C-Mod's ICRF Experimental Capability," Proceedings of the 13th Topical Conference on Radio Frequency Power in Plasmas, Annapolis MD, April 1999, 429-432.

[2] G. Schilling et al., "Upgrades to the 4-strap ICRF Antenna in Alcator C-Mod," Proceedings of the 14th Topical Conference on Radio Frequency Power in Plasmas, Oxnard CA, May 2001, 186-189.

[3] S. J. Wukitch et al., "Results and Status of the Alcator C-Mod Tokamak," Proceedings of the 19th IEEE/NPSS Symposium on Fusion Engineering, Atlantic City NJ, January 2002, 290-295.

[4] S. J. Wukitch et al., "Performance of a Compact Four-Strap Fast Wave ICRF Antenna," presented at the 19th IAEA Fusion Energy Conference, Lyon, France, 14 - 19 October 2002, FT/P1-14.

[5] B. Lipschultz et al., "A Study of Molybdenum Influxes and Transport in Alcator C-Mod," Nuclear Fusion **41**, (2001) 585.

[6] J. W. Myra and D. A. D'Ippolito, "Far Field ICRF Sheath Formation on Walls and Limiters," Proceedings of the Tenth Topical Conference on Radio Frequency Power in Plasmas, Boston MA, April 1993, 421-424.

[7] J. A. Goetz et al., "Operation of the Alcator C-Mod 4-Strap Antenna," presented at the 42nd Annual Meeting of the APS Division of Plasma Physics, Quebec City, Canada, October 2000.

Mode Conversion of High-Field-Side-Launched Fast Waves at the Second Harmonic of Minority Hydrogen in Advanced Tokamak Reactors

R. Sund and J. Scharer

University of Wisconsin-Madison
Department of Electrical and Computer Engineering

Abstract. Under advanced tokamak reactor conditions, the Ion-Bernstein wave (IBW) can be generated by mode conversion of a fast magnetosonic wave incident from the high-field side on the second harmonic resonance of a minority hydrogen component, with near 100% efficiency. IBWs have the recognized capacity to create internal transport barriers through sheared plasma flows resulting from ion absorption. The relatively high frequency (around 200 MHz) minimizes parasitic electron absorption and permits the converted IBW to approach the 5th tritium harmonic. It also facilitates compact antennas and feeds, and efficient fast wave launch. The scheme is applicable to reactors with aspect ratios < 3 such that the conversion and absorption layers are both on the high field side of the magnetic axis. Large machine size and adequate separation of the mode conversion layer from the magnetic axis minimize poloidal field effects in the conversion zone and permit a 1-D full-wave analysis. 2-D ray tracing of the IBW indicates a slightly bean-shaped equilibrium allows access to the tritium resonance.

INTRODUCTION

Generation and control of internal transport barriers is central to the development of advanced tokamak reactors. It is generally believed ICRF waves can generate plasma flows through ion absorption; but given the high electron temperatures expected in a reactor, avoidance of electron Landau damping imposes a very low upper limit on usable $k_\parallel$ for waves near the fundamental cyclotron frequencies. For example, at T_e=15 keV and f=40 MHz (Ω_T at B = 7.9 T), $\omega/k_\parallel > 2v_{th,e}$ requires $k_\parallel < 1.7$ m^{-1}. Use of higher-frequency ion-Bernstein waves (IBWs) mitigates this problem. As an IBW approaches a majority ion resonance its wavelength becomes less than a thermal ion gyroradius, its polarization is nearly electrostatic, and its radial group velocity becomes very small, leading to large amplitudes. In consequence, strong ion absorption occurs even at higher cyclotron harmonics (n=5, 6, …). Use of the IBW is also motivated by theory indicating the IBW drives poloidal flow by a ponderomotive mechanism connected to absorption. In addition, if the absorption layer is on the high-field side (HFS), toroidicity converts HFS perpendicular heating to radially separated counter parallel flows on the low-field side (LFS). Experiments on PLT, Alcator-C, PBX-M, and FTU (waveguide antenna) in which IBWs were launched from the outboard edge have yielded confinement increases consistent with IBW-created sheared-flow transport barriers. But direct IBW launch faces many difficulties, including those arising from sensitivity to density and its gradient, sheath effects, parametric decay, ponderomotive density expulsion, and coaxial mode excitation.

Generating an IBW by mode conversion from a fast magnetosonic wave (FW) within the plasma offers a favorable alternative to direct IBW launch. It is possible to achieve nearly complete conversion of a fast wave incident from the high-field side on the 2nd harmonic resonance of a small minority (5%) hydrogen component ($\omega = 2\Omega_H$) in a D/T reactor plasma. Conversion is negligible without hydrogen, but the amount needed is little more than the residual level present in tokamaks.

CONCEPT

As an example, consider the following model reactor. Assume major and minor radii of $R_0 = 6.5$ m and $a = 2.5$ m, and assume $B \sim 1/R$ with $B = 6.07$ T at R_0. Choosing $f = 200$ MHz places $\omega = 5\Omega_T$ at $R = 5$ m. The conversion layer is then slightly to the HFS of the $2\Omega_H / 4\Omega_D / 6\Omega_T$ resonance at $R = 6$ m, which, assuming a modest Shafranov shift of 0.5 m, is 1 m to the HFS of the magnetic axis. Lower aspect ratio is desirable as it reduces the separation between conversion and absorption surfaces as a fraction of minor radius and increases Shafranov shift, both of which allow greater separation between the magnetic axis and the mode conversion zone, thereby minimizing 2-D complications. In our scheme, a fast magnetosonic wave is launched from a HFS antenna situated near the midplane into a D/T plasma with a minority H component. A comb-line antenna could provide the desired k_ϕ spectrum: sharply peaked at a low value. At this frequency the wave can propagate in vacuum up to $k_\phi = 4.2$ m^{-1}, permitting a large antenna/plasma gap. The launched fast wave is unaffected by the $3\Omega_D$ layer in the low-temperature HFS plasma edge; it next passes through the $5\Omega_T$ resonance where absorption is also negligible (owing to the relatively long wavelength and high harmonic). On reaching the mode conversion layer some absorption may occur, primarily on hydrogen, but it can be minimized by spectral control and adequate hydrogen density. Transmission is negligible. The mode-converted IBW then propagates back toward the $5\Omega_T$ resonance.

In our analysis, the conversion dynamics are modelled by two coupled 2nd-order linear differential equations for E_x and E_y ($E_z = 0$) based on a 2nd-order gyroradius expansion. In the limit $k_\parallel = 0$ these can be written in the form of the standard tunneling equation, with a tunneling factor given by: $\eta = 8.62 \, \eta_H F(\eta_H) n_{e20}^{1/2} \beta L$ where $\beta = .0806 \, n_{e20} T_H / B^2$, $F = (A+B)^{1/2} (B/A - 1)^{5/2}$, $A = \frac{8}{15}\eta_D + \frac{12}{35}\eta_T + \frac{4}{3}\eta_H$, $B = 2 + \frac{2}{15}\eta_D + \frac{2}{35}\eta_T + \frac{2}{3}\eta_H$, $\eta_j = n_j/n_e$, $n_{e20} = n_e$ in 10^{20} m^{-3}, T in keV, B in Tesla, and L is the magnetic scale length in m. The transmitted FW power fraction is $e^{-2\eta}$. For $\eta_D = \eta_T = (1-\eta_H)/2$, F varies almost linearly over the range of interest from $F(0) = 44.3$ to $F(0.1) = 27.2$. Under reactor conditions a small amount of hydrogen has a large effect. Even at $\eta_H = 4\%$, $n_{e20} = 1$, $T = 10$, we have $\eta = 1.4$ and transmission is only 6%. Physically, $\eta \sim k_{\perp\infty} \Delta$ where $k_{\perp\infty}$ is the non-resonant FW wave number and Δ is the width of the conversion zone. Since $k_{\perp\infty} \propto n_e^{1/2}$ it is apparent the bulk plasma acts to broaden the conversion zone, i.e., it causes the mode conversion point to move well away from the resonance toward the high-field side. The effect is proportional to the plasma beta if H is at the bulk temperature. For finite $k_\parallel$, a wave with a not-too-large $k_\parallel$ encounters the mode conversion point before the

Doppler-broadened H absorption zone. As $k_\parallel$ increases, the latter widens, absorption increases and mode conversion diminishes, establishing an upper limit on usable $k_\parallel$. For other parameters fixed, the minimum η_H is established by requiring that the mode conversion fraction exceed some value for some maximum $k_\parallel$. Although the Doppler width increases as $T_H^{1/2}$, the conversion zone widens as T_H, so parasitic H absorption actually decreases with increasing temperature.

NUMERICAL CODE AND RESULTS

We have written a 1-D full-wave code to calculate the fields, power absorption profile and scattering parameters for this scenario. The wave modes are independent at the boundaries of the solution interval. B_z and $k_\parallel$ both vary as $1/(R_0 + x)$. Since the fast wave is well characterized by a single, real $k_\perp$ at the boundaries, the expression for kinetic flux from weak damping theory is valid for that mode. Therefore the mode conversion fraction can be computed by subtracting the transmitted and absorbed powers from the incident (reflection is negligible on the HFS). Although the code computes absorption only for H, this is reasonable because: i) D absorption is smaller by at least a factor of $(n_D/n_H)/400$ for the cases considered, and T absorption is always negligible; and ii) cases of primary interest are those for which absorption is low—for these, any small relative error in absorption will contribute a very small error to the mode conversion coefficient. It must be noted, however, that a number of high-beta cases considered here are beyond the range of high accuracy for equations based on second-order gyroradius expansions and only give qualitative guidance.

The solution interval (xL, xR) is divided in two at xM, a point chosen a bit to the LFS of the mode conversion point. Direct integration yields a transfer matrix relating field values at xL and xM. Invariant embedding is used over the LFS interval (xM, xR) where the IBW modes become evanescent, making straightforward ODE integration numerically unstable in either direction. The original equations are transformed into modal form by a similarity transformation. The mode that grows exponentially to the right, call it IBW$^+$, and the left-propagating FW, call it FW$^-$, are both zero at xR. Each is represented as a

TABLE 1. Various mode conversion results. ($k_{\parallel\,ant}$ in m^{-1} is specified at $R = 4$ m)

case	n_{e20}	% H	T	$k_{\parallel\,ant}$	%Trans	%Abs	%MC
1	1.4	5	5	5	7.2	12	80.7
2	1.4	5	20	5	0.02	1.9	98
3	1	5	10	5	4	37	59
4	2	5	5	5	1.5	14	84
5	2	2.5	10	5	1	37	62
6	2	5	10	5	0.04	1.7	98
7	2	5	10	7	0.05	14.1	85.8
8	2	5	20	7	0	2	98
9	2	2.5	20	7	0.03	61.4	38.5
10	2	2.5	20	5	0.04	14.7	85
11	2	2.5	20	3	0.04	0.6	99
12	2	3.5	20	5	0	1.5	98.5

linear combination—with x-dependent coefficients—of the other two modes, IBW$^-$ and FW$^+$. Substitution into the mode equations yields 4 coupled Riccatti equations for the 4 unknown coefficients that are stable to integration from xR to xM and have zero initial values. Substituting the invariant embedding representations of IBW$^+$ and FW$^-$ into the original mode equations yields a pair of coupled linear ODEs for IBW$^-$ and FW$^+$ stable to integration from xM to xR.

IBW RAY TRACING

As an IBW ray comes close to an ion harmonic, it propagates primarily parallel to B with a sense governed by the sign of $k_\| = \mathbf{k} \cdot \mathbf{B}/B$. Rotational transform and flux-surface poloidal curvature cause up-shifts in $k_\|$ on one side of the midplane and down-shifts on the opposite side. This process is especially sensitive to magnetic geometry near resonance where $|\mathbf{k}|$ becomes very large. In a tokamak with a conventional D-shaped cross section, most ray trajectories on the HFS curve away from the midplane, shifting toward ever increasing values of $|k_\||$. Consequently, the rays are completely absorbed by electron Landau damping (ELD) before reaching the ion resonance. If the equilibrium has a slight bean shape, with poloidal curvature on the midplane reversing sign between the mode conversion and tritium resonance surfaces, a ray near the midplane in the reverse-curvature region will oscillate in poloidal and toroidal angle and in $k_\|$. Thus, $|k_\||$ can remain bounded below the ELD level, allowing the ray to reach the tritium resonance.

We have written a 2-D ray-tracing code based on the full 3x3 hot-plasma dispersion tensor. Figure 2 shows the $k_\|$ evolution of two rays started 20 cm above and below the midplane, just after conversion. ($R_0 = 6.4$ m, $a = 2.2$ m, other parameters as above). Both rays have an initial k_ϕ corresponding to 4 m^{-1} at the antenna. Reverse shear is modelled using a magnetic pitch angle proportional to radius and set to match the pitch of a q=2 surface at the point where the $5\Omega_T$ resonance intersects the midplane. In general, results for reverse-shear equilibria indicate only a very slight bean distortion is needed to assure IBW penetration to the tritium resonance.

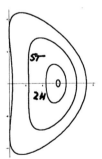

Figure 1. Flux surfaces for slight bean (surfaces through 5T and 2H included)

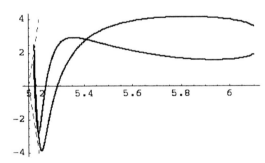

Figure 2. $k_\|$ vs R for rays with $k_{\phi\,ant} = 4$ m and y=±20 cm (strong T absorption begins left of dashed lines)

ACKNOWLEDGMENTS

This work was supported in part by The University of Wisconsin – Madison.

Ion Heating in Field-Reversed Configuration by Radio-Frequency Waves

V. A. Svidzinski, S. C. Prager

Physics Department, University of Wisconsin-Madison, WI 53706

Abstract. A simple modeling of the recent experiment on plasma heating by radio-frequency (rf) waves in the field-reversed configuration (FRC) is made. In the FRC Injection Experiment device [1] the ion heating by rf pulse was observed. Present analysis indicates that the heating can be explained by the Doppler broadening of the ion cyclotron resonance in the low magnetic field device. These results support the suggestion that the direct heating of the ion majority near the fundamental ion cyclotron frequency should be efficient in low field configurations.

The heating of the ion majority at the fundamental ion cyclotron resonance is considered to be inefficient in machines with relatively high magnetic field such as tokamaks. The polarization of the launched rf wave is right-handed (non resonant with ions) at the location of the resonance. We made a more detailed discussion of this subject in our recent theoretical analysis [2] in which we studied the possibility of direct heating of plasma ions in reversed field pinches (RFP) by rf waves at the fundamental ion cyclotron resonance. An RFP operates with a relatively low magnetic field (typically an order of magnitude smaller than in tokamaks). We demonstrated that this heating method should be effective in RFPs for a range of plasma parameters. A more general conclusion was that such heating method can be effective in other configurations with low magnetic field.

When the magnetic field is low the ion cyclotron resonance in a hot nonuniform plasma is more spatially broadened. The condition

$$\left| \frac{\omega - \omega_{ci}}{k_\| v_{Ti}} \right| \lesssim 1 \tag{1}$$

is satisfied in a wider spatial region for smaller magnetic field because the difference $\omega - \omega_{ci}$ is smaller for smaller ω_{ci} (when $\omega \sim \omega_{ci}$). In the vicinity of the cold plasma resonance (location where $\omega = \omega_{ci}$) there is a finite amount of left-handed polarization of the rf wave. This polarization is resonant with ion gyromotion. Generally, with the increase of distance from the resonance the fraction of left-handed polarization increases. In a hot plasma in a low magnetic field the spatial overlap of the broadened resonance and left-handed polarization is relatively large so that ion heating can be efficient.

From the above consideration one can expect that it is generally possible to heat ions at $\omega \approx \omega_{ci}$ in plasma in a low magnetic field. Recent experiment on plasma heating in the field-reversed configuration (FRC) supports this conclusion. The plasma was heated in the confinement section of the FRC Injection Experiment (FIX) device by a low-frequency magnetic pulse [1]. The energy of the pulse was absorbed mainly by ions. In discussion in Ref. [1] it was suggested that the heating is due to excitation of slow (shear) Alfvén wave which is resonant at $\omega = \omega_{ci}$. The slow wave, however, does not propagate efficiently across magnetic field (see, e. g., Ref. [3]) such that the power supplied by an antenna goes into fast (compressional) Alfvén wave which is not resonant at $\omega = \omega_{ci}$. The difficulty of the excitation of the slow wave by an antenna located on the outside of the plasma relates to a poor heating efficiency of the ion majority in tokamaks at $\omega = \omega_{ci}$. Thus the suggested explanation of the ion heating in Ref. [1] seems to be not reliable. We perform simple rf modeling of the plasma in this experiment in order to show that the observed ion heating can be explained by the broadened resonance in the low magnetic field, so that the experimental result supports the suggestion about the effectiveness of the ion majority heating at $\omega \approx \omega_{ci}$ in low magnetic field configurations.

The experimental parameters of FIX experiment are given in Ref. [1]. We use these parameters in our model. We consider a cylindrical deuterium plasma in a vessel of radius $a = 40$ cm. The one-dimensional rigid rotor equilibrium model [4],[5] is adopted. In this model only the z component of equilibrium magnetic field is present. The radial profiles of the plasma density and of the equilibrium field are $n(r) = n_m \text{sech}^2 K \left(r^2/R^2 - 1 \right)$, $B_z(r) = B_0 \tanh K \left(r^2/R^2 - 1 \right)$. In our calculations we use $n_m = 4 \cdot 10^{13} \text{cm}^{-3}$, $B_0 = 400$ G, $K = 0.59$, $R/a = 0.35$, $T_i = T_e = 80$ eV. For these parameters the density and magnetic field profiles are shown on Fig. 1. We assume that the antenna induces rf fields with frequency $f = 8 \cdot 10^4$ Hz and azimuthal and axial wave numbers $m = 0$ and $k_z a = 2$ respectively.

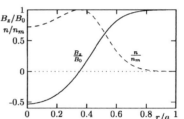

FIG. 1.

An accurate rf modeling of the FRC plasma is a complicated task. In the conditions of the FIX experiment the radial locations where $\omega = \omega_{ci}$ are at $r/a = 0.25$ and $r/a = 0.42$. Between these points and the point of zero magnetic field ($r/a = 0.35$) the higher ion cyclotron harmonics are located. For an accurate treatment of this region finite Larmor radius effects have to be included. Near the point where $B_z = 0$ there are locations of electron cyclotron resonance. Both this resonance and the strong shielding of rf fields in this region result in a very small wave length of rf field. Significantly enhanced numerical resolution of the region near the zero of magnetic field is required. The electron collision rates are approximately three times larger than the wave frequency ω. Thus the electrons are strongly collisional while the ions are collisionless in the experimental conditions. No simple accurate model is available to treat this situation.

We examine here the effects due to the broadened ion cyclotron resonance. In our simple treatment we take the limit $k_\perp \rho_{Li} \to 0$ and do not include the effects due to higher harmonics. We perform a Fourier transform of θ and z coordinates. The ion dielectric response is local with respect to the r coordinate and it is described by the hot plasma dielectric tensor in the limit $k_\perp \rho_{Li} \to 0$. We use two models for the electron dielectric response. In the first case the electrons are described by the collisionless hot plasma dielectric tensor and in the second case they are described by the collisional cold plasma dielectric tensor. These two models for electrons are not very accurate for the parameters of the experiment. On the other hand if these quite different models result in similar conclusions about ion heating then this conclusion can be considered as a reliable one.

The hot plasma dielectric tensor in the limit $k_\perp \rho_L \to 0$ has components [3]

$$K_1 = 1 + \frac{1}{2}\sum_\alpha \frac{\omega_{p\alpha}^2}{\omega^2} \zeta_\alpha [Z(\zeta_{\alpha+}) + Z(\zeta_{\alpha-})], \quad K_2 = -\frac{i}{2}\sum_\alpha \frac{\omega_{p\alpha}^2}{\omega^2} \zeta_\alpha [Z(\zeta_{\alpha+}) - Z(\zeta_{\alpha-})],$$

$$K_3 = 1 - \sum_\alpha \frac{\omega_{p\alpha}^2}{\omega^2} \zeta_\alpha^2 Z'(\zeta_\alpha), \text{ where } \zeta_\alpha = \frac{\omega}{|k_z|v_{T\alpha}}, \quad \zeta_{\alpha\pm} = \frac{\omega \pm \omega_{c\alpha}}{|k_z|v_{T\alpha}}, \quad v_{T\alpha} = \sqrt{\frac{2T_\alpha}{m_\alpha}},$$

α corresponds to ions and electrons, $\omega_{p\alpha}$, $\omega_{c\alpha}$ are the plasma and cyclotron frequency, Z is the plasma dispersion function. The collisional cold plasma dielectric tensor (see, e. g., Ref. [6]) in the second model for electrons has a collision frequency profile $v_e/\omega = 3 \cdot n(r)/n_m$.

We assume that antenna induces an electric field of a unit amplitude with time dependence $\propto \exp(-i\omega t)$ and wave numbers m and k_z on the inner surface of the vessel parallel to the θ direction. We assume a uniform temperature profile in the plasma and find the power absorbed per unit volume as $P = \langle \mathbf{j} \cdot \mathbf{E} \rangle$, see, e. g., Ref. [6]. The Maxwell equations with the described plasma dielectric response are solved numerically by a finite difference method.

Figure 2 shows the radial profiles of the absolute values of arguments of the plasma dispersion function ζ_{i-}, ζ_i, ζ_{i+}. The broadened ion cyclotron resonance [approximately defined by Eq. (1)] covers the region $0 \leq r/a \lesssim 0.5$. If there is a substantial amount of left-handed polarization somewhere in this region, then one should expect ion heating there. There can be also some additional ion heating because ζ_i, corresponding to the Cerenkov resonance, is relatively small as well.

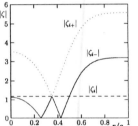

FIG. 2.

Figures 3(a)-(c) are the numerical results for the model with hot collisionless electrons and Figs. 3(d)-(f) are corresponding results for cold collisional electrons. Figures 3(a),(d) show the radial profiles of the total power absorbed per unit volume by ions and electrons P in arbitrary units (solid line). The dashed line is the power absorbed by electrons. In both models the ion heating is dominant. The most power is absorbed by ions in the region of broadened cyclotron resonance. A relatively small part of this ion absorption is due to the Cerenkov resonance. Electron heating is localized near the electron cyclotron resonance ($r/a \approx 0.35$) on Figs. 3(a),(d) and near the Alfvén resonance in the model with cold electrons on Fig. 3(d).

The radial profiles of the electric field amplitudes E_r, E_θ in the two models are presented in Figs. 3(b),(e). The wave length near the point where $B_z = 0$ ($r/a = 0.35$) is very small; the wave structure near this location is hardly resolved on these figures. A large number of grid points is required for the appropriate numerical resolution in this region. Because the shielding of electric fields is strong there, the wave penetration beyond the point of zero magnetic field is suppressed. With these experimental parameters substantial rf power penetrates to the region of ion cyclotron resonance. There is an Alfvén resonance structure near $r/a \approx 0.6$ (at this point $\omega/k_z \approx v_A$) on Fig. 3(e). The absence of a similar structure on Fig. 3(b) is because of the strong Doppler broadening of the resonance ($\omega/k_z v_{Te} \ll 1$). The Alfvén resonance appears in the model with the hot electrons for significantly smaller temperatures.

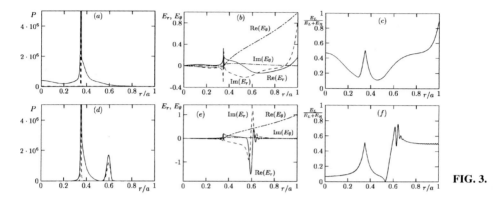

FIG. 3.

Figures 3(c),(f) show the radial profiles of the ratio $E_L/(E_L + E_R)$, where $E_{L,R}$ are the absolute values of left-handed and right-handed parts of wave amplitude (relative to the direction of magnetic field). In the region of broadened ion cyclotron resonance ($0 \leq r/a \lesssim 0.5$) there is a substantial amount of left-handed polarization. When this is combined with the wave accessibility to this region, the ion heating becomes significant.

Both models for electrons result in the same conclusion about ion heating. When the ion temperature is reduced, the wave polarization becomes purely right-handed at the point where $\omega = \omega_{ci}$ (but there is a left-handed component in the vicinity of this point) which means that the slow wave is not directly involved in the ion heating. Reduction of ion temperature or removing the ion cyclotron resonance from the plasma by increasing the wave frequency leads to the suppression of ion heating (both in absolute terms and relative to the electron heating) in both models. Thus, the ion heating observed in the experiment can be explained by the thermal broadening of the ion cyclotron resonance in this low field configuration. This experimental result and our analysis support the conclusion that the heating of the ion majority near the fundamental ion cyclotron frequency should be efficient in low field configurations such as RFPs, FRCs and spherical tokamaks.

REFERENCES

1. K. Yamanaka, S. Yoshimura, K. Kitano, S. Okada and S. Goto, Phys. Plasmas **7**, 2755 (2000).
2. V. A. Svidzinski and S. C. Prager, Phys. Plasmas **9**, 1342 (2002).
3. D. G. Swanson, *Plasma Waves* (Academic, Boston, 1989).
4. W. T. Armstrong, R. K. Linford, J. Lipson, D. A. Platts and E. G. Sherwood, Phys. Fluids **24**, 2068 (1981).
5. R. L. Morse and J. P. Freidberg, Phys. Fluids **13**, 531 (1970).
6. M. Brambilla, *Kinetic Theory of Plasma Waves, Homogeneous Plasmas* (Oxford University Press, Oxford, 1998).

Overview of Alcator C-Mod ICRF Experiments

S.J. Wukitch,[1] P.T. Bonoli,[1] C. Fiore,[1] I.H. Hutchinson,[1] Y. Lin,[1] E. Marmar,[1] A. Parisot,[1] M. Porkolab,[1] J. Rice,[1] G. Schilling,[2] and the Alcator C-Mod Team

[1] *MIT, Plasma Science and Fusion Center Cambridge MA USA*
[2] *Princeton Plasma Physics Laboratory, Princeton NJ USA*

Abstract. In C-Mod, internal transport barriers (ITB) have been formed in both ohmic and off-axis ICRF discharges. With ~0.6 MW of central ICRF, the density and impurity peaking is controlled and allows the ITB to be maintained for the duration of the ICRF. Mode conversion in both D(H) and H(^{3}He,D) has been investigated and the first measurement of D(H) mode conversion has been recently reported. A four-strap ICRF antenna has been shown to operate at high power density in heating phasing. Modifications have reduced the RF electric fields parallel to the magnetic field and shielded the BN-metal interfaces from the plasma. Voltage and power limitations were observed in the two dipole antennas and arc damage was found again where $E \parallel B$. Modifications have been completed and the initial antenna performance is presented.

INTRODUCTION

Ion cyclotron range of frequencies (ICRF) heating is the primary auxiliary heating source on Alcator C-Mod.[1] One of the principle interests is to use the ICRF for modifying and controlling internal transport barrier (ITB) modes plasma through pressure profile modification. These discharges have no central fueling, no external torques, and equilibrated temperatures. These conditions are similar to conditions expected in reactor grade plasmas. With the development of a phase contrast imaging diagnostic to measure density fluctuations at the RF frequency and a sophisticated simulation code, TORIC, to simulate and predict experiments, experiments have been performed to study the mode conversion process in both H(^{3}He,D) and D(H) plasmas. Another area of emphasis is ICRF antenna performance and reliability. With three antennas, two 2-strap and one four-strap fast wave antennas, the voltage, power, and impurity production characteristics of the antennas have been documented with each modification made to the antennas. In the following, the recent results from these research areas are presented.

INTERNAL TRANSPORT BARRIER EXPERIMENTS

In C-Mod, ITB discharges have been observed in both ohmic and ICRF heated H-mode discharges.[2,3,4] For ITB formation in ICRF heated discharges, an EDA H-mode is required, but not sufficient condition, and the cyclotron resonance needs to be located at r/a>0.5 to either the high field or low field side of the plasma center. The ITB forms during the current flat top phase of the discharge and the discharges can remain sawtoothing throughout. An example of an ITB discharge is shown in Fig. 1. As in a typical H-mode, the stored plasma energy and density increase with the transition to H-mode. Unlike a typical H-mode, the density continues to increase and the profile becomes peaked. An ITB discharge is defined as one where the density peaking factor exceeds 1.5 in H-mode. From these signals, improved thermal confinement is inferred from the maintenance of the central temperature with the doubling of the central density. The improved thermal confinement has been confirmed with TRANSP analysis. In addition, the central rotation often reverses sign with respect a nominal H-mode discharge.

These discharges typically show a steep density gradient in the plasma core as shown in Fig 2. With the application of central heating ~ 0.6 MW the density evolution is controlled. Here, the central ICRF power is applied after the barrier is allowed to develop for ~0.3 sec or 10 energy confinement times. The density accumulation is arrested and the profile maintained for the duration of the ICRF. For power levels >0.6 MW, the density profile decays back to a typical H-mode density profile.

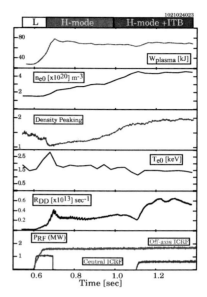

Figure 1, Typical ITB discharge showing the increase in central density and density peaking with approximately constant central temperature with both on and off-axis ICRF heating.

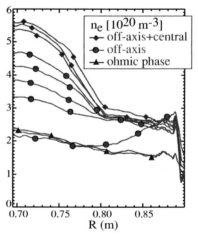

Figure 2, Density profile control with application of central ICRF power during an ITB discharge.

MODE CONVERSION EXPERIMENTS

In multi-species plasmas, ICRF mode conversion (MC) can compete with minority heating. A common ICRF MC scenario is where the fast wave mode converts to a short wavelength ion Bernstein wave at the ion-ion hybrid layer. In addition to competing with minority absorption, MC electron heating has been shown to be localized, FWHM ~0.2r/a, and current drive has been measured on a number of experiments. Recent advances in diagnostics and simulation codes have allowed experiments to directly investigate the MC process. The key diagnostic is a phase

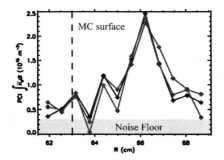

Figure 3 Raw PCI data appears to have wavelength ~1.5 cm and fluctuations appear on the low field side of the ion-ion hybrid layer.

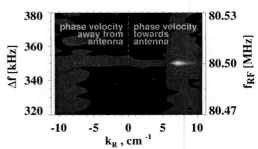

Figure 4 Contour plot of Fourier transformed PCI data showing a wave traveling towards the antenna with a wavenumber between 5-10 cm^{-1}.

contrast imaging system: a line integrated fluctuation diagnostic sensitive to wave numbers 0.5 <|k_R|< 10 cm^{-1}.[5] Furthermore, the addition of a high-resolution ECE radiometer, ~0.7 cm, allows detailed measurements of the electron power deposition profile.[6]

In the H(^{3}He,D) discharges, the PCI data indicated ~1.5 cm wavelength mode located to the low field side of the ion-ion hybrid layer as shown in Fig 3.[7] Simulations with the full wave code TORIC[8,9] and AORSA,[10] completed independently, indicated a wave

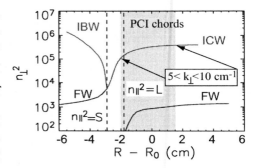

Figure 5 Dispersion curves in the mode conversion region showing the FW, IBW, and ICW.

structure with the appropriate wavelength and location was present. Numerically solving the full electromagnetic dispersion relation showed that an electromagnetic ion cyclotron wave ICW first predicted by Perkins[11] was excited at the ion-ion hybrid layer. The measured wave number was in good agreement with the corresponding wave number from the dispersion relation seen in Figs 4 and 5. The ICW damping could result in a lower MC current drive efficiency due to trapping effects and may make poloidal flow using MC waves possible.

Although D(H) MC scenario has been previously experimentally investigated, direct evidence of localized electron power deposition, as predicted by simulations, has been lacking. Recent experiments with H/D~ 20% on C-Mod have provided the first direct evidence of MC electron heating.[12]

ANTENNA PERFORMANCE

The ICRF system is the only auxiliary heating system on C-Mod. Thus, reliable antenna operation with minimum negative plasma impact is critical to the success of the C-Mod physics program. C-Mod has a compliment of three fast wave antennas, D,

E,[13] and J-port.[14] The primary limitations have been associated with voltage handling and impurity injections from the BN tile-metal interfaces exposed to plasma.

J-port initially suffered from lower than expected voltage handling compared with D and E-port antennas. This limit was experimentally determined to be located to antenna elements where the RF E-field exceeded 15 kV/cm where the RF E-fields were parallel to the tokamak B-field (E∥B). Initial modifications were restricted to orientating the RF E-fields perpendicular to the B-field in the J-port power feeds. The maximum voltage was raised to 25 kV from 17 kV at 78 MHz. Furthermore, the maximum voltage was found to increase to 30 kV at 70 MHz. After modification of the current strap, the J-port antenna has operated up to 3 MW and 35 kV into plasma in heating phasing. In separate experiments, the heating efficiency, density and impurity production were found to be comparable to the D and E-port antennas.

Modifications to the D and E-port Faraday screen resulted in a longer electrical length antenna strap. This moved higher voltage to ~20 Ω section of coaxial vacuum transmission line. Arc damage was found on the coax center conductor where E∥B and the 15 kV/cm empirical breakdown limit was exceeded. This section of line was modified by reducing the center conductor diameter and resulted in a ~30% decrease in the maximum E-field in this region.

The BN material was selected to replace the original Mo protection tiles to minimize the Mo impurity source. Injections however from exposed BN-metal interfaces resulted in large injections that often caused plasma disruptions. The solution was to shield the BN-metal interface from the plasma. Since making these modifications the antennas have been free of BN-metal interface injections.

ACKNOWLEDGMENTS

Work is supported by D.o.E Coop. Agreement DE-FC02-99ER54512. The authors would like to thank the engineering and technical staff of the Alcator C-Mod team for support in antenna modifications and plasma operations.

REFERENCES

[1] I. H. Hutchinson et al., *Phys. Plasmas* **1**, 1511 (1994).
[2] S.J. Wukitch et al., Physics of Plasma **9**, 2149 (2002).
[3] J. E. Rice et al., Nuclear Fusion **42**, 510 (2002).
[4] C. Fiore et al., Bull. of the Amer. Phys. Soc. **47**, 138 (2002).
[5] A. Mazurenko, PhD thesis, Massachusetts Institute of Technology (2001).
[6] J. W. Heard, C. Watts, R. F. Gandy, P. E. Phillips *et al*, Rev. Sci. Instrum. **70** (1), 1011 (1999).
[7] E. Nelson-Melby et al., Phys. Rev. Lett. **90**(15), 155004(2003).
[8] M. Brambilla, Nucl. Fusion **38**, 1805 (1998).
[9] M. Brambilla, Plasma Phys. Control. Fusion, **41** (1), 1(1999).
[10] E.F. Jaeger *et al*, Phys. Rev. Lett. (2003), accepted for publication.
[11] F. W. Perkins, Nucl. Fusion **17**, 1197 (1977).
[12] Y. Lin et al., Plasma Phys. and Control. Fusion (2003), accepted for publication.
[13] Y. Takase et al., Proc. 14th Symp. Fusion Eng., San Diego (Piscataway, NJ):IEEE (1992), p118.
[14] S. J. Wukitch, et al, Proc. 19th IAEA Fusion Energy Conf. (Lyon) FT/P1-14 (2002).

FAST WAVE CURRENT DRIVE

Characterization of Fast Ion Absorption of the High Harmonic Fast Wave in the National Spherical Torus Experiment [1]

A. L. Rosenberg, J. E. Menard, J. R. Wilson, S. Medley, C. K. Phillips, R. Andre, D. S. Darrow, R. J. Dumont, B. P. LeBlanc, M. H. Redi[a], T. K. Mau[b], E. F. Jaeger, P. M. Ryan, D. W. Swain[c], R. W. Harvey[d], J. Egedal[e], and the NSTX Team

[a] *Princeton Plasma Physics Laboratory, Princeton, NJ 08543-0451*
[b] *University of California at San Diego, La Jolla, CA 92093*
[c] *Oak Ridge National Laboratory, Oak Ridge, TN 37831*
[d] *CompX, Del Mar, CA 92014*
[e] *Plasma Science and Fusion Center, Massachusetts Institute of Technology, Cambridge, MA 02139*

Abstract. Ion absorption of the high harmonic fast wave in a spherical torus is of critical importance to assessing the viability of the wave as a means of heating and driving current. Analysis of recent NSTX shots has revealed that under some conditions when neutral beam and RF power are injected into the plasma simultaneously, a fast ion population with energy above the beam injection energy is sustained by the wave. In agreement with modeling, these experiments find the RF-induced fast ion tail strength and neutron rate at lower B-fields to be less enhanced, likely due to a larger β profile, which promotes greater off-axis absorption where the fast ion population is small. Ion loss codes find the increased loss fraction with decreased B insufficient to account for the changes in tail strength, providing further evidence that this is an RF interaction effect. Though greater ion absorption is predicted with lower $k_\parallel$, surprisingly little variation in the tail was observed, along with a small neutron rate enhancement with higher $k_\parallel$. Data from the neutral particle analyzer, neutron detectors, x-ray crystal spectrometer, and Thomson scattering is presented, along with results from the TRANSP transport analysis code, ray-tracing codes HPRT and CURRAY, full-wave code and AORSA, quasi-linear code CQL3D, and ion loss codes EIGOL and CONBEAM.

INTRODUCTION

Interaction between the high harmonic fast wave (HHFW) and energetic particles in a spherical torus (ST) [1] is a new and important research area. A fast ion population of fusion-born alpha particles will be found in a reacting plasma, along with energetic ions from neutral beam injection (NBI) in some scenarios. HHFW is currently being

---

[1] This work was supported by the United States Department of Energy under Contract No. DE-AC02-76CH03073.

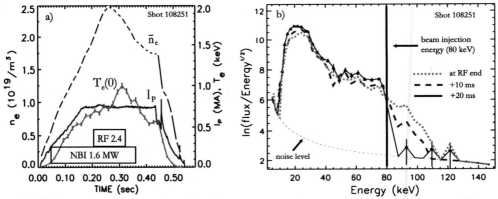

FIGURE 1. a) Typical line-averaged density, temperature, plasma current, NBI, and RF power profiles for scans in recent experiments. b) Neutral particle analyzer data for NSTX shot 108251, launched $k_{\parallel}$ = 14 m^{-1} (heating phasing), after RF turns off at t = 320 ms. Fast ion population above 80 keV beam injection energy collapses to a typical no-RF distribution within 20 ms.

explored as a means of heating and driving plasma current. In NSTX's most recent campaigns, a clear fast ion tail was observed on the neutral particle analyzer (NPA) when HHFW and NBI were active simultaneously. Neutron detector and ion loss probe signals provided further evidence for interaction. This occurred for nearly every shot there was a significant overlap in RF and NBI power traces. Ray-tracing was used to analyze these shots, and found absorption by fast ions to be competitive with electron absorption. Measured neutron rates for similar RF and no-RF shots were also compared with predicted rates, and a significant RF-induced enhancement was found, consistent with the enhanced tail.

EXPERIMENTAL STUDIES

For all shots analyzed, the neutral beam injected deuterium into the plasma at $E_{beam} \approx$ 80 keV, $P_{beam} \approx$ 1.6 MW. Without RF, the energy spectrum observed by the NPA dropped out above ~80 keV. With RF, the energy spectrum extended to ~130 keV. Furthermore, after RF turnoff with NBI remaining active, the tail decayed to the no-RF spectrum on a time scale comparable to that for decay of a beam-only distribution, as seen in Figure 1. The ZnS and fission neutron detectors also saw a significant signal enhancement with RF. This signal began dropping immediately upon RF turnoff as well. As shown in Figure 2a, for similar shots with and without RF, within 25 ms of RF turnoff the enhanced neutron rate decays to the no RF value.

To further examine the fast ion absorption dependence on various parameters, scans in B_0 and $k_{\parallel}$ were performed. These shots all had similar electron temperature and density profiles, so the scan in B_0 effectively became a β_t scan. As demonstrated in Figure 3, the neutron rate was found to decrease with decreasing toroidal field, and the fast ion tail on the NPA dropped to nearly a no-RF spectrum. In Ref. [2], total

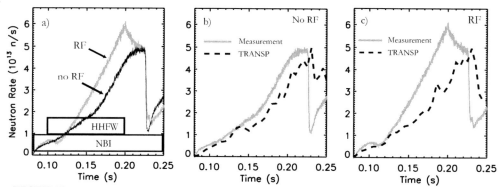

FIGURE 2. a) Neutron rates for NSTX shots 105906 (no RF) and 105908 (RF). b) Neutron rate for no RF vs. TRANSP prediction. After RF turns off, rate decays close to measured and predicted no RF value. c) Measured neutron rate for RF shot significantly exceeds TRANSP prediction without RF input.

absorption is predicted to increase with β, and because the fast ion population is quite centralized, these experimental results are consistent with theory as less power from an antenna on the outboard side of the plasma would be available to core fast ions in a high β shot. Though greater ion absorption is predicted with lower $k_\parallel$ [3,4], surprisingly little variation in the tail was observed, along with a small neutron rate enhancement with higher $k_\parallel$. This discrepancy may be due to different edge coupling conditions at different antenna phasings. The antenna performance and reliability have recently been significantly improved, so this will be reinvestigated in the next NSTX campaign.

THEORETICAL STUDIES

For analysis, the TRANSP [5] transport analysis code was used to calculate fast ion energy and particle density profiles, and this information was used to estimate an effective Maxwellian temperature for the fast ion population. These profiles, along with EFIT and Thomson data, were fed into HPRT, a 2-D ray-tracing code which uses the full hot plasma dielectric with complex k to compute power deposition profiles along the hot electron/cold ion ray path [3]. As shown in Figs. 4a and 5a, fast ion absorption was calculated to be competitive with electron absorption in sustained neutral beam shots, often taking $\sim$35% of the total RF power. Figures 4a-c show that these results and profiles matched those of CURRAY [6], an independently developed ray-tracing code, and AORSA [7], an all orders full wave code, reasonably well for RF+NBI equilibria. CQL3D [8], a quasi-linear code which currently uses a model of the fast ion distribution function rather than an effective Maxwellian, has also calculated 35% ion absorption for this shot, and finds a similar ion deposition profile. A comparison of typical ray paths in HPRT vs. wave front propagation calculated by AORSA is shown in Figure 6. Both codes launch waves from identical antenna geometry on the outboard side of the plasma. The flow of these paths are remarkably similar considering the quite different methods of calculation. It also helps demonstrate the importance of including

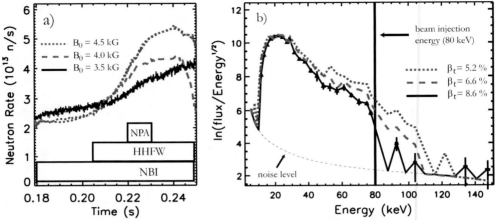

FIGURE 3. a) Neutron rates for otherwise similar NSTX shots at $B_0 = 3.5$, 4.0, and 4.5 kG, which correspond to $\beta_t = 5.2$, 6.6, and 8.6%, respectively. b) NPA signals at $R_{tan} = 70$ cm for B_0 scan, averaged over time window displayed in a). The larger β profile at lower B_0 may promote greater off-axis electron absorption, reducing the fraction of power available to the centralized fast ion population.

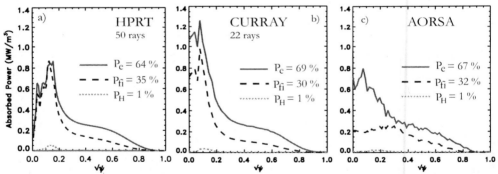

FIGURE 4. Power deposition profiles for NSTX shot 108251, t = 235 ms, $n_\phi = 24$ from a) HPRT, 50 rays, b) CURRAY, 22 rays, and c) AORSA. Good agreement is found in fractional absorption and profiles between codes for RF+NBI shots.

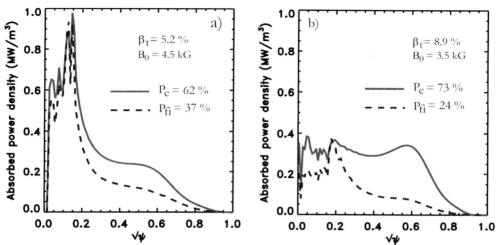

FIGURE 5. a) HPRT power deposition profiles for NSTX shot 108250, t = 230 ms, $B_0 = 4.5$ kG, $\beta_t = 5.2\%$. b) Shot 108252, t = 230 ms, $B_0 = 3.5$ kG, $\beta_t = 8.9\%$. Lower on-axis absorption is calculated for lower B, higher β, in agreement with neutron rate and NPA signals.

2D effects in absorption calculations for ST equilibria. Figure 5 shows HPRT calculated deposition profiles for the experimental magnetic field scan. Lower on-axis absorption is calculated for lower B, or higher β, in agreement with the neutron rate and NPA signals in Figure 3, as the greater off-axis electron absorption in higher β shots would prevent as much RF power from reaching the bulk of the fast ion population near the magnetic axis. An increased fast ion loss fraction at lower B may also contribute to suppression of the observed tail, so loss codes EIGOL [9], which follows the full fast ion orbit, and CONBEAM [10], which follows the guiding center of the orbits and accounts for finite larmor radius effects, were used to determine the significance of this factor. For 120 keV ions, EIGOL calculates a loss fraction of 17% for a $B_0 = 4.5$ kG equilibrium - NSTX shot 108250, t = 235 ms, and 23% for $B_0 = 3.5$ kG - shot 108252, t = 235 ms. CONBEAM calculates a loss fraction of 21% for $B_0=4.5$ kG, and 25% for $B_0=3.5$ kG. According to the NPA data, the tail is suppressed by at least a factor of 8 between high and low field, so the tail reduction is more likely due to an RF effect.

Without providing RF input in either case, TRANSP was also used to calculate the neutron rates for similar RF and no-RF shots. As shown in Figure 2, the measured rate matched the prediction well in the no-RF case, and for the RF shot grew to nearly double the predicted rate. It then decayed to approximately the computed rate after RF turnoff. Figure 7a demonstrates that a single effective Maxwellian, matching fast ion particle and energy density exactly, fits the TRANSP $f(E)$ remarkably well, however it exceeds the neutron rate calculated from the TRANSP distribution function by ~20%. Figure 7b shows that the single Maxwellian has too large of a contribution to $\int_0^E dE f(E)\sigma v$, which is proportional to the neutron rate, above the beam injection energy, 80 keV. To model this neutron rate enhancement, multiple Maxwellians may

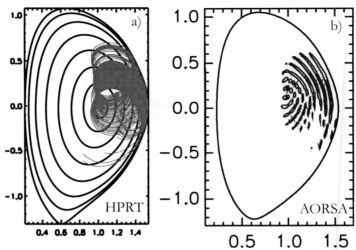

FIGURE 6. a) 50 HPRT ray paths, with initial positions distributed evenly over poloidal range of antenna, for NSTX shot 108251, t = 235 ms, $\beta = 5.2\%$, $n_\phi = 24$. Each ray stops when 99% of its power is absorbed. b) AORSA wave fronts for same shot.

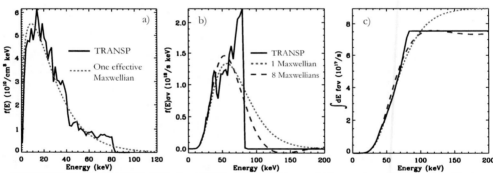

FIGURE 7. a) TRANSP $f(E)$ for NSTX shot 108251, t = 235 ms, r/a = 3% vs. an effective Maxwellian matching its total energy and particle density exactly. The neutron rate from this Maxwellian exceeds TRANSP's by $\sim 20\%$. b) TRANSP $f(E)\sigma v$ vs. one effective Maxwellian, as well as an 8 Maxwellian $f(E)$. c) TRANSP $\int_0^E dE f(E)\sigma v \propto$ neutron rate vs. other two functions. Although $f(E)\sigma v$ cannot be fit well with isotropic, unshifted Maxwellians used in most wave power absorption codes, total neutron rate can, and $\sim 80\%$ of neutrons created below energy range of largest discrepancy, so using this $f(E)$ to estimate neutron rate may have some validity.

provide a more accurate estimate. Although $f(E)\,\sigma v$ cannot be fit well with isotropic, unshifted Maxwellians used in most wave power absorption codes, total neutron rate can. As shown in Figure 7c, ~80% of neutrons are created below the energy range of largest discrepancy in $f(E)\sigma v$, so using this $f(E)$ to estimate neutron rate may have some validity. The next steps in calculating a rough neutron rate enhancement are to compute the fast ion power deposition profiles and confinement times for each Maxwellian, perturb the temperature of each accordingly, and recalculate the neutron rate with the pertubed distribution function. CQL3D may also be used to calculate this enhancement, and will soon be able to read in the TRANSP distribution function to provide more quantitative agreement between experiment and theory.

For equilibria without fast ions, low $n_\phi \approx 6$, and high thermal $T_i(0) \approx 2$ keV, less agreement between wave absorption codes has been found, often differing in bulk deuterium absorption by a factor of ~2, with HPRT consistently calculating less than CURRAY or AORSA. This is surprising considering the level of agreement for the fast ion shots which actually had $T_i(0) \approx 1.5$ keV. CURRAY and HPRT have been found to agree in ray path and deposition profile quite well for RF+NBI and equilibria dominated by electron absorption. A possible explanation is that HPRT is using the complex k in its dielectric tensor, whereas CURRAY does not, and the impact only arises in this particular parameter regime. As outlined in [11], HPRT's equation for absorbed power density

$$\frac{\partial W_{ps}}{\partial t} = \mathbf{E}_1^* \cdot \mathbf{j}_s - \nabla \cdot \mathbf{T}_s \qquad (1)$$

thus includes the latter kinetic flux term:

$$\nabla \cdot \mathbf{T}_s = \frac{\omega}{8\pi} \mathbf{E}_1^* \cdot \left(\mathbf{k}_i \cdot \frac{\partial}{\partial \mathbf{k}} \overline{\overline{\epsilon}}_{Hs} \cdot \mathbf{E}_1 \right) e^{2\phi_i}. \qquad (2)$$

while CURRAY, because it keeps k real in the evaluation of its dielectric tensor, sets $\partial W_{ps}/\partial t = \mathbf{E}_1^* \cdot \mathbf{j}_s$. It should be noted that a rigorous derivation from the energy moment of the Vlasov equation to this form of the kinetic flux term with complex k has yet to be performed. Noting that there is currently discrepancy between theory and experiment in the $k_\parallel$ scan (there has yet to be any experimental evidence of significant bulk ion absorption at all in NSTX), and there is a discrepancy between ray-tracing codes at lower $k_\parallel$, a breakdown of the theory at lower $k_\parallel$ should not be ruled out.

CONCLUSIONS

Analysis of recent NSTX shots has revealed that under some conditions when neutral beam and RF power are injected into the plasma simultaneously, a fast ion population with energy above the beam injection energy is sustained by the wave. In agreement with modeling, these experiments find the RF-induced fast ion tail strength and neutron rate at lower B-fields to be less enhanced, likely due to a larger β profile, which promotes greater off-axis absorption where the fast ion population is small. Ion loss codes EIGOL and CONBEAM find the increased loss fraction with decreased B insufficient to account for the changes in tail strength, providing further evidence that this is an RF interaction effect. Though greater ion absorption is predicted with lower $k_\parallel$, surprisingly little variation in the tail was observed, along with a small neutron rate enhancement with higher $k_\parallel$. Comparison between ray-tracing codes HPRT and CURRAY, full-wave code AORSA, and quasilinear code CQL3D have found good agreement in deposition profile and fractional absorption for RF+NBI shots. Ray paths and wave fronts match remarkably well between the ray tracing codes and AORSA. Less agreement is found between HPRT and CURRAY in lower $k_\parallel$, no-NBI equilibria with high thermal ion temperature. An effective Maxwellian matches the fast ion distribution function below ~60 keV, but to match the neutron rate more Maxwellians are needed. Future work includes attempting to gain more quantitative agreement between experiment and theory by determining how the fast ion distribution function is perturbed by RF and recalculating the neutron rate, either by using several Maxwellians or CQL3D.

REFERENCES

1. Y.-K. M. Peng and D. J. Strickler, Nucl. Fusion **26**, 769 (1986).
2. M. Ono, Phys. Plasmas **2**, 4075 (1995).
3. J. Menard, R. Majeski, R. Kaita, M. Ono, and T. Munsat, Phys. Plasmas **6**, 2002 (1999).
4. C. N. Lashmore-Davies, V. Fuchs, and R. A. Cairns, Phys. Plasmas **5**, 2284 (1998).
5. J. P. H. E. Ongena, M. Evrard, and D. McCune, Fusion Technology **33**, 181 (1998).
6. T. K. Mau, C. C. Petty, M. Porkolab, and W. W. Heidbrink, Modeling of fast wave absorption by beams in DIII-D discharges, in *Radio Frequency Power in Plasmas: 13th Topical Conference*, AIP Conference Proceedings 485, page 148, Annapolis, Maryland, 1999, American Institute of Physics, New York.
7. E. F. Jaeger, L. A. Berry, E. D'Azevedo, D. B. Batchelor, and M. D. Carter, Advances in full-wave modeling of rf heated, multi-dimensional plasmas, in *Radio Frequency Power in Plasmas: 14th Topical Conference*, AIP Conference Proceedings 595, page 369, Oxnard, California, 2001, American Institute of Physics, New York.
8. R. W. Harvey and M. G. McCoy, The CQL3D Fokker-Planck code, in *Proceedings of IAEA Technical Committee Meeting on Advances in Simulation and Modeling of Thermonuclear Plasmas*, Montreal, 1992, available through USDOC, NTIS no. DE9300962.
9. D. Darrow, R. Akers, D. Mikkelsen, S. Kaye, and F. Paoletti, Modeling of neutral beam ion loss from NSTX plasmas, in *Proceedings of the 6th IAEA Technical Committee Meeting on Energetic Particles in Magnetic Confinement Systems*, page 109, Ibaraki-ken, Japan, 2000, Japan Atomic Energy Res. Inst.
10. J. Egedal, M. H. Redi, D. S. Darrow, and S. M. Kaye, Phys. Plasmas **10** (2003).
11. J. E. Menard, C. K. Phillips, and T. K. Mau, Ion absorption effects in high-harmonic fast wave ray tracing theory, in *Radio Frequency Power in Plasmas: 13th Topical Conference*, AIP Conference Proceedings 485, page 345, Annapolis, Maryland, 1999, American Institute of Physics, New York.

Observations of Anisotropic Ion Temperature during RF Heating in the NSTX Edge

T.M. Biewer*, R.E. Bell*, D.S. Darrow* and J.R. Wilson*

Princeton Plasma Physics Laboratory, Princeton, NJ 08543

Abstract. A new spectroscopic diagnostic on the National Spherical Torus Experiment (NSTX) measures the velocity distribution of ions in the plasma edge with both poloidal and toroidal views. An anisotropic ion temperature is measured during the presence of high power HHFW RF heating in He plasmas, with the poloidal T_i roughly twice the toroidal T_i. Moreover, the measured spectral distribution suggests that two populations have temperatures of 500 eV and 50 eV with rotation velocities of -50 km/s and -10 km/s, respectively. This bi-modal distribution is observed in both the toroidal and poloidal views (in both He II and C III ions), and is well correlated with the period of RF power application to the plasma. The temperature of the edge ions is observed to increase with the applied RF power, which was scanned between 0 and 4.3 MW. The ion heating mechanism from HHFW RF power has not yet been identified.

INTRODUCTION

A new spectroscopic diagnostic with both toroidal and poloidal views has been implemented in the edge of NSTX plasmas. This edge rotation diagnostic (ERD) was designed to measure the velocity and temperature of ions in the edge of NSTX. Correlated with HHFW, a strong heating of He II and C III ions is observed. The apparent temperature is anisotropic; the T_i viewing poloidally is roughly twice the T_i measured with the toroidal view. Moreover, a bi-modal distribution of these ions is observed. The presence of two distributions of ions is consistent with calculated ionization and thermalization times for He II. No direct method of edge RF ion heating has yet been identified.

OBSERVATIONS

The edge rotation diagnostic (ERD)[1] measures the intrinsic emission of light from the plasma edge. There are 7 toroidally directed views and 6 poloidally directed views of the outboard plasma edge. The poloidal views are ~ 20 cm (toroidally) from the RF antenna, and the toroidal views are ~ 2 m away. The intersection of the diagnostic sightline with the intrinsic emission shell provides the localization of the measurement. The sightlines are nearly tangent to the flux surfaces. The C III triplet near 4651 Å and the He II line at 4685 Å are measured. In the results presented here, Helium is the bulk "working" ion of the discharge.

The National Spherical Torus Experiment (NSTX)[2] is a large spherical tokamak[3] with a major radius of 0.85 m and a minor radius of 0.65 m. Pulse lengths for these

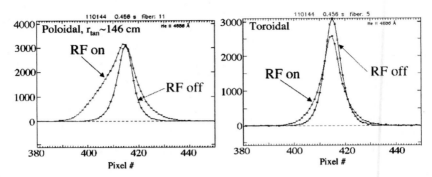

FIGURE 1. The He II spectrum from Shot 110144 during adjacent 10 ms time slices (centered on 455 and 465 ms), showing the difference in the poloidal view and toroidal view when HHFW RF heating is applied to the plasma. RF heating (4.3 MW) ends at 460 ms. Both views are tangent to the flux surface at ~ 146 cm. Error bars indicate statistical uncertainty.

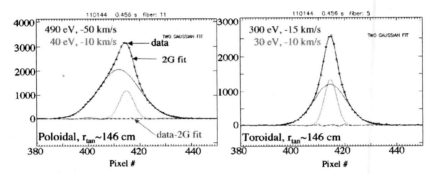

FIGURE 2. The He II spectrum under the influence of RF heating, fit with two Gaussians.

NSTX discharges are $\simeq 600$ ms, with an on-axis toroidal magnetic field of ~ 0.3 T. The plasma current is 500 kA. The on-axis electron temperature and density are ≤ 2 keV and $\sim 2 \times 10^{19}$ m^{-3}, respectively with ≤ 4.3 MW of HHFW RF auxiliary heating[4].

During the application of 30 MHz RF heating, phased to drive current, a distortion to the spectrum of both He II (as shown in Fig. 1) and C III is observed. The distortion is more pronounced in the poloidal view, but it is also present in the toroidal view. Under the influence of RF power, the spectrum is clearly non-Maxwellian. However, fitting the spectrum with two Gaussians yields a very accurate representation of the measured data, as shown in Fig. 2, suggesting "hot" and "cold" components are present.

Fig. 3 shows the time evolution of NSTX Shot 110144. HHFW heating is applied from 60 to 460 ms. The effect of the RF heating is apparent in the edge measured brightness, velocity, and rotation. After the RF is applied, ~ 100 ms are needed to establish a new equilibrium rotation, as indicated by the velocity trace. Once an equilibrium is established, however, the plasma is very robust, as indicated by the near constant level of velocity and temperature. When the RF heating is terminated, the hot component

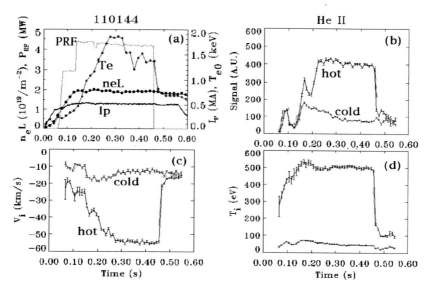

FIGURE 3. Time evolution of NSTX Shot 110144, showing (a) I_p, T_e, n_e, and P_{RF} and the hot and cold components of poloidally measured He II (b) brightness, (c) velocity, and (d) temperature at a tangency radius of 146 cm.

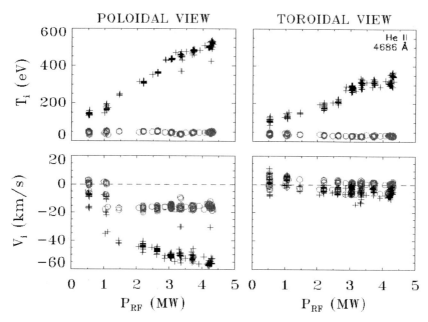

FIGURE 4. The temperature of the "hot" component scales with P_{RF}, whereas the temperature of the "cold" component is not effected by the amount of RF power applied to the plasma. He II data from Shots 110133-110145 are shown.

disappears promptly from the plasma. Fig. 4 shows how T_i and v of the hot and cold components vary with the amount of RF power that is applied to the plasma.

DISCUSSION

The observed anisotropic temperature is consistent with ions having a large perpendicular energy content. One interpretation of the observed bi-modal distribution, is that there are two populations of He ions in the plasma. In this scenario, the RF creates (by some unidentified mechanism) a high-temperature population of He II ions. The emission time scale is ~ 1 ns, implying that light from both populations (hot and cold) would be readily observed, since the time scale for thermalization between two populations of He ions is ~ 10 ms. However, the time scale for ionization is ~ 100 μs. Hence, thermalization between the hot and cold populations would not be observed before ionization occurs. Presumably, thermalization occurs among fully stripped He III ions, which do not emit.

Key to this interpretation of the observed bi-modal distribution is the heating mechanism of the RF which creates the hot component of He ions. The HHFW launched by the NSTX antenna is not expected to heat edge ions[5], but core electron heating is expected and observed. Resonant heating at the ion cyclotron frequency (27th sub-harmonic for He, and 41st for C) is unlikely. One possibility for ion heating is parametric decay of the launched HHFW into a daughter HHFW and an ion quasi-mode at the fundamental ion cyclotron resonance. Ion heating could then occur either directly at the fundamental cyclotron frequency or by stochastic heating. In the near-field of the antenna, a compressional wave could be generated which may stochastically heat He (and C) ions[6]. Attempts to invert the poloidal brightness assuming constant emissivity on a flux surface fail, i.e. negative emissivities are obtained. Localized emission near the midplane can explain the observed brightness profile. Single particle simulations suggest that hot, trapped ions with high perpendicular energies have banana-orbits which are primarily localized to the midplane, increasing emission there.

The authors wish to recognize the many contributions of the NSTX group and collaborators at Oak Ridge National Lab. This research was supported by the U.S. D.O.E. under contract: DE-AC02-76CH03073.

REFERENCES

1. Biewer, T. M., Bell, R. E., Feder, R., Johnson, D. J., and Palladino, R. W., in preparation (2003).
2. Spitzer, J., Ono, M., Peng, M., Bashore, et al., *Fusion Technology*, **30**, 1337 (1996).
3. Peng, Y.-K., and Strickler, D., *Nuclear Fusion*, **26**, 769 (1986).
4. Ono, M., *Phys. Plasmas*, **2**, 4075 (1995).
5. Majeski, R., Menard, J., Batchelor, et al., in *Radio Frequency Power in Plasmas–Thirteenth Topical Conference, Annapolis, MD*, AIP Press, New York, 1999, p. 296.
6. Ono, M., and Peng, M., personal communications (2003).

Fast Wave Current Drive in JET ITB-Plasmas

T. Hellsten[1], M. Laxåback[1], T. Johnson[1], M. Mantsinen[2], G. Matthews[3], P. Beaumont[3], C. Challis[3], D. van Eester[4], E. Rachlew[1], T. Bergkvist[1], C. Giround[3], E. Joffrin[5], A. Huber[3], V. Kiptily[3], F. Nguyen[5], J.-M. Noterdaeme[6], J. Mailloux[3], M.-L. Mayoral[3], F. Meo[6], I. Monakhov[3], F. Sartori[3], A. Staebler[6], E. Tennfors[1], A. Tuccillo[7], A. Walden[3], B. Volodymyr[6] and JET-EFDA contributors.

[1]*Association VR-Euratom, Sweden,*
[2]*HUT, Association Euratom-Tekes, Finland,*
[3]*UKAEA Fusion Association, Culham Science Centre, U.K.,*
[4]*LPP-ERM/KMS, Association Euratom-Belgian State, Belgium,*
[5]*Association Euratom-CEA, CEA-Cadarache, France,*
[6]*Max-Planck-Institut für Plasmaphysik, EURATOM Association,, Garching, Germany.*
[7]*Euratom-ENEA, Italy.*

Abstract. FWCD experiments have been studied on JET in plasmas with electron internal transport barriers, eITB, produced with LHCD. Because of the low single pass damping the current drive is found to be strongly reduced because of parasitic ^{3}He cyclotron damping and that a large fraction of the power is not absorbed in the plasma.

INTRODUCTION

Fast wave current drive, FWCD, is a tool to influence the central current in tokamak plasmas. The weak single pass damping for TTMP/ELD requires high temperatures and large plasmas for efficient current drive. A scenario to test FWCD on JET has been developed based on electron internal transport barriers, eITB, with strongly reversed magnetic shear obtained with LHCD preheat. The high electron temperature should result in a high efficiency and allow studies of how the current drive can influence the transport barrier. In spite of an electron temperature of 8keV the single pass damping for TTMP/ELD was low (1%), which resulted in substantial parasitic absorption by cyclotron damping on residual ^{3}He and in a significant fraction of the power not being absorbed in the plasma.

EXPERIMENTAL RESULTS

Fast wave current drive experiments were carried out in strongly reversed ITB plasmas in deuterium obtained with LHCD preheat in a single-null divertor configuration at a magnetic field of 3.45T and a plasma current of 2MA. Up to 6MW of ICRF power was coupled at a frequency of 37MHz for which the ratio of the phase velocity and the thermal velocity was about 2. The heating performance was compared with heating at 51MHz for which the hydrogen resonance passed through the mid-plane at about 2.9m. For 37MHz the ^{3}He resonance intersected the mid-plane at $R=2.8$m, the hydrogen resonance was located outside the plasma and the deuterium resonance near the plasma edge at the high field side, $R=2.1$m. NBI-, LH- and ICRH-powers, central electron, ion temperature and electron density versus time are shown in Fig. 1 for discharge 58681. Hot plasmas with low density and strongly reversed magnetic shear were created with 2MW LHCD preheat. The LHCD was switched off when the NBI and RF were coupled into plasmas. Due to NBI fuelling the electron temperature decreased, which increased the current diffusion. The initial phase of the discharges, up to 47s, were similar, after that larger discrepancies appeared when an outer strong barrier developed at different times and with different strengths as q_{min} reached 2. This caused a rapid increase in the ion temperature and the density. To avoid disruptions the NBI power was stepped down if the neutron yield exceeded a pre-set value. The dynamics of the discharges were sensitive to small differences in heating, which changed the current diffusion rate, the transport in the barrier, the strength of the sawteeth and bootstrap currents. The increase in the RF-power from 4.5MW to 6MW gave only a small increase in the electron temperature in the initial phase, but the barrier developed rather differently. With 6MW of RF-heating the barrier became weaker and the NBI was not stepped down resulting in a high electron temperature also in the latter phase of the discharge.

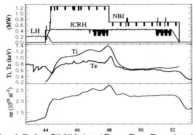
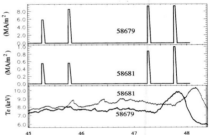

Fig. 1 Pulse 58681 (top)P_{NBI}, P_{RF}, P_{LH}, (top) T_e and T_i (middle) n_e (bottom).

Fig. 2 central current density measured at 4 times and T_e versus time.

The effects of heating and current drive were seen by comparing the pair of discharges 58679 and 58681 with similar heating, but different ICRH phasings; -90° and +90° respectively. The central current density obtained from the motional Stark effect, MSE, diagnostic measured at four different times at the beginning of the ICRH heating is shown in Fig. 2. For reference also the electron temperature measured at 15cm below the magnetic axis is shown in Fig. 2. Initially there is a small difference in the evolution of the current near the magnetic axis consistent with a less fast increase of the current density in 58681 compared to 58679. The evolutions of the current profiles were also subjected to redistribution of the current by sawteeth and development of bootstrap currents in the barrier. The sawteeth appearing at an integer q-surface in the reversed magnetic shear region will throw in current contrary to normal sawteeth for equilibria with positive shear. The central current will also be affected by the increase in bootstrap current in the barrier as the high confinement barrier is formed.

The RF-power deposition on electrons by TTMP/ELD was obtained from Fourier and break-in-slope analyses of the electron temperature measurements during modulation of the RF-power. The total fraction of the ICRF power damped directly on electrons was only 10-15% of the total power coupled to the antennae. A larger fraction was in general seen at the first modulation phase between 47s and 48s than in the second between 51s and 52s, which is consistent with the lower electron temperature at the later time. Although the power density in the outer part of the plasma is low, it can still be significant because of the large volume. However, it will not significantly affect the driven current due of the low current drive efficiency there.

Precautions to avoid ^{3}He were taken in order to avoid parasitic absorption which, because of the weak damping by TTMP/ELD could be considerable. The presence of ^{3}He was detected from a gamma camera and an RF-probe measuring the cyclotron emission. From the anisotropic energy content in the plasma measured by the difference between the diamagnetic energy and MHD equilibrium measurements the presence of fast ions could also be confirmed. The fast energy contents for +90° and −90° phasings were both in the order of 0.2MJ, although for +90° it was slightly higher than for −90°. Also the central electron temperature was slightly higher, Fig. 2, in agreement with the better heating with +90° phasing due to the inward RF-pinch.

To identify the cause of the low fraction of power absorbed by TTMP/ELD we compared the heating at 37MHz with heating at 51MHz, 58682, for which the wave was absorbed by hydrogen cyclotron heating near the centre. Heating at 51MHz produced higher central electron temperatures. An energy balance was carried out by comparing the total energy delivered by the heating system, including ohmic heating, and the sum of the energy delivered into the divertor, measured with 6 thermocouplers, and the integrated radiated power in the main chamber, measured with the bolometer camera. At 37MHz there was larger systematic differences between the energy delivered to the antenna and that radiated from the plasma and delivered to the tiles compared to the 51MHz case, see Table. Thus a large fraction of the RF-power was not absorbed in the plasma.

Table

No	comments	phase	absorbed power %
58679		-90°	32
58680	no NBI	+90°	43
58681		+90°	48
58682	51MHz	+90°	95
58684	6MW	+90°	71

MODELLING OF HEATING AND CURRENT DRIVE

RF-simulations were done with the SELFO code, which calculates the wave field and the distribution function of the resonant ions self-consistently including the effects of RF-induced spatial diffusion and finite drift orbit width [1, 2]. The toroidal mode spectrum was modelled with 80% of the power on $n_\phi=15$ and 20% on $n_\phi=-40$ for +90° and with reversed sign for −90°. Because of the weak single pass damping by TTMP/ELD a significant fraction of the power can be absorbed by ^{3}He. The ^{3}He concentration was low, but difficult to estimate, making predictions uncertain as the absorption increases with the concentration. The uncertainty of the measured fast energy is too large to accurately determine the power absorbed on ^{3}He. Simulations with 0.25% ^{3}He and with 2.25MW absorbed power in the plasma, which for 58681(+90°) equalled the measured averaged power absorbed in the plasma, gave about 65% of the power absorbed by ^{3}He, 28% by TTMP/ELD and less than 10% by D-beams. For the same assumptions on concentration and power we obtained for

58679 (−90°) that 33% is absorbed by ^{3}He, 60% by TTMP/ELD and less than 10% by D-beams. The averaged driven current is about 80kA for 58679 and 35kA for 58681, half of that is driven inside r =0.25m. For this choice of ^{3}He concentration and lost power the fast energy content and the power damped by TTMP/ELD become similar to that measured for 58681. For 58679 the computed power damped by TTMP/ELD is about a factor of 2 larger and the fast energy content a factor of 0.5 smaller than measured. By varying the concentration and absorbed power within the experimental uncertainties simulation and experiment can be brought to agreement also for 58679.

For 58682 with 51MHz, which has strong single pass damping, almost all RF-power was absorbed in the plasma, whereas a considerable fraction is lost for heating with 37MHz with low damping. To study the correlation between the single pass damping and lost power we assume the power partition scales as proportional to the single pass damping. For parameters corresponding to the early phase of the discharges with 0.25% of thermal ^{3}He ions the single pass damping is 0.25% for ^{3}He and 1% for TTMP/ELD. When the ions are heated the single pass damping increases to about 3.3% for 58681(+90°) and about 1.5% for 58679 (−90°). The increase in the damping is due to the increased ^{3}He absorption mainly due to FLR effects and the large ratio of $|E_-/E_+|$. The difference in damping between the discharges is caused by the RF-induced transport. The single pass damping for 58680 (+90) without NBI, which has a much lower electron temperature, 3.8keV, is about 2.0%. By comparing the lost power from the table one finds that for low single pass damping the lost power is correlated with the single pass damping, but with large scattering of the data.

CONCLUSIONS

Of the 4.5MW of power delivered to the antennae only about 15% was absorbed by TTMP/ELD producing currents of the same order as measured with MSE. However, the evolution of the central plasma current was also influenced by current diffusion, redistribution by sawteeth and bootstrap current. Parasitic loss by cyclotron heating of residual ^{3}He ions was large. By comparing the energy delivered by the heating system and time integrated ohmic power with the radiated energy and the energy delivered to the divertor for heating at with hydrogen minority heating with TTMP/ELD and ^{3}He minority heating one can conclude that a large fraction of RF-power was not absorbed in the plasma, which for low single pass damping increased with decreasing damping. Since the lost power was neither radiated nor delivered into the divertor, it cannot come from power being absorbed inside the last closed flux surface or in the SOL region connected to the divertor through magnetic field lines. Thus it had to be absorbed at the RF-antennae or the walls e. g. in rectified RF-sheaths enhanced by weak damping [3]. Increase of RF sheaths effects for weak damping has been seen [4].

REFERENCES

[1] J. Hedin et al in Theory of Fusion Plasmas 1998, Varenna 1999, pp. 467.
[2] M. Laxåback et al 14th Topical Conf on RF Freq. Pow. in Plasmas, Oxnard 2001, pp. 415
[3] T. Hellsten and M. Laxåback, this conference
[4] D. A. D'Ippolitio et al, Nuclear Fusion **42** (202)1357

High-Harmonic Fast-Wave Driven H-Mode Plasmas in NSTX*

B.P. LeBlanc[a], R.E. Bell[a], S.I. Bernabei[a], K. Indireshkumar[a], S.M. Kaye[a],
R. Maingi[b], T.K. Mau[c], D. W. Swain[b], G. Taylor[a], P. M. Ryan[b],
J.B. Wilgen[b], J.R. Wilson[a]

[a] *Princeton Plasma Physics Laboratory, Princeton, New Jersey 08543*
[b] *Oak Ridge national Laboratory, Oak Ridge, Tennessee 37830*
[c] *UCSD, San Diego, California*

Abstract. The launch of High-Harmonic Fast Waves (HHFW) routinely provides auxiliary power to NSTX plasmas, where it is used to heat electrons and pursue drive current. H-mode transitions have been observed in deuterium discharges, where only HHFW and ohmic heating, and no neutral beam injection (NBI), were applied to the plasma. The usual H-mode signatures are observed. A drop of the D_α light marks the start of a stored energy increase, which can double the energy content. These H-mode plasmas also have the expected kinetic profile signatures with steep edge density and electron temperature pedestal. Similar to its NBI driven counterpart – also observed on NSTX – the HHFW H mode has density profiles that feature "ears" in the peripheral region. These plasmas are likely candidates for long pulse operation because of the combination of bootstrap current, associated with H-mode kinetic profiles, and active current drive, which can be generated with HHFW power.

INTRODUCTION

The application of High Harmonic Fast Wave (HHFW) constitutes an important element of the NSTX research program, where it is used to heat bulk electrons and pursue non-inductive drive current [1,2]. Substantial progress has been achieved over the results presented at the previous meeting of this conference [3], and effective heating has been achieved in helium and deuterium plasmas for different antenna $k_{//}$. In particular electron temperature, T_e, up to 3.9 keV has been measured, with profile behavior suggestive of a thermal electron internal transport barrier [4]. NSTX operates naturally at high beta, with parameters entailing wave physics with dielectric constant $\varepsilon \equiv \omega_{pe}^2/\Omega_e^2 \approx 50\text{-}100$, which is large compared to conventional tokamak, where $\varepsilon \approx 1$. For such high ε–value plasmas, an attractive fast-wave window opens in the high harmonic frequency range, $\Omega_i << \omega << \omega_{LH}$, which permits electron heating and current drive [5]. A welcome result has been the observation of H-mode transition during HHFW heating. Such transitions are readily observed when HHFW provides the sole source of auxiliary heating. So far all the H-mode transitions have been observed with lower single null configuration (LSN) and at plasma current lower or equal to 0.5 MA. Attempts made at higher current were not successful, but a systematic study has not been performed to date. In this paper we review some of the parameters of these plasmas and make use of a recent implementation of the ray tracing code CURRAY [6] into the data regression code TRANSP [7] to study the time dependent power deposition under HHFW driven H-mode conditions.

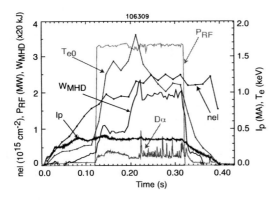

Figure 1. Time evolution of an HHFW driven H-mode discharge. The transition is seen at 0.195 s on the D_α trace; stored energy WMHD doubles. A drop in Te0 occurs during the ELM activity.

HHFW DRIVEN H-MODE PLASMAS

The relevant parameters for a HHFW driven H-mode discharge in deuterium are shown in Fig.1. The plasma current is 0.36 MA, and the magnetic field is 0.45 T. HHFW power of 3.3 MW is applied during interval 0.12-0.32 s. The HHFW frequency is 30 MHz with $k_\parallel = 14$ m^{-1}. T_{e0} rapidly responds to the HHFW power by increasing from 0.3 keV to nearly 1.5 keV in 0.05 s. The H transition occurred at 0.195 s and was accompanied by further heating of the electrons and a doubling in the stored energy. The decrease in central electron temperature observed later on could have resulted from power-coupling losses caused by MHD activity or the ELMs (visible on the D_α trace). TRANSP analysis predicts a 40% bootstrap current fraction for this discharge.

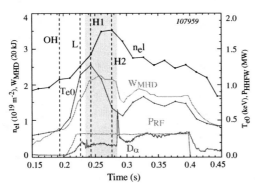

Figure 2. Time evolution of HHFW driven H-mode discharge. Time markers OH, L, H1 and H2 shown with dotted lines.

Kinetic Documentation

We can see in Fig. 2 a temporal overlay of plasma parameters for a deuterium discharge with plasma current of 0.5 MA and toroidal field of 0.45 T. The HHFW power 3.2 MW pulse is applied from 0.2 to 0.4 s and is the sole source of auxiliary power. The antenna $k_\parallel$ is 14 m^{-1} and the frequency 30 MHz. As a result of the HHFW heating, the central electron temperature T_{e0} increases from $\approx$ 0.4 keV to $\approx$ 1.1 keV, before the onset of the H phase occurring during the interval 0.235-0.285 s as can be seen on the D_α trace. Four time points indicated with vertical dotted lines – 0.193, 0.227, 0.243, 0.277 s – correspond respectively to the ohmic phase

(OH), the L phase (L), the early and late H phase (H1 and H2). Also seen in the figure are the line-integrated density, $n_e l$, and the magnetically derived stored energy, W_{MHD}.

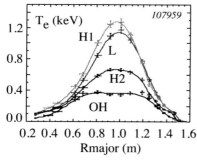 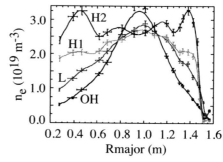

Figure 3. Temporal overlays of $T_e(R)$, and $n_e(R)$ for a HHFW driven H-mode discharge. Four times are shown: ohmic (OH), L-mode (L), early H-mode (H1), late H-mode (H2).

Kinetic profiles of HHFW driven H-mode plasmas show the expected signatures of this high confinement regime. In Fig. 3, we show $T_e(R)$ and $n_e(R)$ profiles for the time points marked OH, L, H1 and H2 in Fig. 2. During the ohmic phase, the T_e profile is flat and limited to 0.3 keV; the density profile is peaked. During the L phase, we observe a T_e increase over the whole profile with the center reaching 1.1 keV; the density profile changes from peaked to triangular shape. There is a hint of a edge profile steepening visible on $T_e(R)$ and $n_e(R)$ outboard data. The early H-mode $n_e(R)$ data show a well established edge gradient. The plasma column has shifted inwards by ≈ 3 cm and the electron temperature is slightly increased. The temperature edge pedestal is ≈ 0.12 keV. During the later H-mode phase, we observe a fully developed edge density gradient with "ears" near the peripheral regions. Meanwhile the central electron temperature has fallen to 0.6 keV.

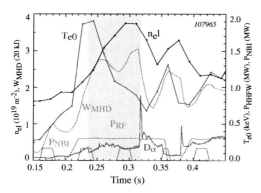

Figure 4. Time evolution of HHFW driven H mode discharge. Short NBI pulses added for T_i measurements

The discharge shown in Fig. 4 has the same nominal parameters as the one just discussed above, but short neutral beam pulses lasting 0.02 s were applied from 0.16 s on to measure the ion temperature profile $T_i(R)$ by charge exchange recombination spectroscopy at 0.06-second intervals. As in the above case, the HHFW power is 3.2 MW. Each beam pulse has a power of 1.7 MW, but TRANSP calculations indicate that only a power level ≈ 0.6 MW contributes to plasma heating. We can see in Fig. 5 plots of the T_i and T_e and n_e profiles during the L and H phases at respectively ≈ 0.230 s and 0.290 s.

Time Dependent Power Deposition with CURRAY

The ray tracing code CURRAY has recently been incorporated into the TRANSP code and we can see in Fig. 6 some preliminary analysis results. In panel (a) we see the predicted power absorbed by the electrons. Besides the total power, we also show the power absorbed in the inner region and in the outer region. One can see that during the H phase, indicated by dotted lines, more power is absorbed in the outer region as a result of the higher peripheral electron density. There are over 150 time points during the HHFW pulse, which gives true temporal information on the power deposition. For example, the drops in the absorbed power occurring, when neutral-beam pulses are present, are caused by wave absorption by fast particles [8]. Panel (b) show the power absorbed by the ions. Absorption by the fast particle constitutes the dominant term and one can see the ion heating staying in sync with the neutral beam pulses. The quick rise in ion heating at the onset of the HHFW pulse – 0.2 s – comes from the residual fast ions generated by the beam blip at 0.16 s.

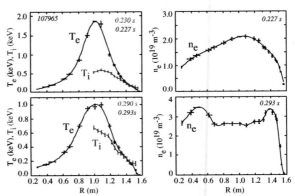

Figure 5. Kinetic profile data at two times near 0.230 s and 0.290 s. T_i from charge exchange recombination spectroscopy. T_e and n_e from Thomson scattering.

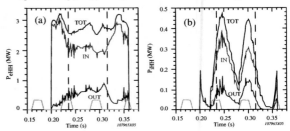

Figure 6. Time evolution of power absorption: (a) power to the electrons; (b) power to the ions. Dotted lines delineate H phase. Neutral beam blips shown for reference.

*This work is supported by U.S. DOE contract DE-AC02-76CH03073.

REFERENCES

1 Wilson, J.R. et al., Phys. Plasmas, Vol. 10, No. 5, (2003) 1733-1738
2 Ryan, P.M, et al. this conference
3 LeBlanc, B.P. et al., R. F. Power in Plasmas, 14th Topical meeting, AIP proc. 595, (2001) p 51
4 LeBlanc, B.P. et al. Submitted to Nuclear Fusion
5 Ono ,M., Phys. Plasmas 2 (1995) 4075
6 Mau, T.K.,el., R. F. Power in Plasmas, 14th Topical meeting, AIP proc. 595, (2001) p.170
7 Ongena J., Evrard M., McCune D., "Numerical Transport Codes", in the Proceedings of the Third Carolus Magnus Summer School on Plasma Physics, (Spa, Belgium, Sept 1997), as published in Transactions of Fusion Technology, March, 1998, Vol. 33, No. 2T, pp. 181-191.
8 Rosenberg, A et al., this conference.

Modeling of HHFW Heating and Current Drive on NSTX with Ray Tracing and Adjoint Techniques*

T.K. Mau[a], P.M. Ryan[b], M.D. Carter[b], E.F. Jaeger[b], D.W. Swain[b],
C.K. Phillips[c], S. Kaye[c], B.P. LeBlanc[c], A.L. Rosenberg[c], J.R. Wilson[c],
R.W. Harvey[d], P. Bonoli[e], and the NSTX Team

[a] *University of California at San Diego, 9500 Gilman Drive, La Jolla, CA 92093-0417, USA*
[b] *ORNL,* [c] *PPPL,* [d] *CompX,* [e] *MIT*

Abstract. In recent HHFW current drive experiments on NSTX, relative phase shift of the antenna array was scanned from 30° to 90° to create $k_\parallel$ spectral peaks between 3 and 8 m^{-1}, for rf power in the 1.1-4.5 MW range. Typical plasma parameters were $I_p \sim 0.5$ MA, $B_T \sim 0.45$ T, and $n_{eo} \sim 0.6$-3×10^{19} m^{-3}, and $T_{eo} \sim 0.6$-3 keV. In this paper, detailed results from the CURRAY ray tracing code at various time slices of some of the earlier discharges are presented. The complete antenna spectrum is modeled using up to 100 rays with different k_ϕ, and k_θ. The rf-driven current is calculated by invoking the adjoint technique that is applicable to toroidal plasmas of all aspect ratios and beta values. In these low β ($\sim$2-3%) discharges, the rf-driven current is peaked on axis and minority ion absorption displays a tendency to increase at lower $k_\parallel$. Reasonable agreement with inferred results from the voltage measurements has been obtained that points to evidence of current drive, while the calculated power deposition profiles agree very well with the HPRT ray code for these discharges. The use of the adjoint method will become more important in future high β NSTX discharges.

INTRODUCTION

During the past two years, fast waves in the high-harmonic range ($\omega/\omega_{ci}\sim$8-10) were launched in the National Spherical Tokamak Experiment (NSTX) to measure wave heating and current drive in plasmas with high dielectric constant ($\omega_{pe}^2/\omega_{ce}^2 \gg 1$). This campaign also provides the basis for advancing the plasma performance to high β and long pulse lengths in future experiments. The wave launcher consists of a toroidal array of twelve antenna elements that were phased to excite wave spectra centered at 3 m$^{-1} \leq k_{\parallel 0} \leq 8$ m^{-1} at a frequency of 30 MHz and a launched power of up to ~4.5 MW. Due to the absence of direct current measurement (with MSE), the driven current was inferred from the measured loop voltage difference between two identical discharges with co- and counter-CD antenna phasing. On the other hand, both ray tracing and full-wave modeling codes were used to calculate the driven currents with the measured plasma profiles and equilibrium. Agreement between the measured and calculated results for a number of well analyzed discharges points to the evidence of HHFW current drive. A normalized CD efficiency of ~0.035×10^{19} A/W-m^2 was recorded at T_{eo} = 1.5 keV and $k_\parallel \sim$ 8 m^{-1}. In this paper, we focus on the calculation of driven current in recent experiments, based on the CURRAY ray tracing code that makes use of the linear adjoint technique to evaluate the local CD efficiency.

Modeling of HHFW Current Drive in NSTX Discharges

The CURRAY ray tracing code [1] was used to model heating and current drive by HHFW in recent NSTX discharges. In this code, hot electron, cold ion dispersion relation is used in solving for the ray paths. High harmonic ion cyclotron, and electron Landau and TTMP damping are calculated using the fully kinetic E-field polarization factors, to all orders in $k_\perp^2 \rho_i^2$, with the local complex $k_\perp$ derived from an order reduction approximation. Separate calculations have verified that, for the fast wave, the ray paths and complex $k_\perp$ obtained this way are good approximations to those obtained from a fully hot dispersion relation. The plasma geometry is described by an EFIT-reconstructed equilibrium, while the density and temperature profiles are numeric fits to radial data points from Thomson scattering measurements. The full launched spectrum, calculated by RANT3D, is represented typically by 50-100 rays, with one dominant k_ϕ and usually $k_\theta = 0$ for each significant power lobe, and rays starting from points along the antenna poloidal extent just inside the separatrix. The power deposition profiles for both electrons and ions (thermal and energetic) from CURRAY are found to agree with those from HPRT and GENRAY using typical NSTX data points, while there are minor variations in the species power absorption fractions.

We first analyze three earlier discharges that have the following antenna phasing: (1) co-CD with $\Delta\theta = -\pi/2$ and $k_\phi = -8$ m^{-1}; (2) counter-CD with $\Delta\theta = \pi/2$ and $k_\phi = 8$ m^{-1}; (3) $00\pi\pi00$ phasing with spectrum peaks at $k_\phi = \pm 8$ m^{-1}. Here $\Delta\theta$ is the phase shift between adjacent antenna elements, and all three spectra have dominant poloidal mode $m = 1$. The plasma discharge has R=0.85 m, a=0.66 m, B_o=0.45 T, I_p=0.5 MA, and the wave frequency is 30 MHz that corresponds to f ~$9f_{cD}$ on axis. Density and temperature profiles with $n_{eo,13}$ ~ 0.6-1.1 and T_{eo} ~ 0.6-1.5 keV, and equilibrium data are obtained at five time slices in each discharge during the rf pulse. These are then used as input to CURRAY to calculate the power deposition to each species and the driven current. In these cases, the ion species concentrations are taken to be f_D=0.931, f_H=0.046, f_C=0.04, and f_C=0.003, giving Z_{eff}=3.25, and T_i=0.68T_e is taken as our best estimate. In Fig. 1 are shown the power deposition and driven current profiles with an input power of 2.1 MW and co-CD antenna phasing. The single pass absorption fraction is relatively strong (~40%) with more than 99% of the power absorbed by electrons and the rest to minority hydrogen. Both absorption and driven current are peaked on axis with the CD efficiency in the 0.071-0.082 A/W range. Reversing to counter-CD phasing at 1.1 MW produces very similar absorption profiles and current drive in the opposite direction with essentially the same range of CD efficiencies. At symmetric antenna phasing, currents were driven in opposite directions that add to very low levels of total driven current. A time history of the total driven current during the rf pulse for the three discharges, but at different input power, indicates almost constant driven current throughout the rf pulse with roughly constant profiles peaked on axis, which is expected of a plasma of 2-3% β.

The time slices at 391 ms in shots corresponding to co- and counter-CD phasing are analyzed in some detail by experimental inference from the measured loop voltage difference [2], and by ray tracing (CURRAY) and full wave (TORIC) codes. The results of I_{CD}/P (A/W) = 0.056, 0.046, and 0.075 for experiment, TORIC and CURRAY, respectively, indicate quite reasonable agreement, pointing to evidence of current drive.

Another series of discharges with antenna phasing of $\Delta\theta = \pm\pi/4$ have also been investigated in detail, under similar plasma conditions. In Fig. 2 we show the results for the case of $-\pi/4$ co-CD phasing, corresponding to $k_\phi = 3.2$ m^{-1}, at 1.9 MW of launched power for two time slices. At these two points (301 ms, 403 ms), 56 and 60 kA of current were driven respectively. It is noted that the electron absorption and driven current are peaked slightly off axis at $\rho = 0.1$ mainly due to refractive effect at a lower k_ϕ, causing the main power flow to skirt the axis. In addition, the fraction of power absorbed by minority H is found to increase significantly (to ~10%).

Current Drive Calculations Using the Adjoint Technique

We have examined the regime of validity for the Ehst-Karney empirical fit [3] in calculating the CD efficiency in a spherical tokamak because the plasma is paramagnetic at high β, i.e., the magnitude of the total magnetic field is non-monotonic with the major radius R, and $R_o/a < 3$. In this situation the inherent assumption that the inverse aspect ratio ε for a circular plasma in the E-K fit for $j/p(w,\varepsilon,\theta)$ can be replaced by $(B_{max}-B_{min})/(B_{max}+B_{min})$ on each non-circular flux surface is no longer valid. Instead, for each equilibrium, $j/p(w,\varepsilon,\theta)$ has to be evaluated using the adjoint technique [4], and this is carried out in CURRAY to create a table of j/p for use during ray tracing. In a previous work [5], it was found that at high β (~20%), the E-K formula under-estimates the CD efficiency by ~20% at $\varepsilon = 0.2$ and by ~100% at $\varepsilon = 0.4$ for $w = v_\parallel/v_{te} < 2.5$. We have compared the driven current profiles calculated using the adjoint technique, the E-K formula and without neoclassical trapping effect for two NSTX equilibria, and the results are displayed in Fig. 3. It is clear from Fig. 3(a) that at $\beta \sim 2\%$, the E-K formula gives a good approximation to the adjoint result as the current is driven on axis ($\varepsilon < 0.3$), and the trapping effect accounts for 30-50% reduction in the driven current. However, in an anticipated NSTX $\beta = 25\%$ equilibrium, the current is driven mainly off-axis ($\varepsilon \geq 0.3$) as shown in Fig. 3(b) and the E-K formula under-estimates the driven current by a substantial amount. At the same time trapping effect is seen to be much stronger than at lower β, resulting in much more than 50% reduction in driven current. We therefore conclude that as NSTX moves into high β discharges, HHFW current drive must be analyzed with the adjoint method or by solving the quasilinear Fokker Planck equation.

ACKNOWLEDGMENTS

This work at UCSD is supported by US DOE Grant DE-FG03-99ER54526.

REFERENCES

1. Mau, T.K., et al., Proc. 14th Top. Conf. RF Power in Plasmas, Oxnard (2001) 170.
2. Ryan, P.M., et al., 19th Int. Conf. on Plasma Phys. and Controlled Nucl. Fusion Res., paper IAEA-CN-94/EX/P2-13, Lyon, France (2002).
3. Ehst, D.A., Karney, C.F.F., Nucl. Fusion 31 (1991) 1933.
4. Antonsen, T.M., Chu, K.R., Phys. Fluids 25 (1982) 1295.
5. Chiu, S.C., et al., Nucl. Fusion 29 (1997) 1515.

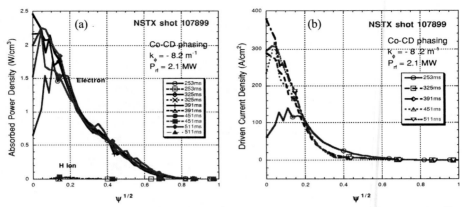

Fig. 1. Calculated (a) power absorption and (b) rf-driven current profiles for five time slices with $-\pi/2$ co-CD antenna phasing, in NSTX shot 107899.

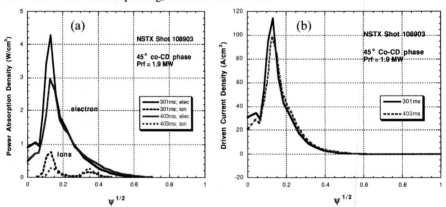

Fig. 2. Calculated (a) power absorption and (b) rf-driven current profiles for two time slices with $-\pi/4$ co-CD antenna phasing, NSTX shot 108903.

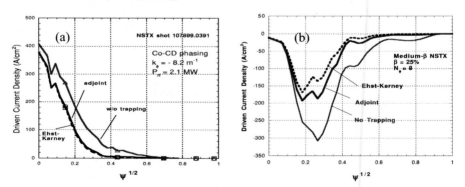

Fig.3. Calculated driven current profiles using Ehst-Karney formula, adjoint method, and without trapping effect for (a) low β [2.4%] and (b) high β [25%].

High Harmonic Fast Wave Current Drive Experiments on NSTX[*]

P. M. Ryan[a], D. W. Swain[a], J. R. Wilson[b], J. C. Hosea[b], B. P. LeBlanc[b],
S. Bernabei[b], P. Bonoli[c], M. D. Carter[a], E. F. Jaeger[a], S. Kaye[b],
T. K. Mau[d], C. K. Phillips[b], D. A. Rasmussen[a], J. B. Wilgen[a]
and the NSTX Team

[a]*Oak Ridge National Laboratory*
[b]*Princeton Plasma Physics Laboratory*
[c]*Massachusettes Institute of Technology*
[d]*University of California–San Diego*

Abstract. Phased-array experiments were performed with the 30-MHz, 12-element HHFW antenna array on NSTX for rf powers in the 2-4.5 MW range. Relative phase shift of the array was scanned from $\pi/6$ to $\pi/2$ to create k_z spectral peaks between 3 and 7.6 m^{-1}; it was also possible to switch from $\pi/2$ to $\pi/4$ phasing during a shot. Typical plasma parameters were $I_p \sim$ 0.5 MA, $B_T \sim$ 0.45 T, and $n_e(0) \sim 1-3 \times 10^{19}$ m^{-3}, and $T_e(0) \sim$ 1-3 keV. Electron heating was observed even at the highest wave phase velocities, reaching 1.3 keV with 2.2 MW at $-\pi/6$ phasing and 2.7 keV with 2.9 MW at $-\pi/3$. Lower plasma loading and higher central heating efficiency are observed for counter-CD phasing. MSE measurement of the current profile is not yet available; therefore, the rf-driven portion of the plasma current is calculated from the difference in loop voltage between co- and counter-CD phasing. The normalized current drive efficiency for $k_z = 7.6$ m^{-1} is $\gamma \sim 0.035 \times 10^{19}$ A•W^{-1}•m^{-2} for both D and He plasmas.

THE HHFW SYSTEM ON NSTX

The HHFW antenna array consists of twelve evenly spaced, identical current strap modules connected in pairs 1-7, 2-8, etc.[1] Each pair forms a resonant loop with the strap currents 180º out of phase. Each loop is powered by a single 1 MW, 30 MHz transmitter, allowing arbitrary, rapid phase shift between transmitters to be made for real-time spectral control. Each loop is isolated from its nearest neighbors by a shunt decoupling circuit that cancels the mutual inductance between adjacent straps. A new digital system has been implemented that provides active feedback control of the transmitter phase based on phase measurements near the vacuum feedthroughs, allowing full control of the array phase during a shot.

The NSTX plasma presents a relatively large dielectric constant at 30 MHz operation due to its high density and low magnetic field ($\varepsilon = \omega_{pe}^2/\omega_{ce}^2 \sim 1$). Under these conditions the high harmonic fast waves (HHFW), with $\omega \sim$ 10-20 ω_{ci}, do

---
[*]Oak Ridge National Laboratory, managed by UT-Battelle, LLC, for the U.S. Dept. of Energy under contract DE-AC05-00OR2272

not couple efficiently to the thermal ions but will readily damp on the electrons via Landau damping and Transit Time Magnetic Pumping (TTMP) [2]. A directed wave spectrum launched from the HHFW array uses this strong coupling to the electrons for non-inductive current drive (CD).

PHASED ARRAY OPERATION

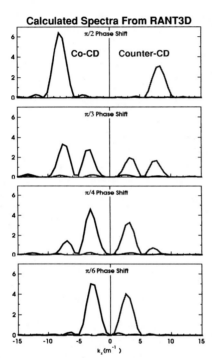

FIGURE 1. Calculated wave spectra in plasma as the relative phase between array elements is varied.

Because the array elements 1-7, 2-8, ... are connected in pairs, the NSTX 12-element array acts as two adjacent 6-element arrays that are permanently 180º out of phase. This results in a correct phase relation between elements 6 and 7 only when the relative phase shift between elements is either $\pi/6$ or $\pi/2$, resulting in a monochromatic spectrum peaked at toroidal wave number $k_z = 3$ m^{-1} or 7.6 m^{-1}, respectively. Other phase shifts result in bifurcated spectra in which the power is divided between peaks centered around these two wave numbers, as illustrated in Fig. 1. It also indicates that more power is launched in the co-CD direction than in the counter-CD direction for equal currents (or voltages) on the straps, as calculated by the RANT3D code[3]. This imbalance is due to the large poloidal field in NSTX, and the resulting large pitch angle of the magnetic field in front of the antenna (up to 45°). The propagation asymmetries in the direction perpendicular to field lines, arising from radial gradients, are rotated into the toroidal direction [4]. A loading analysis made with a coupled-transmission line model of the power distribution system and the plasma coupling calculated with the RANT3D code show good agreement with the measurements of the loading imbalance [5]. The plasma loading for counter-CD phasing at 7.6 m^{-1} was typically half that for co-CD (8 Ω vs 17 Ω) and limited the power that could be delivered when running in the counter-CD phasing.

CURRENT DRIVE EXPERIMENTS

Phased array operation to produce directional wave spectra for HHFW current drive (CD) has concentrated primarily relative phasing of $\pm\pi/2$ ($k_z = \pm 7.6$ m^{-1}) with limited operation at higher phase velocities (for example, k_z primarily at ± 3 m^{-1} for

$\Delta\phi = \pm\pi/4$). Current density profiles could not be measured during these phased array experiments; future operation will have a Motional Stark Effect diagnostic available for this purpose. In the absence of this crucial measurement, the HHFW driven current can be estimated by establishing identical plasma conditions for co-CD and counter-CD phasing and assuming:

1. Steady-state (t > L/R) conditions have been achieved.
2. The plasma ohmic resistance R_p is the same for both co- and counter-CD phasings (same T_e).
3. The HHFW driven current is proportional to the rf power.

Under these assumptions a 0D calculation of the driven current can be made from the measured loop voltage, after subtracting the calculated bootstrap current and compensating for the portion of the loop voltage which drives the changing magnetic stored energy. The non-inductive current drive (I_{CD}) portion of the total plasma current I_p (kept constant at 0.5 MA) can be estimated from

$$I_p = (V_{loop} - 0.5 \bullet I_p \bullet dL_i/dt)/R_p + I_{BS} + I_{CD}$$

where V_{loop} is the measured loop voltage, L_i is the internal inductance calculated by EFIT, and I_{BS} is the bootstrap fraction calculated by TRANSP.

Figure 2 shows two comparison shots for $k_z = \pm 7.6$ m^{-1}, D_2 plasma, $B_T(0) = 0.445$ T, $I_p = 0.5$ MA. The HHFW power needed to obtain comparable core density and temperatures was 2.1 MW for co-CD and 1.1 MW for counter-CD; Thomson scattering measured $n_e(0) = 1.1 \times 10^{19}$ m^{-3}, $T_e(0) = 1.4$ keV. The 0.22 V difference in loop voltage (Fig. 3) for the two cases gives an estimated $\Delta I = 180$ kA (110 kA co-CD, 70 kA counter-CD). The current drive efficiency based on the launched power is 0.05 A/W for a normalized efficiency ($\gamma = I_{CD}<n_e>R/P_{RF}$) of 0.034×10^{19} A$\bullet$W$^{-1}\bullet$m^{-2}. Note however, that the assumption of steady-state is not well satisfied; internal MHD activity leads to a decrease in voltage difference with time.

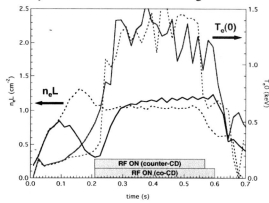
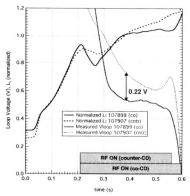

FIGURE 2. $T_e(0)$, n_eL, I_p, and P_{RF} vs time for co-CD (shot 107899, solid) and counter-CD (shot 107907, dotted) with $k_\parallel = 7.6$ m^{-1}.

FIGURE 3. The calculated internal inductance (EFIT) and measured loop voltage for co/counter-CD cases.

HHFW current drive calculations have been made using the 2D full-wave code TORIC [6] and the ray-tracing code CURRAY[7]; results are shown in Table 1. Both analyses indicate strong central power absorption and negligible damping on the H$^+$ minority.

TABLE 1. Comparison of code calculations of current drive with experimental values

	Co-CD (kA)	Counter-CD (kA)	Total ΔI (kA)	I_{CD}/P (A/W)
0D Calculation	110	70	180	0.056
TORIC	96	50	146	0.046
CURRAY	162	79	241	0.075

After modifying the vacuum feedthroughs where arcing was limiting high power operation, current drive experiments were performed in He plasmas. The rf power was applied earlier in the shot in order to keep q(0) from dropping below unity. For I_p = 0.5 MA, B_0 = 0.45 T, and line average densities of 0.8-1×10^{19} m^{-3}, central electron temperatures of 2.3 keV were achieved with 4.3 MW at $\Delta\phi = -\pi/2$ and 2.3 MW at $\Delta\phi = +\pi/2$. The normalized current drive efficiency estimated from the 0D analysis was somewhat lower ($\gamma = 0.03 \times 10^{19}$ A•W^{-1}•m^{-2}) at this T_{e0} than that obtained for D plasmas at T_{e0} = 1.4 keV. This could indicate a mismatch between the wave phase velocity and the electron thermal velocity. Limited operation in D plasmas at $\pm\pi/4$ relative phasing, where most of the power is at 3 m^{-1}, gave $\gamma \sim 0.043 \times 10^{19}$ A•W^{-1}•m^{-2} for T_{e0} = 1.2 keV ($<n_e>$ = 1.8×10^{19} m^{-3}, P_{RF} = 1.9 MW). However, plasma conditions were too transient for a steady-state analysis to be applicable and these observations remain speculative at this time.

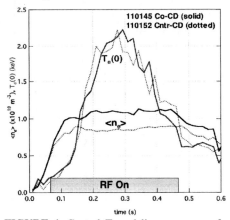

FIGURE 4. Central T_e and line average n_e for co/cntr-CD comparison at ± 7.6 m^{-1}, He plasma.

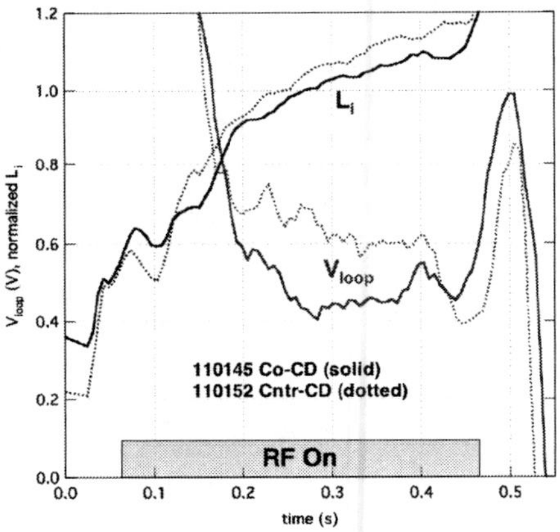

FIGURE 5. Normalized internal inductance and loop voltage for co/cntr-CD at ± 7.6 m^{-1}

REFERENCES

[1] Ryan, P. M., et al., *Fusion Engineering and Design* **56-57**, 569-573 (2001)
[2] Ono, M., *Physics of Plasmas* **2**, 4075-4082 (1995)
[3] Carter, M. D., et al., Nucl. Fusion 36, 209-224 (1996)
[4] Jaeger, E. F., et al., "Co-counter Asymmetry in Fast Wave Heating and Current Drive", in *12th Top. Conf. on Radio Frequency Power In Plasmas*, edited by P. M. Ryan and T. Intrator, AIP Conf. Proc. 403, New York, 1997, pp. 285-289
[5] Swain, D. W., et al., *Fusion Science and Technology* **43**, 503-513 (2003)
[6] Brambilla, M., *Plasma Physics and Controlled Fusion* **41**, 1-34 (1999)
[7] Mau, T. K., et al., this conference

Investigations of Low and Moderate Harmonic Fast Wave Physics on CDX-U [1]

J. Spaleta[†], R. Majeski[†], C. K. Phillips[†], R. J. Dumont[*†], R. Kaita[†], V. Soukhanovskii[†], Z. Zakharov[†]

[†] *Princeton Plasma Physics Laboratory, Princeton, NJ 08543-0451*
[*] *Association Euratom-CEA sure la Fusion Contrôlee, F-13108 St Paul lez, Durance, France*

Abstract. Third harmonic hydrogen cyclotron fast wave heating studies are planned in the near term on CDX-U to investigate the potential for bulk ion heating. In preparation for these studies, the available RF power in CDX-U has been increased to 0.5 MW. The operating frequency of the CDX-U RF transmitter was lowered to operate in the range of 8 – 10 MHz, providing access to the ion harmonic range $2\Omega \sim 4\Omega$ in hydrogen. A similar regime is accessible for the 30 MHz RF system on NSTX, at 0.6 Tesla in hydrogen. Preliminary computational studies over the plasma regimes of interest for NSTX and CDX-U indicate the possibility of strong localized absorption on bulk ion species.

INTRODUCTION

The physics of the low harmonic fast wave regime has been studied extensively in tokamak environments, but is largely unexplored for ST's which typically have a high dielectric constant and a higher ion beta than tokamaks. In this regime mode conversion to the IBW may become significant [1]. It has been suggested that third harmonic fast wave heating could be used for bulk ion heating in reactor scale tokamak plasmas [2]. However, little work has been done to examine the applicability of third harmonic ion cyclotron heating in the ST geometry. The two strap phasable antenna on CDX-U provides an excellent opportunity to investigate the low ion cyclotron harmonic regime of ST wave physics. Figure 1 provides an overview of the CDX-U geometry and typical operational parameters. The CDX-U operation parameter space allows for the investigation of second, third and fourth hydrogen cyclotron harmonics in both on-axis and off-axis scenarios. A similar third hydrogen harmonic regime is accessible in NSTX at 0.6 Telsa toroidal field.

---

[1] Work supported in part by D.o.E Fusion Science Fellowship.

CDX-Parameters:
- $R_o = 34$ cm
- $a = 22$ cm
- $b = 35 \sim 38$ cm
- $\kappa = b/a = 1.55 \sim 1.7$
- triangularly $= 0.2 \sim 0.4$
- $B_{T0} = 2.3$ kG
- $I_p \sim 80$ kA
- $n_{e0} \sim 4 \times 10^{19}$ m^{-3}
- $T_{e0} \sim 100$ eV

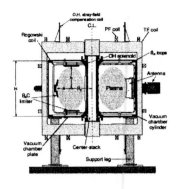

FIGURE 1. Cut-away view of CDX-U, with typical operational parameters

NUMERICAL SIMULATIONS

Two numerical simulation codes, CRF [3] and METS [4], were used to explore the feasibility of low harmonic fast wave power absorption. Both codes are 1-D slab plasma models which include a toroidal geometry induced radial k_z up shift. CRF calculates solutions to the full hot plasma dispersion relation through the entire plasma region. METS solves for the wave fields to all orders in $k_\perp \rho_i$ and produces a power absorption profile for each species present in the plasma.

Figure 2 provides an example of the CRF generated plasma dispersion relation for typical CDX-U plasma parameters, with the radial location of the hydrogen and deuterium cyclotron resonant layers highlighted. Using a simple model for wave attenuation of a ray launched radially inward, $P_{loss}(r) = 1 - exp\left[-2 \int_0^r Im\left(k_\perp(x)\right) dx\right]$, one can estimate the power attenuation profile. The attenuation profile in Figure 3 indicates the potential for localized absorption in the typical CDX-U 90% H / 10% D plasmas near the hydrogen ion cyclotron harmonics. Results in a pure hydrogen plasma have similar localized behavior near the hydrogen cyclotron harmonics, suggesting that the minority species plays a negligible role in the fast wave damping process near the hydrogen harmonics.

METS was subsequently employed to obtain more detailed information about the absorbed power split between species, and to determine how much of the fast wave energy propagated onto the IBW branch. Figure 4 shows the power absorption profiles calculated by METS for a CDX-U relevant plasma, with the third hydrogen harmonic located on-axis. METS has been used to explore the applicability of third ion cyclotron harmonic heating in larger spherical tori. Figure 5 shows the fast wave power absorbed on the third hydrogen harmonic located at the magnetic axis as a function of ion beta for a target 90% H / 10% D plasma in NSTX and NSST relevant parameter regimes. The β_{ion} scans indicate that fast wave power absorption is strongly dependent on β_{ion}. Because of the smaller physical size of the CDX-U plasma, the METS code has difficulty converging for some parameter choices. Work is underway to update the numerical algorithm in METS. Upon completion, similar β_{ion} studies will be performed

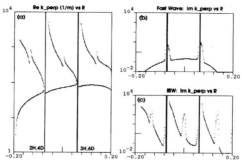

FIGURE 2. 1-D dispersion relation for a 90% H / 10% D plasma in CDX-U under typical operational parameters. Shown here is $k_\perp$ as a function of CDX-U minor radius, with the ion harmonic locations indicated and labeled. (a) Real part of $k_\perp$ for both fast wave and IBW. (b) Imaginary part of $k_\perp$ for the fast wave. (c) Imaginary part of $k_\perp$ for the IBW. Similar IBW and fast wave coupling features near the hydrogen harmonics occur for 100% H plasma simulations as well, suggesting that the minority species is not critical to the ion cyclotron heating mechanism

for CDX-U.

CONCLUSIONS

Numerical simulations, using CRF and METS 1-D codes, indicate that 3rd harmonic fast wave heating could lead to strongly localized single pass absorption on bulk ion species in ST geometries. Furthermore, simulations indicate power absorption is strongly dependent on β_{ion}. Discrepancies between the different types of β_{ion} scans, temperature, density and magnetic field variations, suggests a more complicated relationship between these primary plasma parameters. Experimental studies on CDX-U to investigate third harmonic hydrogen absorption are planned to begin in the near future. Future work also

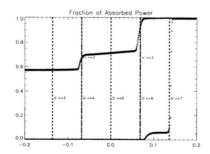

FIGURE 3. Fast wave attenuation calculation for a typical CDX-U plasma, based on the simple propagating ray approximation, $P_{atten}(r) = 1 - exp\left[-2\int_0^r Im(k_\perp(x))\,dx\right]$. Drops in wave power are localized to hydrogen harmonic regions, where the dispersion relation in Figure 2 indicates close coupling of IBW and fast wave branches.

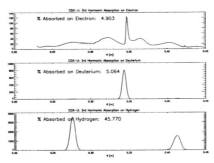

FIGURE 4. METS calculated by species power absorption profile in a 90% H / 10% D plasma for typical CDX-U operational parameters. Strong absorption on the dominant hydrogen at the 3rd and 2nd hydrogen harmonics.

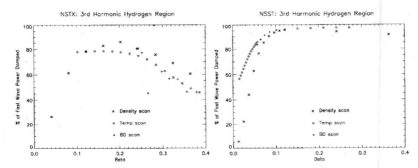

FIGURE 5. β_{ion} scan comparisons between NSTX and NSST. Both show a strong dependence on β_{ion}. ($\times$ variation in density, $\Diamond$ variation in T_i, $+$ variation in B_0)

includes continued investigations of absorption parameter dependence via numerical simulations, as well as work towards an understanding of the applicability of analytical models for third harmonic heating such as [2] to the the ST regime.

REFERENCES

1. M. Ono, Phys. Plasmas **2** (1995).
2. D. G. Swanson, Phys. Fluids **28**, 1800 (1985).
3. D. W. Ignat and M. Ono, Phys. Plasmas **2** (1995).
4. D. Smithe, Plasma Physics and Controlled Fusion **31**, 1105 (1889).

LOWER HYBRID CURRENT DRIVE

Hybrid and Steady-State Operation on JET and Tore Supra

A. Bécoulet, the Tore Supra Team[*]
and the Contributors to the EFDA-JET Workprogramme[**].

*Association EURATOM-CEA sur la Fusion, CEA Cadarache
13108 St Paul-lez-Durance cédex, FRANCE*

Abstract : Producing fusion energy requires to simultaneously sustain in a tokamak environment fully non inductive regimes at the highest Q-values and a "significant" fusion performance level under MHD-stable conditions, while insuring a satisfactory confinement of the fast alpha particles. This ambitious goal is being investigated on many devices worldwide, particularly focusing on the role played by the current density profile. The paper reports on the recent experimental progress of both the JET and Tore Supra devices towards i) long to very long pulse operation relying on a careful use of lower hybrid current drive under various current profile tailoring conditions (namely so-called "hybrid" peaked current density profiles and so-called "steady-state" hollow current density profiles) and ii) discharges performed with real-time controlled pressure and/or current density profiles. Such discharges are detailed and interpreted using the CRONOS integrated modelling suite. Its fully predictive capability, including real time control features, is used to provide keys to future experiments.

INTRODUCTION

Recently, the International Tokamak Physics Activity (ITPA) Group on Steady-State Operation and Energetic Particles proposed a classification of the tokamak discharges which aim at long pulse operation, in present and next step devices. This classification is based on the commonly admitted result that the current profile tailoring is the key issue for this type of discharges. Thus, regardless to the absolute long pulse technical capability of the concerned devices, two major target discharges were defined. The first one, so-called "hybrid" regime, relevant to the extended burn phase of ITER, consists in discharges where i) some inductive flux is saved mostly by a non inductive additional H&CD source and ii) the central current density profile is tailored in order to avoid the main q=1 MHD sawtoothing activity. The resulting q-profile, depicted in Fig 1a, is monotonic, with an edge value not larger than ~ 5 and a central value in the vicinity of 1. The second target regime, called "steady-state", is more ambitiously preparing the fully stationary operation of devices like ITER. The baseline is to take full advantage of the self-consistent bootstrap current in order to set-up a regime with the capability of providing actual stationarity to the plasma core. The resulting plasma current density is broadened by the large contribution of the bootstrap current (say above 50%), even leading to a magnetic shear reversal in the plasma core. The regime thus operates at vanishing loop voltage,

---

[*] *see J. Jacquinot, 19th Fusion Energy Conf., Lyon, France, 2002*
[**] *see J. Paméla, 19th Fusion Energy Conf., Lyon, France, 2002*

with large values of the normalized beta. Here again, considerations on overall MHD stability and fast ion confinement bound the central q-value around a typical value of 3, and the minimum q-value in the range 1.5-2.5. Though the value of the magnetic shear inversion radius is not specified for this target, the recent results by many devices on the existence of internal transport barriers (ITBs) in such regimes tend to push towards the largest possible values. The typical "steady-state" target q-profile is depicted in Fig 1b.

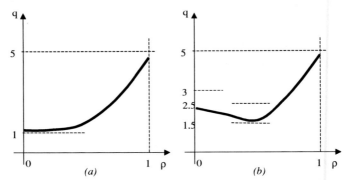

Fig1: Typical safety factor profile for a) the hybrid target and b) the steady-state target

Referring to this classification, the ITPA then encourages each tokamak device to document the "hybrid" and "steady-state" scenarios in terms of existence, stability and performance, investigating in the largest density / plasma current / configuration parameter space. Obviously, the performance has to be sustained over the longest pulse duration allowed by the devices. The development and implementation, in parallel, of real-time control algorithms is strongly supported as part of the optimisation process.

The present paper locates the Tore Supra and JET devices in this respect, underlining the key role often played by lower hybrid current drive (LHCD), either when bringing a dominant non inductive additional current source and/or when tailoring the current density profile prior to the high power phase. Section 1 illustrates the "hybrid" and "steady-state" regimes with recent discharges from both devices, and discusses the present related performance and limits; section 2 presents some specific issues linked to the real time control at vanishing loop voltage operation; it finally proposes an alternative procedure of real time algorithms, performance oriented, that are presently developed using the predictive capability of the CRONOS[1] integrated modelling suite.

HYBRID AND STEADY-STATE TARGETS

The Tore Supra experimental programme is focused on long pulse plasma operation, progressively integrating all the technology and physics aspects of a multi-megawatt tokamak discharges lasting over hundreds of seconds. The constraint on the scenarios makes the vanishing loop voltage a pre-requisite. The circular limiter configuration of the device (R=2.42m, a=0.72m) prevents from investigating the ITER-relevant geometries or the compatibility between the edge barrier and a plasma core with a strong current density profile shaping. On the other hand, Tore Supra's long pulse radio-frequency power capability (LHCD, ion and electron cyclotron resonant heating (ICRH, ECRH) systems) allows to produce and study such "hybrid" and "steady-state" current profiles over time durations much larger than the resistive current diffusion time, and under vanishing or controlled-zero loop voltages. The most investigated regime so far is the low density fully non inductive discharge in which a dominant fraction (~80-90%) of the plasma current is driven thanks to LH waves, the small remaining fraction being provided by the bootstrap effect. The typical current drive efficiency obtained in these fully stationary discharges ($\eta_{vloop=0} \sim 0.7 \; 10^{19}$ A/W/m^2) allows long pulse operation with the following typical parameters: I_p=0.55 MA, B=4 T, $q_a \sim 9$, P_{LHCD}=3 MW, $n_{//inj} \sim 1.8$, n_{e0}=2.5 10^{19} m^{-3}. The propagation/absorption conditions[2] of LHCD are such that the resulting current profile is centrally peaked ($l_i \sim 1.6$), naturally producing an "hybrid-like" target. It is to be noted that the LHCD system is presently being upgraded (both through new 700kW klystrons, and a new antenna[3]) and will bring the operating point closer to $q_a \sim 5$ in the near future. The present longest discharge of Tore Supra[4] (#30414) reached 4minutes25s, thanks to 0.75GJ injected. The regime has also been consolidated adding up to 3 MW ICRH (D(H) central minority heating) at somewhat higher density (Fig2).

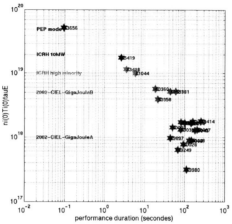

Fig2: Tore Supra: $n_i(0)T_i(0)\tau_E$ (keV.s.m-3) as a function of the performance duration (s)

Accessing a steady-state target is somewhat more complex. For a machine like Tore Supra, the constraint of 50% of bootstrap current requires the poloidal beta value to exceed 1.5. Such values were obtained applying a high power level (>6MW) of ICRH in the so-called fast wave electron heating (FWEH) mode[5]. The scenario was performed at B=2.15T, and a wave frequency of 48MHz. In that case, the only ion cyclotron resonance present in the plasma is the 3rd harmonic of deuterium, with a negligible absorption compared to the direct absorption of the wave by the electrons under the combined electron Landau damping and transit time magnetic pumping effects. The resulting discharges exhibit the expected 40-50% of bootstrap current, a significant confinement enhancement factor ($H_{ITER97-L}=1.6$), thanks to a highly enhanced electron stored energy (more than twice the Rebut-Lallia-Watkins zero-D prediction). It is to be noted that, so far, no signature of shear reversal or ITB has been observed in such discharges. Here again, the on-going upgrade of the ICRH capability of Tore Supra, through CW sources and an improved antenna concept more resilient to variations of load will help investigating further this route in the coming years[6]. Meanwhile, hollow current profile targets are produced and studied, at lower bootstrap current fractions, using the RF heating and current drive systems. Two main routes are followed on Tore Supra for the production of such reversed magnetic shear discharges. The first one follows the conventional proven recipe of many tokamaks, relying on a fast plasma current ramp-up phase, which naturally produces a transient hollow current profile[7]. In the present case, a low current stationary plasma is first produced (I_p<400kA for 8-10s), and the plasma current is then rapidly ramped-up, at a typical rate of 1.6MA/s, while additional ICRH power and density are raised. Target q-profile relevant of the "steady-state" are then temporarily produced over 1 to 2 seconds, as reconstructed from magnetics and polarimetry data. The regime is only transiently produced as the inductive current fraction remains dominant. It is to be noted that an ITB on electron temperature is then observed, the normalized radius for the foot of the barrier being as broad as 0.5-0.6. One must also point out that the superposition of LHCD power to this sequence in order to prolong the high performance phase failed so far, as LHCD is then deposited mostly centrally and counter-acts the reversed shear configuration. More successful in terms of duration are the attempts to reverse the magnetic shear directly from a monotonic situation using LHCD alone and/or a combination of LHCD and electron cyclotron current drive (ECCD) power. The initial situation is similar to the hybrid long pulse operation, but i) the LHCD deposition profile is broader due to a proper choice of parameters[2] (higher-$n_{//}$ launched spectrum, larger plasma current, ...) and/or ii) a local perturbation is applied off-axis on the current density profile by counter-ECCD[8]. In such situation(s), a so-called lower hybrid enhanced performance (LHEP) mode is triggered, exhibiting an ITB on electron temperature. Due to the modest volume occupied by this barrier, which foot point lies between 0.2 and 0.3 in normalised radius, it is difficult to conclude on the related shear reversal, but through 1-D current diffusion reconstructions which give good confidence in its presence. The on-going increase of the available long pulse ECRH capability of Tore Supra allows us to expect producing broader ITBs with more off-axis ECCD deposition[9]. The massive role played by LHCD eases the stationarity of the phenomenon. LHEP targets lasting more than a minute (LHCD only), and/or as long as the ECCD pulse, have already been observed this way, routinely. Interestingly, the transition between the hybrid target and a hollow current density profile may lead to a stationary solution oscillating at very low frequency[8][10] (observed at a few Hz for more than 100 s). Such oscillatory behaviour is presently identified as an interplay between the current density profile evolution and the associated local transport properties, and viewed as an incomplete transition towards the core ITB.

Complementarily, the JET device allows to investigate both "hybrid" and "steady-state" targets integrating ITER-relevant configuration aspects and compatibility with an H-mode edge. JET also provides an alternative set of additional heating and current drive systems. Though high power phases include large fractions of positive neutral beam power, the ICRH and LHCD systems of JET appear to also play key roles in setting-up the regimes[11]. The development of the "hybrid" target is pursued in the frame of the so-called "JET/Asdex-Upgrade identity" experiments[12][13]. In both machines, the additional heating power waveform is carefully timed so that the current density is tailored with the identical "hybrid" features described in introduction. The core region exhibits a rather flat q-profile, clamped in the vicinity of q=1 by a regular fishbone activity which prevents the current profile from further peaking. The plasma current, still dominated by ~70% of ohmic current, typically contains 20% of bootstrap current and 10% of neutral beam current drive, responsible for this core current density flattening. Though differences may occur, the comparison between the two machines is carried out extensively, including the match of the ρ^* and edge q-values at various triangularities. The present JET performance reaches $H_{89}.\beta_N > 5$ at 1.4MA/1.7T (q_{95}~4, δ up to 0.45), the compatibility with an H-mode edge is satisfactory and the possible limitation due to the role of the central fishbone activity on fast ion confinement seems presently modest.

JET provides also major contributions to the viability of the "steady-state" target. The reversed magnetic shear configuration is provided routinely by imposing LHCD power during the fast plasma current ramp-up phase (~0.4MA/s). A careful dosing of the power and the ramp-up rate gives access to a large range of current profile targets from monotonic to extremely hollow (so-called "current hole" configurations[14]). The target current profile is then carefully adjustable, very close to the stationary solution obtained during the high power phase, which combines bootstrap current, NBCD and LHCD. This way of proceeding allows JET to access and study the steady-state targets despite the limited high power pulse duration. Steady-state discharges mixing 50% of bootstrap current to 25% of NBCD and 25% of LHCD have been maintained up to typically 10 seconds. This represents about one current diffusion time for JET. $H_{89}.\beta_N$ values in such discharges reach 4 at present[15] (Fig3).

It is noticeable that such discharges are always associated so far with ITBs: ITBs on electron temperature are present from the prelude LHCD phase, ITBs on electron and ion temperature, but also on plasma density and toroidal velocity then develop when the high NBI+ICRH+LHCD power phase is triggered. As a common feature to all these developments, steady-state targets in JET are carefully designed to last, and not to explore the performance limits (fusion, MHD, ...). To that respect, they do not really bring new information, but on long time scale phenomena. The most noticeable result here concerns the high-Z impurity behaviour. It has been demonstrated[16] that impurities follow the neo-classical expectation, i.e. accumulate in the plasma core all the more that the density gradient is large and the temperature gradient is weak. Core radiative collapses have even been observed in long lasting strong ITBs, though it must be noted that the regime recovered after such collapses, as the current density profile was still under control.

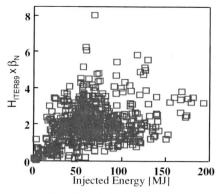

Fig3: JET: present performance ($H_{ITER89}\beta_N$ vs injected energy) in steady-state target discharges.

CONTROLLING THE REGIMES, IN VIEW OF STATIONARY OPERATION

A careful control of the various profiles is thus necessary for the viability of these regimes. An intensive work is being conducted on both devices in the direction of a real-time control of the advanced discharges.

To alleviate the high-Z impurity accumulation on long pulse discharges with ITBs, JET uses for instance a real-time feedback loop between the measured characteristic gradient length of the ITB and the input power. The ITB is detected[17] and then followed in location and strength through the maximum value of the local ion Larmor radius at the sound speed ρ_{se} ($\rho_{se}^2 = m_i T_e/eZ_i^2 B^2$) normalised to the local electron temperature gradient length (deduced from electron cyclotron emission measurement). The best actuator appears to be the ICRH power.

To better match the requested current profile targets, JET is also equipped with a real-time determination of the q-profile, based on real-time equilibrium reconstruction constrained by polarimetry data, and soon by MSE data. The control algorithm is a model-based control[18] which presently uses a pre-determined matrix link between five values along the safety factor profile and the LHCD power actuator. It insures a least square minimisation between the actual q-values and five prescribed ones. The matrix can either be computed from models or deduced from previous open-loop experiments. Successful prescribed monotonic or reversed q-profile have been obtained and maintained this way, combining current ramp-up and LHCD power. It is important to note that the model based algorithm can accepts much more sophisticated actions through any desired link between measured q-profile, pressure profile, ... values and LHCD, ICRH, NBI, ... actuators. Such algorithms are under intense development and tests presently on JET[19].

Tore Supra has also a long tradition of real time control developments and operation[20]. One can remind the routine operation at constant primary flux consumption combined with a plasma current feed back control by the LHCD power, which allows most of the long pulses, feedback loop between the internal inductance and the LHCD antenna phasing, which allows a certain control on the current profile width, ... But Tore Supra is also equipped with an increasing series of feedback loops relevant to a safe long pulse operation, i.e. very accurate plasma shape controller (R and Z to be controlled within millimetres, triangularity within percents), feedback loops involving radiated power, local surface temperature of plasma facing components (infrared measurements), water cooling flows, ... The electron cyclotron emission and the hard-X ray tomography signals will also be available in real time during the coming campaign, allowing determination of the electron temperature and lower hybrid current profiles during the pulses.

The success of advanced tokamak physics is intimately linked to our capability to access key information and to react in real time on the major plasma parameters, including some radial profiles. The difficulty to develop performant algorithms can be summarize under three major bullets:

i) The relevant physics addresses detailed profile characteristics, and thus imposes to access and treat a very large amount of information in real time. Some of this information is even hardly accessible by diagnostics and/or requires long computation time.

ii) The actuators mostly act on global parameters and/or upon several profiles simultaneously. One must also note their few number and the tendency to become "weaker" in the large size next step devices (lower power densities).

iii) Our present level of predictability is low, in the sense that the routes towards the desired stationary targets are still difficult to model satisfactorily.

Thus, it appears crucial to consider in the control algorithms a part of self-optimisation from the discharge itself. In fact, the real time control process is then conceived more like an online help to decision, rather than like a permanent constraint along the discharge. One can then envisage to run a tokamak discharge the following way (note that this is naturally relevant of long duration discharges):

i) Global plasma/machine parameters are prescribed: would it be the toroidal magnetic field, the plasma shape, …

ii) Global constraints are imposed: in the case of a stationary discharge, for instance, the primary flux consumption might be imposed to zero. The discharge must then run constantly under these constraints. Note that one might envisage to suppress this constraint during given phases of the plasma scenario, if necessary. Constraints concerning the safety of the subsystems are also to be included.

iii) The series of actuators is identified: one mostly recognizes here the additional power systems, with their usual parameters of power, phase, …, but also the gas fuelling (puff, pellets, …) and pumping or, on limited time duration, the primary flux consumption. Each actuator is given an operating window.

iv) The physics limits are pre-programmed: this part plays a key role in the self-optimisation process. These limits have either a zero-D character (as the density limit, the maximum radiated power fraction, …), or a more sophisticated one-D character (as the MHD stability domain dependent upon the actual pressure and current density profiles, the transport enhancement conditions, …).

v) The algorithm is then in charge of running the plasma discharge satisfying an optimisation condition, say the maximisation of the fusion power, or of the stored energy for instance. After a plasma initiation phase, a snapshot of the discharge is thus taken from the real time data, and the actual value(s) of the quantity(ies) to optimise is stored. The algorithm then guesses whether a given step in one of the actuators is compatible with the physics limits. This guess might be strict ("the plasma density when increased by one step will remain below the Greenwald density limit" for instance) or more fuzzy ("the plasma density when increased by one step will still remain "far enough" from the Greenwald density limit", "far enough" to be quantified). Of course, the more constraints and prescribed limits, the more sophisticated the decision step will be. All the physics and knowledge are located at that step, would they come from first principles or from fits or various learning processes. If a green light is given by the controller, the action is taken ("the density is increased"), the discharge evolves following the constraints and the controller waits for a new state to establish. The duration here might either be pre-programmed or determined by the controller itself. It is essentially determined by the characteristic time of the (local) current profile relaxation. The following snapshot of the discharge is taken from the real time data, and the quantity to optimise is compared to its previous value. If better, the same actuator is used again under the same procedure, if worse the controller comes back one step and restarts the procedure on the following actuator.

The controller then progressively brings the plasma discharge from its initial state to the "best" situation in terms of fusion power, or stored energy, and on a safe route (if not always the fastest). The procedure described here above somewhat decorrelates the plasma engineering procedures from i) the physics governing the discharge and ii) the choice of constraints and limits. For instance, even if several local optima exist, possibly depending upon current density or velocity shear profiles for instance, the controller will cope with such a situation, if it is properly taught. The procedure also gives room to a given device to follow its own learning curve, along a given discharge and/or from discharge to discharge.

As a matter of illustration, such a real time central controller has been implemented in the CRONOS suite of code[21]. The code is not used as an interpretative tool for an existing discharge, but as a "virtual tokamak" enabling us to adjust the engineering part of the controller, and test the best strategy in terms of steps in the actuators, delays between snapshots, ... The Tore Supra device is chosen, as this first demonstration will be proposed for an actual experiment.

The global parameters are fixed as follows: circular shape R=2.40m, a=0.72m, B_T=3.9T. The constraint forces to operate at constant primary flux (V_{loop}=0 V). Only two actuators are selected: i) the LHCD power, bounded between 0 and 7 MW, with steps of 0.3MW, and ii) the line averaged density (actuated by gas fuelling and pumping), bounded between 0 and 10^{20}m^{-3}, with steps of 2.10^{18} m^{-3}. A single zero-D limit condition is pre-programmed: the density must not be expected to overcome 95% of the Greenwald limit. The optimisation condition is chosen to be the maximisation of the fusion power. The algorithm was developed with three different delays between snapshots (0.5, 1 & 2s), resulting in three different routes. The controller was then able to bring itself the discharge from a low density – low current hybrid discharge to the highest combination of density/power/current compatible with the imposed limits and constraints, as expected in such a simplistic demonstration case. It is interesting to note that once the vicinity of the optimum point is reached, the algorithm carries on oscillating in terms of actuators, in a stationary way. The time requested to reach the optimum region ranges from 20 to 100 seconds depending on the delay between snapshots (Fig4).

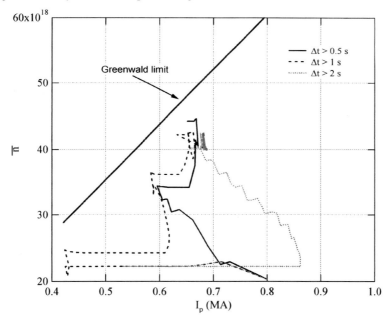

Fig4: Tore Supra: self optimisation trajectories in the line averaged density / plasma current operating space, for three different delays imposed between snapshots.

Of course, when transposing this type of algorithm to larger ignited stationary discharges, constraints and limits will significantly increase and gain in complexity, actuators will change, physics will change, but the procedure should basically remain the same. From now on, it can be developed and tested on existing long pulse devices, with the help of integrated codes, as the CRONOS suite.

CONCLUSION

Besides the considerable technological actions needed to bring tokamaks from the plasma performance exploration to the stationary production of fusion energy, the integration of physics aspects linked to stationarity requires a dedicated effort. Various regimes are candidate, either to an extended pulse length regime or to a truly steady-state solution. Many devices, like JET and Tore Supra, are dedicating an increasing experimental time to generate and study stationary plasma targets. This issue is intimately linked with the capability to diagnose the discharge in real time, and then to react through the relevant actuators in order to drive the plasma towards the desired operating point, and/or to sustain this operating point. Real time feedback controls are now routinely used on Tore Supra and JET, from simple one-to-one actions to more sophisticated non-diagonal model-based controls, mixing physics quantities to control and actuators. The safety of the devices (plasma facing components, disruptions, ...) are progressively also included in the algorithms. Finally, it is proposed to use real time controls also as an online help to decision in a self-optimisation process, rather than as a permanent constraint along the discharge.

ACKNOWLEDGEMENTS

It is a great pleasure to acknowledge many fruitful discussions among the members of the ITPA Group on Steady-State Operation and Energetic Particles, in particular with Drs Claude Gormezano, Tim Luce and Shunsuke Ide.

REFERENCES

[1] V. Basiuk, J.F. Artaud, F. Imbeaux, et al, accepted in Nucl. Fus.
[2] Y. Peysson, « Status of lower hybrid current drive », in *Proceedings of the 13th Topical Conference on Radio Frequency Power in Plasmas*, AIP Conf. Proceedings 485, Melville, New York, 1999, pp183-192
[3] B. Beaumont et al, "Lower Hybrid system upgrade on Tore Supra", this conference
[4] J. Jacquinot, "Recent development in Cadarache towards steady-state operation", in Proceedings of the 19th Fusion Energy Conference, IAEA, Lyon, France, 2002
[5] Hoang, G. T., Saoutic, B., Guiziou, L., et al., Nucl. Fusion **38**, 117 (1998)
[6] K. Vulliez, G. Bosia, G. Agarici, et al., "Tore Supra ICRH antenna prototype for next step devices", in Proceedings of the 22nd Symposium on Fusion Technology, Helsinki, Finland, 2002.
[7] Hoang, G.T., Bourdelle C., Garbet X., *et al.*, Phys. Rev. Letters **87**, 125001-1 (2001)
[8] F. Imbeaux, JF Artaud, V. Basiuk, et al., « Separate and combined Lower Hybrid and Electron Cyclotron current drive experiments in non-inductive plasma regimes on Tore Supra », this conference
[9] R. Prater, "ECCD in DIII-D: Experiment and Theory", this conference.
[10] G. Giruzzi, F. Imbeaux, JL Ségui, et al., submitted to Physical Review Letter
[11] A. Ekedahl, et al., this conference.
[12] Wolf R et al, Plasma Physics and Controlled Fusion, **41**, B93-107 (1999)
[13] E. Joffrin, et al., Plasma Physics and Controlled Fusion, **44**, p.1203-1214 (2002)
[14] N. Hawkes, C. Giroud, E. Joffrin, et al., Phys. Rev. Letters, vol.87 n°11 p.115001-1/115001-4 (2001)
[15] X. Litaudon, A. Bécoulet, F. Crisanti, R. Wolf et al., "Progress towards steady-state operation and real time control of ITBs in JET", in Proc. of the 19th Fusion Energy Conference, IAEA, Lyon, France, 2002
[16] R. Dux, C. Giroud, et al., "Impurity behaviour in ITB discharges with reversed shear on JET", in Proceeding of the 28th EPS conference on Contr. Fusion and Plasma Physics, Madeira, Portugal, 2001, ECA, vol25A, p505.
[17] G. Tresset, X. Litaudon, D. Moreau, et al., Nuclear Fusion, **42** p. 520-526 (2002)
[18] D. Moreau, et al., "Real Time control of the current density profile in JET", this conference
[19] E. Joffrin, "Integrated scenarios at JET using real time profile control", to appear in Proceeding of the 30th EPS conference on Contr. Fusion and Plasma Physics, St Petersburg, Russia, 2003.
[20] G. Martin, J. Bucalossi, A. Ekedahl, et al., "Real time plasma feed-back control : an overview of Tore Supra achievements", in Proceedings of the 18th Fusion Energy Conference, Sorrento, Italy, 2000
[21] V. Basiuk, JF Artaud, A. Bécoulet, et al., "Towards predictive scenario simulations combining LH, ICRH, and ECRH heating", this conference

Long Distance Coupling of Lower Hybrid Waves in ITER Relevant Edge Conditions in JET Reversed Shear Plasmas

A. Ekedahl[1], G. Granucci[2], J. Mailloux[3],
V. Petrzilka[4], K. Rantamäki[5], Y. Baranov[3], K. Erents[3], M. Goniche[1],
E. Joffrin[1], P.J. Lomas[3], M. Mantsinen[6], D. McDonald[3],
J.-M. Noterdaeme[7,8], V. Pericoli[9], R. Sartori[10], C. Silva[11], M. Stamp[3],
A.A. Tuccillo[9], F. Zacek[4] and EFDA-JET Contributors[*]

[1] *Association Euratom-CEA, CEA/DSM/DRFC, CEA-Cadarache, 13108 St Paul-lez-Durance, France*
[2] *Associazione Euratom-ENEA sulla Fusione, IFP-CNR, Via R. Cozzi, 53 - 20125 Milano, Italy*
[3] *Euratom/UKAEA Fusion Association, Culham Science Centre, Abingdon, OX14 3DB, UK*
[4] *Association Euratom-IPP.CR, Za Slovankou 3, P.O.Box 17, 182 21 Praha 8, Czech Republic*
[5] *Association Euratom-Tekes, VTT Processes, P.O.Box 1608, FIN-02044 VTT, Finland*
[6] *Helsinki Univ. of Technology, Association Euratom-Tekes, P.O.Box 2200, FIN-02015 HUT, Finland*
[7] *Max-Planck-Institut für Plasmaphysik, Euratom Association, D-85748 Garching, Germany*
[8] *Gent University, EESA Department, B-9000 Gent, Belgium*
[9] *Associazione Euratom-ENEA sulla Fusione, CR Frascati, C.P. 65, 00044 Frascati, Rome, Italy*
[10] *EFDA Close Support Unit (Garching), Boltzmann-Str. 2, D-85748 Garching, Germany*
[11] *Associaçao Euratom-IST, Centro de Fusao Nuclear, 1049-001 Lisboa, Portugal*

Abstract. A significant step towards demonstrating the feasibility of coupling Lower Hybrid (LH) waves in ITER has been achieved in the latest LH current drive experiments in JET. The local electron density in front of the LH launcher was increased by injecting gas (D_2 or CD_4) from a dedicated gas injection module magnetically connected to the launcher. P_{LHCD}=3MW was coupled with an average reflection coefficient of 5%, at a distance between the last closed flux surface and the launcher of 10cm, in plasmas with an internal transport barrier (ITB) and H-mode edge, with type I and type III ELMs. Following a modification of the gas injection system, in order to optimise the gas localisation with respect to the LH launcher, injection of D_2 proved to be more efficient than CD_4. A D_2 flux of $5-8\times10^{21}$el/s provided good coupling conditions at a clearance of 10cm, while when using CD_4, a flux of 12×10^{21}el/s was required at 9cm. The plasma performance (neutron rate, H-factor, ion temperature) was similar with D_2 and CD_4. An additional advantage with D_2 injection was found, as it reduced the amplitude of the ELMs, which further facilitated the LH coupling. Furthermore, preliminary results of the study of the behaviour of electron density profile in the scrape-off layer during injection of C_2H_6 and C_3H_8 are reported. Finally, the appearance of hot spots, resulting from parasitic absorption of LHCD power in front of the launcher mouth, was studied in the long distance discharges with near gas injection.

---

[*] See annex of Paméla, J., et al., in *Proc. of the 19th IAEA Fusion Energy Conf.*, Lyon, France, 2002.

INTRODUCTION

Coupling of radio frequency (RF) waves in H-mode plasmas with large amplitude ELMs is a problem in existing tokamaks, both for Lower Hybrid (LH) Current Drive (CD) systems, as well as for Ion Cyclotron Resonance Frequency (ICRF) systems. The steep density gradient associated with the H-mode can bring the density in front of the antennas below the cut-off densities needed for efficient coupling. In addition, the presence of ELMs causes perturbations at the edge that give very rapid variations in loading in front of the antennas. Furthermore, the next step tokamak, ITER, is foreseen to operate at a distance between the Last Closed Flux Surface (LCFS) and the first wall of >12cm. Large efforts are therefore placed on finding methods for improving the coupling of these RF systems in these conditions, in view of operation in ITER.

For the LHCD system in JET, gas puffing in the vicinity of the launcher has proven efficient for increasing the local electron density, thereby improving the coupling of the LH waves. The launcher is equipped with a dedicated gas injection pipe, located in the outer wall 1.2m from the launcher. Before the experiments reported here, the gas injection system was changed back to its original design, as used in the experiments before 1997 [1]. In the experiments carried out with D_2 injection in 1997-1999 [2, 3] and with CD_4 injection in 2000-2001 [4, 5], the gas injection system had two extra holes added in the upper part of the pipe. When comparing the efficiency of improving the LH coupling using D_2 injection with the two designs, it appeared that the original design (with eight holes, rather than ten holes) gave a better localisation of the gas injection with respect to the launcher.

EXPERIMENTAL SCENARIO

A dedicated experiment was carried out in the JET tokamak in spring 2003, in order to demonstrate the feasibility of coupling LH waves under ITER relevant conditions, i.e. at large distance between the LCFS and the launcher, and with type I ELMs. The plasma scenario was based on a configuration at $I_P=1.5MA$ and $B_T=3T$, used in experiments aimed at achieving internal transport barriers (ITBs) in steady state conditions [6]. The scenario has a fast current ramp-up with 2-2.5MW LHCD, in order to produce a target q-profile with reversed shear, before the application of Neutral Beam Injection (NBI) and ICRF. $P_{NBI}\sim16MW$ was requested in these experiments in order to produce type I ELMs. The ICRF power was maintained at a moderate level, and the ICRF antenna located next to the LH launcher was not used, as this antenna in particular can degrade the LH coupling when operating simultaneously [7].

The LH coupling obtained in these discharges without near gas injection is shown in Fig. 1. The launcher was retracted 2cm behind the poloidal limiter while the LCFS was held at 5cm from the limiter. The LHCD power was stepped down at 44s and the re-applied at 45s, due to diagnostic purposes. However, the low level of LHCD power coupled at 45s and onwards was due to poor coupling, i.e. too low density in front of the launcher. The low coupled power level and the ragged power signal is caused by a launcher protection system that switches off the power on a klystron if the difference in the reflected powers in the two multijunctions powered by the klystron is too large. Fig. 2 shows the result when adding CD_4 injection near the LH launcher, at a rate of 8×10^{21}el/s, in an otherwise identical discharge. P_{LHCD}=2.5MW could be maintained during type I and type III ELMs. At the end of the discharge the gas injection rate was stepped down. This resulted in an increase in average power reflection coefficient (RC) and a decrease in coupled LHCD power, as this rate was not sufficient to maintain sufficient electron density in front of the launcher.

All discharges in this experiment had a second current ramp-up in the later phase of the pulse. This was programmed in order to be able to study the appearance of hot spots in the divertor region, caused by the acceleration of particles in front of the LH launcher, as described later. It can be noted that at the start of the second current ramp-up which takes place at 47.0s (the plasma is initiated at 40.0s) in Fig. 1 and Fig. 2, the ELM behaviour changes from type I to type III and the neutron yield increases, due to the formation of an ITB. It has been found that a large fraction of current at the edge, as obtained during a current ramp, can have some effect on the ELM activity [8].

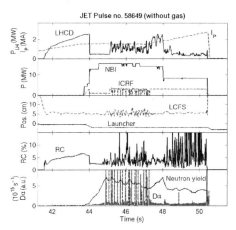

FIGURE 1. Poor LH coupling on type I and type III ELMs without near gas injection.

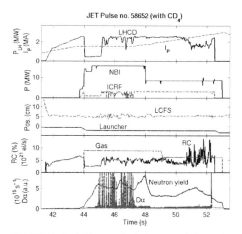

FIGURE 2. LH coupling improved by using injection of CD_4 near the LH launcher.

LONG DISTANCE LH COUPLING WITH CD_4 AND D_2

In the subsequent discharges, the LH launcher was placed closer to the poloidal limiter (0.5cm in the shadow of the limiter), while the LCFS was moved further away from the limiter. The result of a discharge with CD_4 injection near the launcher, in which the LCFS was 9cm from the limiter, is shown in Fig. 3. The second current ramp-up was delayed until 48.0s in this discharge, in order to maintain the type I ELM activity, as required for the LH coupling experiment. The long distance discharges were then repeated with injection of D_2 from the gas injection pipe near the launcher. The result of one of these discharges is shown in Fig. 4. $P_{LHCD}=3MW$ was coupled with an average reflection coefficient of 5%, at a distance between the LCFS and the launcher of 10.5cm. When using D_2 injection, the amplitude of the ELM was smaller compared to the discharges with CD_4 injection, which was beneficial for coupling, both of LH waves and ICRF waves.

The decrease of the ELM amplitude was not the main factor for reducing the LH reflection coefficient in the discharges with D_2, compared to CD_4. A detailed comparison of the reflected power signals during CD_4 and D_2 injection, at similar ELM amplitude, revealed that the reflected power signals were lower and had less fluctuation in the pulses with D_2 injection. This suggests that the electron density in front of the launcher was higher when using D_2 injection, even though the injected rate of electron/s was lower in the discharges with D_2. The measurements of the electron density profile in the scrape-off layer (SOL), taken by a reciprocating Langmuir probe magnetically connected to the LH launcher, confirm that the electron density in the SOL was higher when using 8×10^{21}el/s D_2 compared to 12×10^{21}el/s CD_4. This can probably be explained by the fact that deuterium recycles at the plasmas edge, which further increases the electron density.

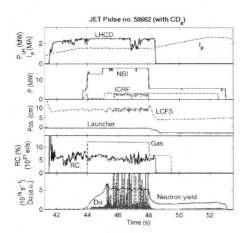
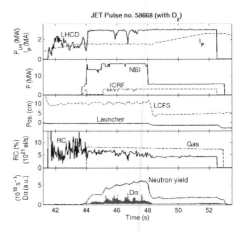

FIGURE 3. Long distance coupling during type I ELMs, using CD_4 injection (12×10^{21}el/s) near the LH launcher.

FIGURE 4. Long distance coupling with D_2 injection near the LH launcher. The amplitude of the ELMs reduced when using D_2.

PLASMA BEHAVIOUR WITH CD_4 AND D_2

Several pulses were carried out with CD_4 injection at $10\text{-}12\times10^{21}$el/s and with D_2 injection at $7\text{-}8\times10^{21}$el/s. Similar plasma performance was obtained with the two gases, as inferred from plasma parameters, such as neutron yield, ion temperature, and H-factor. Comparisons of the electron density profiles and the ion temperature profiles from otherwise identical discharges, one with CD_4 injection and one with D_2 injection, are shown in Fig. 5 and Fig. 6. Both figures show that an ITB was produced in the discharge with D_2 injection. This behaviour is in contrast to the results obtained in the earlier Optimised Shear experiments in 1998-1999, in which D_2 injection at 8×10^{21} el/s rather provoked type I ELMs and caused a degradation of the ITB [2, 3]. In addition, the LH reflection coefficient could not be maintained at a sufficiently low level without reducing the distance between the LCFS and the launcher to 5cm, or less. These results initiated the use of CD_4 injection near the LH launcher in order to improve LH coupling and still maintaining the ITBs [4, 5].

The reason to which D_2 injection has become efficient for improving the LH coupling in these latest experiments is most likely the change in the gas injection design, from ten holes to eight holes, which improved the localisation of the gas injection in the area magnetically connected the launcher. What concerns the behaviour of the ITB with D_2 injection in the latest and in the earlier experiments, it should be noted that the experiments were carried out in different plasma configurations. The earlier experiments (in the Optimised Shear configuration) used a flat q-profile, while the present ITB experiments use a q-profile with reversed shear. It has been found that this latter configuration is robust towards gas injection and gas injection has been used in recent ITB experiments in JET to modify the ELM-activity.

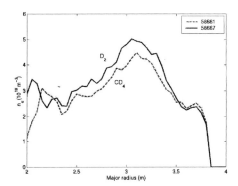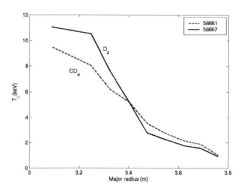

FIGURE 5. Electron density profiles in a discharge with D_2 injection (solid line) and with CD_4 injection (dashed line).

FIGURE 6. Ion temperature profiles in a discharge with D_2 injection (solid line) and with CD_4 injection (dashed line).

ELECTRON DENSITY IN THE SCRAPE-OFF LAYER DURING INJECTION OF DIFFERENT GASES

CD_4 was chosen as an alternative gas to D_2 to be used for LH coupling improvement, because its ionisation cross section is higher than that of D_2 and because

CD$_4$ does not cause an increase in the separatrix density or in the main plasma density [9]. C$_2$D$_6$ or C$_3$D$_8$ should be even more efficient for increasing the electron density in the SOL, since their ionisation cross sections are even higher than that of CD$_4$. This was investigated in a specific experiment using the same plasma scenario as presented above, and using injection of C$_2$H$_6$ and C$_3$H$_8$ near the LH launcher. In order to compare the results obtained with the four gases (D$_2$, CD$_4$, C$_2$H$_6$ and C$_3$H$_8$), the corresponding electron density profiles were normalised to the same rate of electrons/s (4×10^{21}el/s). This is shown in Fig. 7. CD$_4$, C$_2$H$_6$ and C$_3$H$_8$ seem to follow their respective ionisation cross sections at ~25eV. The density obtained with D$_2$ is higher than what expected from its ionisation cross section, probably due to the recycling effect.

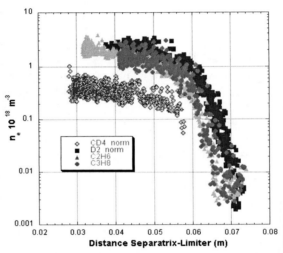

FIGURE 7. Electron density in the scrape-off layer normalised to the same electron/s rate (4×10^{21}el/s) during injection of different gases from the gas injection pipe near the LH launcher.

STUDY OF LH GENERATED OF FAST PARTICLES IN FRONT OF THE LH GRILL

Parasitic absorption of LH power in front of the launcher can lead to acceleration of fast particles, which travel along the magnetic field lines and impinge on components magnetically connected to the launcher, causing localised hot spots. Since the power involved in the parasitic absorption depends on the electron density in front of the launcher [10, 11], it is possible that this effect can become significant when using local gas injection near the launcher. The second part of the discharges in the LH coupling experiments was therefore dedicated to this study. A current ramp-up was programmed in order to change continuously the angle of the magnetic field line so as to ensure that, at some point during the ramp, there would be a magnetic field line connection between the launcher and the specific region of the divertor, which was monitored by CCD cameras. The CCD images of the LH produced hot spots were then analysed in order to obtain a qualitative picture of the dependence of the hot spots. A preliminary analysis of the data obtained in this experiment and in earlier LHCD experiments using local gas injection is shown in Fig. 8. It can be seen that the relative brightness of the hot spots decreases when increasing the distance between the LCFS and the launcher, which is an encouraging result in view of long distance operation with LHCD, as required in ITER. Note that, since only CCD cameras were used, no quantitative estimate of the power involved was made in the JET experiments. However, experiments in Tore Supra have shown that only <2% of the LHCD power is involved in this parasitic absorption, in conditions where the electron density in front of the launcher is less than $1 \times 10^{18} m^{-3}$ [11]. Finally, it is important to note that, although these hot spots can be potentially dangerous in present experiments with conventional LH launchers, as they can involve very localised heat loads, this effect will be negligible for the Passive Active Multijunction (PAM) launcher, as foreseen in ITER [12].

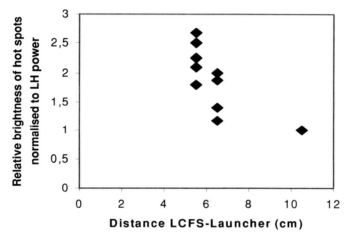

FIGURE 8. Relative brightness of hot spots as a function of the LCFS-launcher distance for several discharges with CD_4 or D_2 injection near the launcher, and LHCD power in the range 1-3MW.

SUMMARY

Very encouraging results regarding coupling of LH waves under ITER relevant edge conditions have been obtained in recent LHCD experiments in JET. 2.5MW LHCD power was coupled at a distance between the LCFS and the launcher of 9.5cm during type I ELMs, while 3MW was coupled at a distance of 10.5cm during type III ELMs.

At comparable injected electron/s rate, D_2 injection near the launcher proved more efficient than CD_4 for improving the LH coupling. This can probably be attributed to the recycling effect of the deuterium, which further increases the electron density, as confirmed by the reciprocating Langmuir probe measurements. In addition, D_2 injection reduced the amplitude of the ELMs, which is beneficial for the coupling of LH waves, as well as ICRF waves.

Internal transport barriers could be maintained in the discharges with D_2 injection, and the plasma performance was comparable in the discharges with CD_4 injection and D_2 injection.

Furthermore, the reciprocating Langmuir probe measurements in the experiments with injection of C_2H_6 and C_3H_8 near the LH launcher show that higher electron densities in the SOL are obtained with C_2H_6 and C_3H_8, compared to CD_4. This is in agreement with their corresponding ionisation cross sections.

Finally, preliminary results of the study of fast particle generation in front of the LH launcher indicate that the brightness of the produced hot spots becomes weaker when operating at large LCFS-launcher distance.

ACKNOWLEDGMENTS

The authors acknowledge the support of the UKAEA JET Operator of the JET-EFDA Facility. In particular, the support of the UKAEA Heating and Fuelling Department during the execution of the experiments is gratefully acknowledged and especially that of the members of the LHCD team, i.e. P. Finburg, C. Fleming, P. Haydon and G. Platt. This work has been performed under the European Fusion Development Agreement.

REFERENCES

1. Ekedahl, A., et al., in *Proc. of the 12th Top. Conf. on RF Power in Plasmas*, Savannah, USA, 1997.
2. Joffrin, E., et al., "LH-Power Coupling in Advanced Tokamak Plasmas in JET", JET Internal Report, JET-R(99)09.
3. Söldner, F.X., et al., in *Proc. of the 13th Top. Conf. on RF Power in Plasmas*, Annapolis, USA, 1999.
4. Pericoli, V., et al., in *Proc. of the 14th Top. Conf. on RF Power in Plasmas*, Oxnard, USA, 2001.
5. Sarazin, Y., et al., in *Proc. of the 28th EPS Conf.*, Madeira, Spain, 2001.
6. Litaudon, X., et al., *Nuclear Fusion* **43**, 565 (2003).
7. Ekedahl, A., et al., This Conference.
8. Bécoulet, M., et al., Invited Paper at *the 30th EPS Conf.*, St Petersburg, Russia, 2003.
9. Strachan, J., et al., *Journal Nuclear Materials* **290-293**, 972 (2001).
10. Mailloux, J., et al., *Journal Nuclear Materials* **241-243**, 745 (1997).
11. Goniche, M., et al., *Nuclear Fusion* **38**, 919 (1998).
12. Bibet, Ph. and Mirizzi, F., "Report on ITER FEAT LHCD launcher", Contract FU05-CT-2001-00019 (EFDA/00-553) 2001.

Profile Control and Plasma Start-up by RF Waves Towards Advanced Tokamak Operation in JT-60U

Yuichi Takase for the JT-60 Team*

Graduate School of Frontier Sciences, The University of Tokyo
Hongo 7-3-1, Bunkyo-ku, Tokyo 113-0033 Japan
*Japan Atomic Energy Research Institute, Naka Research Establishment
Naka-machi, Naka-gun, Ibaraki-ken 311-0193 Japan

Abstract. A central electron temperature of 23 keV was obtained by a combination of ECH and ECCD. A highly localized current density profile was obtained by central ECCD, indicating a diffusion coefficient for resonant electrons of at most 1 m^2/s. A comparison of HFS and LFS off-axis ECCD confirmed the theoretically expected trapped particle effect on the current drive efficiency. LHCD was used to broaden the q_{min} radius and the ITB foot radius. N-NBCD was used to shrink the current hole radius. An integrated scenario consisting of (1) a novel plasma start-up method using RF plasma production and induction by vertical field and shaping coils, (2) a noninductive ramp-up stage, and (3) a transition to a high-density, bootstrap-dominated, high-confinement plasma, without the use of the OH solenoid, has been demonstrated. The plasma created by this technique had both internal and edge transport barriers, with $\beta_p = 3.6$ ($\varepsilon\beta_p = 1$), $\beta_N = 1.6$, $H_{H98y2} = 1.6$ and $f_{BS} \approx 90\%$ at $I_p = 0.6$ MA and $\bar{n}_e = 0.5 n_{GW}$.

INTRODUCTION

Advanced tokamak operation with high β, high confinement, and high bootstrap current fraction is necessary to improve the economic competitiveness of a tokamak fusion reactor. Advances were made in controlling the current density profile and the location of the internal transport barrier (ITB) by LHCD and N-NBCD. ECCD is a promising candidate for providing current density profile control, because highly localized current can be driven. Both on-axis and off-axis current drive were investigated, and current drive efficiencies are compared with theoretical expectations.

In conventional tokamak operation, an Ohmic heating (OH) solenoid is used to ramp up the plasma current (I_p) by induction. If I_p ramp-up and sustainment could be accomplished without the use of the OH solenoid, a substantial improvement can be achieved in the economic competitiveness of a fusion reactor by enabling a more compact design with higher magnetic field [1,2]. In particular, elimination of the OH solenoid is a necessity for a low aspect ratio spherical tokamak (ST) reactor [3]. An advanced tokamak plasma with $H_{H98y2} = 1.6$ and $f_{BS} \approx 90\%$ was formed without the use of the OH solenoid. This result opens up the possibility of OH-less operation, which is a requirement for ST reactors, and can also make a substantial improvement in the economic competitiveness of conventional aspect ratio tokamak reactors.

ECH AND ECCD

A fourth unit with toroidal (± 20°) and poloidal steering capability has been installed in addition to the three existing units of 110 GHz ECRF system with a poloidally steerable antenna at a fixed toroidal injection angle of 20° (corresponding to a parallel index of refraction of $N_\parallel = 0.5$ at the plasma center). The fundamental O-mode is launched from the low field side of the torus. Each unit is equipped with one gyrotron capable of delivering 1 MW of output power, and a transmission line with an overall transmission efficiency of 70-80%. Up to 3 MW of power has been injected into the plasma for 2.7 s, and 1.5 MW for 5 s. The total injected energy reached 10 MJ. Further improvements in both injected power (up to 4 MW) and longer pulse length (up to 10 s) are anticipated.

ECCD is a promising candidate for providing current density profile control, because of its ability to drive a highly localized current. Both on-axis and off-axis current drive were investigated, and current drive efficiencies were compared with theoretical expectations. A central electron temperature of 23 keV was maintained for 0.8 s by on-axis EC injection (Fig. 1) [4]. The duration of the high temperature phase was limited by an instability which occurred when q_{min} reached 2.5. This experiment was carried out to investigate ECCD in a high T_e regime typical of a fusion reactor. A highly peaked T_e profile was obtained. The EC driven current density profile was also highly localized (Fig. 1). An upper bound of $D = 1$ m²/s can be placed on the diffusion coefficient of resonant electrons. The spatially integrated EC driven current was 0.74 MA (the total current was 0.6MA), corresponding to a current drive figure of merit of $\eta_{CD} = 4.2 \times 10^{18}$ A/W/m². This figure of merit, obtained at an average T_e of 21 keV, is a factor of two higher than $\eta_{CD} \equiv \bar{n}_e I_{EC} R / P_{abs} = 2.1 \times 10^{18}$ A/W/m² obtained at 7 keV, indicating a less than linear dependence on T_e. A linear calculation based on ray tracing and a relativistic adjoint evaluation of the driven current (without

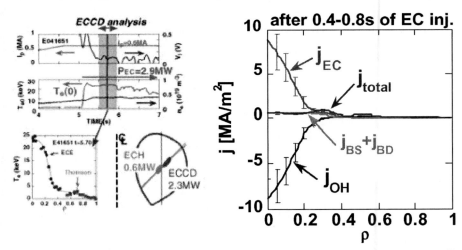

FIGURE 1. Electron temperature of 23 keV maintained for 0.8s (left). Current density profiles (right). The current is strongly overdriven by ECCD near the plasma center.

the effect of the inductive electric field) [5] predicted $\eta_{CD} = 6.3 \times 10^{18}$ A/W/m² for the experimental condition. This discrepancy may be due to the negative loop voltage of 6 mV (8.6% of the Dreicer field) induced in the plasma core because of local overdrive. Several factors are expected to improve η_{CD} for ITER. According to the same calculation, higher density (1.5×10^{20} m⁻³ instead of 0.4×10^{19} m⁻³), larger $N_\parallel$ (0.8 instead of 0.5), and lower Z_{eff} (1 instead of 3) result in increases of 40%, 70%, and 60% in η_{CD}, respectively, resulting in an overall improvement in η_{CD} by a factor of 4.

A comparison of high field side (HFS) and low field side (LFS) off-axis ECCD was made to test the theoretically expected trapped particle effect on the current drive efficiency. The normalized current drive efficiency, $\zeta \equiv (e^3/\varepsilon_0^2)(\eta_{CD}/T_e)$, for the LFS case was about a factor of two lower than that for the HFS case (Fig. 2), as expected theoretically, confirming that the trapped particle effect is important. Note that both the back emf effect and the nonlinear effect are negligible for off-axis CD, because of lower power densities compared to the on-axis case. However, the predicted degradation of ζ with $\varepsilon^{0.5}$ (a measure of the trapped particle fraction) was not confirmed by the present experiment. Data at larger values of ε are needed.

A complete stabilization of the $m = 3$, $n = 2$ neoclassical tearing mode by real time feedback control of the poloidal steering mirror was demonstrated. Details of this experiment are reported by Isayama [6] at this Conference.

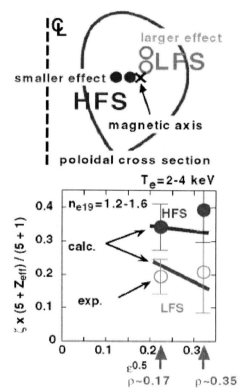

FIGURE 2. Comparison of normalized current drive efficiency for HFS and LFS ECCD.

CURRENT DENSITY PROFILE CONTROL

In addition to the improvements in the ECRF system, performance of the negative ion based neutral beam (N-NB) has also been improved. Up to 6.2 MW of power was injected at 381 keV for 1.7 s, and 2.6 MW at 355 keV for 10 s. These improved capabilities enabled studies of the current density profile evolution. In particular, modification of the current density profile in a reversed shear plasma with a large bootstrap current fraction was achieved by LHCD and NBCD.

N-NB and LH were injected into a reversed shear plasma with a large ITB radius, established by P-NB heating during I_p ramp-up. The loop voltage was slightly negative (0 – -0.2 V) over the entire profile, indicating noninductive overdrive (Fig. 3 left). The bootstrap current fraction was 62%, with the rest provided by LHCD and NBCD [7]. High confinement (H_{H98y2} = 1.4) and high beta (β_N = 2.2) were maintained at a high density ($\bar{n}_e$ = $0.8 n_{GW}$). The electron and ion temperatures were nearly equal at 4 keV.

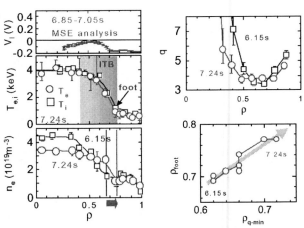

FIGURE 3. V_l, T_e, T_i, n_e profiles (left). q profiles at t = 6.15 s and 7.24 s, and evolutions of ρ_{foot} and ρ_{qmin}.

Normally, without off-axis current drive, the normalized minor radius of the q_{min} location (ρ_{qmin}) decreases with time due to penetration of the inductive current. In the present case, with LHCD just outside ρ_{qmin}, ρ_{qmin} increased gradually during LHCD. The location of the ITB foot (ρ_{foot}), located slightly outside of ρ_{qmin}, also increased as ρ_{qmin} increased (Fig. 3 right). A large current hole [8] already existed prior to injection of N-NB and LH. This is not favorable for confining high energy particles, including fusion produced α particles. Central current drive by N-NB has resulted in shrinkage of the current hole. These results demonstrate that it is possible to modify the current density profile, even when the current is largely self-driven.

PLASMA CURRENT START-UP

An integrated scenario consisting of (1) a novel plasma start-up method combining RF plasma production and induction by vertical field and shaping coils, (2) a noninductive ramp-up stage, and (3) a transition to a high-density, bootstrap-dominated, high-confinement plasma with β_p = 3.6, β_N = 1.6, H_{H98y2} = 1.6 and f_{BS} 90% has been demonstrated on JT-60U [9]. The poloidal field coil configuration of JT-60U is shown in Fig. 4 In these experiments the current in the F-coil, which corresponds to the OH solenoid, was kept constant at zero throughout the entire discharge (Fig. 5). The contributions of the inboard VT (triangularity control) coil, outboard VT coils, and the VR (main vertical field) coils to the vertical field B_v and the poloidal flux Ψ (evaluated at a nominal major radius R = 3.4 m) are:

$$B_v (T) = (-0.537 + 1.948) I_{VT} (MA) + 8.720 I_{VR} (MA)$$
$$\Delta\Psi(Wb) = (30.1 + 88.1) \Delta I_{VT} (MA) + 257.6 \Delta I_{VR} (MA)$$

The two coefficients in the parentheses for the VT coil correspond to contributions from the inboard and outboard turns of the VT coil, respectively. In the experiment described in this paper, the inboard VT coil provided about 20% of the total poloidal flux. This fraction is approximately 70% in the usual start-up scenario using the OH solenoid.

In the example shown in Fig. 5 ($B_T R$ = 13.45 Tm), a plasma with I_p = 0.2 MA was formed by a combination of preionization by EC (110 GHz) and LH (2 GHz) waves and induction by VR and VT coils. The VR and VT

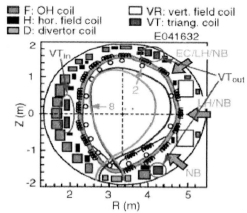

FIGURE 4. Poloidal coil configuration of JT-60U.

coil currents were ramped linearly from +0.1 to +1.1 kA and from −7.3 kA to +6.5 kA, respectively (positive current is defined in the direction that produces B_v required for equilibrium), from t = 2.10 to 2.25 s. Such an operation is necessary because if both coils were ramped from zero, the resultant B_v would become too high to hold the plasma in equilibrium. These current ramps provided a loop voltage of up to 6 V at loop 8 (inboard midplane) and 12 V at loop 2 (close to the upper outboard VT coil). The VT coil set produces poloidal field minima (poloidal field "nulls") at two locations, at the inboard midplane and the outboard midplane. The VT and VR current ramps shift the field minima towards the outboard side. The existence of a field null facilitates I_p start-up.

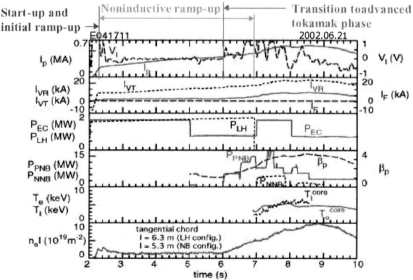

FIGURE 5. An integrated scenario of producing an advanced tokamak plasma without using the OH solenoid.

In a discharge in which VT and VR coil currents were ramped over 70 ms, B_v was initially negative (wrong direction to hold the plasma in equilibrium) and did not reverse sign until 90 ms after the start of ramp, but I_p started to ramp up just 5 ms after the start of ramp. In a discharge that had neither EC nor LH, I_p did not start rising until 90 ms after the start of ramp, when a field null has formed. Therefore, it can be concluded that a strong source of plasma is required for I_p to start up in the absence of proper B_v for establishing a toroidal equilibrium.

For a typical average B_v of 10 mT in a 4T toroidal field, the length along the field line from the vacuum vessel center to the vacuum vessel wall is approximately 600 m, which corresponds to about 30 toroidal revolutions. When plasma current starts to flow, the negative B_v pushes the plasma outward. The eddy current induced in the vacuum vessel by this motion acts to push the plasma back, but this alone is not sufficient. During this time, a continuous source of plasma by EC and/or LH is needed in order to maintain or increase I_p [10]. When B_v becomes positive and large enough, it becomes possible to maintain a toroidal equilibrium. In the example shown in Fig. 5, plasma current started to ramp up at 2.11 s. Magnetic configurations at several time slices, reconstructed using the FBI filament code (which takes into account the vacuum vessel eddy currents) [11], are displayed in Fig. 6. The plasma moves outward during the start-up phase until 2.20 s and becomes limited by the outboard wall. Plasma is shifted to the center of the vacuum vessel again as B_v is increased.

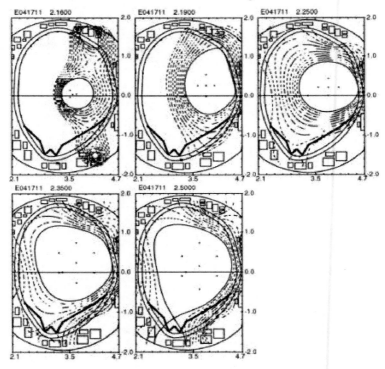

FIGURE 6. Magnetic configurations at several time slices during the rampup phase, reconstructed using a filament code.

It has also been demonstrated that it is possible to start up the plasma current by EC alone. I_p was ramped up inductively by VT and VR coils, as in the case shown in Fig. 5, but without LHCD. It was possible to maintain a constant I_p at 200 kA for 300 ms, but the injected power was not sufficient to ramp up I_p further. The termination of the discharge in this case was caused by a slow positional drift (radially inward, and downward) of the plasma because plasma position was not feedback controlled. This can easily be remedied, and it should be possible to ramp up I_p further with higher EC power.

A transition to a diverted configuration starts at 2.4 s and is accomplished by 2.5 s (Fig. 5). Thereafter, the plasma configuration (plasma position, X-point, etc.) is feedback controlled. Further ramp-up to 0.4 MA was achieved by 6 s, by a combination of electron heating and current drive by EC and LH waves. A current hole is already formed during this phase. The conversion efficiency from the total external noninductive input energy $\int P_{NI}\, dt$ to the total poloidal magnetic field energy $W_m = (L_{ext} + L_{int}) I_p^2/2$ is 3.6%, averaged over the time interval 2.6 to 5.0 s. Here, $P_{NI} = P_{LH} + P_{EC}$ is the total noninductive input power. (Because EC and LH powers were nearly the same, the conversion efficiency would be larger by a factor of two if only the LH power is considered to be useful for I_p ramp-up). The input power from the poloidal field coils P_{ext} was approximately 40% of dW_m/dt. Therefore, the usual definition of current ramp-up efficiency was $(dW_m/dt - P_{ext})/P_{NI} = 2.2\%$. This is a rather low efficiency, and points out that it is desirable to make the maximum use of induction by outboard PF coils. However, it should be possible to raise I_p arbitrarily (limited only by the available power) by extending the noninductive current ramp-up period.

A transition from a low-density noninductively driven phase to a high density, nearly self-sustained (bootstrap dominated) phase begins at 6 s when the current becomes high enough (0.4 MA) to confine the injected beam ions. Density was increased by gas puffing from 5.8 to 7 s, and 85 kV NB injection was started from 6 s. In addition to the noninductive current drive effect, I_p ramps up due to the flux provided by the current increase in VR and VT coils (the latter effect is dominant). Addition of the 376.5 kV negative ion based neutral beam (NNB) contributes to further ramp-up by current drive and β_p increase (NNB dropout at $t = 7.8$ s was not intentional). The plasma generated by this scenario had an internal transport barrier (ITB) and an edge transport barrier (H mode). The current density in the plasma core is nearly zero ("current hole"), and the q profile is deeply reversed with $q_{min} = 5.6$ at $r/a = 0.7$ and $q_{95} = 12.8$. A conservative evaluation of the bootstrap current fraction yielded $f_{BS} = 90\%$ as a lower bound. At $t = 8.5$ s (time of maximum stored energy), $\beta_p = 3.6$ ($\varepsilon\beta_p = 1.0$), $\beta_N = 1.6$, and $H_{H98y2} = 1.6$ were achieved at $\bar{n}_e = 0.5 n_{GW}$. In order to ramp up I_p further by heating under the same condition, it is necessary to increase the β limit (e.g., by wall stabilization).

CONCLUSIONS

A central electron temperature of 23 keV was obtained by a combination of ECH and ECCD. A highly localized current density profile was obtained by central ECCD, indicating an upper limit on the diffusion coefficient for resonant electrons of 1 m^2/s. The experimentally obtained current drive figure of merit was lower than theoretical prediction, possibly because of the negative loop voltage induced by overdrive. A comparison of HFS and LFS off-axis ECCD confirmed the theoretically expected trapped particle effect on the current drive efficiency. Off-axis LHCD was used to broaden the q_{min} radius and the ITB foot radius. N-NBCD was used to shrink the current hole radius.

Plasma start-up, I_p ramp-up, and transition to a bootstrap dominated advanced tokamak with high β and high confinement ($\beta_p = 3.6$, $\beta_N = 1.6$, $H_{H98y2} = 1.6$ and f_{BS} 90%) was demonstrated in JT-60U. This result gives confidence in reducing, and eventually eliminating the OH solenoid in ST and tokamak fusion reactors. In the present experiment, the triangularity control coil with turns on the inboard midplane was used to control the plasma shape. The inboard turns of this coil contributed about 20% of the total poloidal flux input. Demonstration of this start-up technique without using any coils on the inboard midplane is a remaining task. Extension of I_p ramp-up to higher plasma currents (*i.e.*, lower q) and achievement of higher β_N without compromising the bootstrap current fraction is also a topic of future research. Since B_v ramp down (caused for example by a stored energy loss) will ramp down I_p due to the same mechanism, and therefore degrade confinement, a more serious issue is the development of a control algorithm that can react to abnormal events such as a β collapse.

ACKNOWLEDGMENTS

This work was performed as JAERI-University collaboration. Contributions of international collaborators and engineering and technical staff at JAERI are gratefully acknowledged.

REFERENCES

1. Nishio, S., et al., in *Fusion Energy 2000* (Proc. 18th Int. Conf., Sorrento 2000) IAEA-CN-77/FTP2/14.
2. Nishio, S., et al., in *Fusion Energy 2002* (Proc. 19th Int. Conf., Lyon 2002) IAEA-CN-94/FT/P1-21.
3. Stambaugh, R.D., et al., Fusion Technology **33**, 1 (1998).
4. Suzuki, T., et al., in *Fusion Energy 2002* (Proc. 19th Int. Conf., Lyon 2002) IAEA-CN-94/EX/W-2.
5. Hamamatsu, T., Fukuyama, A., Fusion Engineering and Design **53**, 53 (2001).
6. Isayama, A., et al., "Stabilization of neoclassical tearing mode by electron cyclotron wave injection in JT-60U," this conference.
7. Ide, S., et al., in *Fusion Energy 2002* (Proc. 19th Int. Conf., Lyon 2002) IAEA-CN-94/EX/C3-3.
8. Takase, Y., et al., in *Fusion Energy 2002* (Proc. 19th Int. Conf., Lyon 2002) IAEA-CN-94/PD/T-2.
9. Fujita, T, et al., Phys. Rev. Lett. **87**, 245001-1 (2001).
10. Zakharov, L.E., Pereverzev, G.V., Sov. J. Plasma Phys. **14**, 75 (1988).
11. Matsukawa, M., et al., Fusion Engineering and Design **21**, 341 (1992).

Control and Data Acquisition System for Lower Hybrid Current Drive in Alcator C-Mod

N.P. Basse, J. Bosco, T.W. Fredian, M. Grimes, N.D. Kambouchev, Y. Lin, R.R. Parker, Y.I. Rokhman, J.A. Stillerman, D.R. Terry, S.J. Wukitch

MIT Plasma Science and Fusion Center

Abstract. A lower hybrid current drive (LHCD) system is being installed on the Alcator C-Mod tokamak. Initially, 12 klystrons operating at 4.6 GHz will deliver a total power of 3 MW to the coupler. The LHCD system will make it possible to modify the current density profile in the outer half of the plasma. This implies that the q-profile can be manipulated, thereby enabling the study of advanced tokamak regimes. In this paper we describe the overall structure of the control and data acquisition system for LHCD. The acquisition setup collects data from the active controller, the transmitter protection and the coupler protection systems. Long pulse tests of the klystrons are presented and a monitor camera is introduced.

INTRODUCTION

Active control of the current density profile has been shown to enable the attainment of advanced tokamak operation in several magnetic confinement devices (see Ref. [1] and references therein). This has motivated the installation of a lower hybrid current drive (LHCD) system on the Alcator C-Mod tokamak [2, 3]. A combined model for current profile control and magnetohydrodynamic stability analysis has identified stable operating modes for $\beta_N \simeq 3$ and a high bootstrap current fraction $f_{BS} = 0.70$ [4]. For this scenario, the q-profile is controlled by LHCD in the outer half of the plasma, resulting in a non-monotonic q-profile, where $q_{\min} > 2$.

The LHCD system on Alcator C-Mod will initially consist of 12 klystrons operating at 4.6 GHz, supplying a total of 3 MW to a single coupler. In the second phase, 4 klystrons and a second coupler will be added. A schematic view of the coupler is shown in Fig. 1. The plasma position is indicated on the left-hand side, the coupler on the right-hand side. Note the 3 dB divider forming an 'X' close to the center of the coupler. The construction of the LHCD system is a collaboration between Princeton Plasma Physics Laboratory (PPPL) and MIT. PPPL is responsible for the front-end of the system (grill, waveguide run); status of that part of the hardware can be found in Ref. [5].

CONTROL SOFTWARE

The LHCD control software is made up of two separate parts: The operator system (OS) and an active controller system (ACS). The ACS has been described in Ref. [6], so we

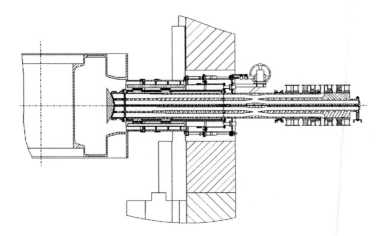

FIGURE 1. Side view of the coupler, taken from [5].

focus on the OS. This has 4 main purposes:

- To construct amplitude and phase waveforms for the klystrons
- To establish the timing sequence
- To acquire raw data
- To calibrate raw data

The main component of the OS is a graphical user interface made using the interactive data language (IDL) running on a Linux platform. It is called the lower hybrid control system (LHCOSY); a preliminary version of the main panel is shown in Fig. 2. The 'RF gate' is controlled from the left-hand column and determines the start time and duration of the klystron pulse. The RF gate is displayed in the central column. Amplitudes and phases supplied to the klystrons can be altered by clicking on the buttons above the RF gate plot (this opens new widgets). Below the plot, timing information can be viewed and modified. The timing information - along with amplitudes, phases and other data - is written to the Alcator C-Mod MDSplus tree. The timing gates are transferred to Jorway 221 modules, which in turn sends the signals to the klystron systems and the ACS. A number of options in LHCOSY makes it easy to re-use waveforms and other settings from previous discharges or test shots. The status of LHCOSY is that all main features are implemented, except the treatment of calibrations.

DATA ACQUISITION AND KLYSTRON TESTS

Data acquisition (and calibration, when completed) is done through the MDSplus tree. The data acquisition hardware consists of a number of compact PCI (cPCI) modules, each sampling 32 channels at 250 kHz with 16 bit resolution [7]. Data is acquired both

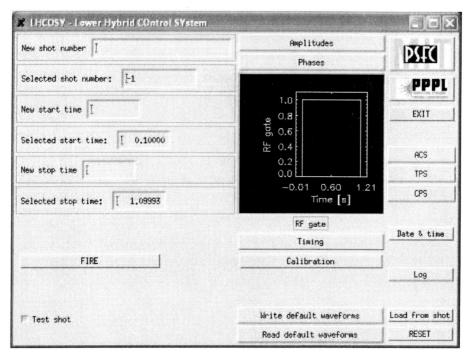

FIGURE 2. Main panel of the lower hybrid control system (LHCOSY).

from the ACS and from the transmitter (TPS) and coupler (CPS) protection systems. The measurements can be viewed either through LHCOSY (from buttons in the right-hand column, see Fig. 2) or using scopes, an MDSplus plotting utility.

To illustrate some of our first results using the cPCI system, we discuss measurements made of TPS signals. These measurements were conducted to test the high voltage power supply and the klystrons [8]. The klystrons are mounted on 3 carts, 4 klystrons on each one; in Fig. 3, we show waveforms from test shot 138. Here, we tested klystron 4 on cart 2 for long pulse, high power operation: The discharge lasted 1 second and the forward power (2nd trace from top) was 256 kW (rating: 250 kW). The body current settled at 20 mA (bottom trace). All 12 klystrons have now been successfully tested for long pulse, high power usage. Note that only voltage measurements are shown in Fig. 3, the calibration part of the software is not yet complete.

The CPS will be used to monitor and protect the waveguide run, for phase calibration purposes and to measure the coupling to the plasma. Since the coupling efficiency is affected by density fluctuations, the CPS detectors will indirectly measure those fluctuations. Therefore we have ordered a 16 channel fast (10 MHz sampling rate) cPCI board to be able to study high frequency density fluctuations in front of the grill.

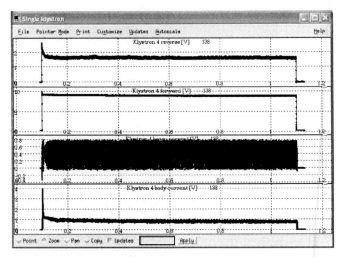

FIGURE 3. Waveforms from a long pulse test of a klystron, top to bottom: Reverse power, forward power, beam current and body current.

MONITOR CAMERA

An important diagnostic in the initial phase of LHCD operation will be a camera. This will allow us to monitor the occurrence of e.g. arcs at the grill. A camera observing a limiter is already installed in a port from where the grill can be viewed. Since this camera will be used alternately for observing the limiter and the grill, we only needed to design and machine a modified camera holder. This has now been done, and the view will be tested during the current Alcator C-Mod campaign. The camera is mounted in a re-entrant port, allowing us to change the camera holder without breaking vacuum.

ACKNOWLEDGMENTS

The authors would like to thank B. Lipschultz, H. Savelli and R.F. Vieira for their help in designing the camera holder and for providing the camera.

REFERENCES

1. Mailloux, J. *et al.*, *Physics of Plasmas*, **9**, 2156–2164 (2002).
2. Parker, R. *et al.*, *Bulletin of the American Physical Society*, **46**, KP1 10 (2001).
3. Bernabei, S. *et al.*, *14th Topical Conference on RF Power in Plasmas*, **1**, 237–240 (2001).
4. Bonoli, P. T. *et al.*, *Nuclear Fusion*, **40**, 1251–1256 (2000).
5. Bernabei, S. *et al.*, *Fusion Science and Technology*, **43**, 145–152 (2003).
6. Terry, D. *et al.*, *19th IEEE Symposium on Fusion Engineering*, **1**, 23–26 (2002).
7. Stillerman, J. A. *et al.*, *Fusion Engineering and Design*, **60**, 241–245 (2002).
8. Grimes, M. *et al.*, *19th IEEE Symposium on Fusion Engineering*, **1**, 16–19 (2002).

A proposal for a planar Lower Hybrid launcher

G. Bosia[1], S. Kuzikov[2], P. Testoni[3]

[1] *Association EURATOM-CEA, CEA/DSM/DRFC, CEA-Cadarache, 13108 St Paul-lez-Durance, (F)*
[2] *Institute of Applied Physics (IAP), Niznhy Novgorod (RF)*
[3] *Università di Cagliari (I)*

Abstract. The concept of a planar Lower Hybrid (LH) wave guide using resonant power division in slotted waveguide, and featuring potential desirable features, such as a simpler construction, less obtrusive dimensions and weight, adaptive matching, and eide range $N_{//}$ modulation capability is presented

INTRODUCTION

The purpose of this paper is to investigate the concept of a planar Lower Hybrid (LH) wave guide array featuring the potential practical advantages of being only few centimetres thick and end fed.. The array could be integrated in the first wall of a large tokamak, in locations other than a main port, and be installed on large, otherwise unused surfaces (such as the inner vacuum vessel wall), so that directivity could be enhanced and average power density minimized.

In the current LH launchers, a multistage binary division is used to bring the input power of the RF source (typically 500-750 kW from a single klystron tube) to a power output of 30-45 kW per waveguide element, consistent with an acceptable electric field at the plasma end. This process requires a rather massive construction that usually constrains the LH launcher to a main tokamak port.

For an overall input power of 20 MW, such as in the ITER case, the number of array elements is large (> 2000) and the LH launcher accordingly complex and difficult to build, also taking into account the required construction tolerances (~ 0.1 mm) and weight (tens of tons). It is obvious that a simpler layout could be useful.

PRINCIPLE OF OPERATION

Current drive requirements (e.g. $N_{//}$ ~ 2, f = 3.7 GHz and $\Delta\phi = 90°$) usually lead to an array period δ ~ 1 cm. In the proposed scheme, a rectangular cross section wave guide cavity is used as a standing wave fed, linear array of N (with N ~ 10) radiating slots, acting as a simple in-phase 1/N power divider.

Different radiating slot geometries can be used, Two examples of resonantly spaced ($\lambda_g/2$) slots radiating in phase in rectangular wave guide geometry, are shown in Figure 1a) and b). In the first case (1a) the slots are cut in the broad face and alternatively offset from wave guide mid plane. In the second, the slots are located in the mid plane of the narrow face, and may be tilted at an angle. In both cases a linear array of elements radiating in-phase is obtained.

If a number of linear arrays of the type shown in Figure 1 a) and b) are placed side by side (Figure 1c and 1d), and the relative phase of the radiated waves electronical

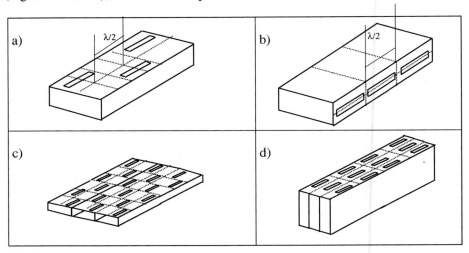

FIGURE 1. Wave-guide arrays : a) slotted in the in the broad face; b) slotted in the narrow face. c) Planar broadside array d) Planar narrow side array.

controlled, the result is a planar array with a governable radiation pattern. If the planar array is positioned on the Tokamak vacuum vessel, and the electric field in the slots is directed in the toroidal direction, to couple to the slow wave, a launcher topology similar to the usual lower hybrid "grill" is obtained..

To act as a power divider, the slotted sections are made resonant by adjustable short circuit plunger(s) and excited in a suitable point at one and both ends.

If in the section of N_s resonant slots, copper losses are neglected, and the slot impedance is adjusted at Z_0/N_s by position and size, the input impedance is Z_0.

As the slot impedance varies with plasma coupling, the input impedance varies accordingly. The end plungers can be used to actively match the input(s) of the structure against resistive and reactive variations, using techniques well developed for Ion Cyclotron heating.

As the power division involves one single array linear element, and each element is individually powered, large and rapid variations of $N_{//}$ can be obtained by electronically varying the phase between two sources.

Field patterns in the slotted waveguide sections

The toroidal spacing d of the LH wave-guide array used to drive plasma current is related to the desired parallel refraction index $N_{//}$, to the incremental toroidal phase step between adjacent elements $D\!f$ (radians) and to the frequency f by $d = (D\!f\ c_0)/(2\pi N_{//} f)$, where c_0 is the speed of light in vacuum. For $N_{//} = 2, f = 3.7$ GHz, and $D\!f = \pi/2$, $\delta = 1.01$ cm.

If a *narrow side* launch is selected, a reduced-width wave guide with b = δ can be used as basic linear array. At 3.7 GHz, the array of reduced-width wave guide elements would be typically ~ 100 mm "thick", and perhaps difficult to to install in the first wall of a small tokamak, but easily integrated in the first wall of ITER blanket modules.

The pattern of the electric field module in a waveguide section with 10 resonantly spaced, $\lambda_g/4$ long slots, radiating from the narrow side, at 5 GHz is shown in Figure 2.

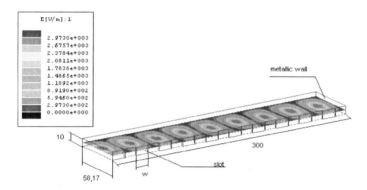

FIGURE 2. Electric field pattern for a 10 longitudinal slot wave guide radiating on the narrow side, fed by the same power at 5 GHz.

If a *broad side* launch, with a = δ is selected, the arr ay is in cut-off at 3.7 GHz, as the vacuum wavelength $\lambda_0 = 81$ mm and $\lambda_c = 20.2$ mm. In this case, we choose $\Delta\phi = 3/2\pi$ in order to increase the wave guide broad side dimension, at the expense of a

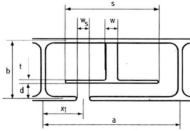

FIGURE 3. Definition of broadside slotted wave guide dimensions

somewhat reduced array directivity (but still maintaining more than one sample per cycle on the toroidal slow wave), and obtain $\delta = a = 30.38$ mm, $\lambda_c = 60.77$ mm, and b = 0.45a = 13.67 mm. A T-shaped ridge geometry shown in Figure 3 can be used to remove the still existing wave guide cut-off (1).

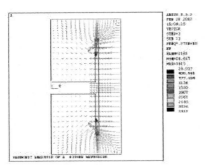 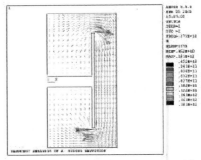

FIGURE 4. E-field and B-fiels distribution in the ridged waveguide

It is found that, for typical normalized values of the ridge dimensions (d/b= 0.25, w/a = 0.1, t/b = 0.1, s/a = 0.7 and d/b = 0.25, the cut-off and guide wavelengths are λ_c = 5.2a =15.18 cm, and $\lambda_g/2$ = 4.71 cm. respectively. For the same set of parameters, the value of the gap impedance at infinite frequency $Z_{g(\infty)}$ = 40 Ω and the gap impedance at 3.7 GHz, is Z_g= 57.3 Ω. In Figure 4 the E-field and B-field distribution in the ridged waveguide section are shown.

Conclusions

Based on the above parameters, an assessment of the electric parameters in vacuum of both broad side and narrow side slotted arrays can be performed. The details of this analysis are not discussed here for lack of space, but in the companion poster.

In short, we find that for slotted section of typically 10 slots, at 10 GHz, the electric field at the slot, is generally below $1kV/m/(W)^{1/2}$ i.e. it is comparable with the one of "grill"launchers. Here however the value depends on the slot width, slot offset and angle respect to waveguide midplane, and can be easily modified to tune the array to the proper plasma loading.

Each linear array is terminated at both ends by an adjustable short circuit plunger, positioned at $\lambda_g/4 + n\ \lambda_g/2$ from the last slot. The resonant condition is maintained by acting on the plunger position and on small frequency modulation. Both plunger and feeder can be located outside the tokamak vessel for easier maintenance and connected to the radiating section by non-slotted sections.

References

1) G.G. Mazumder, P.K. Saha, IEEETrans.MTT, **35-**2, 201 (1987).

Lower Hybrid System upgrade on Tore Supra

B.Beaumont[1], A.Beunas[2], P.Bibet[1], F.Kazarian[1]

[1] *Association EURATOM-CEA, CEA/DSM/DRFC, CEA-Cadarache, 13108 St Paul-lez-Durance, France*
[2] *Thales Electron Devices, 2 rue Latécoère, BP23, 78141 VELIZY Cedex, France*

Abstract. An upgrade of the heating systems is in progress on Tore Supra. It is focused on the manufacturing of a new coupler, using the PAM concept as foreseen for ITER launcher, and the upgrade of the existing transmitter by replacing the 16 existing 500 kW, 210 s klystrons with new 750 kW, 1000s klystrons. Main features of the launcher and klystron are presented.

INTRODUCTION

The LHCD system was implemented on Tore Supra from its start, in 1988. The initial specifications were aiming at plasma duration up to 30s, which was considered as a long discharge at that time. The need of extending the pulse length has been stressed by the studies on plasma wall interaction: erosion, deposition, wall equilibrium and fuelling, etc... Tore Supra, thanks to its supraconducting toroïdal coil system, has a unique opportunity to study these fields during long plasma duration, several minutes, and thus to play a role in preparing the next step.

The inner components of the torus have been just upgraded and partially commissioned: they allow extraction of power and particles on very large time scale. Long pulses experiments have been already initially explored with the existing system, for which power flow is limited, and high energy plasmas up to 0.7 GJ have been performed. The LHCD system, which is the main tool to sustain the plasma current, now requires a major upgrade to reach plasma duration in the range of 1000 s.

This upgrade is being carried out in the 'CIMES' project[1]. The main components of the project are a new LH launcher based on the PAM (Passive Active Multijunction) concept (proposed for the ITER array), a new high power klystron (700 kW CW on VSWR up to 1.4, any phase), and a new quasi-continuous pellet injector (12000 pellets, 16Hz max, 800m/s max.).

LHCD LAUNCHER DESIGN

Experience of long pulse launcher. A first long pulse launcher has been installed in Tore Supra in 1999, using a conventional multijunction array of 6 rows of 48 waveguides (70 x 8 mm²). The geometric period has been chosen in order to radiate N//

---

[1] Composants pour l'Injection de Matière et d'Energie en Stationnaire = steady state power and particle fuelling

with the main peak between 1.7 and 2.3, depending on the feeding phases. Radiating surface is 0.16 m², corresponding to 4 MW coupled power at a reliable flux of 25 MW/m², experimentally obtained on long pulse operation. The results obtained with this launcher have confirmed its design parameters and construction choices.

New design compatible with ITER choices. A second launcher has been designed, following the PAM concept proposed for ITER: an array of alternating active and passive waveguides. This new array is as the former one composed of 6 rows of waveguides, allowing the use of the already designed rear part of the antenna, such as mode converters, vacuum waveguides and vessel. The passive waveguides allow to couple at low density -near (or even lower than) the electron cutoff density-, to improve the neutron shielding properties, and to incorporate more cooling at the antenna mouth (to extract neutrons "heat"), – 3 features highly desirable for ITER.

Each horizontal row will consist of 16 active guides (76 x 14.6 mm²) and 17 passive guides (76 x 13 mm²). The toroï dal pitch is then 35.3 mm, and the N// ranges from to 1.44 to 2. See the Fig.1 and 2 for the lay out of this launcher.

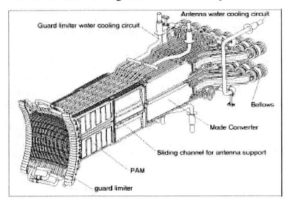

FIGURE 1. General view of the PAM launcher   **FIGURE 2.** View of PAM array

Design features. Each klystron power is divided into 2 poloï dal circuits in the upper and lower part of the launcher. Each part incorporates a DC-break, a bi-directional coupler, and a vacuum window. The following RF circuit in the vacuum is represented in the Fig.3., where can be seen the S shaped vacuum line, the mode converter from TE10 to TE30, and the PAM module featuring the bijunction toroï dal splitting and the large cooling capabilities through the thick passive guide walls.

All components parameters have been optimised, such as bijunction length, taper and mode converter geometry,

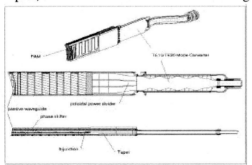

FIGURE 3. View of one PAM circuit.

using finite element electromagnetic field computation.

The expected directivity is 70%, and the coupled power at 25MW/m² is 2.7 MW with a power reflection below 5%. The design allows CW operation. Thermo mechanical calculations, including effects induced by disruptions, predict a fatigue limit in the range of 10000 shots.

The manufacturing of most pieces is under way. The mode converters will be delivered in early 2004, and the PAM modules will be ready at the end of next year, so that the first experiments on plasma are foreseen in 2005.

LHCD GENERATOR UPGRADE

The generator power and availability need to be improved, in consistence with development of launchers. The power level will be increased by the use of a new high power klystron, the TH2103C, which is today in the prototype development phase. The goal is to build a klystron not only able to deliver a higher power than previously obtained with the TH2103, that is to say 750 kW instead of 500 kW, but also to be able to deliver this power on a load presenting a VSWR up to 1.4 for duration up to 1000s. High VSWR on tube output requires special optimisation of output cavity, and long pulse duration requires special development on the thermal control of the cavities and collector.

TABLE 1. Klystron Specifications		TH2103C	TH2103
Output power VSWR<1.1	kW	750 min	500 min
Output power VSWR=1.4 any phase	kW	700 min	
Pulse length	s	1000	60
Duty cycle		1/6	0.25/0.35
Frequency	MHz	3700 ± 5	3700 ± 5
Cathode voltage	kV	76	60
Cathode current	A	22	19.5
Anode voltage	kV	NA	45
Perveance	$\mu AV^{-1.5}$	1.05	2.04
Efficiency	%	45 min	42
Gain	dB	50 min	50 min
Collector power	kW	1680	1300
RF window		1 x TE11 BeO	2 x TE11 BeO
Output guide		1 x WR284	2 x WR284

The new tubes will be installed in TS transmitter in place of the existing tubes, requiring compatibility with all existing equipment: oil tanks, focaliser magnet, power supplies, etc... In addition, compatibility with the supplier test facilities is also mandatory to perform the qualification of the tube and all factory tests. The retained design parameters are listed in the table 1, with the existing TH2103 parameters as reference.

The increased RF output power is obtained by an increase of the cathode voltage and current. As a consequence, the collector load can reach nearly 1.7 MW, and it is designed to keep the wall heat flux bellow the flux that was used on the former TH2103. The tube setup has been simplified: there is no modulated anode, and only one output window, with a specially designed S-shaped vacuum guide (see Fig.5). The suppression of the modulated anode will avoid instabilities on the beam current control experienced on the TH2103: depending on surface conditions and beam interception, secondary electrons can be reemitted by the anode and then invert the anode current. In our set-up, which is using single tetrode modulator, this situation can only be stopped by switching off the high voltage power supply, leading to the loss of 4 klystrons, since one power supply feeds 4 klystrons at the same time. In the new configuration with the TH2103C, a specially developed solid-state switch will allow to switch on or off each of the klystrons. This switch will replace the anode modulator in the oil tank.

FIGURE 4. First TH2103C before baking

FIGURE 5 . Cavities assembly and S-shaped output guide

Mock-ups of output guide and cavity allowed testing fabrication process and heat exchange coefficient of output cavity. Specific tests on the RF window will allow validating the thermo mechanical calculations.

The first demonstrator tube is finished, see Fig.4, and now under test at the factory. The prototype will be delivered at Cadarache in December 2003. The final acceptance tests will take place at the beginning of 2004 and will include performance tests on matched and unmatched load for duration up to 1000s. The test stand is presently being overhauled in preparation of these experiments (control system (PLC), line cooling, RF switches, 1MW CW load).

The first series of tubes will then be produced at a rate of one tube per month from October 2004 on. A first coupler will be equipped at the end of 2005 with its set of 8 tubes for the experiments in 2006.

SUMMARY

Main components of LHCD upgrade are manufactured to allow installation of the PAM launcher in the end of 2004 and increased transmitter power at the fall of 2005.
First experiments with the Tore Supra PAM launcher are foreseen in 2005.

Modeling of Lower Hybrid Current Drive (LHCD) and Parametric Instability (PI) for high performance Internal Transport Barriers (ITBs)

R. Cesario, A. Cardinali, C. Castaldo, F. Paoletti[1], C. Challis[2], J. Mailloux[2], D. Mazon[3], and the EFDA-JET Contributors

Associazione EURATOM-ENEA sulla Fusione, Centro Ricerche Frascati c.p. 65, 00044 Frascati, Italy
[1] PPPL Princeton, USA
[2] Euratom-UKAEA fusion association, Culham Science Centre, Abingdon, Oxfordshire, OX 14 3DB, England
[3] Association Euratom-CEA, CEA Cadarache 13108, St Paul lez Durance, France

ITBs with high performance in time duration (4 seconds) were produced at JET in plasma discharges operating at the plasma current of 2.4 MA and toroidal magnetic field of 3.45T using Lower Hybrid (LH) radiofrequency power (2.3MW) for heating and current drive (CD) [1]. The coupling of the LH radiofrequency power started during the plasma current ramp-up and was maintained during the main heating phase (neutral beam injection (NBI) and ion cyclotron radiofrequency heating (ICRH) combined). The ITB's feature with longer time duration in shots performed by coupling LH power during the main heating phase was attributed to the low magnetic shear located in the outer half of plasma, as effect of the local LH current drive [2].

The first results of the modeling devoted to calculate the LH power deposition and current density profiles for ITB plasmas are presented, considering the same plasma discharges of Ref. 1 and Ref.2. In shot 53432, the LH power coupled only in the early phase of discharge. In shot 53429, the LH power coupling is continued also during the main heating phase. The q-profiles obtained with EFIT-equilibrium code constrained by the polarimetry and the Motional Stark effect (MSE) data are compared with the q-profiles obtained from the JETTO code [4]. The experimental kinetic profiles and the EFIT magnetic reconstruction data of the aforementioned plasma discharges are considered. In the modeling, the LH power deposition and the current density profiles are calculated both by the 1-D LH ray tracing module [5] included in the JETTO code, and by the LH ray-tracing utilizing a 2-D relativistic Fokker-Planck solver [6]. The 2-D code can simulate the effect of the non-linear scattering of LH waves by parametric instability driven by ion sound quasi-modes, producing some LH power to be redistributed in a wide spectrum of the refractive index parallel to the toroidal field direction ($n_{//}$) [7]. The LH power deposition and driven current density profiles are calculated considering the magnetic reconstructed equilibrium provided by the EFIT code. The LH power density profile shown in Fig. 1 is calculated considering the nominal LH power $n_\parallel$ spectrum launched by the antenna ($n_{\parallel 0}$=1.84, width 0.43). A substantially centrally deposition (at $\rho \approx 0.3$, ρ is the square root of the normalized toroidal flux) is obtained. Many passes (>10) are necessary for producing a significant fraction of the coupled LH power to be absorbed. Some broadening (20%) of the launched $n_\parallel$ power spectrum, simulating the effect of a non-linear wave scattering, is considered for obtaining the LH power density deposition profile shown in Fig. 2. The analysis of the parametric instability (which is beyond the scope of the present contribution) is necessary for calculating the realistic propagating $n_\parallel$ power spectrum.

Fig. 1 – LH deposition profiles with nominal $n_\parallel$ power spectrum (see the text)

Fig. 2 - LH deposition profiles with broad $n_\parallel$ power spectrum (see the text)

The obtained LH power density deposition profile is shown in Fig. 2. Most of the LH power is deposited at the first pass, mainly in the outer half of plasma (at $\rho \approx 0.7$). If no broadening is considered (by both the 1-D or the 2-D LH ray tracing modeling), an unrealistic number of passes (>10) is necessary for obtaining significant LH deposition in the outer half of plasma. The LH current density profile is not significantly affected considering the self consistent magnetic equilibrium in the ray-tracing calculation, as a

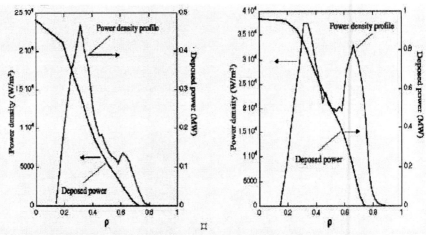

low fraction of the LH driven current occurs in the considered high plasma current discharges. The simulation gives, indeed, a moderate amount (60%) of non-inductive current, including 30% of LHCD fraction. The q-profiles from polarimetry and from MSE at the beginning of the main heating phase are shown in Fig. 3. In shot 53429 (Fig 3a), at the time t=4.4s (before the application of the LH power occurring at t=5.8s), the MSE and polarimetry q-profiles have both a reversed shaping. To be noted the different q-minimum in the profiles at that time: q_{min}=2.8 from MSE and q_{min}=3.2 from polarimetry. The q-profile simulated by the JETTO code has a good agreement with the EFIT-MSE profile at t=4.4 s. The simulation is performed considering the LHCD current density profile obtained by the broad $n_{//}$ spectrum

FIGURE 3 - q-profiles at the beginning of the main heating phase.

FIGURE 4. q-profiles during the main heating phase

In shot 53429, during the ITB phase, the evolution of the q-profile shows a wide region (R=3.6m) with low magnetic shear, (see Fig. 4a). The q-minimum value is lower than the integer value q=3. The ITB persists during the whole time window considered in the figure. In shot 53432 (see Fig. 4b), a well pronounced reversed shaped q-profile occurs around the time of the ITB collapse (t=6.3s). The collapse may be due to MHD starting as the consequence of the crossing of the q-profile with the integer value q=3.

Both the q-profiles simulated by the JETTO code and those obtained by polarimetry for shot 53429 show a low magnetic shear in a wide region off-axis and the q minimum lying between the integer values of 3 and 2. For shot 53432, both profiles show a well-pronounced reversed shape. However, the profile crossing around q=3 is

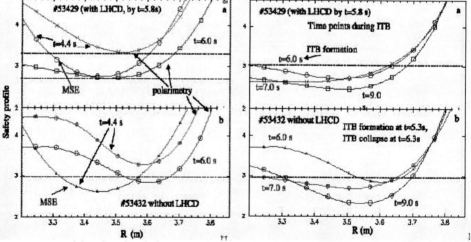

not recovered by the simulation. In a previous work [2], the ITB collapse was related to the inward migration of the layer with negative shear (consistent with the polarimetry q-profile trend), in region dominated by neoclassical transport.

Non-linear plasma edge phenomena allow propagation of some LH power with large $n_{//}$. Such effect should be retained for a realistic LHCD modeling of ITB plasmas. The consequent enhanced off-axis LHCD is consistent with the observed large ITBs and the obtained large region with low magnetic shear. The LH power might provide a powerful tool for controlling the q-profile for ITB at high plasma current, for potential application to the advanced tokamak regimes.

Acknowledgements

The authors acknowledge L.M. Carlucci for her editorial comments.

References
1. J. Mailloux, B. Alper, Y. Baranov et al., Phys. Plasmas **9**, 5 (2002) 2156
2. C. Castaldo, R. Cesario, A. Cardinali, et al., Phys. Plasmas in press
3. G. Cenacchi, A. Taroni, in Proc. 8th Computational Physics, Computing in Plasma Physics, Eibsee 1986, (EPS 1986), Vol. 10D, 57
4. L.L. Lao, J.R. Ferron, R.J. Groebner, et al. Nucl. Fusion **30**, (1990) 1035
5. A.R. Esterkin and A.D. Piliya, Nucl. Fusion **36** (1996) 1501; A. Saveliev et al., private communication
6. A. Cardinali, in Recent Results in Phys Plasmas **1**, (2000) 185
7. R. Cesario, A. Cardinali, Nucl. Fusion **29**, 10 (1989) 1709

Density Convection near Radiating ICRF Antennas and its Effect on the Coupling of Lower Hybrid Waves

A. Ekedahl[1], L. Colas[1], M.-L. Mayoral[2],
B. Beaumont[1], Ph. Bibet[1], S. Brémond[1], F. Kazarian[1], J. Mailloux[2],
J.-M. Noterdaeme[3,4], A.A. Tuccillo[5] and EFDA-JET Contributors

[1] *Association Euratom-CEA, CEA/DSM/DRFC, CEA-Cadarache, 13108 St Paul-lez-Durance, France*
[2] *Euratom/UKAEA Fusion Association, Culham Science Centre, Abingdon, OX14 3DB, UK*
[3] *Max-Planck-Institut für Plasmaphysik, Euratom Association, D-85748 Garching, Germany*
[4] *Gent University, EESA Department, B-9000 Gent, Belgium*
[5] *Associazione Euratom-ENEA sulla Fusione, CR Frascati, C.P. 65, 00044 Frascati, Rome, Italy*

Abstract. Combined operation of Lower Hybrid (LH) and Ion Cyclotron Resonance Frequency (ICRF) waves can result in a degradation of the LH wave coupling, as observed both in the Tore Supra and JET tokamaks. The reflection coefficient on the part of the LH launcher magnetically connected to the powered ICRF antenna increases, suggesting a local decrease in the electron density in the connecting flux tubes. This has been confirmed by Langmuir probe measurements on the LH launchers in the latest Tore Supra experiments. Moreover, recent experiments in JET indicate that the LH coupling degradation depends on the ICRF power and its launched $k_{//}$-spectrum. The 2D density distribution around the Tore Supra ICRF antennas has been modelled with the CELLS-code, balancing parallel losses with diffusive transport and sheath induced E×B convection, obtained from RF field mapping using the ICANT-code. The calculations are in qualitative agreement with the experimental observations, i.e. density depletion is obtained, localised mainly in the antenna shadow, and dependent on ICRF power and antenna spectrum.

INTRODUCTION

Lower Hybrid Current Drive (LHCD) is the main tool for achieving non-inductive current drive, as well as current profile control, with the aim to studying paths to achieve steady state operation in several tokamaks. In recent Tore Supra experiments, a plasma current of 0.5MA has been sustained for more than 4 minutes by the means of LHCD at a power level of 3MW [1]. In JET, LHCD plays an important role both for pre-forming the q-profile in the Internal Transport Barrier (ITB) experiments, as well as for providing an off-axis current during the main heating phase with Neutral Beam Injection (NBI) and ICRF waves [2]. Both scenarios and both tokamaks require application of LH waves simultaneously as ICRF. Effects on the LH coupling observed under these conditions are described in this paper.

COUPLING OF LH WAVES IN THE PRESENCE OF ICRF

Experimental observations in Tore Supra. Tore Supra has two LH launchers, located in adjacent ports, and three 2-strap ICRF antennas. In most experimental conditions the ICRF antenna Q1, which is located close to the LH launchers, is magnetically connected to at least one of the launchers. For efficient LH non-inductive current drive, the long pulse discharges in Tore Supra are carried out at low electron density ($n_e < 2.5 \times 10^{19} \text{m}^{-3}$), which generally lead to poor coupling conditions for ICRF [3]. If ICRF is applied simultaneously as LHCD, the coupling on the part of the LH launchers magnetically connected to the antennas decreases, indicating a decrease in the electron density in the connecting flux tubes. Fig. 1 shows a discharge in which the ICRF antennas were powered, using dipole phasing, in sequence during LHCD. When antenna Q1 was powered, the reflection coefficient on the upper part on the nearby LH launcher increased, while when antenna Q4, which was not connected, was used there was no effect on the LH coupling. Fig. 2 shows the recent results in the CIEL configuration, in which the ICRF antennas are at different locations. However, antenna Q1 is still in the same location and still connected to the LH launchers. Fig. 2 shows the reflection coefficients on the upper and lower waveguide rows on the LH launcher closest to the ICRF antennas. In agreement, the density deduced on the Langmuir probe on the upper part on this launcher decreases as ICRF is switched on, while the density on the lower part stays unaffected. The degradation in LH coupling is mainly seen when the LH launchers are retracted behind the ICRF antennas, indicating that the density decrease takes place in the shadow of the antennas, whici is also in agreement with observations in ASDEX-Upgrade [4].

Experimental observations in JET. JET has four 4-strap ICRF antennas, located in pairs in opposite ports of the torus. The antenna pair A+B is located next to the LH launcher, separated by a poloidal limiter. A series of experiments was conducted in order to studying the effect of mainly antenna B, which is located immediately next to the LH launcher, on the LH coupling. The ICRF power, antenna spectrum, and the position of the LH launcher relative to the antenna were varied. Fig. 3a and Fig. 3b show a comparison of the effect of the ICRF antenna spectrum. When using dipole spectrum, the LH power remained constant as ICRF is switched on. The same result was obtained when using ICRF with +90° phasing. However, when applying ICRF with monopole spectrum, the LH coupled power dropped, indicating an increase in the reflected power levels. The drop in power is caused by a launcher protection system that switches off the power on a klystron if the difference in the reflected powers in the two multijunctions powered by the klystron is too large. Similar effect as monopole phasing was obtained with –90° phasing. The effect of the launcher position relative to the ICRF antennas and the poloidal limiter is demonstrated in Fig. 3b and Fig. 3c. At the position –15mm, <1MW from antenna B affects the LH coupling, while at –5mm, higher power on this antenna can be used. In order to increase the density to allow better coupling of the LH waves, injection of gas (CD_4 or D_2) near the LH launcher is used routinely in JET. The discharge shown in Fig. 3d is identical to the one of Fig. 3c, except that local CD_4 injection was added. As a result, the electron density increased sufficiently to allow maintaining the LH power level. Using local gas injection, up to 3MW of LH power has been coupled in H-mode and ITB-plasmas with type I and type III ELMs, at a plasma-launcher distance of 10cm [5].

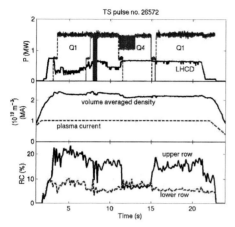 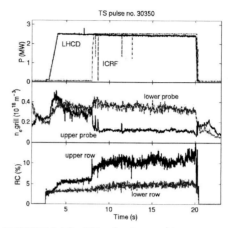

FIGURE 1. Behaviour of the LH reflection coefficients on the upper and lower parts of the launcher as antenna Q1 (connected) and antenna Q4 (not connected) are powered.

FIGURE 2. The LH reflection coefficient on the upper part of the launcher increases as ICRF is switched on. Accordingly, the electron density on the corresponding Langmuir probe decreases.

MODELLING OF ICRF DENSITY CONVECTION

The 3D electric field pattern of the Tore Supra antennas has been calculated with the ICANT-code, which uses the realistic geometry of the Tore Supra antenna Q1 [6]. The output of the ICANT-code is then coupled with the CELLS-code, which calculates the 2D density distribution around the antenna, by taking into account cross-field diffusion, parallel transport losses and the sheath induced E×B-convection from ICRF [7]. Fig. 4 shows the radial dependence of the electron density profile, averaged over the poloidal extent of the Q1 antenna, for the experimental parameters of a discharge in the Tore Supra CIEL configuration (TS pulse no. 30463, $n_e=2.0\times10^{19}m^{-3}$, $I_P=0.8MA$, $B_T=3.8T$), having similar behaviour as the discharge in Fig. 2. The resulting electron density profiles in front of the antenna, assuming 1.5MW with dipole and with monopole spectrum, are shown, as well as the initial ohmic electron density profile. In Fig. 4, the radial positions of the side protections of the ICRF antennas and of the LH launchers are indicated. A stronger density decrease is obtained when using monopole spectrum compared to dipole, because the electric field is stronger with monopole phasing [7]. This is in qualitative agreement with the JET observations (Fig. 3).

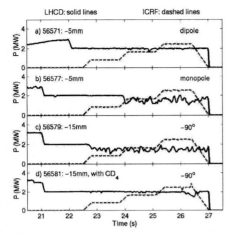
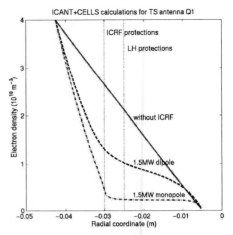

FIGURE 3. Results from JET: Effect of ICRF antenna B on the LH coupled power, with different antenna spectra and different positions of the LH launcher relative to the antenna.

FIGURE 4. Radial electron density profiles averaged over the poloidal extent of the ICRF antenna Q1, for Tore Supra pulse no. 30463. The separatrix is at –0.045m.

SUMMARY

Due to the formation of convective cells, application of ICRF waves can cause a decrease in the electron density near the antenna, which in turn can degrade the coupling of LH waves if the LH launcher is magnetically connected to the ICRF antenna. Avoiding placing an LH launcher next to an ICRF antenna is therefore recommended. The density depletion is stronger in the shadow of the ICRF antennas and, therefore, locating the LH launcher and the ICRF antenna at the same radial position reduces the effect. Being electric field dependent, the density depletion will depend on $P_{ICRF}^{1/2}$, as well as on the antenna spectrum, i.e. monopole spectrum has a stronger effect than dipole spectrum.

REFERENCES

1. Equipe Tore Supra, in *Proc. of the 19th IAEA Fusion Energy Conference*, Lyon, France, 2002.
2. Litaudon, X., et al., in *Proc. of the 19th IAEA Fusion Energy Conference*, Lyon, France, 2002.
3. Goniche, M., et al., *Nuclear Fusion* **43**, 92-106 (2003).
4. Noterdaeme, J.-M., et al., in *Proc. of the 23rd EPS Conference on Controlled Fusion and Plasma Physics*, Kiev, Ukraine, 1996.
5. Ekedahl, A., et al., This Conference.
6. Pécoul, S., et al., *Computer Physics Communications* **146**, 166-187 (2002).
7. Bécoulet, M., et al., *Physics of Plasmas* **9**, No. 6 (2002).

Lower Hybrid Experiments in MST

J.A. Goetz, M.A. Thomas, P.K. Chattopadhyay, M.C. Kaufman, R. O'Connell, S.P. Oliva, and P.J. Weix

Department of Physics, University of Wisconsin, Madison, WI 53706

Abstract. Current drive using RF waves has been proposed as a means to reduce the tearing fluctuations responsible for anomalous energy transport in the RFP. A traveling wave antenna that operates at 800 MHz is being used to launch lower hybrid waves into MST to assess the feasibility of this approach. The power flowing through the antenna has been measured with pickup loops installed in the antenna backplane. The measured power damping length of 2-4 parallel wavelengths is consistent with analytic calculations. The damping length decreases with increasing plasma density and increases in improved confinement discharges. Localized hard x-ray emission has been observed when energy flows in either direction through the antenna.

INTRODUCTION

Experimental and theoretical results indicate that anomalous energy and particle transport observed in the reversed field pinch (RFP) is due to tearing fluctuations. These magnetic fluctuations are reduced when parallel current drive is added in the edge of a standard RFP plasma [1]. Recent inductive current drive experiments on MST [2] have produced a nine-fold improvement in the energy confinement, a tripling of the electron temperature, and a doubling of beta [3]. However, inductive current drive is an inherently transient technique and is non-local in nature. RF poloidal current drive has been proposed to provide steady control and more localized current deposition. Studies have indicated that the lower hybrid slow wave should be a good tool for current density profile control in the RFP [4].

An antenna had been designed and built to launch lower hybrid slow waves into MST in order to address the goal of improving transport in an RFP [5]. Due to the stringent restraints of the MST vacuum vessel, a traveling wave antenna based on an interdigital line [6] was chosen. RF power enters the structure at one end and then propagates to the other end; along the way some power is radiated as a lower hybrid slow wave. The input power is fed to one end of the antenna (port 1 for co-current or port 2 for counter-current drive) and the output end is connected to a dummy load.

Experiments with the prototype antenna showed that port 1 had a power limit of 2-3 kW, above which the antenna was fully reflecting. Port 2 was fully reflecting at all power levels. This antenna had 1 cm diameter vacuum feedthrus and a VSWR ~ 6. Bench measurements of the wave spectrum showed that the desired $n_{||}$ = 7.5 was being produced. It was also noted that for some external tunings, a standing wave resulted instead of a traveling wave. With these limitations identified, a second antenna was designed, built, and installed in MST in May 2002.

EXPERIMENTAL OBSERVATIONS

Improved computer modeling of second MST LH antenna led to better impedance matching and a lower VSWR (~2) and lower Ohmic losses inside the structure. In addition, a larger diameter (2 cm) vacuum feedthru was designed in order to achieve a better impedance match at the coaxial feed and to obtain better power handling. Instrumentation was added to the antenna backplane to measure power flow near the center of the antenna.

Experiments designed to investigate coupling and loading issues were carried out with the second antenna. The present antenna can handle up to 50 kW (for the duration of the 10ms power supply pulse) for both feed directions (see Fig. 1). This corresponds to a power density at the vacuum feedthru of >160 MW-m^{-2}. It is also in contrast to the first generation antenna that had a power limit of ~3 kW. One major improvement to the antenna operation was the implementation of arc detection and suppression circuitry. This circuit detects arcs in the antenna (reflection coefficient > 50%) and interrupts the rf drive to the klystron in 5 μs with a dead time of 100 μs. This technique allows for vastly improved conditioning compared to the first antenna.

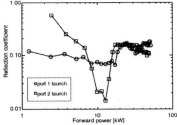 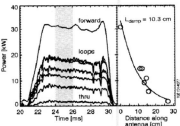

FIGURE 1. The second MST LH antenna handles 50 kW with a good impedance match between the antenna and plasma for both port 1 (circles) and port 2 (squares) feed directions.

FIGURE 2. The power flow through the antenna as measured with the sampling loops is shown on the left side for a typical shot. The power damping length (right side) is calculated from the data in the shaded region.

There is a good impedance match between the antenna and the plasma up to the present transmitter power limit. The power reflection coefficient is low, between 10 and 15%, and is symmetric at high power with respect to power feed direction (Fig. 1). The power reflection coefficient increases only slightly with edge plasma density. The antenna has a good impedance match with plasma present over a wide range of electron density. The power transmission coefficient has been observed to be ~10% when the antenna is well conditioned. It appears now that power flow is determined by plasma loading and not breakdown in the antenna cavity or feedthru.

To evaluate antenna performance, it is desirable to measure the power density along the structure. Thus, power sampling loops were installed behind the center five radiating elements of the second antenna. Each loop consists of a wire inserted in a slot cut in the backplane terminated in 50 Ohms, thus sampling the current density. The power damping length is calculated by fitting an exponential function to the data measured by these five loops together with the input and output power (Fig. 2).

Damping lengths of 5 – 15 cm have been measured in a variety of RFP plasma conditions. The antenna spectrum peaks at $\lambda = 4.8$ cm. Thus, a power damping length of > 10 cm should not significantly diffract the $n_{\parallel}$ spectrum. The measured power damping length is a combination of the lengths due to vacuum losses and plasma interaction. Given that the measured damping length in vacuum is about 56 cm and a total damping length of 15 cm, the damping length due to plasma interaction would be 20 cm. This is consistent with that calculated by analytic means [7].

The power damping length decreases as the edge density increases, implying that antenna loading can be controlled with edge density. The power damping length increases with forward power but the dependence is not very strong. This is expected because the increasing ponderomotive force should decouple the plasma from the antenna. In enhanced confinement plasmas [8] the antenna becomes unloaded. The damping length is much longer, presumably from the measured change in the density profile and the increase in the vacuum gap. In addition, the power damping length depends on the angle of the magnetic field lines at the edge of the plasma (Fig. 3). The coupling efficiency in standard RFP plasmas is best when the field line is within five degrees of the antenna. Future antenna optimization may require rotating the antenna to match the field. Further investigations of antenna performance in enhanced confinement discharges (field line angle ≤ 45 degrees) are planned.

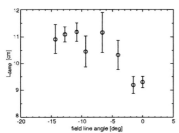

FIGURE 3. The power damping length in standard RFP plasmas is shortest when the magnetic field line angle is within five degrees of being aligned with the antenna (0 degrees).

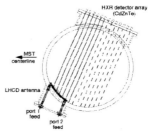

FIGURE 4. The layout of the MST HXR detectors is shown (poloidal plane). The views shown with solid lines observe emission when there is power flowing through the antenna.

Hard x-rays (HXR) are generated when lower hybrid waves are launched into MST. MST has an array of CdZnTe detectors that detect HXRs with energies of 10 – 100 keV [9]. HXRs have been generated when power is fed in either direction through the antenna. However, only viewing chords that intersect the antenna have detected HXR flux (Fig. 4). The HXR flux correlates closely with rf propagation along the antenna. The energies of the HXRs produced is in the range of 10 – 25 keV which is ~10 times the estimated rf potential in front of the antenna rods and greater than that of the ~4 keV electrons resonant with the wave at $n_{\parallel} = 8$. The generation mechanism for these x-rays is still unclear and under further investigation.

DISCUSSION

Great strides have been made in upgrading the MST lower hybrid antenna. The present antenna can handle 50 kW, which is the limit of the klystron given the capabilities of the pulse forming network and high voltage power supply. The impedance match to the antenna is good over a range of plasma parameters and input power when fed from both ports. Instrumentation added to measure power flow along the antenna structure has been used to calculate power damping length. It scales as expected with respect to analytic theory. Hard x-rays with energies of 10 – 25 keV have been produced when there is sufficient power flowing through the antenna in either direction.

In order to carry the MST lower hybrid program forward there will be several upgrades performed to the transmitter system and antenna. The pulse forming network that drives the klystron will be upgraded to a 30 ms – 50 kV pulse. This should allow for 200 kW tube operation. To handle this increased power, the antenna will be redesigned with larger diameter feedthrus (4 cm) and full 3D computer modeling will be used for even better impedance matching of the antenna to the feeds. The next antenna will have power sampling loops along the whole structure, not just the center, to constrain the damping length calculations. In addition, full instrumentation should identify whether or not the feed end of the antenna is unloaded in high power operation. Phase and amplitude electronics are under construction and will be used to look for changes in the antenna dispersion with loading. During this period, current drive efficiency calculations will be performed.

ACKNOWLEDGMENTS

This work is supported by U.S. DoE Contract DE-FG02-96ER54345.

REFERENCES

1. Ding, W.X., Brower, D.L., Terry, S.D. et al., Phy. Rev. Lett. **90**, 035002-1 (2003).
2. Dexter, R.N., Kerst, D.W., Lovell, T.H., Prager, S.C., and Sprott, J.C., *Fus. Tech.* **19**, 131 (1991).
3. Chapman, B.E., Anderson, J.K., Biewer, T.M. et al., Phy. Rev. Lett. **87**, 205001-1 (2001).
4. Uchimoto, E., Cekic, M., Harvey, R.W., *et al., Phys. Plasmas* **1**, 3517 (1994).
5. Goetz, J.A., et al., "Design of a Lower Hybrid Antenna for Current Drive Experiments on MST", in *Proceedings of the 14th Topical Conference on Radio Frequency Power in Plasmas* (2001).
6. Matthaei, G.L., *IRE Transactions on Microwave Theory and Techniques*, **Vol MTT-10**, 479 (1962).
7. Golant, V.E., *Sov. Phys.-Tech. Phys.* **16**, 1980 (1972).
8. Sarff, J.S., Almagri, A.F., Cekic, M., *et al., Phys. Plasmas* **2**, 2440 (1995).
9. O'Connell, R., Den Hartog, D.J., Forest, C.B., and Harvey, R.W., Rev. Sci. Ins. **74**, 2001 (2003).

Effect of the launched LH spectrum on the fast electron dynamics in the plasma core and edge

M.Goniche[1], V.Petržílka[2], A.Ekedahl[1], V.Fuchs[2],
J.Laugier[1], Y.Peysson[1], F.Žáček[2]

[1]*Association EURATOM-CEA, CEA/DSM/DRFC, CEA-Cadarache,*
13108 Saint Paul-lez-Durance (France)
[2]*Association EURATOM / IPP.CR, Czech Academy of Sciences, Za Slovankou 3,*
182 21 Praha 8, Czech Republic

Abstract. The lower hybrid current drive efficiency in the Tore Supra tokamak was investigated in various cases of launched $N_{//}$ spectra. By varying the number of energized waveguides, the broadening of the $N_{//}$ spectrum is varied by a factor 3.5. Weak effect of this broadening is found. The effect of the central value $N_{//0}$ and the MHD activity is also documented. The parasitic losses of fast electrons in the scrape-off layer are analyzed from infrared images of the antenna protection limiter. For these experiments performed at constant LH power, the heat flux scales mainly with the RF electric field. The effect of the number of powered waveguides is discussed.

INTRODUCTION

The lower hybrid current drive system is routinely used on the Tore Supra tokamak in order to drive most of the plasma current and allow long pulse operation. Moreover the ability of the LW waves to modify the current density profile gives access to enhanced confinement regime. In order to minimize the installed RF power, the current drive efficiency is a key parameter, which is addressed in this paper. Specific experiments were carried out in order to vary the width of the $N_{//}$ spectrum lobes for different values of the main peak $N_{//0}$.

Parasitic absorption of LH power near the antennas leads to particle acceleration in the scrape-off layer and localized heat flux deposition on the first wall components magnetically connected to the LH antennas [1,2] which may limit the operational domain of the LHCD system. For various cases of the antenna power feeding, this heat flux was measured on the guard limiter of one of the antenna and compared to results of simulation.

CURRENT DRIVE EFFICIENCY

Two series were performed at the same LH power (1 MW per antenna): the first one at medium plasma current and line-averaged plasma density ($0.8MA/1.8\pm0.15\times10^{19}m^{-3}$) with a significant DC electric field ($V_L\sim150mV$ at 2 MW), the second one at lower current and density ($0.5MA/1.45\pm0.15\times10^{19}m^{-3}$) allowing quasi full current drive ($|V_L|<30mV$) when both antennas are launching power. The Zeff of serie 1 (resp.2) was in the range of 2-2.5 (resp. 2.9-3.4). The current drive efficiency was computed taking into account the plasma conductivity and the hot conductivity [3]. The efficiency from this 0-D modelling was compared, for some shots, to the efficiency from the CRONOS code [4]: when the bootstrap current fraction (10%) and the re-constructed line-averaged density $\int_0^1 n(r/a)d(r/a)$ (7% lower than the direct measurement from the interferometer

$\int_{-a}^{a} n_l dl / 2a$) are taken into account, good accordance is found. For these LH-heated only plasmas, we will include the bootstrap current in the RF-driven current and the direct measurement is quoted for the line-averaged density.

The two LH antennas, MarkI and MarkII have a different number of waveguides in a row: 32 for MarkI and 57 (including passive waveguides) for MarkII, leading to more narrow $N_{//}$ lobes for the latter. By varying the number of klystrons energizing the antennas from 3 to 8, the width ΔN normalized to the standard width of MarkI, ΔN_0, was varied between 0.6 and 2.2. This scan was performed for two $N_{//0}$ values. The current drive efficiency of each antenna is investigated from discharges with residual DC electric field ($\Delta V_L/V_L$=0.66-0.86). At the high $N_{//}$ (1.84) value, for the MHD-free discharges, the same weak dependence of the CD efficiency with $\Delta N_{//}$ leading to a maximum efficiency of 1.2. The same efficiency was found for MarkI and MarkII although for this latter, only two points are available (figure 1). Efficiency of MHD discharges is lower by 10-20%. Series 2 was performed in such conditions that the loop voltage with 2 MW (two launchers) was very similar to the loop voltage with 1 MW of series 1 ($\Delta V_L/V_L$=0.70-0.78). Although . the $N_{//0}$ value of MarkII was slightly higher (2.02) and the spectrum of MarkI was strongly distorted by wrong setting of the klystron phasing, leading to two wide lobes centred at $N_{//0} \approx$ 1.5 and 2 and a lower weighted directivity $D_n = \int_{Nacc}^{\infty} 4\frac{P(N_{//})dN_{//}}{N_{//}^2} - \int_{-\infty}^{-Nacc} 4\frac{P(N_{//})dN_{//}}{N_{//}^2}$ (~ -15%), the efficiency normalized to Zeff=1 is found to be higher for these shots performed at higher plasma (0.8MA) with the same dependence on the normalized width $\Delta N_{//}$.

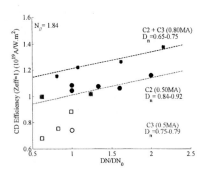
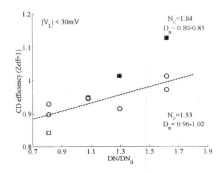

Fig.1 CD efficiency as a function of the normalized $N_{//}$ spectrum width (P_{LH}=1MW, $\Delta V_L/V_L$=0.76±010). Open symbols are for plasmas with MHD activity.

Fig.2 . CD efficiency as a function of the normalized $N_{//}$ spectrum width (P_{LH}=2MW, $|\Delta V_L/V_L|$<0.05). Open symbols are for plasmas with MHD activity.

At the lower $N_{//0}$ value (1.53), the CD efficiency of MarkII was found smaller than of MarkI by ~10%, this is likely to be related to a significant part of the $N_{//}$ spectrum which is non accessible to the plasma, since a strong negative phasing of the modules is needed on this launcher to obtain such a low $N_{//0}$ value. Calculation of the Dn directivity indicates a lower value (~-15%) for MarkII with respect of MarkI at $N_{//0}=1.53$.

For the quasi-full current drive just two MHD free pulses were obtained (figure 2). From the pulses with MHD activity, the same trend for the effect of the width of the spectrum is obtained. For the broader spectrum, an efficiency of 0.85 is measured for the high $N_{//0}$ MHD-free pulse as for the low $N_{//0}$ non MHD-free pulse. However, taking into account the Zeff, we end up with a higher efficiency for the MHD-free pulse by a factor 1.1.

Hard X-ray (30-150keV) measurements indicate, for MHD-free pulses, a slightly more peaked deposition profile for the $N_{//0}=1.53$ pulses, indicating that for these low density values, the accessibility of the wave does not limit the wave penetration [5]. When the $N_{//}$ spectrum is broadened, the normalized deposition profile is almost unchanged, but the integrated X-ray emission increases, at all energies, consistently with the increase of the current drive efficiency.

HEAT FLUX CARRIED BY FAST ELECTRONS IN THE EDGE

Infra-red images of the water-cooled graphite guard limiters of MarkI indicate peaked heat flux deposition at a location where the flux tubes passing in front of the waveguide row terminates. From 2-D simulations, the parallel heat flux, $F_{//}$, was estimated, assuming an e-fold decay length for the flux of 5mm from the grill aperture. Electron acceleration in front of the LH antenna is known to depend on the amplitude of the RF electric field but also on the number of waveguides, the stochastic limit being reached after about 20 waveguides [1]. The average electric field, computed from the SWAN code [6], was varied from 3.2 and 4.6 kV/cm and the number of waveguides between 12 (wide lobes) and 32 (narrow lobes). The two series were performed at different edge densities. Langmuir probes embedded in MarkI, indicates saturation current in the range of 17-23 mA (resp.8-11 mA) for series 1 (resp. series2). As expected, for the same electric field, the heat flux is larger for series 1. When the heat flux is normalized to the saturation current, the results are consistent and strong dependence with the electric field ($\sim E_{RF}^4$) is found (figure 2).

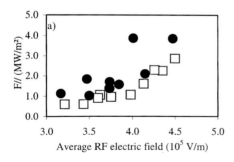
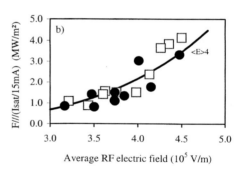

Fig. 3-a. Parallel heat flux on the MarkI guard limiter as a function of the average RF electric field ($P_{LH}=1$MW). The line indicates a E_{RF}^4 law. 3-b same with the heat flux normalized to the saturation current $I_{sat}=8-11$mA (open squares), $I_{sat}=17-23$mA (closed circles).

In order to go further in the analysis, from the electric field mapping, the mean energy gained by thermal electrons was calculated from 1-D code. No correlation is found between the mean electron energy (200-300and the heat flux: for the high electric field cases, achieved with a lower number of energized waveguides, the computed mean energy is indeed smaller (or equivalent) to the one from the low electric field cases for which the full antenna is coupling power.

CONCLUSION

Although for non-single pass absorption process the absorbed LH spectrum is broadened (and up-shifted) from the launched spectrum, a weak beneficial effect of the broadening of the launched spectrum is measured. When the directivity (including the wave accessibility) is taken into account no difference in CD efficiency is found between MarkI and MarkII. MHD activity reduces the efficiency by no more than 10% in most cases. At low plasma current (0.5 MA), full current drive with a normalized efficiency $\eta(Z_{eff}=1) = 1.1$ ($N_{//0}=1.84$) was achieved. For lower $N_{//0}$ value, higher efficiency was not obtained because of the loss of directivity and/or MHD activity. However, plasma with residual electric field indicates a higher efficiency (~10%) at $N_{//0}=1.53$.

The stochastic nature of the acceleration process leads to non predictable results for the mean energy of the electrons when the number of active waveguides is changed at constant launcher power. Whereas we found a reasonable agreement between the mean RF electric field and the heat flux, two explanations could explain this discrepancy between the computed mean energy gained by the electrons and this measured heat flux. First, in order to maintain plasma quasi-neutrality, a static potential arises and 3D convection flows occur [7]. The particle flux is inhomogeneous and locally underestimated. Secondly, velocity of fast particles may be enhanced by random fields resulting from wave diffusion on plasma fluctuations [8]. This process is expected to be more important at high electric field.

ACKNOWLEDGMENTS

This work has been partly supported by the Czech grant project GA AV 1043101.

REFERENCES

1. V.Fuchs et al., Phys. Plasmas **3**, 4023 (1996).
2. M.Goniche et al., Nuclear Fusion, **28** (1998) 919.
3. N.J.Fisch, Phys. Fluids **28**, 245 (1985).
4. V. Basiuk, J.F. Artaud et al, Nuclear Fusion, to be published in 2003 in a Greifswald special issue.
5. Y.Peysson et al, "Tore Supra X-Ray pulse height analyzer diagnostic" in *Proc of the 18th EPS conference on Cont. Fus. and Plasma Phys.,Berlin,* edited by European Phys. Soc., Genève, 1991,Vol.15D, pp345-348.
6. D.Moreau, T.K.N'Guyen, Rep.EUR-CEA-FC-1246, Centre d'Etudes Nucléaires de Grenoble.
7. V.Petržílka. et al. "A 3-D model of the plasma vortex in front of LH grills" "in *Proc of the 28th EPS conference on Cont. Fus. and Plasma Phys.,Funchal,* edited by European Phys. Soc., Genève, 2001, Vol.25A,pp289-292.
8. V.Petržílka et al. Czech. Journ. Phys. **S3**, 127 (1999).

Absorption of Lower Hybrid Waves by Alpha Particles in ITER

F. Imbeaux, Y. Peysson, L.G. Eriksson

Association EURATOM-CEA sur la Fusion
DSM / Département de Recherches sur la Fusion Contrôlée
CEA/Cadarache, 13108 St. Paul-lez-Durance (FRANCE)

Abstract Absorption of Lower Hybrid (LH) waves by alpha particles may reduce significantly the current drive efficiency of the waves in a reactor or burning plasma experiment. This absorption is quantified for ITER using the ray-tracing+2D relativistic Fokker-Planck code DELPHINE [1]. The absorption is calculated as a function of the superthermal alpha particle density, which is constant in these simulations, for two candidate frequencies for the LH system of ITER. Negligible absorption by alpha particles at 3.7 GHz requires $n_{\alpha,supra} \leq 7.5 \; 10^{16}$ m^{-3}, while no significant impact on the driven current is found at 5 GHz, even if $n_{\alpha,supra} = 1.5 \; 10^{18}$ m^{-3}.

INTRODUCTION

The absorption of lower hybrid (LH) waves by alpha particles produced by the fusion reactions may reduce the current drive efficiency significantly in a reactor or burning plasma experiment like ITER. To avoid absorption by the alpha particles, a high wave frequency should be used. A previous study, involving 1D quasi-linear calculations for both electron and alpha particle distributions in a simplified cylindrical geometry [2], suggested that the minimum frequency should be chosen as 5 GHz. Nevertheless, this high frequency is a source of technological constraints, and it would be much easier to build a LH system at a slightly lower frequency, i.e. 3.7 GHz, for which RF power sources already exist. Therefore it is an important issue to assess the choice of the frequency for the possible LH system of ITER, taking advantage of more recent tools (i.e. the DELPHINE package [1], wich includes ray-tracing in arbitrary geometry, 2D relativistic Fokker-Planck solver for the electron distribution function, and in which linear damping on alpha particles has been implemented).

EQUILIBRIA AND ALPHA PARTICLE DISTRIBUTION

Two different equilibria have been used in the simulations. Both correspond to the ITER non-inductive scenario. They have the same density and temperature profiles (see Fig. 1), but slightly different plasma current (respectively 9 and 7.6 MA). The q-profile is weakly reversed with a minimum at toroidal flux coordinate $\rho = 0.6$. This corresponds to the foot of an internal transport barrier, as shown by the high electron and ion temperature gradient inside this position. In the 9 MA case, 10 MW of LH waves are launched at a peak refractive index $n_{//0} = 2.0$. In the 7.6 MA case, 30 MW of LH waves are distributed among a co-current peak centered on $n_{//0} = 1.9$ (87 % of the power) and a counter-current peak centered on $n_{//0} = -3.8$ (13 % of the power) [3].

The alpha particle distribution follows the classical slowing down expression, $f(p) = \dfrac{c_0}{p^3 + p_c^3}$ where p is the alpha particle momentum, c_0 a normalisation constant, and the critical momentum is given by $\dfrac{p_c^2}{2m_\alpha}[keV] = 14.8 T_e[keV] \left(\dfrac{A_\alpha}{n_e} \sum_{j=ions} \dfrac{n_j Z_j^2}{A_j} \right)^{\frac{2}{3}}$

[4]. The distribution function drops to zero above an energy of 3.5 MeV, which is the kinetic energy of the alpha particles just after they are created by the fusion reaction. Since we do not take into account quasilinear effects for the alpha particles (i.e. a possible increase in the alpha particle energy due to the absorption of LH waves), we estimate only the lower bound of the absorption.

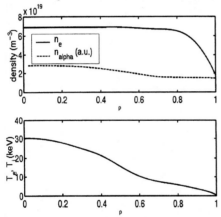

FIGURE 1. Electron density (m^{-3}), alpha particle density (arbitrary units), and temperature profiles.

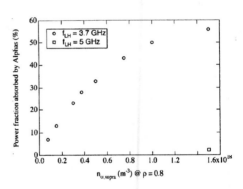

FIGURE 2. LH power fraction absorbed by alpha particles, 9 MA case, for LH frequency of 3.7 GHz (circles) and 5 GHz (square), as a function of the superthermal alpha particle density at the edge.

RESULTS

Though the details of the rays propagation are different between the two equilibria, the results on the LH power fraction absorbed by alpha particles is very similar (see Fig. 2). Significant absorption by alpha particles occurs for LH frequency of 3.7 GHz, except if $n_{\alpha,supra} \leq 7.5 \; 10^{16}$ m^{-3}, while it is only of the order of a few percents for 5 GHz frequency and $n_{\alpha,supra} = 1.5 \; 10^{18}$ m^{-3}. The power fraction absorbed by alpha particles for $f_{LH} = 3.7$ GHz increases linearly with their superthermal density for $n_{\alpha,supra} \leq 5\ 10^{17}$ m^{-3}, then the slope of the curve flattens for higher densities. The amount of driven current is reduced approximately in the same ratio. The exception to this rule is when the absorbed power corresponds to the counter-current lobe, which drives in fact very few current (only 80 kA in the 7.6 MA case for $f_{LH} = 5$ GHz and $n_\alpha = 0$, to be compared to the 2.1 MA driven by the co-current lobe). Therefore, even if in the 7.6 MA, 5 GHz case, the LH power fraction absorbed by alpha particles reaches 15 % for $n_{\alpha,supra} \leq 1.5\ 10^{18}$ m^{-3}, it corresponds mostly to power in the counter-current lobe, and at the end there is a negligible reduction of the LH driven current.

The LH waves being damped on alpha particles by perpendicular Landau damping, the minimum alpha particle perpendicular velocity required to absorb LH waves is related to their maximum $n_\perp$ through : $v_{\perp min} = c/n_{\perp max}$, where c is the speed of light in vacuum. Since $n_{\perp max}$ is higher at low LH frequency (in the electrostatic approximation, $n_\perp/n_{//} \propto 1/\omega_{LH}$) the interaction of the waves with the alpha particle distribution function may occur at lower energies. As shown on figure 3, for LH frequency of 3.7 GHz in the 9 MA case, there is a significant number of rays above the energy cut-off of the alpha particle distribution function (3.5 MeV, horizontal line on the figure). Absorption on superthermal alpha particles occurs, provided their density is not negligible. Conversely, since almost no ray propagates above the energy cut-off in the $n_\perp$ space at 5 GHz, the absorption by alpha particles is negligible (see Fig. 4). Figures 3 and 4 also show that the LH waves propagation in ITER is not purely single-pass, even though the electron temperature is high in the plasma core. For this reason, it is critical to take into account the toroidal $n_{//}$-upshift in a realistic geometry, in order to assess correctly the absorption of LH waves by alpha particles.

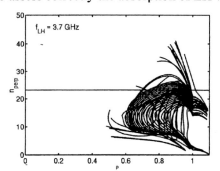

FIGURE 3. Perpendicular refractive index of the rays versus toroidal flux coordinate, 3.7 GHz and 9 MA case. The horizontal line corresponds to the perpendicular index required to have a resonance with alpha particles at their energy cut-off (3.5 MeV).

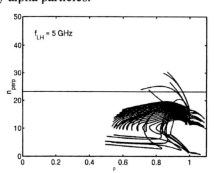

FIGURE 4. Perpendicular refractive index of the rays versus toroidal flux coordinate, 5 GHz and 9 MA case. The horizontal line corresponds to the perpendicular index required to have a resonance with alpha particles at their energy cut-off (3.5 MeV).

As shown on figure 5, for the 7.6 MA case, the local power deposition on alpha particles is of the same order of magnitude as the deposition on electrons in the low electron temperature region ($\rho > 0.7$), while electron absorption clearly dominates in the internal transport barrier region. In the 9 MA case, the power deposition profile on electron is broader in reason of a faster toroidal $n_{//}$-upshift, but still the damping of LH power on alpha particles is restricted to the low electron temperature region $\rho > 0.7$ (see Fig. 6).

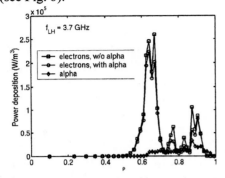 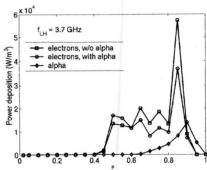

FIGURE 5. LH power deposition on electrons with no alpha particles (squares), on electrons (circles) and alpha particles (diamonds) for $n_{\alpha,supra} = 1.5\ 10^{17}\ m^{-3}$, 3.7 GHz and 7.6 MA case.

FIGURE 6. LH power deposition on electrons with no alpha particles (squares), on electrons (circles) and alpha particles (diamonds) for $n_{\alpha,supra} = 3.0\ 10^{17}\ m^{-3}$, 3.7 GHz and 9 MA case.

CONCLUSIONS

In spite of slight differences due to the plasma current, both cases show the same trends : a LH frequency of 3.7 GHz may yield significant absorption of LH power by alpha particles, even at low superthermal alpha particle density, while at 5 GHz, there is a negligible current drive reduction due to superthermal alpha particles, even at high density. The counter-current lobes of the LH power spectrum drive very few current in the absence of alpha particles. Moreover they are easily absorbed by alpha particles, since their power deposition is localized in the low electron temperatue region. Ray-tracing in realistic geometry is a important tool to calculate the LH power deposition in ITER, since the LH waves propagation is not purely single-pass. The results presented here also show that the absorption on alpha particles occurs at the edge of the plasma ($\rho > 0.7$), where it is difficult to estimate the alpha particle distribution function. In this work, we used fixed alpha particle distribution function, and showed that the results are sensitive to its parameters, like the superthermal density and energy cut-off. Therefore, the next step is to develop a detailed modeling of the alpha particle distribution function in a burning plasma experiment, taking into account their specific trajectories and interactions with the various heating systems and the plasma.

REFERENCES

1. F. Imbeaux, report EUR-CEA-FC-1679 (1999).
2. E. Barbato and F. Santini, Nucl. Fus. **31**, 673 (1991).
3. P.T. Bonoli et al, ITPA Meeting, Cadarache, France (2002).
4. T.H. Stix, Nucl. Fusion **15**, 737 (1975).

Separate and Combined Lower Hybrid and Electron Cyclotron Current Drive Experiments in Non Inductive Plasma Regimes on Tore Supra

F. Imbeaux, J.F. Artaud, V. Basiuk, G. Berger-By, F. Bouquey, C. Bourdelle, J. Clary, C. Darbos, G. Giruzzi, G.T. Hoang, G. Huysmans, M. Lennholm, P. Maget, R. Magne, D. Mazon, Y. Peysson, J.L. Ségui, X.L. Zou, and the Tore Supra Team

Association EURATOM-CEA sur la Fusion
DSM / Département de Recherches sur la Fusion Contrôlée
CEA/Cadarache, 13108 St. Paul-lez-Durance (FRANCE)

Abstract During the 2002 campaign of the TORE SUPRA tokamak, long pulses with a high fraction of Lower Hybrid Current Drive (~80 %) have been obtained, leading to a new record of injected energy of 0.75 GJ. Specific phenomena are related to these experiments, like the onset of periodic oscillations of the core electron temperature, which is ascribed to a coupling between current drive and heat transport. In combination with LHCD, a modest amount of Electron Cyclotron (EC) Current Drive power can modify the q profile in the central part of the plasma, which has a significant impact on the behaviour of the discharge. When used in the counter-current direction, ECCD triggers an electron transport barrier.

INTRODUCTION

Taking advantage of its new actively cooled inner vessel components and non-inductive current drive systems (Lower Hybrid (LH) waves at 3.7 GHz and Electron Cyclotron (EC) at 118 GHz) the TORE SUPRA tokamak can realise steady state discharges over long durations (several minutes). During the 2002 campaign, new results on the physics of steady state tokamak plasmas have been obtained. Among them, a new regime with periodic oscillations of the core electron temperature, and the triggering of internal transport barriers in combined LH and EC current drive experiments.

PURE LHCD DISCHARGES

Using LHCD only, at a power level of 3 MW, discharges up to 4mn 25s have been obtained, setting a new world record of injected energy of 0.75 GJ. The main parameters of the longest discharge (#30414) are : plasma current 0.51 MA, toroidal magnetic field 3.8 T, edge safety factor 9.3, central density $2.5 \; 10^{19} m^{-3}$, central electron temperature 4.8 keV, central ion temperature 1.6 keV, effective ion charge 2, radiated power fraction 20%, loop voltage below 10 mV, see Fig. 1. In the first 3 minutes of this discharge, the LH waves drive 82 % of the total current, 14 % is driven by bootstrap current, the remaining 4 % are driven by inductive means. All plasma parameters, including the density, remain perfectly constant during this 3 minutes phase. The typical safety factor profiles obtained in such discharges during the full LH power phase, are flat or slightly reversed in the plasma core (Fig. 2). This flat shear in the core ($\rho < 0.2$), combined with a high shear at mid-radius, is close to MHD tearing mode limits. Hence, in some discharges with high internal inductance ($l_i \sim 1.6$), a core MHD activity around q = 3/2 triggers in cascade a (2,1) and a (3,1) mode (Fig. 3).

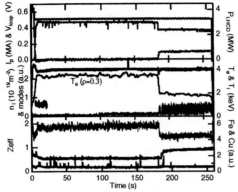

FIGURE 1. Times traces for discharge #30414 (0.75 GJ injected power, 4mn 25s)

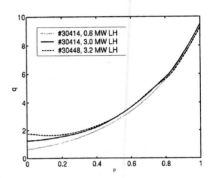

FIGURE 2. Safety factor profiles (toroidal flux coordinate) of typical pure LHCD shots at high power (solid and dashed lines), and quasi-ohmic (0.6 MW LHCD, dots).

PERIODIC OSCILLATIONS OF CORE TEMPERATURE

In several discharges with loop voltage close to zero (< 200 mV) and moderate plasma current (I_p < 700 kA), the spontaneous onset of stable and periodic oscillations of the central electron temperature T_e has been observed (Fig. 5). They have amplitudes ranging from 100 eV to 1 keV, frequencies from 4 to 20 Hz, and are localized in a very central region (toroidal flux coordinate $\rho < 0.2$). Their spatial structure is purely radial (no helicity), so they do not correspond to MHD activity. Many clues suggest a link between the oscillations and the current profile. The amplitude of the T_e oscillations increases with the width of the Hard X-rays emission, which is representative of the LH power deposition [1]. Also, a small amount of ECCD (350 kW, co-ECCD on axis), added on top of 3 MW of LHCD provides surprising changes in the core energy confinement and trigger temperature oscillations (Fig. 4). As shown on figures 4 and 5, the central temperature increases just before the beginning of the oscillations. This reminds of the hot core Lower Hybrid Enhanced Performance (LHEP) transition, in which the electron transport is reduced in the central plasma region (typically, $\rho < 0.2 - 0.3$), owing to turbulence suppression associated with a negative magnetic shear [2-4]. The experimental evidence of a link between the oscillations and a flat central q-profile leads to the plausible interpretation of the oscillations as an incomplete transition to the LHEP regime, with a cyclic behavior of electron temperature and current density.

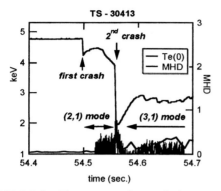

FIGURE 3. Time traces of central electron temperature deduced from ECE, and MHD activity measured by Mirnov coils.

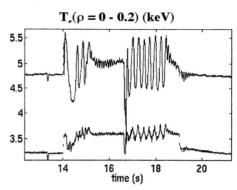

FIGURE 4. Time traces of T_e at $\rho = 0$ and 0.2. The dashed lines indicate the ECCD pulse (between 14 and 19 s).

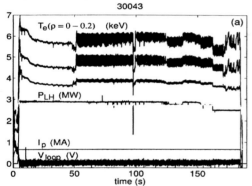

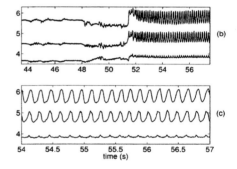

FIGURE 5. Time traces for discharge #30043. a) From top to bottom: ECE channels, LH power, line-averaged density, plasma current and loop voltage. b), c): enlarged views of the ECE channels.

In order to support this hypothesis, simulations have been carried out with the CRONOS transport code [5], introducing specific mechanisms coupling T_e and the current density j. The physical basis for the coupling phenomena are : i) the turbulence responsible for heat transport is suppressed or reduced by effect of negative magnetic shear [6] and/or of the reduced rational surface density around a main rational value q_r [7,8], ii) the LH-driven current density j_{LH} is an increasing function of both temperature and current, as experimentally observed [3,9]. By introducing such coupling mechanisms ($j_{LH} \propto j, T_e$ in the region $\rho < 0.3$ and χ_e reduced in the region of negative magnetic shear or when q is close to q_r) in solving the transport equations for j and T_e, simultaneous stationary oscillations of T_e and q have been obtained (Fig. 6). Their amplitude, frequency, and radial structure are similar to the experimental observations. This result supports the interpretation of the oscillation as an incomplete transition of the plasma core to a regime of enhanced energy confinement.

INTERNAL TRANSPORT BARRIER USING ECCD

When ECCD is used in discharges with low loop voltage, it can significantly change the q profile in the plasma core, in spite of the low power ratio between the two waves (300-400 kW ECCD / 1-3 MW LHCD). For instance, a narrow internal transport barrier can be produced by slightly off-axis counter-current ECCD ($\rho \sim 0.2$), obtained with 400 kW ECCD, 1 MW LHCD, $n_{e0} = 2.1 \; 10^{19}$ m^{-3}, $I_p = 550$ kA, $V_{loop} \sim 130$ mV, see Fig. 7.

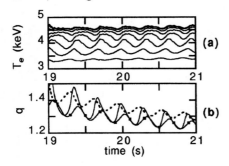

FIGURE 6. Results of simulations by the CRONOS code. (a) T_e vs time at various radial positions $0 \leq \rho \leq 0.15$; (b) q vs time for $\rho = 0$ (solid), $\rho = 0.04$ (dashed), $\rho = 0.1$ (squares).

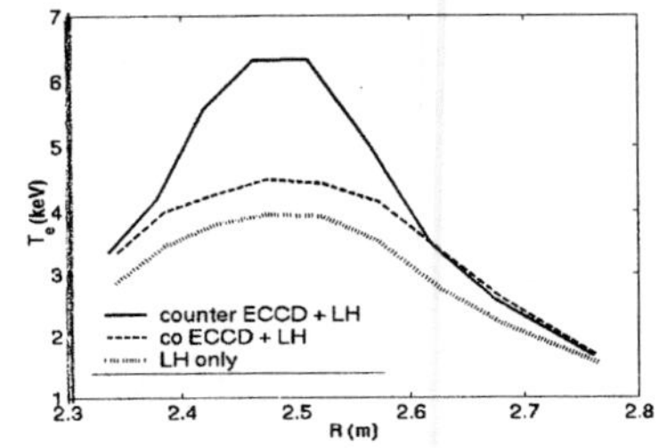

FIGURE 7. Electron temperature profiles of shots with same density, LH power (1 MW), but different ECCD schemes.

CONCLUSIONS

The LHCD plasmas optimised for long pulse operation have a flat q-profile in the core ($\rho < 0.2 - 0.3$). Because of the link between energy confinement and q-profile, these plasmas are just at the limit of transition to high core confinement, which results in new regime with stable oscillations of the central electron temperature. A small amount of ECCD in the core can trigger a transition to a fully established electron transport barrier. Such phenomena may be relevant for the "hybrid" scenario of ITER, which corresponds to a wide flat q profile.

REFERENCES

1. Y. Peysson and F. Imbeaux, Rev. Sci. Instrum. 70, 3987 (1999).
2. G.T. Hoang et al., Nucl. Fusion **34**, 75 (1994).
3. Y. Peysson and the Tore Supra Team, Plasma Phys. Contr. Fus. **42**, B87 (2000).
4. X. Litaudon et al., Plasma Phys. Contr. Fus. **43**, 677 (2001).
5. V. Basiuk et al., Nucl. Fus. (to be published, 2003).
6. J. F. Drake et al., Phys. Rev. Lett. **77**, 494 (1996).
7. F. Romanelli and F. Zonca, Phys. Fluids B **5**, 4081 (1993).
8. X. Garbet et al., Phys. Plasmas **8**, 2793 (2001).
9. K. Ushigusa, Plasma Phys. Contr. Fus. **38**, 1825 (1996).

Real-Time Control of the Current Profile in JET

D. Moreau[1,2], F. Crisanti[3], X. Litaudon[2], D. Mazon[2], P. De Vries[4],
R. Felton[5], E. Joffrin[2], L. Laborde[2], M. Lennholm[2], A. Murari[6], V. Pericoli-Ridolfini[3], M. Riva[3], T. Tala[7], G. Tresset[2], L. Zabeo[2], K. D. Zastrow[5]

[1] *EFDA-JET CSU, Culham Science Centre, Abingdon, OX14 3DB, U. K.*
[2] *Euratom-CEA Association, CEA-DSM-DRFC Cadarache, 13108, St Paul lez Durance, France.*
[3] *Euratom-ENEA Association, C.R. Frascati, 00044 Frascati, Italy.*
[4] *Euratom-FOM Association, TEC Cluster, 3430 BE Nieuwegein, The Netherlands.*
[5] *Euratom-UKAEA Association, Culham Science Centre, Abingdon, U. K.*
[6] *Euratom-ENEA Association, Consorzio RFX, 4-35127 Padova, Italy.*
[7] *Euratom-Tekes Association, VTT Processes, FIN-02044 VTT, Finland.*

Abstract. New algorithms using a truncated singular value decomposition of a linearised model have been implemented in the JET control system. Using three heating and current drive actuators (LHCD, NBI, ICRH), successful control of the safety factor profile has been achieved in quasi steady state conditions where the loop voltage was small and a large fraction of the plasma current was carried by the bootstrap current.

INTRODUCTION

In order to control the current and pressure profiles in high performance tokamak plasmas with internal transport barriers (ITB), a multi-variable model-based technique has been proposed which offers the potentiality of retaining the distributed character of the current and heat diffusion (distributed-parameter system) [1, 2].

Here, we describe first experiments using the simplest, lumped-parameter, version of this technique. In section 3 only one actuator (LHCD) is used. In section 4, the first experiments using a multiple-input-multiple-output controller to control the q-profile with 3 heating and current drive actuators (LHCD, NBI, and ICRH) are reported.

TRUNCATED SINGULAR VALUE DECOMPOSITION

Let $\mathbf{Q}(s)$ be the linearized Laplace response around the target equilibrium:

$$\mathbf{Q}(s) = \mathbf{K}(s) \cdot \mathbf{P}(s) \qquad (1)$$

where $\mathbf{Q}$ represents a safety factor difference vector and $\mathbf{P}$ an input power difference vector. The kernel $\mathbf{K}(s)$ can be identified from power modulation experiments around the target steady state, or by simulations using a predictive transport code.

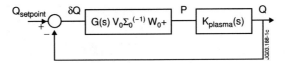

FIGURE 1. Control diagram used for the feedback control of the safety factor.

For the experiments described below, the steady state gain matrix $\mathbf{K}(0)$ was sufficient. A singular value decomposition is performed, yielding :

$$\mathbf{K}(s) = \mathbf{W}(s) \, \mathbf{\Sigma}(s) \, \mathbf{V}(s)^+ \qquad (2)$$

and this defines decoupled modal inputs and modal outputs :

$$\alpha(s) = \mathbf{V}(s){+}. \, \mathbf{P}(s) \quad \text{and} \quad \beta(s) = \mathbf{W}(s){+}.\mathbf{Q}(s) \qquad (3)$$

related by :

$$\beta(s) = \mathbf{\Sigma}(s) \, . \, \alpha(s) \qquad (4)$$

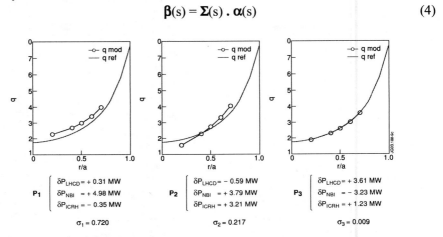

FIGURE 2. Respective influence of the 3 identified singular input vectors on the q-profile when using simultaneously LHCD, NBI and ICRH as actuators. The full profile which appears in all frames corresponds to the reference equilibrium obtained with P_{LHCD} = 3MW, P_{NBI} = 7MW, P_{ICRH} = 3MW, around which the system is linearized. The open circles on the left, central, and right frames show the q-profiles estimated from the model when adding the power combinations P_1, P_2, and P_3, respectively, to the reference case powers. P_1, P_2, and P_3, are proportionnal to the singular vectors, $\mathbf{V}_1$, $\mathbf{V}_2$, and $\mathbf{V}_3$, but normalized to 5MW. σ_1, σ_2, and σ_3 are the corresponding singular values.

Pseudo-modal control techniques can be used by taking the steady state limits, $\mathbf{V}_0$, $\mathbf{\Sigma}_0$, $\mathbf{W}_0$, of $\mathbf{V}(s)$, $\mathbf{\Sigma}(s)$, $\mathbf{W}(s)$, and by inverting the diagonal steady state gain matrix, $\mathbf{\Sigma}_0$, where $[\mathbf{\Sigma}_0]_{i,j} = \sigma_i.\delta_{i,j}$. In order to obtain a simple PI feedback control with minimum (least square) steady state offset, we choose the controller $\mathbf{G}(s)$ as (Fig. 1) :

$$\alpha(s) = \mathbf{G}(s) \, . \, \beta(s) = g_c[1 + 1/(\tau_i.s)] \, \mathbf{\Sigma}_0^{(-1)} \, \beta(s) \qquad (5)$$

where g_c is the proportional gain and (g_c/τ_i) is the integral gain.

The SVD expansion of $\mathbf{K}$ is truncated to retain the most significant singular values. In the example below (Fig. 2), 3 actuators are used (LHCD, NBI, ICRH), but σ_3 is small which indicates that the family of accessible profiles is only a 2-parameter family spanned by the first two singular vectors, $\mathbf{W}_{0,1}$ and $\mathbf{W}_{0,2}$. We then use the truncated operator with only σ_1 and σ_2 : $\mathbf{K}_{0,T} = \sigma_1 \mathbf{W}_{0,1}.\mathbf{V}_{0,1}^+ + \sigma_2 \mathbf{W}_{0,2}.\mathbf{V}_{0,2}^+$.

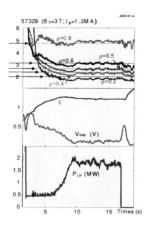 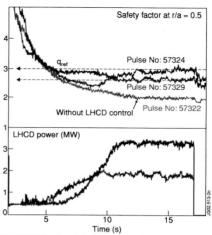

FIGURE 3a. Real-time control experiment with LHCD only (pulse #57329, $B_T = 3T$, $I_p = 1.3$ MA). Top : Safety factor at r/a = [0.2 0.4 0.5 0.6 0.8]. Centre : Internal inductance parameter, l_i, and loop voltage, V_{loop}.(Volts). Bottom : LHCD power (MW).

FIGURE 3b. Measured and requested q-values at a r/a = 0.5, and LHCD power for 2 controlled pulses (#57329 and #57324 $B_T=3T$, Ip=1.3MA). A pulse without control is shown for comparison (#57322). Control starts at 2s and stops at 17s.

CONTROL OF THE SAFETY FACTOR PROFILE WITH LHCD

The first, simplest - and in some sense trivial - application of the lumped parameter SVD control scheme was to reach a predefined q-profile with only one actuator, namely LHCD, but 5 q-setpoints. The accessible targets are restricted to a one-parameter family of profiles and must therefore be chosen reasonably. Applying an SVD technique with 5 points, rather than only one, can be an advantage if the q-profile "rotates" when varying the power. It may not allow to reach any one of the setpoints exactly, but minimizes the error on the profile shape (e.g. weak or reversed shear), contrary to a control of l_i. Such an experiment was performed during an extended LHCD preheat phase [3], a usual prelude to the formation of ITB's in JET. The central line-integrated density was $2.7 \times 10^{19} m^{-2}$ to allow for efficient LHCD. The toroidal field was 3T and the plasma current was 1.3 MA so as to approach a full non-inductive regime. A linearized model which links the values of q(r) at r/a = 0.2, 0.4, 0.5, 0.6, 0.8 to the input LH power was identified from simple step power changes. The q-profile reconstruction uses the real-time data from the magnetic measurements and from the interfero-polarimetry, and a parameterization of the magnetic flux surface geometry [4, 5]. The output signals are available every 50 ms.

The effectiveness of the controller in achieving, and maintaining in steady state, various q-profiles can be seen on Fig. 3.

CONTROL OF THE q-PROFILE WITH LHCD, NBI AND ICRH

The second set of experiments conducted so far and using the proposed SVD technique (Fig. 3) was a first attempt at using the 3 available heating and current drive systems to control the q-profile during a strong heating phase, in an ITB scenario with moderate bootstrap current. The toroidal magnetic field was 3T and the scenario started with a reversed shear 2.5 MW LHCD preheat phase during which the plasma current was ramped up to 1.8 MA, at a line-integrated plasma density around 3×10^{19} m^{-2}. At 4.3 s, 12.5 MW of NBI and 3 MW of ICRH power were applied resulting in the triggering of an ITB and, starting at 7 s, the plasma current was ramped down to a final 2-second-long 1.5 MA plateau. The determination of the steady state responses to variations of the heating and current drive powers was made from the analysis of four dedicated discharges, including the reference discharge around which the system is to be linearized. The result is shown on Fig. 4.

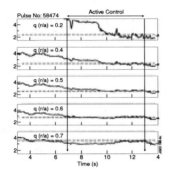

FIGURE 4a. Measured and requested time traces of q(r/a) at the 5 radii selected for the control experiment with LHCD, NBI and ICRH (pulse #58474, $B_T = 3T$, $I_p = 1.5$ MA).

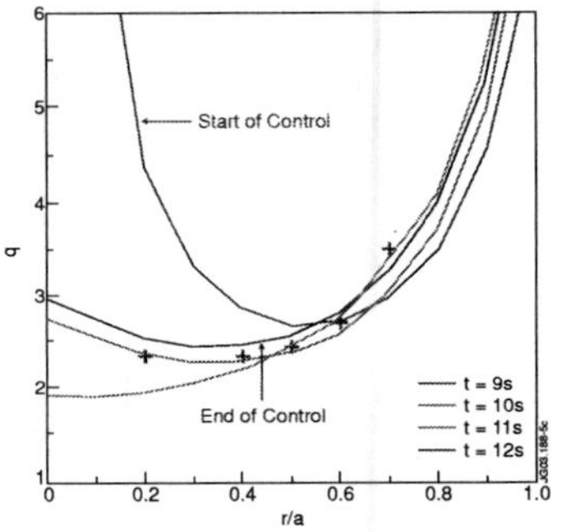

FIGURE 4b. Same as in Fig. 4a (pulse #58474, $B_T = 3T$, $I_p = 1.5$ MA) : q-profiles at four times between 47 s and 52 s. Pluses represent the 5 setpoints.

CONCLUSION

A successful control of the safety factor profile was obtained with a lumped-parameter SVD algorithm. This provides an interesting starting basis for a future experimental programme at JET, aiming at the sustainement and control of ITB's (q(r) and ρ_T^* criterion) in fully non-inductive plasmas with a large bootstrap current.

REFERENCES

1. Moreau, D., Litaudon, X., Mazon, D., et al., in *Proc. IAEA TCM on Steady State Operation of Magnetic Fusion Devices*, Arles (France) 2002.
2. Moreau, D., Crisanti, F., Litaudon, X., Mazon, D., et al., to be published in Nucl. Fus. (2003).
3. Mazon, D., Litaudon, X., Moreau, D., et al., Report JET PR(02) 21, EFDA-JET, Culham Science Centre, Abingdon, U. K. ; to be published in Plasma. Phys. Control. Fusion (2003).
4. Zabeo, L., Murari, A., Joffrin, E., et al., Plasma. Phys. Control. Fusion 44 (2002) 2483.
5. Riva, M., Zabeo, L., Joffrin, E., et al., in *Proc. 22th Symp on Fusion Tech. (SOFT)*, Helsinki, Finland (2002).

Construction, Calibration and Testing of a Multi-Waveguide Array for the Application of LHCD on Alcator C-MOD

J.R. Wilson, S. Bernabei, E. Fredd, N. Greenough, J.C. Hosea, C.C. Kung, G.D. Loesser

Princeton Plasma Physics Laboratory, Princeton University, Princeton, NJ 08540, USA

Abstract. A collaborative project between MIT and Princeton has been undertaken to fabricate a high power system for application of lower hybrid current drive (LHCD) on the Alcator C-MOD device. The ultimate goal of the project is to have the ability to apply up to 3 MW of 4.6 GHz rf power to drive current for the study of advanced tokamak operation on C-MOD. As part of the PPPL portion of the collaboration a high power wave-guide launching array has been designed and constructed to support 1.5 MW application to the plasma based on previous experimental power limits (and possibly supporting 2 MW application with advanced conditioning). In this paper we will describe the construction, calibration and testing techniques utilized and the results obtained during the construction phase.

INTRODUCTION

Lower Hybrid Current Drive (LHCD) is a proven technique for efficient rf current drive in tokamak plasmas. The so-called Advanced Tokamak (AT) scenarios require carefully tailored current profiles to obtain maximum benefit, high plasma beta and bootstrap current fraction. A program of AT physics studies has been proposed for the Alcator C-MOD tokamak. As part of this program an LHCD system is to be installed on the machine. As part of the project to build and install such a system a waveguide launcher has been designed and constructed. The geometric constraints of C-MOD as well as the physics requirements necessitated a novel launcher design [1]. As part of the construction process a number of issues had to be resolved by prototype construction, as well as testing and calibration of the final assembly.

LAUNCHER DESIGN AND VERIFICATION

The C-MOD launcher design consists of four vertically stacked arrays, each array consisting of twenty-four toroidally adjacent 0.55 cm x 6.0 cm waveguides. Each 250 kW, 4.6 GHz, klystron feeds eight waveguides. These guides form a two wide by four high subset of the array. The power is first into four feeds using commercial power dividers. The final division takes place in a stacked plate assembly via a custom designed internal splitter. The stacked plate assemblies form the 4 x 24-waveguide channels providing 48 inputs and 96 outputs. The assemblies come in two parts with the rear section including the internal splitter from aluminum and the forward section from copper plated stainless steel. The stainless steel is required to withstand the electromagnetic forces induced by plasma disruptions. The two sections, as well as the individual vacuum waveguide couplers, are connected with gaskets, which allow for individual tolerance mismatches of up to 0.01 in. The gaskets are fabricated from aluminum be plunge EDM. Aluminum was chosen for its greater compliance and lack of work hardening. The stacked waveguide assemblies are attached to four vacuum couplers. Each coupler is machined from a solid piece of titanium and is 24-waveguides wide each containing an individual alumina vacuum window brazed into the waveguide channel. Power is monitored in the stacked plate assemblies with probes that consist of two coaxial loop couplers spaced 1/4 wavelength apart and combined in a hybrid to give either forward or reflected power.

Stacked Waveguide and Power Splitter

The stacked waveguide concept was employed in the PBX-M lower hybrid system [2]. Here it has been extended because of the confined port geometry. Each waveguide

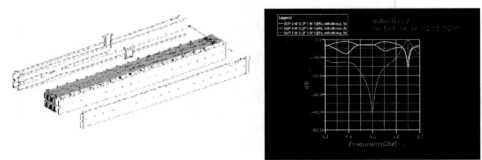

FIGURE 1. 3dB splitter and rear stacked waveguide assembly. (a) Drawing of splitter stacked assembly (b) electrical properties of splitter with shorted fourth leg.

is formed by a groove cut into the plate butted against the back of the next plate, Fig.1. The plates are held together with a series of bolts. In the forward waveguide the stainless steel plates are copper plated to reduce the rf losses. Even with the plating losses were high, ~ 1dB/ft, until the plates were polished, whereupon the losses fell to ~0.1 dB/ft. The polishing effectively reduced the surface area at the scale of the rf skin depth, ~ 1×10^{-5} in. In the rear stacked waveguide a power division was required. A first design utilized a simple slot. This slot coupler then acts like a 90-degree hybrid junction. This was adequate provided the four ports, one input, two output and one rejection port, are terminated in non-reflective loads. Space constraints dictated that the rejection port be terminated with a shorting plate. With this termination it was found that the splitter was very sensitive to the exact dimensions and the phase shift back to the shorting plate. A new design was made with a double slot coupler, the two slots separated by a small post. This design proved to be compatible with the shorting plate and dimensional tolerances of the slots providing a 3-dB split in practice, Fig 1(b).

Vacuum Window Assembly

The vacuum waveguide coupler is electrode discharge machined (EDM) from a solid block of titanium 6242. The 24 waveguide slots are machined to a precision of 0.02-mm. The vacuum windows are alumina oxide bricks brazed into the channels. To achieve low reflection the windows must be cut accurately to be exactly a half

 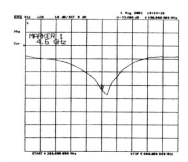

FIGURE 2. Vacuum window test fixture and measurement result. (a) fixture with window in place. (b) network analyzer measurement of window resonance

wavelength long at the rf frequency. Since the manufacturer cannot guarantee the exact dielectric constant of the material, only that a given billet will be uniform, a production scheme had to be devised that would allow precise cutting of the bricks. Samples from the raw billet were sent to Princeton where they were placed into a jig (Fig 2a) and the half wavelength resonance measured (Fig 2b). The billets were found to be quite uniform and return losses greater than 30 dB were easily obtained at the operation frequency. In order for this to work the bricks needed to be coated around their circumference with a conductive coating. Three coatings were tried, conductive paint, vacuum vapor deposited silver and gold leaf. All three techniques yielded acceptable results. The dielectric constant obtained from these measurements allowed specification of the exact brick length to be given to the manufacturer who proceeded to cut the 96 bricks and then metalize them. The bricks were subsequently brazed into the coupler at Princeton.

Power Probes for Stacked Waveguide

This design allows flexible phasing of the waveguide array to produce a wide range of $n_{\|}$. In order to control this spectrum the phase in the stacked waveguide must be monitored. A probe has been designed for this function. Additionally it can be used to

FIGURE 3. Power Probe for measuring forward or reflected power in stacked waveguide array (a) finished probe, (b) Electrical schematic of probe function

measure the reflection coefficients in the individual guides and verify the correct operation of the system. The narrow dimension, 0.55 mm, of the waveguide requires a compact design. In additional the design must insure directionality. Two coaxial loops inserted into the top of the waveguide channel spaced one-quarter wavelength apart can be combined in a hybrid to measure either forward or reflected power. By swapping the hybrid output port and a termination the hybrid output will be either forward or reflected power. By monitoring forward power via directional couplers further back in the line and reflected power in the stacked waveguide a complete characterization of the spectrum can be obtained

SUMMARY

A novel coupler for application of LHCD has been constructed for the Alcator C-MOD tokamak. Measurements during the construction project have verified that the launcher performs as designed at low power levels and that diagnostics in place will allow characterization of its performance in high power testing and operation on the tokamak.

ACKNOWLEDGMENTS

This work supported by U.S. Department of Energy Contract DE-AC02-76CH03073.

REFERENCES

1. Bernabei, S., Hosea, J.C., et al., *Fusion Science and Tech.*, **43**,145-152 (2003).
2. Greenough, N. et al. *Proc. Symp. Fusion Engineering* IEEE San Diego CA 1992 pp. 127.

ELECTRON CYCLOTRON HEATING
AND CURRENT DRIVE

Optimization of Electron Cyclotron Heating in LHD

S. Kubo*, T. Shimozuma*, H. Idei[†], Y. Yoshimura*, T. Notake**,
K. Ohkubo*, Y. Mizuno*, S. Ito*, S. Kobayashi*, Y. Takita* and
LHD Experimental Group*

National Institute for Fusion Science, 322-6 Oroshi cho, Toki 509-5292, Japan
[†]*Adv. Fusion Research Center, Research Inst. for Appl. Machanics, Kyushu Univ., Kasuga, 816-8580, Japan*
**Dept. of Energy Science, Nagoya Univ., Furo cho, Nagoya 464-8603, Japan*

Abstract. Formation of electron transport barrier and achievement of central electron temperature of more than 10 keV are confirmed by electron cyclotron heating (ECH) in LHD. These are the result of highly concentration of the deposited power. The system are described which enables these experiments and clarifying the mechanisms of the formation of electron transport barrier. The formation of the transport barrier in LHD is limited in rather low density of below 5×10^{18} m^{-3} so far. The efforts and their results of further optimization of ECH system for the better understanding and for the formation in wider plasma parameters of the transport barrier are also discussed.

INTRODUCTION

Large Helical Device is the world largest $l = 2$ and $m = 10$ heliotron type plasma confinement device to explore the confinement properties of fusion relevant plasma parameters. Electron cyclotron resonance heating (ECRH) had been used as a plasma production and electron heating method since the first plasma production of LHD in 1998. ECRH system have been upgraded step by step[1, 2]. In 2001, central electron temperature of more than 10 keV is achieved using ECRH at the plasma density of $5 - 6 \times 10^{18}$m^{-3}[3]. These results are attributed to the power of nearly 2 MW at fundamental and second harmonic heating concentrated near the magnetic axis. This concentration can be achieved by its strong focusing antenna system and adjusting the magnetic field. In order to achieve higher electron temperature at higher electron density, further optimization of the ECRH system have been performed. The plasma confinement experiments using this system show the advantage of the local heating character. The present status and performance of the system are discussed in this paper.

ELECTRON CYCLOTRON HEATING SYSTEM IN LHD

Birds eye's view of ECH system in LHD operated during last experimental campaign is shown in Fig. 1.

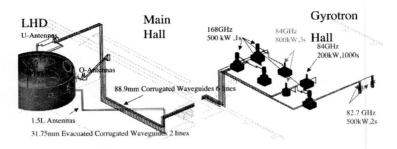

FIGURE 1. Total ECRH System for LHD in 7th cycle experimental campaign

In order to provide the high voltage to the gyrotrons, two types of gyrotron power supply are used. One is for conventional type gyrotrons and can operate two gyrotons in parallel. The others are for collector potential depression. These power supplies consist of three sets of collector power supply. Body and anode voltages are supplied by independent inverter type power supplies. Two of the collector power supply is connected to triode type CPD gyrotrons (#1,2 ,3 and 7) at 168 GHz mainly used for the second harmonic heating. The other one is connected to the diode type CPD gyrotrons. These combinations are listed in TABLE 1. For the non-CPD type gyrotron, non-coolant type cryo-magnets are introduced. Those can be kept super conducting state without feeding liquid helium. One more advantage of this magnet is the capability to operate persistent current mode. By removing the magnet power supply during the gyrotron operation, but keeping the magnetic field with persistent current, the miss operation of the magnet power supply could be prevented. As the result, these magnet worked stably and reliably during the last experimental campaign without any problems.

Oversized non-evacuated corrugated waveguide with the diameter of 88.9 mm is introduced at first. Six lines of this type transmission with total length of about 100 meters are in operational. From the beginning phase of the ECH system of LHD, we are suffered from arcing problems in the waveguides[1, 2]. The experience indicated that the threshold power of arcing increases as the purity of the HE_{11} excited in the waveguide. All matching optics unit (MOU) mirrors are optimized using the phase retrieval from the measurement output beam from each gyrotron. The accuracy of the alignment of the output beam from MOU to the waveguides is enhanced. Figure 2 shows the ratio of the power lost at MOU and transmission line and that coupled into LHD to that from gyrotron window for each system. It should be noted that the transmission efficiency for #11 and 12 exceeds 90 %,although the number of miter bends and total path length are large. After this optimization, the threshold power for the arcing inside the waveguide appreciably increased. Relatively low efficiency for 168 GHz systems (#1-3,#7) may be due to the sensitivity of the alignment of the input beam axis to that of waveguide. Low transmission efficiency for 31.75 mm diameter evacuated waveguides (#4 and 5) may be attributed to the low purity of the coupled HE_{11} mode. Since the axis alignment is less critical in the small diameter waveguide system, further optimization of the MOU mirrors or adjustment of the position of the waveguide mouth might be necessary.

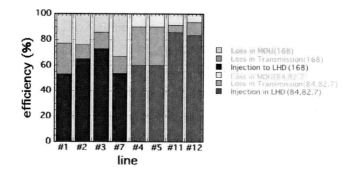

FIGURE 2. Power injected to the LHD and loss at MOU normalized by output measured at the gyrotron window for each line. Here, #1,2,3 and 7 are for 168 GHz, #4,5 are for 84 GHz and #11,12 for 82.7 GHz.

TABLE 1. List of operated gyrotron and transmission lines in LHD

Gyrotron No.	#1	#2	#3	#7	#4	#5	#11	#12
frequency (GHz)	168	168	168	168	84	84	82.7	82.7
manufacturer	Toshiba	Toshiba	Toshiba	Toshiba	Gycom	Gycom	Gycom	Gycom
	Triode	Triode	Triode	Triode	Diode	Diode	Diode	Diode
spec. power (kW)	500	500	500	500	800	800	500	500
pulse width (s)	1	1	1	1	3	3	2	2
Power Supply	#1 CPD	#1 CPD	#1 CPD	#3 CPD	#2 CPD	#2 CPD	Non CPD	Non CPD
waveguide dia.(mm)	88.9	88.9	88.9	88.9	31.75	31.75	88.9	88.9
	dry air	dry air	dry air	dry air	evacuated	evacuated	dry air	dry air
total length (m)	92	92	78	94	65	72	116	115
No. of Bends	18	21	16	21	10	10	19	15
Antenna position	$5.5U_{in}$	$9.5U_{in}$	$2O_{left}$	$2O_{right}$	$1.5L_{out}$	$1.5L_{in}$	$5.5U_{out}$	$9.5U_{out}$
Max. Inj. Power (kW)	180	212	186	160	383	408	254	286
Max. Inj. width (s)	1.0	1.0	1.0	0.9	1.5	1.5	1.5	1.5

The power from each gyrotron is transmitted through a corrugated waveguide system and injected by a quasi-optical antenna system. Two sets of U (upper) port antenna consist of two sets of mirrors for 82.7 and 168GHz. The antenna mirrors are designed using the phase constant method assuming Gaussian optics[4]. Designed beam waist sizes on the midplane of LHD are 15 and 50 mm in radial and toroidal direction. These values and its steerability on the midplane of the LHD are confirmed by low power test. Similar but symmetric and less focussed beams are formed in L (lower) and O (outer) ports. In each antenna system, final plasma facing mirror is plane mirror and can be steered by remote controlled super sonic motors around two axis (azimuth and elevation angle).

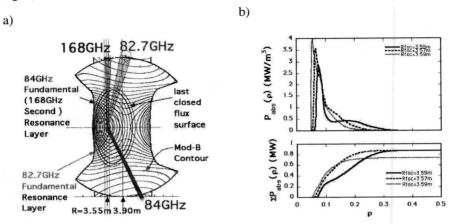

FIGURE 3. a) Flux surfaces and mod-B contours in LHD at vertically elongated cross section. Injected microwave beam from upper and lower antennas are shown with the beam waist size in scale. b)Expected power deposition profile calculated from ray tracing. Power from horizontal antenna is excluded in this calculation.

Fig. 3 a) shows the injection beams, mod-B contours, and flux surfaces in the vertically elongated poloidal cross section. Two sets of upper beams, one lower beam from a vertical antenna, and two beams from horizontal antennas are used. In order to attain high electron temperature at center, the magnetic field strength and the confinement

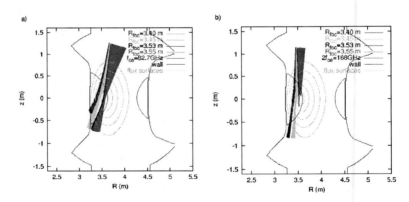

FIGURE 4. Ray tracing results for a) 82.7 GHz beam and b) 168GHz beam. Each beam is represented by bundle of 351 rays, equally distributed within 1.5 times of waist size at the antenna. All rays are projected on the vertically elongated cross section.

magnetic field configuration are selected to have a power deposition as nearly on axis as possible. One of the optimum combinations is the magnetic axis at 3.53 m and the toroidally averaged magnetic field strength on the axis at 2.951 T. The expected power deposition profile estimated by ray tracing described below indicates that almost all of the injected power from the upper and lower antennas are concentrated within an averaged minor radius of $\rho \approx 0.2$ as shown in Fig. 3 b) , here, three cases of different injection angle is plotted to see the allowances and errors in antenna setting. Lower traces are integrated absorption power. Almost 100% power absorption can be expected in the density, and temperature regime discussed in this paper, provided that the beams crosses the resonance in the plasma confinement region. Examples of the raytracing calculation for 82.7 GHz and 168GHz are shown in Fig. 4. Here, projected bundles of rays are shown for the injection angles corresponding to target focal positions on the mid plane of LHD, R_{foc}=3.40, 3.45, 3.53 and 3.55 m. Starting points of each ray are distributed within 1.5 times of waist size with the distance between each other so as to represent an equal area on the starting plane. The injection beam is strongly focused by the last second mirror and steered by the last plane mirror. Initial wave-vector of each ray is set normal to the equi-phase surface defined by the Gaussian beam parameter of each antenna. Normally, a ray is traced using a cold dispersion relation, and the absorption is estimated from quasi perpendicular, weakly relativistic absorption formula[5]. Due to the diffraction of the beam, result tend to be narrower than the real one, especially in the low refractive (density) plasma. The injection polarization, so far is assumed to be pure X or O mode for the raytracing calculation. The injection polarization can also be controlled by setting a set of rotation angles of $\lambda/4$ and $\lambda/8$ plates installed in the waveguide transmission system. These setting angles for optimum focal point and polarization have been controlled independently using the remote control system. The polarization monitor has been developed to reduce real injection polarization in real time. This monitor utilizes the mirror with multi-hole coupler sensitive to the linear polarizations parallel and perpendicular to the incidence plane. Outputs from both polarizations are combined and phase relations are analyzed by quadrature circuits to reduce the polarization states at the surface of the mirror. The principle and result from high power test using gyrotron will be presented elsewhere[6].

OPTIMIZATION OF THE HEATING SYSTEM

Square wave modulation experiments are performed at the averaged electron density of 1.0×10^{19} m^{-3} with flat profile, in order to get detailed power deposition profile when the focal position is scanned for 82.7 GHz and 168 GHz. The electron temperature of parabolic profile and $T_{e0} \approx 2.5$ keV is sustained by NBI. The magnetic field is set at 2.951 Tesla and magnetic axis at 3.50 m. The fundamental and second harmonic resonance layers just cross the magnetic axis under this condition as shown in Fig. 4. The direction of the beam injection is labeled by the radial focal position, R_{foc}, where the center of the injected beam crosses the mid-plane. Under the same magnetic field, the radial position and shape of the cross section between resonance layer and injected beam can be varied by changing the R_{foc} as shown in Fig. 4. The modulated power of both the fundamental and the second harmonic heating is applied on this target plasma. The boxcar technique is used for ±3 ms data points at every turn on and off timings [1, 3] to deduce the first order power deposition profile. The power deposition profile deduced from this method and that calculated from ray tracing well coincides, especially in the case of fundamental resonance heating (82.7GHz), although discrepancies in 168 GHz case are rather enhanced may be due to the limitation of the ray tracing calculation using geometrical optics. These results indicate that the control of focal point by steering mirrors work well as designed, but the reduction of power deposition using ECE data and calculation using geometrical optics needs more optimization[3].

Due to the complexity of the configurations of the transmission system and quasi-optical antenna system, the setting of the antenna angles and polarizer rotation angles for a desired polarization are not straightforward. Furthermore, the definition of the injection parameters on the frame of the LHD magnetic configuration is necessary. In order to be able to do a quick selection and setting of these angles, depending on the various experimental purposes, a software to pick up optimum angle combination of the two polarizers is developed. The desired polarization state to excite the effective mode is determined by the angle between injection and magnetic field directions at the interface of the injected beam to the plasma (near the surface). The polarization state at the interface is controlled by the combination of rotation angles of two polarizers and antenna angles. Given the magnetic field configuration and the magnetic field strength, the optimum or desired antenna setting angles can be determined using a geometrical configuration of the steering antenna. Using such derived magnetic field strength, the angle and the mode, the necessary polarization state can be calculated by the cold plasma dispersion relation. This polarization state is projected back to the polarizers as (α_{opt} and β_{opt}). Here, α and β denotes the angle of the long axis and fatness of polarization ellipse. β =0 means linear polarization, and β=+(-)π/4 indicates right (left) hand circular polarization. A set of rotation angles of polarizers is selected so as the output polarization state (α and β)to maximize

$$\eta = cos^2(\alpha - \alpha_{opt})cos^2(\beta - \beta_{opt}) + sin^2(\alpha - \alpha_{opt})sin^2(\beta + \beta_{opt}) \quad (1)$$

for the given polarizer configuration and the reasonably linear polarization input from the gyrotron. The real polarization state at the far end of transmission line is checked by the newly developed polarization monitor described in the previous section. The results indicates that the the polarization state changes as designed.

To check the polarization effect on the actual plasma heating, the angle α is scanned during the additional heating on NBI target plasma. Figure 5 shows an example of the effect of polarization control. Here, the power from 168GHz is turned on to the NBI heated plasma in LHD. The power deposition profile here is again deduced using the change of local temperature rise at the time of power turned on. The change of the power deposition profile in the polarization α angle is shown as contour plot. The expected optimum angle in this case is -60 degree. It is shown that the localized heating at normalized minor radius of 0.3 is achieved near the optimum polarization (X-mode). It is also noted that broad power deposition appears near the orthogonal polarization at 30 degree. These results indicate that the the algorism to determine the optimum injection angle and polarization works well and ECRH absorption mechanism is just as expected

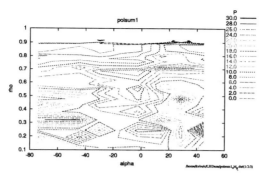

FIGURE 5. Power deposition profile during angle scan for 168 GHz. Here, the power deposition profile is deduced by the local temperature rise at turning on phase.

from cold dispersion relations. Detailed study for the effect of mode scrambling due to the shear [7] are left for the future work.

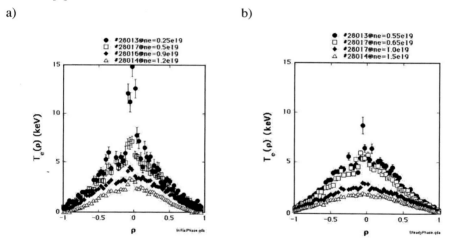

FIGURE 6. Electron temperature profiles measured by YAG-Laser Thomson scattering system for various density a) just after power increased and b) 200 ms after.

OPTIMIZATION FOR HIGHER PLASMA PARAMETERS

The central electron temperature of more than 10 keV is achieved in the experimental campaign in 2001. Before that, it had not been possible to do efficient heating by ECH on LHD due to the lack of the resonance condition on the axis. The magnetic field at the magnetic axis is increased by shifting the axis inward (R=3.5 - 3.6) from the standard position(R = 3.75 m) to locate the cyclotron layer across the axis. The achievement of high electron temperature is clearly the result of the concentration of ECH power near the center. It is also clarified that the density range of such achievement is limitted blow 6×10^{18} m^{-3} with the power available. The non-linear dependence of the centeral electron temperature on the density suggests the action of of radial electric field induced by the ECH in the low collisional regime[8]. The formation of the electron internal transport barrier is confirmed during ECH on the NBI target plasma. Similar structures are also observed in the ECH only plasma as shown in Fig. 6. Here, a) show electron temperature profiles at the initial phase of ITB formation during the scan of the density at 0.25, 0.5, 0.9 and 1.2×10^{19}m^{-3}. In this initial phase, ITB feature is observed just near the center at $\rho \approx 0.1$. On the other hand, temperature profiles at the final phase of high power ECH during this density scan are shown in Fig. 6 b). It is shown that the temperature profile similarity in the corresponding density, but ITB features are weaken. This means that there is a hysteresis feature in the formation of ITB. The well developed high temperature profile in ECH only case do not show a apparent foot of ITB but high $\nabla T_e / T_e$ is formed almost over the confinement area. The electron temperature profile shows ITB like feature even at the initial phase of applying high power with a relatively low temperature even at the density more than 1×10^{19}m^{-3} as shown in Fig.6 a). The foot point of ITB point within here the sharp reduction of the transport appears and spreads outside as the target electron density decreases. These results indicates that the transient and dynamic behavior of the electron temperature profile are closely related to the balance between narrow power deposition and high temperature gradient formation. In this sense, further optimization of the ECH power deposition profile is important to attain high electron temperature in wider density range and in space and longer in time.

CONCLUSION

The power deposition profile is the essential key in analyzing and understanding the transport mechanisms. Optimization of the system has been performed to achieve well defined power deposition to clarify the mechanism of high electron temperature and formation of ITB in LHD. Further optimizations of the combination of power deposition and therefore each injection polarization are necessary to achieve high electron temperature at higher density with wider area for longer duration.

REFERENCES

1. S. Kubo, *et al.*, "RF Experiments in LHD," in *Radio Frequency Power in Plasmas*, edited by S. Bernabei and F. Paoletti, AIP Conference Proceedings 485, American Institute of Physics, New York, 1999, pp. 237–244.
2. T. Shimozuma, *et al.*, *Fusion Engineering and Design*, **53**, 525–536 (2001).
3. S. Kubo, *et al.*, *J. Plasma Fusion Res.*, **78**, 99–100 (2002).
4. S. Kubo, *et al.*, *Fusion Engineering and Design*, **26**, 319–324 (1995).
5. M. Bornatici, *et al.*, *Nucl. Fusion*, **23**, 1153– (1983).
6. T. Notake, *et al.*, "Polarization Monitor of High Power mm-wave for ECH/ECCH in LHD," in *Conference Digest of 28th Int. Conf. on Infrared and Millimeter waves*, 2003.
7. H. Idei, *et al.*, *Fusion Engineering and Design*, **53**, 329–336 (2001).
8. A. Fujisawa, *et al.*, *Physics of Plasmas*, **7**, 4152– (2000).

Physics Studies with ECH/CD in the TCV Tokamak

A.Pochelon, S.Alberti, C.Angioni, G.Arnoux, R.Behn, P.Blanchard, Y.Camenen,
S.Coda, I.Condrea, T.P.Goodman, J.Graves, M.A.Henderson, J-Ph.Hogge, E.Nelson-Melby, P.Nikkola, L.Porte, O.Sauter, A.Scarabosio, M.Q.Tran, G.Zhuang, TCV Team

*Centre de Recherches en Physique des Plasmas CRPP EPFL, Association EURATOM-Confédération Suisse,
Ecole Polytechnique Fédérale de Lausanne EPFL, CH-1015 Lausanne, Switzerland*

Abstract. The TCV tokamak program is centred on flexible plasma shaping matched by a flexible auxiliary-heating system based entirely on ECH at 2nd and 3rd harmonics. This paper reviews some of the recent highlights of TCV ECW (Electron Cyclotron Waves) results. Very high elongation high q plasmas, which would be vertically unstable in Ohmic conditions, are stabilised by broadening the current profile making use of far off-axis ECH or ECCD. Central power deposition in such a configuration can be achieved at high central density (beyond 2nd harmonic cut-off) using 3rd harmonic heating. The 3rd harmonic power deposition meets expectations with coupling of up to 100% and feedback is developed to maximise coupling in transient conditions. Experimental measurements and Fokker-Planck modelling suggest that radial diffusion of suprathermal electrons play a crucial role in determining the ECCD efficiency at the maximum EC power density of TCV. Fully sustained off-axis ECCD scenarios have allowed the creation of plasmas with broad steady-state electron ITBs, high bootstrap fractions (80%) and reversed central shear. Reversed central shear and eITBs have also been developed in discharges dominated by inductive current by combining off-axis heating with central counter CD. Shortening of sawtooth period, useful in NTM avoidance schemes, is produced with local power deposition just inside the q=1 surface.

1. INTRODUCTION

An essential part of the TCV tokamak program is dedicated to the study of the effect of plasma shape on plasma confinement and stability. ECW, offering power coupling independent of the plasma-launcher distance, was an obvious choice. Moreover, the most important property that ECWs offer is localised power deposition in real space and phase space, a crucial property for the control of instabilities (sawtooth or NTM), or for the control of the current profile (enhanced confinement or vertical stability).

The TCV ECW system is composed of six 2nd harmonic X2 gyrotrons and LFS launchers (3MW, 2s, 82.7GHz, $n_{e\ cut-off\ X2} = 4.3 \times 10^{19} m^{-3}$) and three 3rd harmonic X3 gyrotrons using one top launcher (1.5MW, 2s, 118GHz, $n_{e\ cut-off\ X3} = 11.5 \times 10^{19} m^{-3}$). All launchers have two degrees of freedom and can be steered in real time [1]. This represents high power density with potentially more than 30MW/m^3 in central X2 injection. The relative alignment of the different launcher mirrors is accomplished using the highly localised sawtooth period response at the $q=1$ resonant surface. The main codes used are TORAY-GA for the ray tracing and linear CD [2], Fokker-Planck CQL3D (X2&X3) with hard X-ray module [3], equilibrium LIUQE and CHEASE [4] and transport PRETOR [5] including also sawtooth stability, PRETOR-ST [6].

2. CURRENT PROFILE BROADENING WITH OFF-AXIS ECW

Vertical stability of high elongation κ plasmas requires broad current profile $j(\rho)$, as in Ohmic low-q high-current discharges. High κ operation has been extended to lower currents at constant quadrupole field using far off-axis ECW to broaden the current profile, thereby creating high κ discharges that would otherwise be ohmically unstable [7], as shown in Fig. 1a. This enables us to study the stability and confinement properties of a wider range of elongations and current profiles. The current profile broadens on a resistive time scale through T_e-profile broadening, which is caused by an ECH deposition at ρ_{dep}~0.5-0.7, as shown in the discharge evolution of phase 1 of Fig. 1b. The radial co-ordinate used is $\rho=(V/V_{LCFS})^{1/2}$, where V is the volume inside the flux surface considered.

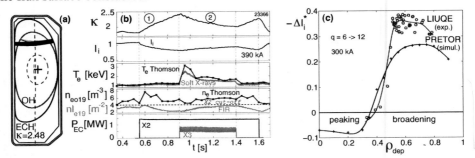

FIGURE 1. a) Increase of elongation using far off-axis ECH deposition, **b)** Parameter evolution in a discharge first elongated with far off-axis X2 (phase 1), then centrally heated with X3 (phase 2, see section 3). **c)** Δl_i vs ρ_{dep}, showing the narrow layer of effective j-broadening.

The change in amplitude of l_i (and κ) depends strongly on the deposition location [8], as shown in Fig. 1c. Modifications are very effective in a narrow radial layer, typically $0.55 < \rho_{dep} < 0.7$. Inside a pivot point at $\rho_{dep}=0.4$ the deposition generates current peaking. This pivot point moves to a larger radius for higher plasma current, pushing the effective layer further out. These results are confirmed by PRETOR transport code simulations. Increasing the power deposited off-axis increases the broadening effect. At higher power, it increases less, which suggests an increase of the thermal diffusivity χ_e outside ρ_{dep} with absorbed power, and is again confirmed by PRETOR transport simulations. Injecting the EC beam at a toroidal angle of typically $\varphi=\pm15°$ appreciably improves effectiveness by 15-20% over ECH conditions ($\varphi=0°$). The improvement is not due to ECCD since injection occurs totally in the trapped electron region of phase space as demonstrated by CQL3D calculations, but is due to the fast electrons generated, effectively measured on the HFS ECE diagnostics, and decreasing the loop voltage symmetrically for both $\varphi=\pm15°$.

Operation of X2 at high-density, even at densities above cut-off in the core, improves current profile broadening effectiveness. This is both due to the usual improvement of electron confinement time with density, which is reflected by a large reduction of χ_e outside the deposition location, and by the increased P_{ei} transfer in the core due to peaked density profiles, flattening the central T_e-profile measured from Thomson scattering. The presence of high harmonics MHD modes with $q=m/n=1$ (1/1, 2/2, 3/3) measured from a 64-channel multi-wire proportional soft X-ray camera [9] is also indicative of a flatter central current profile [10]. At high-density (n_{eo}~6×10^{19} m^{-3}), 1MW was sufficient to raise the elongation from 1.7 to 2.5 at 300kA, using the narrow deposition layer for effective current profile broadening.

3. CENTRAL X3 HEATING IN ELONGATED PLASMAS

Third harmonic X3 heating is exploited for central heating in conditions overdense to X2. Taking the above elongated discharges as an example, X3 power absorption is maximised by launching the wave from the top with a shallow incidence of the beam on the resonance layer, Fig. 2 [11]. Here, the X3 launching mirror angle was optimised to maximise the electron temperature T_{eo} rise soon after the start of X3 (t=0.95s) since the density and equilibrium transients did not allow optimisation of performance throughout the discharge, see phase 2 of Fig. 1b. With far off-axis deposition, X2 can be operated with a central density up to typically twice X2 cut-off density, (i.e. half to two-thirds X3 cut-off density) [7], with significant refraction of X2 and X3 beam paths (Fig. 2). The central region in cut-off of X2 is indicated.

When X3 heating is applied, the elongation starts to decrease due to X3 central power deposition peaking the current profile. The rigid body growth rate γ_{RB} of the vertical instability during X2 increases with elongation (initial phase 1), but starts to decrease at the onset of X3, probably due to a predominance of the decrease of κ over the peaking of $j(\rho)$. Efforts at maintaining constant high elongation during X3 heating by increasing the equilibrium quadrupole field, otherwise maintained constant in these experiments, leads to loss of vertical equilibrium, unless X2 far off-axis power deposition is increased to oppose the X3 current peaking effect.

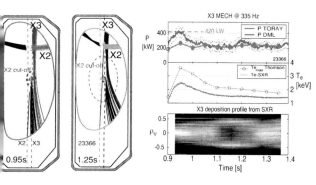

FIGURE 2. Central X3 heating on X2 far off-axis elongated plasma at two different times; the position of the cold resonances and cut-offs are indicated.

FIGURE 3. X3 modulated ECH DML power compared with power absorbed from ray tracing, central temperature evolution from Thomson and soft X-rays absorber method, power deposition profile from soft X-rays.

The total X3 power coupled and its deposition profile are measured by modulation techniques [12]. The time evolution of the measured and predicted X3 absorbed power (from diamagnetic loop (P_{DML}) and ray tracing TORAY-GA (P_{TORAY}), respectively), the central electron temperature (by Thomson and soft X-ray), and the emissivity profile measured with a 64 channel soft X-ray wire chamber (MPX) are shown in Fig. 3. Both the DML and MPX yield reliable results with a response phase close to 90° at an optimal X3 power modulation frequency (337Hz). The power deduced from the DML is identical to that from ray tracing for most of the time slices, which implies a fully thermal plasma. For these times the deposition profile, similar to the one deduced from ray tracing, is monotonic and peaked on axis. The fact that $P_{DML}>P_{TORAY}$ indicates absorption on suprathermal electrons. The presence of suprathermal electrons is confirmed by $T_{ECE}/T_{Thomson}>1$, where T_{ECE} is the radiation temperature measured with a high-field side radiometer. The suprathermal electrons are created during the X2 phase ($\varphi \neq 0°$) and persist over most of the X3 heating phase. At the beginning of X3, full absorption is measured by the DML, the power being shared between bulk and suprathermal absorption, since this full absorption is correlated with an increase of the T_{ECE}-profile (with $T_{rad} > T_{Thomson}$).

4. X3 POWER DEPOSITION AND ABSORPTION

The experiments described in section 3 confirm that X3 absorption is very sensitive to X3 launcher angle. The assessment of X3 heating and power deposition is made in more steady-state plasmas [13]. An example of a top launcher angle sweep during a medium density discharge is shown in Fig. 4a. The plasma soft X-ray response and the absorption calculation from ray tracing are compared, Fig. 4b. For all densities up to 3/4 of cut-off density, there is good overlap showing that ray tracing appears adequate to describe the absorption. It shows that differences of the order of 0.2° in the mirror angle, or variations of the plasma parameters can significantly modify the absorbed power, motivating the development of a feedback system to maximise absorption throughout the discharge.

Experiments were performed to test real-time data analysis methods and to determine the best signal for use in real-time feedback control of the launching mirror angle. In these experiments the launcher mirror angle was swept linearly in time through the resonance layer with a low amplitude sinusoidal oscillation superimposed. The oscillation allows us to determine the maximum response by examining the phase and amplitude of the perturbations measured on a soft X-ray camera with a line of sight through the plasma centre. Figure 5 shows the applied perturbation and the plasma response, as well as the amplitude and phase of the analysed signals [14]. Using the phase response as a feedback signal it will be possible to adjust the launcher angle in real time to maintain maximum absorption throughout the discharge. Algorithms that can be implemented in a digital signal processing system are being developed.

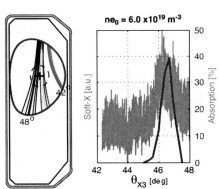
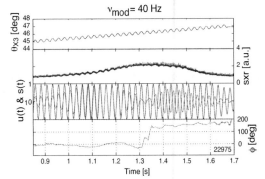

FIGURE 4. X3 mirror angle sweep:
a) X3 mirror angles θ_{X3} of Fig. 4b),
b) soft X-ray plasma response and TORAY absorbed power fraction.

FIGURE 5. X3 mirror angle sweep with a 40Hz dithering, soft X-ray response, phases of soft X-ray and mirror angle, phase difference exhibiting a phase jump at maximum coupling location.

5. SUPRATHERMAL ELECTRON TRANSPORT WITH STRONG ECCD

Steady-state fully sustained ECCD discharges are used for the study of ECCD efficiency in the absence of any inductive component ($E_\parallel=0$). At the high power densities in TCV (even >30MW/m^3), non-linear effects in Fokker-Planck calculations were predicted to enhance markedly current drive efficiencies above normalised power densities of $Q_e[\text{MW/m}^3]/n_e^2[10^{19}\text{m}^{-3}] \sim 0.5$ [15]. However, with normalised power densities in excess of 10 in the same units, non-linear effects were not observed as strong as predicted, requiring a better understanding of suprathermal electron transport. A hard X-ray camera (15 spatial chords each with 8 energy bins between 10 and 100keV [16]) and a HFS ECE [17] radiometer provide fast electron measurements. The hard X-ray emission profiles after inversion show an emission going from peaked to hollow when the deposition is moved from the core to edge, but are always broader than the ECCD calculated deposition location, Fig. 6a, thus providing a first evidence of radial transport [18]. The time delay of the ECE response to an ECCD pulse increases with distance from deposition, Fig. 6b, a second evidence of transport.

The hard X-ray emissivity spectrum along different plasma chords was simulated with CQL3D with a central fast electron diffusivity $1.5<D_o<5\text{m}^2/\text{s}$, Fig. 7 (loop voltage $V_l\neq0$, central co-CD deposition $\rho_{co}\sim0.2$, L-mode plasma). The diffusion coefficient was chosen to match the observed to the simulated I_p-I_{BS} value, where the bootstrap current is obtained from Thomson electron pressure profiles. The simulation of the photon temperature yields a flat radial profile in agreement with experiment [19], whereas zero diffusion would yield a photon temperature peaked on axis. In fully ECCD sustained discharges, matching $I_{CD}=I_{tot}-I_{BS}$ with the simulated driven current I_{CD-FP} yields both the diffusivity, $0.5<D_o<1$ m^2/s, and the driven current profile $j(\rho)$, Fig. 8 ($V_l=0$, $\rho_{co}\sim0.45$, eITB plasma). The results are fairly independent of the transport model used [20].

The simulated I_{CD-FP} without diffusion can easily be ten times higher than with the diffusivity suggested by suprathermal electron measurements, and a deposition off-axis can generate current deposition profiles peaked on-axis. The non-linear current drive enhancement is strongly reduced by such high suprathermal electron diffusion at these high power densities, but it still increases the CD efficiency above the linear efficiency at the power level used.

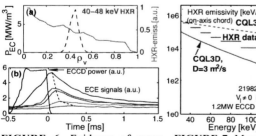
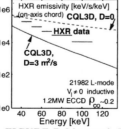
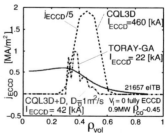

FIGURE 6. Evidence of suprathermal diffusion, **a)** local deposition leads to delocalised hard X-ray emission, **b)** HFS ECE signals indicating time delays increasing with distance to ECCD pulse.

FIGURE 7. Measured and simulated hard X-ray spectra for a central co-ECCD power deposition.

FIGURE 8. Driven current from TORAY (linear), CQL3D (quasi-linear) with and without radial diffusion. Both the I_{CD} value and j-profiles are strongly affected by radial diffusion.

6. FULLY SUSTAINED ECCD SCENARIOS WITH EITBs

Different ECW scenarios are used to form electron internal transport barriers (eITBs) in fully ECCD sustained discharges [21]. The eITBs are formed in the current plateau, without exploiting the help of the strong current rise and with no momentum input. The standard schemes used are: 1) co-ECCD deposited off-axis with ECH on-axis to obtain fully ECCD sustained discharges, 2) ECH off-axis with counter on-axis to form eITBs in inductive discharges [22].

The first approach is described here. The eITB develops with co-ECCD applied off-axis at $\rho_{co} \geq 0.35$ (0.9MW) and typically with ECH on-axis (0.45MW) with $V_l=0$ set by I_{OH}=const., I_p=90kA, $n_{eo}=10^{19}$m^{-3}, κ=1.55, δ=0.5. Co-ECCD broadens the current and pressure profiles. The pressure gradient steepens, which increases the bootstrap current fraction I_{BS}/I_{tot}. A hollow total current density profile j_{tot} develops on a few current redistribution times τ_{CR} with non-monotonic q-profile, carried by j_{BS}, since the j_{CD} is rather flat (see section 5). The addition of central ECH or counter-CD enhances the pressure gradient increasing the bootstrap fraction, increasing the reverse shear, Fig. 9a, and enhancing the barrier performance. This is confirmed by an increase of τ_{Ee} with increased central power deposition.

Displacing ρ_{co} outwards improves the confinement $H_{RLW}=\tau_E/\tau_{RLW}$ for $\rho_{co}>0.35$, Fig. 9b. The increase of confinement is associated with an enlargement of the high confinement volume which scales as ρ^2_{q-min} and ρ^2_{eITB} [23] (defined as the location ρ_{p*} of the maximum of $\rho^*_T(\rho) \equiv \rho^*_{Tmax}$) [24]. Giving a toroidal angle on the central ECH beam, φ=0°, to launch counter-CD (co), while maintaining the same deposition for the off-axis co-ECCD, further enhances (degrades) the confinement, Fig. 9c. The counter-CD creates a hollower current profile, which enhances the eITB. The confinement enhancement is attributed to a more reversed shear with q_0/q_{min} and ρ^*_{Tmax} increasing.

The confinement enhancement at the eITB can be controlled either by increasing the volume enclosed by the barrier (ρ^2_{eITB} or ρ^2_{q-min}) by increasing the off-axis co-ECCD deposition radius, or by increasing the shear reversal factor q_0/q_{min} or ρ^*_{Tmax} with on-axis counter-CD. The global figure of merit for the eITB can be expressed as $H_{RLW} \sim q_0/q_{min} \times \rho^2_{q-min}$ [23] or with Thomson data, $H_{RLW} \sim \rho^*_{Tmax} \times \rho^2_{eITB}$, Fig. 9d [25].

In inductive discharges, with ECH deposition off-axis, increasing the power of counter-CD in the core is another way to create reverse shear and enlarge ρ_{q-min} and ρ_{eITB}. This scenario ($P_{ECH\_off-axis}$=0.9MW and $P_{counter-CD\_central}$=1.35MW) allowed the achievement of record electron temperatures of 15-18keV, with β_N close to 1.5.

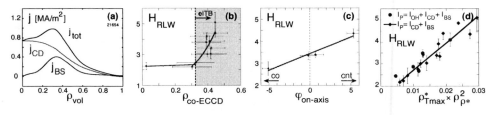

FIGURE 9. eITBs in fully sustained ECCD discharges: **a)** off-axis co-CD deposition at ρ=0.4 leads to non-monotonic $j(\rho)$. The hollow total current j_{tot} is due to the bootstrap current. Improvements of H_{RLW} achieved by **b)** displacing ρ_{co}>0.3, **c)** changing central ECH to counter-CD; **d)** confinement enhancement scales with the barrier strength × barrier volume, i.e. $\rho^*_{Tmax} \times \rho^2_{eITB}$ (also some inductive data).

7. SAWTOOTH CONTROL WITH LOCALISED POWER DEPOSITION

Large sawteeth can seed NTMs, reducing their onset β_N [26]. Shortening the sawtooth period (destabilising sawteeth) by means of adequate local power deposition is therefore of interest. Recent sawtooth control experiments with ECW [6] have been undertaken in several machines, also in the presence of ion heating [27]. Sawtooth period maximisation using a poloidal mirror sweep through the location of the q=1 surface ρ_1 has been regularly used as the most efficient means of cross-calibrating the alignment of the different launchers, to an accuracy of 2% radius (~5mm). The sawtooth period is a nearly linear function of the power density [28, 29, 6], thereby allowing also an alternate measurement of the beam width [29]. The sawtooth period response to localised deposition can be simulated with the PRETOR-ST [6] code. For destabilisation, that is to make the q=1 shear $s_1=dq_1/dr$ steeper, both heating and co-CD inside ρ_1, or counter-CD outside ρ_1 can be used. For stabilisation, that is to flatten s_1, both heating and co-CD outside ρ_1, or counter-CD inside ρ_1 can be used [6].

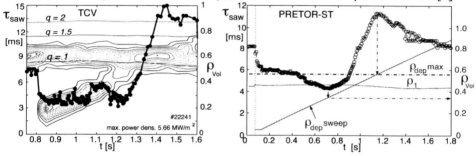

FIGURE 10. Experimental sawtooth period, position of main rational q-surfaces (LIUQE) and TORAY-GA ECH power density contours: 0.9MW ECH (slight co-CD) placed for optimal stabilisation; 0.45MW of swept ECH (slight co-CD). Localised heating is found to destabilise sawteeth inside $q=1$.

FIGURE 11. Simulated sawtooth period in the case of a full radial sweep of localised 0.45MW of ECH, while stabilising sawteeth with 1.35MW at the fixed ρ_{max}.

A case with multiple beams and high total power is shown for illustration. Several beams (2 in the experiment, Fig. 10, and 3 in the predictive simulation, Fig. 11) deposit the power at the location maximising the sawtooth period, ρ_{max}, (ρ_{dep}~0.6). Another single beam sweeps the power deposition from the plasma centre to the edge. The stabilising power deposited beforehand at ρ_{max} increases the sensitivity of the destabilising effect of the swept beam. In the present case, both experiment and simulation yield optimal stabilisation for ρ_{max}/ρ_1~1.2-1.3. Simulations show that the location of most efficient sawtooth stabilisation depends on the deposition width $\Delta\rho$, such that ρ_{max}~$\rho_1+\Delta\rho/2$. Since the stabilisation/destabilisation effect is proportional to the power density, this confirms the advantage of narrow and intense power deposition offered by ECW techniques. Optimal destabilisation is obtained for a deposition totally inside ρ_1 with discrepancies between simulation and experiment presumably due to finite-island size. Figures 10 and 11 provide evidence that modelling of sawteeth is in close agreement with experiment.

8. CONCLUSIONS

High power far off-axis locally deposited X2 ECW allows broad current profiles and improved vertical stability. At the maximum ECCD power densities in TCV, suprathermal diffusion, assessed by hard X-ray and HFS ECE measurements, plays a central role in the resulting ECCD efficiency and radial j_{CD} profile. The local deposition of co-, counter-ECCD and ECH powers are used as tools to shape eITB in fully ECCD sustained discharges and inductive discharges, making TCV an ideal resource for various current profile tailoring experiments. Sawtooth control is more effective with high power density and sawtooth destabilisation is obtained with deposition inside $q=1$. Initial X3 heating has demonstrated full power absorption in high-elongation high-density discharges. The X3 launcher feedback is being developed to couple efficiently during evolving target plasmas.

ACKNOWLEDGMENTS

The authors recognise the competent support of the entire TCV team. This work was partly supported by the Swiss National Science Foundation.

REFERENCES

1. Goodman T.P. et al., *Proc. of the 19th Symposium on Fusion Technology*, Lisbon, (1996) 565.
2. Matsuda, K., *IEEE Transactions on Plasma Science* **PS-17** (1989) 6.
3. Harvey, R.W., et al., *Proc. of the IAEA Tech. Conf. on Advances in Simulation and Modelling of Thermonuclear Plasmas*, Montreal, 1992, (IAEA, Vienna, 1992).
4. Lütjens H., Bondeson A., Sauter O., *Computer Physics Communications*, **97** (1996) 219.
5. Boucher D., Rebut P.-H., *Proc. of the IAEA Tech. Conf. on Advances in Simulation and Modelling of Thermonuclear Plasmas*, Montreal, 1992, (IAEA, Vienna, 1992) 142.
6. Angioni C., et al., *Nuclear Fusion* **43** (2003) 455.
7. Pochelon A., et al., *19th IAEA Fusion Energy Conf.*, Lyon 2002, France, IAEA-CN-94/EX/P-14.
8. Camenen Y. et al., *Proc. of 12th Joint Workshop on ECE and ECRH (EC-12)*, Aix-en-Provence 2002, (World Scientific, Singapore, 2002) 407.
9. Sushkov A. et al, *29th EPS Conf. on Plasma Phys. and Contr. Fusion*, Montreux 2002, ECA Vol. **26B** (2002) Paper P4-118.
10. Duperrex P.-A., Pochelon A., et al., *Nuclear Fusion* **32** (1992) 1161.
11. Hogge J.-P., et al., *EC-12*, Aix-en-Provence 2002, p. 371, accepted for publ. in Nuclear Fusion.
12. Manini A., Moret J.-M., et al., *Plasma Phys. Control. Fusion* **44** (2002) 139.
13. Alberti S., et al., *29th EPS*, Montreux 2002, ECA Vol. 26B (2002), Paper P0-273.
14. Porte L., et al., *19th IAEA Fusion Energy Conference*, Lyon 2002, France, IAEA-CN-94/EX/P5-15.
15. Harvey R.W., *Physical Review Letters* **62** (1989) 426.
16. Peysson Y., Coda S. Imbeaux F., *Nucl. Instrum. and Methods in Phys. Res. A* **458** (2001) 269.
17. Blanchard, P., et al., *Plasma Phys. Control. Fusion* **44** (2002) 2231.
18. Coda S., et al., *19th IAEA FEC*, Lyon 2002, France, IAEA-CN-94/EX-W5, to appear in NF.
19. Harvey, R.W., Sauter O., Prater R., Nikkola P., et al., *Physical Review Letters* **88** (2002) 205001.
20. Nikkola, P., et al., to be publ. in *Nucl. Fus.*; *EC-12*, Aix-en-Provence (2002) 257.
21. Sauter O., et al., *19th IAEA FEC*, Oct. 2002, Lyon, France, IAEA-CN-94/EX/P5-06.
22. Behn R., et al., *30th EPS Conf. on Plasma Phys. and Contr. Fusion*, St. Petersburg 2003.
23. Henderson M., et al., *44th APS Plasma Div.*, Orlando 2002, to be publ. in *Phys. of Plasmas*, 2003.
24. Tresset G., Litaudon X., Moreau D., Garbet X., et al., *Nuclear Fusion* **42** (2002) 520.
25. Martin Y., et al., *30th EPS Conf. on Plasma Phys. and Contr. Fusion*, St. Petersburg 2003.
26. Sauter O., Westerhof E., Mayoral M.L., et al., *Phys. Rev. Lett.* **88** (2002) 105001.
27. Goodman T., Mück A. et al., *19th IAEA-FEC*, Lyon 2002, France, IAEA-CN-94/EX/P5-11.
28. Goodman T.P. et al, *26th EPS*, Maastricht 1999, ECA Vol. 23J (1999), 1101.
29. Henderson M.A., Goodman T.P., et al., *Fusion Engineering and Design*, **53** (2001) 241.

Electron Cyclotron Current Drive in DIII-D: Experiment and Theory

R. Prater, C.C. Petty, T.C. Luce, R.W. Harvey,[a] M. Choi, R.J. La Haye, Y.-R. Lin-Liu, J. Lohr, M. Murakami,[b] M.R. Wade,[b] K.-L. Wong[c]

General Atomics, P.O. Box 85608, San Diego, California 92186-5608
[a]*CompX, Del Mar, California.*
[b]*Oak Ridge National Laboratory,, Oak Ridge, Tennessee.*
[c]*Princeton Plasma Physics Laboratory, Princeton, New Jersey.*

Abstract. Experiments on the DIII-D tokamak in which the measured off-axis electron cyclotron current drive has been compared systematically to theory over a broad range of parameters have shown that the Fokker-Planck code CQL3D provides an excellent model of the relevant current drive physics. This physics understanding has been critical in optimizing the application of ECCD to high performance discharges, supporting such applications as suppression of neoclassical tearing modes and control and sustainment of the current profile.

INTRODUCTION

Electron cyclotron heating (ECH) and current drive (ECCD) have been intensively studied in experiments on the DIII-D tokamak [1-3]. The key objectives are to develop and validate a predictive computational model which will be applicable to next-step devices as well as present-day experiments and to use that model to optimize the many applications of EC power in DIII-D as part of the Advanced Tokamak program. Full validation of the model requires that the relevant physics parameters be recognized and varied in a systematic way in experiments, and that the measurements of the magnitude and profile of driven current be determined and compared with the predictions of the model. The physics model is implemented in the ray tracing code TORAY-GA [4,5] and in the quasilinear Fokker-Planck code CQL3D [6]. This model has been used successfully to obtain the best possible performance of the ECH system on DIII-D in experiments on stabilizing m=3/n=2 [7-9] and m=2/n=1 [10] neoclassical tearing modes and to improve and sustain the current profile which has led to improved tokamak performance [11].

EXPERIMENTAL APPROACH

Detailed current drive experiments were carried out on the DIII-D tokamak [12]. For these experiments plasmas have major radius 1.7 m, minor radius 0.6 m, elongation typically 1.8, and toroidal field up to 2.1 T. The EC system uses up to five gyrotrons in these experiments to generate up to 2.5 MW incident on the plasma for pulses up to 2 s [13]. The 110 GHz frequency of the gyrotrons is resonant with the second harmonic EC frequency at a field of 1.96 T. The EC launcher is placed at the outboard side about 0.7 m above the midplane, and most of the launchers can steer the beam independently in both the vertical (poloidal) and horizontal (toroidal) directions [14]. This launching system provides great experimental freedom to control the location and width of the profile of driven current, as both the radial location and the $n_\parallel$ can be determined by the launch angles for a given equilibrium and kinetic profiles. The physics of ECCD is studied in these experiments through orthogonal scans, such as of minor radius with $n_\parallel$ fixed or of $n_\parallel$ with the minor radius fixed.

The critical diagnostic for measurement of the magnitude and profile of ECCD is the motional Stark effect system [15], which measures the pitch angle of the magnetic field at ≈5 cm intervals across the plasma outer midplane. Time-dependent plasma simulations are used for a range of radial locations, magnitudes, and profile widths to determine the profile of driven current which best matches the actual measurements [2]. Thus, the effects of the back emf generated by the localized ECCD can be taken into account, since in many cases the toroidal electric field is not fully equilibrated.

The dimensionless current drive efficiency $\zeta = (e^3/\varepsilon_0^2)\, n_e\, I_{EC}\, R/T_e\, P_{EC}$ is used as a figure-of-merit to characterize the effectiveness of the EC power in driving current [3]. This efficiency follows from a natural normalization of the current and power density, and it includes the theoretical dependencies on plasma density and temperature [aside from the slowly-varying $\ln(\Lambda)$ factor] which would be expected of any current drive scheme which acts on the electron distribution near the thermal velocity. Hence, deviations of ζ from a constant are due to interaction with a different part of velocity space or to an effect like trapping of electrons in the magnetic well. Plotting ζ in a radial scan illustrates the dependence of the physics of off-axis ECCD without the expected but uninteresting fall-off in current due to decreasing T_e.

The driven currents can be compared with the currents theoretically expected. The relevant theory has been encapsulated in the ray tracing code TORAY-GA and in the quasilinear Fokker-Planck code CQL3D. TORAY-GA uses the cold plasma dispersion relation to determine the ray trajectories and relativistic models for the wave polarization and absorption. It also uses the Cohen Green's function model [5] for calculating the driven current. The ray trajectories and polarizations from TORAY-GA can be input to CQL3D to calculate currents including quasilinear effects due to higher power densities and the effect of $E_\parallel$ on the distribution function. CQL3D uses a collision operator which preserves momentum in electron-electron collisions, unlike that in TORAY-GA. In practice, all of these effects can be important.

COMPARISON OF EXPERIMENT AND THEORY

Excellent agreement of the measured current with the calculated current is found over the DIII-D database of about 80 discharges [1]. This agreement is illustrated in Fig. 1, which shows the measured current vs the current calculated by the CQL3D code, including the effects of $E_\parallel$. The dashed line in Fig. 1 represents perfect agreement, and the error bars of most data points intersect that line. Note that the data base includes many cases with negative driven current which are also in good agreement with theory. In those cases the quasilinear effects on the resistivity given finite positive $E_\parallel$ increase the calculated ECCD rather than decrease it, but still the agreement with theory is excellent.

The data points illustrated in Fig. 1 represent a range of temperatures, densities, minor radii ρ, $n_\parallel$, and θ_p, where the θ_p is the angular variable along a flux surface going from 0 at the outboard midplane to 180 deg at the inboard midplane, and all quantities are measured at the location of the wave-particle interaction. Subsets of these data can then be replotted as single parameter scans, where all parameters but the independent parameter are held fixed. In Fig. 2 the dimensionless current drive efficiency is plotted against $n_\parallel$ for two such scans. In these scans, the density and temperature dependences are categorized jointly as β_e, which is a good characterization of the behavior even though the physics of the dependences on n_e and T_e is slightly different. In Fig. 2 the two curves are for low β_e L-mode discharges, the principle difference being that the upper curve is for an interaction near the inboard midplane where the magnetic well depth is minimum, while the lower curve is for an interaction near the top of the flux surface where the well depth is moderate, both cases being for $\rho=0.3$. Comparing these cases, it can be seen that the efficiency saturates for sufficiently large $n_\parallel$, but that the saturation level is larger for the case with smaller well depth (i.e., $\theta_p = 175$ deg). The agreement between the data points and the CQL3D calculations, including $E_\parallel$ effects, shows that the model has very little systematic variance from the measurements for this cut through the data.

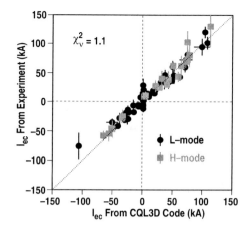

FIGURE 1. The measured ECCD current for the entire database of discharges in DIII-D versus the current calculated by the CQL3D Fokker-Planck code including the effect of the parallel electric field. The circular data points are for L-mode and the squares are for H-mode discharges.

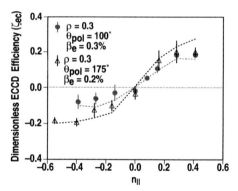

FIGURE 2. Dimensionless current drive efficiency versus parallel index of refraction for discharges with ECCD at $\rho=0.3$, low β_e (<0.3%), and $\theta_{pol} = 175$ deg (triangles) and $\theta_{pol} = 100$ deg (circles). The dashed lines represent the values from calculations using CQL3D.

The data in Fig. 2 are consistent with the physically intuitive idea that trapping of the current-carrying electrons reduces the efficiency of ECCD. Figure 3 shows another case illustrating this behavior. The unfilled points in Fig. 3 show that the efficiency at low fixed β_e, fixed θ_p near 90 deg, and fixed $n_\parallel$ drops as the normalized minor radius increases and the well depth increases. However, the filled points for the same conditions but higher β_e show that the decrease with radius is much smaller, and at ρ=0.4 the dimensionless efficiency for the higher β_e case is double that of the lower β_e case.

The behavior of the efficiency with β_e must be understood in the context of the competition between the Fisch-Boozer current (FB) and the Ohkawa current (OK), the two elements comprising the generation of ECCD [14]. The FB current arises when electrons with one sign of $v_\parallel$ are accelerated by the EC waves to higher energy. At the higher energy, the electrons pitch-angle scatter more slowly than the electrons at the lower energy. In the limit of a steady-state process, this leaves an excess of electrons with that sign of $v_\parallel$. The OK current arises when electrons with one sign of $v_\parallel$ are accelerated in $v_\perp$ into the trapped region of velocity space. These electrons in the trapped region symmetrize rapidly, leaving behind a deficit at the sign of $v_\parallel$. So the toroidal direction of the two current components is opposite, and the net current depends on the details of the balance. For this reason, no simple scaling is possible for estimating the decrement of the net current drive efficiency of off-axis ECCD.

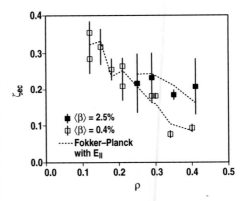

FIGURE 3. Dimensionless current drive efficiency versus rho for low beta plasmas with average beta of 0.4%, θ_{pol} = 95 deg and $n_\parallel$ = 0.4 (unfilled symbols) and for high beta plasmas with average beta of 2.5%, θ_{pol} = 90 deg and $n_\parallel$ = 0.3 (filled symbols). The dashed curves are calculations by the CQL3D code.

The TORAY-GA code calculates this net current drive. For the relevant geometry of Fig. 4(a), the data along the central ray are plotted in Fig. 4(b). This figure shows the normalized power deposited in the plasma per unit ray length (1/P)dP/ds and the quantity (ω/2Ω_e) - 1, where ω is the applied frequency and Ω_e is the local electron cyclotron frequency. Also shown is the calculated driven toroidal current per unit ray length, dI/ds. The case shown in Fig. 4 is for low β_e, and hence relatively weak damping, as Fig. 4(b) shows that the ray crosses the cold resonance where (ω/2Ω_e) - 1 = 0 with significant power still remaining.

The behavior of dI/ds in Fig. 4(b) is easily understood from the discussion of FB and OK current drive discussed above. As the ray propagates, it enters the region close enough to the cold resonance that power is absorbed but still with (ω/2Ω_e) - 1 > 0. To satisfy the relativistic cyclotron resonance for the applied ω and wavenumber $k_\parallel$,

$$\frac{v_\perp^2}{c^2} = 1 \pm \frac{v_\parallel^2}{c^2} - \left(\frac{\omega}{2\Omega_e}\right)^2 \left[1 \pm 2n_\parallel\left(\frac{v_\parallel}{c}\right) + n_\parallel^2\left(\frac{v_\parallel^2}{c^2}\right)\right] , \qquad (1)$$

the $v_\parallel$ must be large relative to the thermal velocity. This resonance in velocity space corresponding to this location in physical space is shown in Fig. 4(c) for the central ray of the ray bundle. The wave-particle interaction takes place along this resonance curve, which does not closely approach the T-P boundary. Because of this relatively weak interaction with the trapped region of velocity space, the FB current is strongly dominant and dI/ds is large and positive. As the wave approaches closer to the resonance,

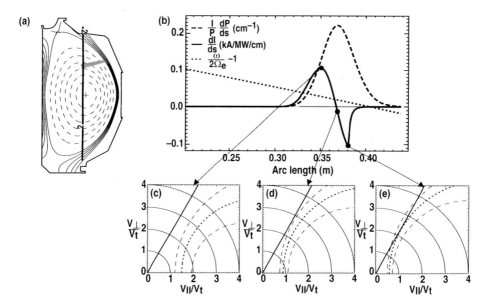

FIGURE 4. (a) Geometric arrangement of rays; (b) $(1/P)(dP/ds)$, dI/ds, and $\omega/2\Omega_e -1$ as a function of ray arc length for the central ray of the bundle shown in (a) as calculated by the TORAY-GA code; (c-e) resonance curve (dotted) and trapped-passing boundary (solid curve) in normalized velocity space, for the locations along the ray path indicated. The dashed curves in (c–e) represent the range of the resonabce given the spred in $k_{\parallel}$ in the beam. For this discharge, the wave power is deposited near $\rho=0.5$ and θ_p near 90 deg, with local density 1.4×10^{19} m^{-3} and local $T_e = 2.2$ keV. The electron beta $\beta_e=0.33\%$ and the estimated optical depth is 10.

smaller $v_{\parallel}$ is needed to satisfy Eq. (1). Hence, the resonance moves closer to the T-P boundary as calculated for the location shown corresponding to Fig. 4(d). For this location the FB and OK currents are approximately equal, so that dI/ds is near zero while $(1/P)dP/ds$ remains large. Power deposited in this part of the ray has almost no ability to drive current. Finally, at the location shown corresponding to Fig. 4(e) the T-P boundary is osculatory to the resonance and the OK current dominates strongly, so the net dI/ds is negative. Further along the ray dI/ds rapidly drops to zero because the wave power is predominantly deposited in trapped electrons for which neither the FB nor the OK current drive is effective. Integrated along the ray, the net current is small because of the cancellation of the current driven along the different parts of the ray. In practice the ray bundle has a spread in $k_{\parallel}$ and in physical location, so the parts of the rays with positive and negative dI/ds average to a small value on any particular flux surface under most conditions.

At higher β_e the situation is quite different. The absorption rate and optical depth at the second harmonic are proportional to β_e, which implies that at higher β_e the wave will be absorbed further from the cold resonance. This reduces the trapping effects. Fig. 5 shows data from a case nearly identical to that of Fig. 4 except that the density and temperature have been doubled. The peak of absorption takes place further from the cold resonance with $(\omega/2\Omega_e) - 1=0.029$ as compared with 0.018 for Fig. 4. This strongly shifts the resonance away from the T-P boundary, reducing the degradation in driven FB current due to the trapping effects of the OK current.

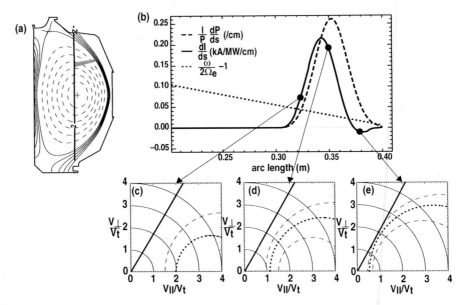

FIGURE 5. Same as Fig. 4, but the local density is 2.7×10^{19} m^{-3} and $T_e=4.2$ keV. For this case $\beta_e=1.24\%$ and the estimated optical depth is 38.

For a large device like ITER the optical depth will be very large, well above 100 even for interactions near the plasma boundary where T_e is reduced. Much of the increase in optical depth comes from the large physical size of ITER, as optical depth is proportional to B/(dB/ds) which varies like the major radius. Therefore, ITER will have very little OKCD compared to FBCD. Even so, trapping effects cannot be neglected due to convective cells which set up in velocity space in response to the EC power deposition.

APPLICATION OF ECCD TO AT DISCHARGES IN DIII–D

The developing understanding of ECCD which was described above has been applied to optimize the applications of ECCD to Advanced Tokamak discharges in DIII-D. Two important applications which require off-axis ECCD are stabilization of neoclassical tearing modes and control and sustainment of current profiles with optimized magnetic shear.

Stabilization of the neoclassical tearing mode can lead to significant improvement in the performance of discharges [9]. Such a case is illustrated in Fig. 6. Here, 2 MW of ECH power is used to drive a highly localized current near the q=3/2 surface. The aiming of the launcher optimizes the current density driven at that flux surface rather than the total current, since the criterion for stabilization is that the quantity $j_{ECCD}/j_{BOOTSTRAP}$ be greater than a number of order unity rather than a condition on the total driven current. For the discharge in Fig. 6 the ECCD is applied at 3 s when the 3/2 NTM is fully grown. The mode shrinks and disappears over a 400 ms period. After the mode disappears the neutral beam heating power is raised in three steps from 4 MW to 7 MW, raising β_N from less than 2 when the mode is saturated to 3. The neutron rate, a measure of plasma performance, doubles as in the last panel.

As the beta of the discharge in Fig. 6 increases, the plasma Shafranov shift increases, and the q=3/2 surface moves relative to the current drive location which is fixed by the toroidal field. So in the higher beta phase of the discharge, the ECCD is no longer well aligned with the rational surface and the 3/2 mode eventually restarts. (Algorithms which keep the location well aligned even when the mode is not present are under development.) This observation illustrates the accuracy with which the ECCD must be aligned, and it further suggests that the actual width of the j_{ECCD} must be not much greater than the displacement which loses the stabilizing effect, or 1.5 cm in this case. This is consistent with the calculations of j_{ECCD} by TORAY-GA which give 2.8 cm in FWHM. Note that this high degree of localization is observed even in a discharge with H-mode edge with large ELMs and sawteeth [Fig. 6(d)] and with a large tearing mode present.

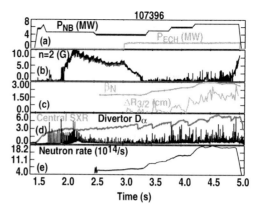

FIGURE 6. Discharge in which the m=3/n=2 neoclassical tearing mode was stabilized by the application of ECCD near the q=3/2 surface at ρ=0.54. (a) Neutral beam and ECH power, (b) amplitude of the n=2 magnetic field fluctuation, (c) β_N and $\Delta R_{3/2}$, the displacement between the q=3/2 surface from its location when the mode is stabilized at 3.5 s, (d) central soft x-ray emission and D_α emission from the divertor region, and (e) the neutron emission rate, as a function of time.

A second critical application of off-axis ECCD is control of the current profile as needed to obtain or sustain discharges with desired current profiles, as required by the Advanced Tokamak program. Recent results from DIII-D have demonstrated the ability to modify the current density in a high beta plasma [11]. In Fig. 7 two discharges are presented, one of which has ECCD driven by 2.5 MW of EC power while the other discharge has heating by the same EC power but without the current drive (that is, the EC power is launched radially rather than obliquely). In both cases the EC interaction takes place near ρ=0.4. The evolution of the q profile is quite different in the two cases, with the co-ECCD case having much larger q_0 and hence stronger central negative shear. This stronger central shear results in improved transport properties in the plasma, causing the central ion temperature to rise strongly as shown in Fig. 7(c. For this 1.2 MA discharge the fraction of the current supported noninductively is about 90%, comprising 10% ECCD, 20% NBCD, and 60% bootstrap current. These results validate a key motivation for the development of the ECH system for DIII-D: that control of the current profile will lead to improved and sustained performance.

ACKNOWLEDGMENT

Work supported by the U.S. Department of Energy under Contracts DE-AC03-99ER54463, DE-AC05-00OR22725, DE-AC02-76CH03073, and DE-FG03-99ER54463.

REFERENCES

1. Petty, C.C., Prater, R., Luce, T.C., et al., "Physics of Electron Cyclotron Current Drive on DIII-D," in Proc. of the 19th IAEA Fusion Energy Conference, Lyon, France, 2002 (International Atomic Energy Agency, Vienna).
2. Petty, C.C., Prater, R., Lohr, J., et al., *Nucl. Fusion* **42**, 1366 (2002).
3. Luce, T.C., Lin-Liu, Y.-R., Harvey, R.W., et al., *Phys. Rev. Lett.* **83**, 4550 (1999).
4. Matsuda, K., *IEEE Trans. Plasma Sci.* **17**, 6 (1989).
5. Cohen, R., *Phys. Fluids* **30**, 2442 (1987).
6. Harvey, R.W., and McCoy, M.G., in Proc. of the IAEA Technical Committee Meeting, Montreal, 1992 (International Atomic Energy Agency, Vienna, 1993) 498.
7. Prater, R., Austin, M.E., Burrell, K.H., et al., in Proc. of 18th IAEA Fusion Energy Conference, Sorrento, Italy (International Atomic Energy Agency, Vienna, 2001) paper EX8/1.
8. La Haye, R.J., Günter, S., Humphreys, D.A., et al., *Phys. Plasmas* **9**, 2051 (2002).
9. Prater, R., La Haye, R.J., Lohr, J., et al., "Discharge Improvement Through Control of Neoclassical Tearing Modes by Localized ECCD in DIII-D," General Atomics Report GA-A24179 (2003) and submitted to *Nucl. Fusion*.
10. Petty, C.C., La Haye, R.J., Luce, T.C, et al., "Complete Suppression of the m=2/n=1 Neoclassical Tearing Mode Using Electron Cyclotron Current Drive on DIII-D," General Atomics Report GA-A24223 (2002) and submitted to *Nucl. Fusion*.
11. Murakami, M., Wade, M.R., DeBoo, J.C., et al., *Phys. Plasmas* **10**, 1691 (2003).
12. Luxon, J.L., Nucl. Fusion **42**, 614 (2002).
13. Lohr, J., Gorelov, Y.A., Callis, R.W., et al., in Proc. of the 12th Joint Workshop on Electron Cyclotron Emission and Electron Cyclotron Resonance Heating, Aix-en-Provence, France, 2002 (World Scientific, Singapore, 2003), 283.
14. Ellis, R., Hosea, J., Wilson, J., et al., in Radio Frequency Power in Plasmas, 14th Topical Conf., Oxnard, California 2001 (American Institute of Physics, Melville NY, 2001), 318.
15. Prater, R., Harvey, R.W., Lin-Liu, Y.-R., et al., *ibid.*, 302.
16. Rice, B.W., Burrell, K.H., Lao, L.L., Lin-Liu, Y.R., Phys. Rev. Lett. **79**, 2694 (1997).

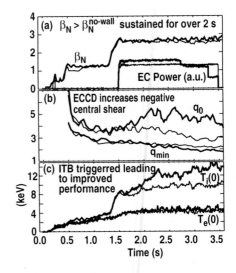

FIGURE 7. Traces from a high performance discharge in which the effects of ECCD (heavier lines) and ECH (finer lines) are compared. (a) β_N and EC power (a.u.), (b) central safety factor q_0 and minimum of the q profile q_{min}, and (c) central ion and electron temperatures.

Toroidal Rotation in ECH H-modes in DIII-D

J.S. deGrassie, K.H. Burrell, D.R. Baker, L.R. Baylor,[a] J. Lohr,
R.I. Pinsker, and R. Prater

General Atomics, P.O. Box 85608, San Diego, California 92186-5608
[a]*Oak Ridge National Laboratory, Oak Ridge, Tennessee*

Abstract. In ECH H-mode discharges with essentially no toroidal momentum input, counter toroidal rotation is measured in the interior, $\rho < 0.5$, while co rotation is measured outside this region. This is contrasted with Ohmic H-modes where the rotation is everywhere in the co direction. A simple parameterization is suggested to facilitate comparison between machines.

INTRODUCTION

Toroidal rotation exists in tokamak discharges with essentially no deliberately injected external torque, in ion cyclotron radiofrequency (ICRF)-heated discharges [1-3], and Ohmically-heated discharges [4]. Understanding the generation of this rotation is important because of the role of toroidal rotation in the stabilization of resistive wall modes [5] and the sheared E×B flow stabilization of microturbulence [6].

H-modes in DIII-D driven by electron cyclotron heating (ECH) also have nonzero toroidal rotation. The rotation in the interior region is in the counter direction, while in the outer region, $\rho > 0.6$, it is in the co direction. Co and counter are relative to the direction of plasma current, I_p, and ρ is the normalized toroidal flux coordinate. This is in contrast to rotation profiles measured in DIII-D for Ohmic H-mode discharges, which are in the co direction for all ρ.

The injection of short neutral beam (NB) pulses, or blips, is required to measure the velocity and temperature of the intrinsic carbon ion impurity in DIII-D using charge exchange recombination (CER) spectroscopy [7]. These blips do inject toroidal torque into the discharge. We find that using only the first 2 ms of the initial blip does give a good measure of the unperturbed velocity profile, as well as temperature profile. The toroidal velocity, U_ϕ, increases in the beam direction (co) during the 10 ms blips utilized in this experiment and the resultant toroidal momentum appears to have a very long confinement time relative to the energy confinement time of the discharge.

An effect of the long time momentum confinement in the NB blip format is that the toroidal velocity "stacks" up in DIII–D from blip to blip, even if the time interval is 200 or 400 ms. This was not known in a previous ECH experiment on the electron heat pinch in which NB blip velocity measurements were also used to infer the effect of

ECH [8]. Co core velocity was measured in that experiment, but this is now understood to be likely driven by a single NB blip 600 ms before this measurement. That experiment used L-mode discharges, rather than the H-modes considered here.

ECH AND OHMIC H-MODES

A comparison was made between ECH and Ohmic H-modes using the same CER measurement technique. For the ECH case, power from four gyrotrons was launched from the outboard side along the vacuum rays shown in Fig. 1, at 110 GHz [9]. The intersection with the vertical line indicates the second harmonic resonance. Gyrotrons 1-3 were launched radially ($k_\phi = 0$), while gyrotron 4 launched at a nonzero toroidal wavenumber. The nominal power from each was 500 kW. Raytracing with the TORAY-GA code indicates that the total EC driven current is less than 10 kA. For the ECH H-mode, $I_p = 1.3$ MA, $B_T = -1.75$ T, with line-averaged target electron density $\bar{n}_e = 3\times10^{19}$/m^3. For the Ohmic H-mode case I_p was increased to 1.5 MA, with no ECH power applied.

Profiles of toroidal rotation frequency, ω, are shown in Fig. 2, for both H-mode cases. In Ohmic H-mode the rotation is in the co direction everywhere, while for the ECH H-mode it is in the counter direction in the interior coincident with the region of ECH power deposition, as indicated in Fig 2. We did not have the opportunity to vary the deposition profile, so it is not yet known if there is a causal relation here. The lines shown merely connect the points to guide the eye.

The electron and ion temperature profiles, T_e and T_i, accompanying these rotation profiles are shown in Fig. 3. The higher temperature trace in each plot goes with the ECH H-mode (dashed curves). The density profiles for each are relatively flat, with a density within the edge pedestal of approx. 6×10^{19}/m^3. T_e is measured with Thomson scattering and T_i with the CER. The smooth curves are spline fits to the data points.

In both of these types of discharge the toroidal rotation profile evolves with time. For the ECH H-mode there is a relatively long ELM-free period during which the

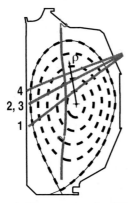

Figure 1. Vacuum rays for EC power launched with four gyrotrons in DIII-D.

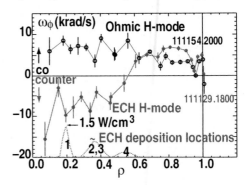

Figure 2. Toroidal rotation profiles for Ohmic (open circles t = 2000 ms) and ECH H-mode discharges (t = 1800 ms). The ECH power deposition profile is shown.

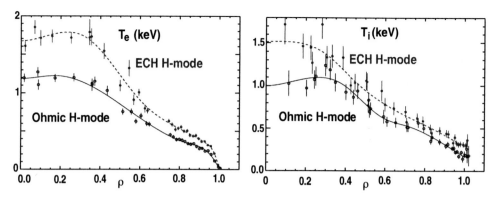

Figure 3. (a) Electron temperature, T_e, for the ECH and Ohmic H-mode discharges at the time of the rotation profiles shown in Fig. 2. (b) Ion temperature, T_i.

density rises until the ECH power is cut-off from depositing at the nominal locations indicated in Fig. 2. This is confirmed by cut-off in electron cyclotron emission from the plasma at a nearby frequency, and by ray tracing with TORAY-GA. Once there is no longer ECH power being deposited in the core, the density rise ceases, ELMs begin and the counter rotation in the core has died away. Also with the density rise T_e and T_i become equal, unlike the profiles shown in Fig. 3. It is not clear which of these phenomena, if any, are responsible for the loss of counter rotation in the core. For the Ohmic H-mode, ELMs begin relatively soon after the H-mode transition and the density rises, albeit less than in the ECH H-mode.

The rotation profiles shown in Fig. 2 were taken at the measurement times of maximum rotation magnitude near the magnetic axis for each case. The relative timing of these profiles with the density rise is shown in Fig. 4, for the ECH H-mode and the Ohmic H-mode. The first 2 ms of the first NB blip in each shot was used for the rotation measurement. For the Ohmic H-mode case the NB blip generated a short ELM-free period. Clearly it is necessary to do more experiments in steady conditions to find the cause of the counter core rotation with ECH power.

COMPARISON PARAMETER

It is useful to have some parameter with which to compare toroidal rotation between devices in discharges with little or no toroidal momentum input. For the toroidal velocity we propose scaling by a characteristic electric field divided by the poloidal magnetic field that is, define $\Delta \equiv U_\phi/(E_\rho/B_\theta)$. This characteristic electric field has a magnitude determined by the radial pressure gradient, which is taken simply to be the total energy density on axis divided by the minor radius. (The sign of the velocity will not be given by this scaling.) Thus Δ can be written as

$$[\omega/(3/2)(T_i+T_e)]_0 = \Delta\, q_{95}/a^2 B_\phi \tag{1}$$

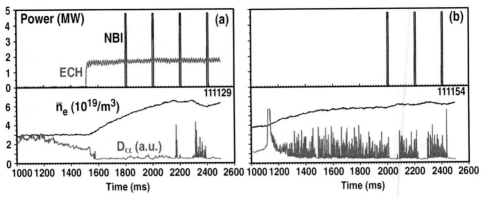

Figure 4. Density evolution in (a) ECH H-mode (1800 ms) and (b) Ohmic H-mode (2000 ms) discharges.

where we have cast this relation in terms which are readily scalable between devices. Here q_{95} is the safety factor at the 95% poloidal flux surface, a the minor radius, and B_ϕ the toroidal field on axis. For a simple circular approximation, Eq. (1) shows that $\Delta \sim (a/R_0)\Omega_\theta(L/W)_0$, where Ω_θ is the poloidal gyrofrequency and L and W are the angular momentum and thermal energy densities on axis, respectively. This scaling is consistent with that found on C-Mod [4], that $U_\phi \sim$ (total stored energy)$/I_p$. We have not yet tested this scaling in DIII-D.

For this Ohmic H-mode in DIII-D, $\Delta=0.5$, while for a C-Mod example $\Delta \sim 0.8$ [4]. A JET ICRH example has $\Delta \sim 1$ [3], while for the ECH H-mode described here $\Delta = 0.4$ (but negative). We note that for standard NB driven ELMing H-modes in DIII-D this same parameterization yields $\Delta \sim 1.5$–2 [10].

ACKNOWLEDGMENT

Work supported by the U.S. Department of Energy under Contracts DE-AC03-99ER54463 and DE-AC05-76OR00033.

REFERENCES

[1] Eriksson, L-G, et a.l, *Plasma Phys. and Contr. Fusion* **39**, 27 (1997).
[2] Rice, J.E. et al., *Nucl. Fusion* **38**, 75 (1998).
[3] Noterdaeme, J-M, et al., *Nucl. Fusion* **43**, (2003) 274.
[4] Hutchinson, I.H., et al., *Phys. Rev. Lett.* **84**, 3330 (2000).
[5] Garofalo, A.M., et al., *Nucl. Fusion* **41**, 1171 (2001).
[6] Burrell, K.H., *Phys. Plasmas* **4**, 1499 (1997).
[7] Burrell, K.H., et al., *Rev. Sci. Instrum.* **72**, 1028 (2001).
[8] deGrassie, J.S., et al., "Co-Toroidal Rotation with Electron Cyclotron Power in DIII-D," *Proc. 14th Top. Conf. on RF Power in Plasmas*, edited by T.K. Mau and J.S. deGrassie, AIP Conference Proceedings, Oxnard, 2001, p. 294.
[9] Callis, R.W., et al., "Maturing ECRF Technology for Plasma Control," *Proc. 19th IAEA Fusion Energy Conf.*, Lyon, France (International Atomic Energy Agency, Vienna, 2002) CD-ROM.
[10] deGrassie, J.S., et al., *Nucl. Fusion* **43**, 142 (2003).

Neoclassical Tearing Mode Suppression with ECRH and Current Drive in ASDEX Upgrade

G. Gantenbein*, A. Keller[§], F. Leuterer[§], M. Maraschek[§], W. Suttrop[§], H. Zohm[§], ASDEX Upgrade-Team[§]

*Institut für Plasmaforschung, Universität Stuttgart, Pfaffenwaldring 31, D-70569 Stuttgart,
[§] Max-Planck-Institut für Plasmaphysik, Boltzmannstr. 2, D-85748 Garching, EURATOM Association

Abstract. Neoclassical tearing modes (NTMs) occurring in high β discharges have a substantial influence on the performance of a tokamak. The suppression and avoidance of these instabilities is considered as a primary task for future machines like ITER. In ASDEX Upgrade, ECRH power is used to stabilise NTMs in high confinement discharges by replacing the 'missing' bootstrap current. For the m=3, n=2 modes complete stabilisation at high $β_N$ was successfully demonstrated in ASDEX Upgrade. Additional features in the case of 2/1 NTM are the vicinity of the q= 2 surface to the vacuum vessel which triggers undesirable interaction, leading to mode locking, and the lower plasma temperature and hence lower current drive efficiency compared to the q= 3/2 surface.

INTRODUCTION

Neoclassical tearing modes (NTMs) occurring in high β discharges have a substantial influence on the performance of a tokamak and can even lead to disruptions. The suppression and avoidance of these instabilities is considered as a primary task for future machines like ITER. Investigations have shown that the ITER plasma will have a strong tendency to develop a 2/1 NTM. NTMs are characterised by magnetic islands in the plasma which are driven by a flattening of the profiles (temperature, pressure) and a subsequent reduction of the bootstrap current. They are localised at the rational surface q= m/n. NTMs are observed in several tokamaks (e.g. ASDEX Upgrade, DIII-D, JT-60U) and experiments have been conducted to demonstrate the stabilisation of these modes by ECCD [1, 2, 3]. In ASDEX Upgrade, ECRH power is used to stabilise rotating NTMs in high confinement discharges by replacing the 'missing' bootstrap current. To be effective it is crucial to deposit the ECH power accurately at the rational surface. For the m=3, n=2 modes complete stabilisation at high $β_N$ (~2.6) was successfully demonstrated in ASDEX Upgrade [4]. Following these results experiments have been performed to suppress NTMs with poloidal mode number m=2 and toroidal mode number n=1. Additional features, compared to the q= 3/2 surface, are the vicinity of the q= 2 surface to the vacuum vessel which triggers undesirable interaction, leading to mode locking, and the lower current drive efficiency caused by several effects (e.g. lower plasma temperature, trapped particles).

SUPPRESSION OF m=2, n=1 NTM

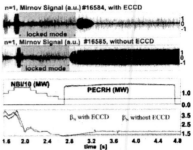

Figure 1: Comparison of two ASDEX Upgrade discharges and the effect of ECCD (#16584) on the 2/1 NTM. The discharges are identical exept the fact that P_{ECRH} was switched off in #16585. Shown are signals of the Mirnov coils for 2/1 mode, NBI (divided by 10) and ECCD power and β_N versus time

In ASDEX Upgrade four gyrotrons operating at 140 GHz deliver up to 1.9 MW of RF power to the plasma. The injection of the RF beam is from the low field side, in the scenario described here, the absorption takes place at the high field side. The deposition of the power in the plasma can be optimised by steering the launching beam in poloidal and toroidal directions. The localisation of the absorbed power, the position and the width of the island can be measured by the ECE diagnostics. A set of Mirnov coils at different poloidal and toroidal positions measures dB/dt in the plasma. In the discharges described, ASDEX Upgrade has been operated in a lower single null high confinement mode (H-mode) with edge localised modes. The plasma current has been kept constant to I_p= 0.8 MA, the electron density was fixed to $<n_e> \approx 5 \times 10^{19}$ m^{-3}, the toroidal field $B_t \approx 2$ T.

This plasma configuration is strongly heated by an additional NBI system with up to 15 MW. If β_N is raised above the onset value of a rotating 2/1 NTM the instability grows to its saturated width (see Fig. 1). Due to the nearness of the q=2 surface to the vacuum vessel interaction of the mode with the surrounding structure occurs which results in many cases in a locking of the mode and a large reduction of β_N. In the example shown in Fig. 1 β_N is decreased after mode locking to approximately half of the value before the mode onset. Since the Mirnov coils monitor only dB/dt which can be due to variation of the width (amplitude of B) of a periodically rotating island or/and a change of the rotating frequency care must be taken with the interpretation of these data.

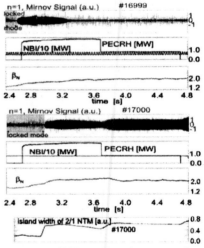

Figure 2: #16999, top: Complete suppression of 2/1 NTM. #17000, bottom: higher average NBI power, ECCD power is not sufficient for complete suppression.

To obtain a clear indication of the influence of the ECCD the NBI power is reduced, well below the onset value, before ECCD injection. Due to the low β_N value(≈ 1.2) the island restarts to rotate with ECCD, sometimes also without ECCD.

If ECCD is applied (#16584 in Fig. 1) the suppression of the 2/1 mode occurs within few 100 ms and is accompanied by increasing β_N up to 30 % higher compared to the case without ECCD (#16585 in Fig. 1).

POWER THRESHOLD FOR SUPPRESSION

In previously conducted experiments on the suppression of the 3/2 NTM complete stabilisation in a steady state regime at $\beta_N \approx 2.6$ has been demonstrated. TORBEAM [5] calculations predict a driven current at the q= 3/2 surface of up to 40 kA with 1.6 MW RF power. The total driven current and the current drive efficiency is considerably reduced in the case of 2/1 NTM at the q= 2 surface. In Fig. 2 time traces showing the n=1 perturbation of the magnetic field, the heating power and β_N are given for typical discharges. The discharge #16999 (Fig. 2, top) shows a successful suppression of the instability by ECCD. A continuous increase of β_N can be observed until the signal of the island vanishes. The mode does not reappear if ECCD is reduced (t= 3.7 s), also not if ECCD is completely switched off (t= 4.7 s) because β_N is below the onset value.

In discharge #17000 (Fig. 2, bottom) the same parameters have been chosen, but the average NBI power and thus β_N has been increased by a small amount. In that case, the available ECCD power (here 1.9 MW, driven current up to 23 kA) is not sufficient to completely suppress the instability, the island size shrinks only to 65 %. Moreover it is obvious that the mode increases if ECCD is reduced (t= 3.7 s) and switched off (t= 4.7 s), demonstrating the direct impact on the island size.

INCREASING β AFTER 2/1 NTM SUPPRESSION

In the case of earlier experiments at ASDEX Upgrade on 3/2 NTM stabilisation it had been shown that once 3/2 mode has beeen suppressed β could be increased above the onset value of the mode in the presence of ECCD. In Fig. 3 a discharge is given where NBI power has been increased after the 2/1 mode has been suppressed. Although β_N exceeds the threshold for the 2/1 mode it does not reappear during ECCD injection. At t≈ 3.8 s the 3/2 mode is triggered at $\beta_N \approx 3.5$. Due to the high β the Shafranov effect pushes the plasma towards larger R (major radius) and the deposition of the RF power, determined by the resonant magnetic field, is shifted to a larger ρ (minor radius). The result with respect to the location of the current drive density (j_{CD}) is shown in Fig. 4 where TORBEAM simulations are plotted. At t≈ 3 s the ECCD power is located at the q=2 surface, the small shift at that time is caused by a feed forward variation of the magnetic field. At t≈ 4 s the ECCD power is moved to a position outside the q= 2 surface which makes the development of

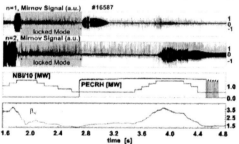

Figure 3: Time history of 2/1 (n=1) and 3/2 (n=2) NTM. The 3/2 mode is triggered at high β_N, where the deposition is shifted towards outside the q= 2 surface.

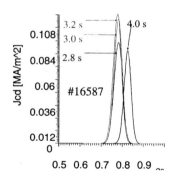

Figure 4: Current drive density as calculated by TORBEAM for the discharge given in Fig. 3.

a 3/2 mode easier. In such a situation it may be necessary to suppress both the 2/1 and 3/2 NTM.

ENHANCEMENT OF 2/1 NTM

A different situation with an increase of the island size during ECCD is given in Fig. 5. In this discharge the ECCD power (here 1.6 MW) is not sufficient to suppress the 2/1 NTM and we observe that the corresponding Mirnov signal shows a decreasing behaviour around $t \approx 4.0 - 4.4$ s. A closer analysis, taking the rotation frequency into account shows that the rotation slows down and the island is actually growing in this region. TORBEAM calculations result in a shift of the location of current drive density for this example. This shift is due to the feed forward variation of the magnetic field and moves the deposition inside the q= 2 surface. This mismatch can also lead to a positive Δ' (matching index, describing the jump of the radial derivative of the magnetic perturbation) and hence to an additional drive of the 2/1 island [6].

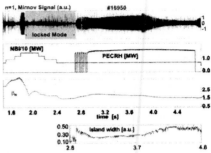

Figure 5: Growing island due to deposition of ECCD inside the q= 2 surface. The ECCD power is below the threshold for stabilisation.

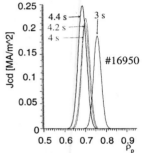

Figure 6: Inward shift of location of current dive as calculated by TORBEAM according to the feed forward variation of the magnetic field

CONCLUSIONS

In this paper we described experiments and discussed results on the stabilisation of m=2, n=1 NTMs with ECCD at ASDEX Upgrade. It has been shown that the suppression of this mode is possible and the required ECCD power is higher than in case of 3/2 mode stabilisation. Increasing β above the onset value is possible after 2/1 NTM suppression in the presence of ECCD, but the 3/2 NTM may possibly occur. Misalignment of the beam and the deposition of the ECCD power inside the q=2 surface showed a increase of the NTM, probably caused by a unfavorable influence on current distribution around the respective surface.

REFERENCES

[1] H. Zohm et al., *Phys. Plasmas*, **8**, 2009 (2001)
[2] R.J. LaHaye et al., *Phys. Plasmas*, **9**, 2051 (2002)
[3] A. Isayama et al., *Plasma Phys. Control. Fusion*, **42**, L37 (2000)
[4] G. Gantenbein et al., *Proc. 29th EPS Conf. Controlled Fusion and Plasma Physics*, P1.036 (2002)
[5] E. Poli et al., *Phys. Plasmas*, **6**, 5 (1999)
[6] A. Pletzer et al., *Phys. Plasmas*, **6**, 1589 (1999)

Stabilization of Neoclassical Tearing Mode by Electron Cyclotron Wave Injection in JT-60U

A. Isayama, K. Nagasaki*, S. Ide, T. Fukuda[†], T. Suzuki,
M. Seki, S. Moriyama, Y. Ikeda and JT-60 team

Japan Atomic Energy Research Institute, Naka, Ibaraki 311-0193, Japan
*Institute of Advanced Energy, Kyoto University, Uji, Kyoto 611-0011, Japan
[†] Present address:Osaka University, Ibaraki, Osaka 567-0047, Japan

Abstract. Real-time neoclassical tearing mode (NTM) stabilization system has been developed in JT-60U. In this system, mode location is first identified by calculating the plasma shape using the Cauchy condition surface method and evaluating electron temperature perturbation profile in real time. Subsequently, steerable mirror angle is changed so that electron cyclotron (EC) wave is absorbed at the mode location. By using this system, a 3/2 NTM has been completely stabilized, and an increase in the beta value and H-factor has been observed after the stabilization. Effects of EC injection before the NTM onset ('early EC injection') have been investigated. It is found that the NTM is more strongly suppressed for the early injection compared with the case of EC injection after the saturation of the NTM growth ('late EC injection'). It is also found that the suppression effect for the early injection strongly depends on the EC injection angle, as in the case of the late EC injection.

INTRODUCTION

Stabilization of neoclassical tearing modes (NTMs) by localized heating and/or current drive by electron cyclotron (EC) wave is important to sustain high beta plasmas. In JT-60U, it was demonstrated that the m/n=3/2 NTM can be completely stabilized by injecting the fundamental O-mode EC wave to the center of the magnetic island [1]. Here m and n are poloidal and toroidal mode numbers, respectively. At that time, EC injection angle was fixed after determining the optimum injection angle, and EC wave was injected after the saturation of the growth of the NTM. However, in a future device such as ITER, NTMs should be detected and stabilized in real-time, since the mode location changes in time. In addition, the NTM should be stabilized with smaller amount of EC power. In this paper, results from real-time NTM stabilization and effects of early EC injection on the NTM are described.

REAL-TIME NTM STABILIZATION

In JT-60U, a real-time NTM stabilization system has been developed [2]. The procedure of this system is shown in Fig. 1. At the first step, the plasma shape is first calculated in real-time with the Cauchy condition surface (CCS) method [3], and the mode location is coarsely estimated. At the second step, fine tuning is performed by evaluating the electron temperature perturbation profile, utilizing the fact that an M-shaped perturbation profile is obtained at the magnetic island and that the center of the island corresponds to the local minimum point of the profile. In obtaining the

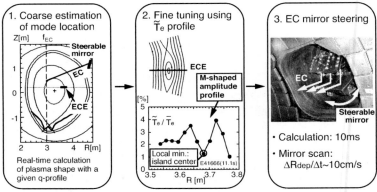

FIGURE 1. Schematic procedure of real-time NTM stabilization system.

perturbation profile, standard deviation of the ECE heterodyne radiometer signal is evaluated. This evaluation method has the advantage that the mode amplitude can be obtained without evaluating the mode frequency. At the third step, angle of the EC steerable mirror is changed so that EC wave is absorbed at the center of the island. The calculation time is less than 10 ms, which is much shorter than typical timescale of the evolution of NTMs (~several hundred milliseconds).

Typical waveforms of the NTM stabilization experiment are shown in Fig. 2. Plasma parameters of this discharge are as follows: $I_p=1.5$ MA, $B_t=3.7$ T, $R=3.3$ m, $a=0.78$ m, $q_{95}=3.8$. A 3/2 NTM was destabilized at $\beta_N\sim1.5$ by the neutral beam (NB) injection power of ~20 MW. The mode amplitude gradually decreased after the EC injection at $t=7.56$ s ($H_{89PL}=1.8$, $HH_{y2}=1.0$), and the 3/2 mode was completely stabilized at 8.8 s. Even after the turn-off of the EC injection at 9.5 s, the 3/2 mode did not appear and β_N continued to increase to 1.67. Since NB injection power was fixed, this shows confinement improvement. In fact, H_{89PL} and HH_{y2} increase to 1.9 and 1.1, respectively. The Fokker-Planck code and the ACCOME code show that the maximum

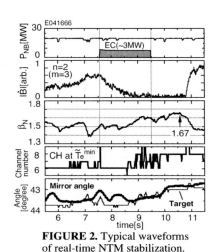

FIGURE 2. Typical waveforms of real-time NTM stabilization.

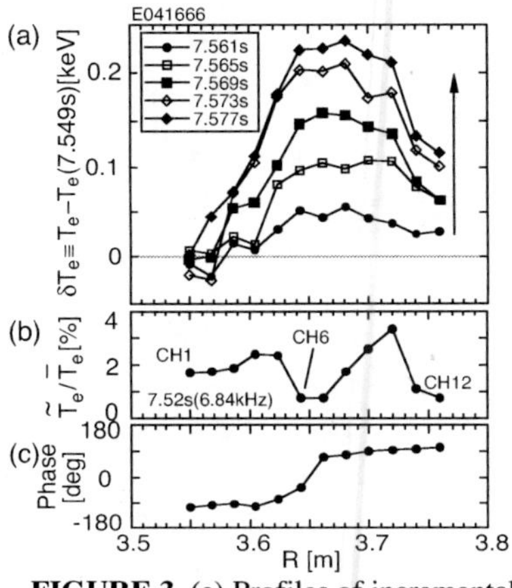

FIGURE 3. (a) Profiles of incremental electron temperature; profiles of (b) amplitude and (c) phase at the mode frequency.

EC-driven current density at the mode rational surface is comparable to the bootstrap current density (~0.25 MA/m^2). This shows that the stabilization was effectively accomplished.

Deposition location can be experimentally estimated from the increment of electron temperature after the EC injection, and the mode location can be identified from an electron temperature perturbation profile. Profiles of the incremental electron temperature after the EC injection are shown in Fig. 3(a). It can be seen that the peak position is located at $R \sim 3.67$ m. Profiles of the amplitude and phase of the 3/2 mode are shown in Figs. 3(b) and 3(c), respectively. The M-shaped structure in the amplitude profile and the jump in the phase profile are clearly observed, which shows that the center of the magnetic island is located at $R \sim 3.65$ m. Note that the the minimum point of the amplitude profile is located at channel 6 or 7, which is consistent with the results from the real-time system (see $t=7.5$ s in Fig. 2). Result from the Fokker-Planck code [4] shows that the full-width half-maximum of the EC driven current profile is about 0.1 m in volume-averaged minor radius. Thus, most of the EC power is deposited in the island region.

EFFECT OF EARLY EC INJECTION

It is important to suppress the NTMs with smaller amount of EC power. Theoretical analysis shows that required EC power can be reduced if the EC wave is injected in the growth phase of the NTM [5]. In JT-60U, the effects of early EC injection on NTM have been investigated experimentally [6].

Typical waveforms in the early and the conventional late injection cases are shown in Fig. 4. Plasma shape and discharge condition are almost the same as those in Fig. 2. In the early injection case, EC wave was injected from 4.6 s with a fixed injection angle. In the late injection case, EC wave was injected from 7.5 s, and the injection angle was changed using the real-time stabilization system. As shown in Fig. 4(a), in the early injection case, the growth of the 3/2 NTM is suppressed during the EC phase. It is notable that the amplitude of magnetic perturbations in the early injection case is smaller than that in the late injection case throughout the discharge. This suggests that EC injection time also affects the evolution of NTMs, possibly

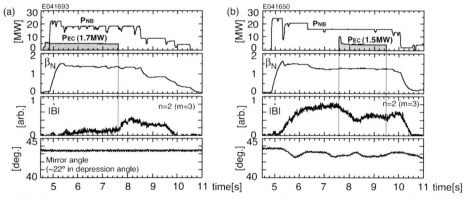

FIGURE 4. Typical waveforms for the cases of (a) early EC injection, and (b) late EC injection.

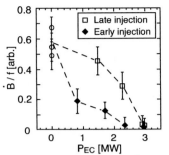
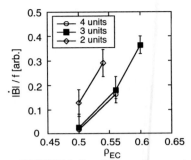

FIGURE 5. Dependence of magnetic perturbation on EC injection power.

FIGURE 6. Dependence of magnetic perturbation on EC deposition location.

through the change in the current and temperature profiles.

In Fig. 5, amplitude of magnetic perturbations during EC injection is plotted as a function of EC injection power. In the late injection case, the amplitude gradually decreases with increasing EC power, and the NTM is completely stabilized at four-unit EC injection (~3 MW). In the early injection case, the mode amplitude significantly decreases even at one-unit EC injection (~0.8 MW), and complete stabilization is achieved at three-unit injection (~2.3 MW). This shows that EC power required for complete stabilization can be reduced by early injection, which is promising in NTM stabilization in ITER.

It was reported that in late injection case the suppression effect strongly depends on the EC deposition location, and that the effect is significantly reduced if the deposition location is misaligned by about half of the island width (~several centimeters) [1]. Dependence of the amplitude of magnetic perturbations on EC deposition location for early injection is shown in Fig. 6, where the magnetic island is located at $\rho \sim 0.5$. It can be seen that the suppression effect significantly decreases with the deviation from the mode location, as in the case of late injection. This shows that the injection angle must be precisely adjusted also in the early injection case.

CONCLUSIONS

Real-time NTM stabilization system has been developed. By using this real-time system, a 3/2 NTM has been completely stabilized, and an increase in beta value and H-factor has been observed. Effect of early EC injection has been investigated. It has been found that NTM can be more efficiently suppressed by the early injection compared with the conventional late EC injection case. It has been also found that the suppression effect strongly depends on the EC deposition location, as in the late EC injection case.

REFERENCES

[1] Isayama A. *et al.*, *Plasma Phys. Control. Fusion* **42**, L37–L45 (2000).
[2] Isayama A. *et al.*, Proc. 19th IAEA Fusion Energy Conf., Lyon, (2002) IAEA-CN-94/EX/C2-2.
[3] Kurihara K. *et al.*, *Fusion Eng. Design* **51-52**, 1049–1057 (2000).
[4] Hamamatsu K. *et al.*, *Fusion Eng. Design* **53**, 53–58 (2001).
[5] Pustovitov V.D. *et al.*, Proc. 18th IAEA Conf. Sorrento (2000), IAEA-CN-77/ITERP/07.
[6] Nagasaki K. *et al.*, submitted to Nucl. Fusion.

Launcher Performance in the DIII-D ECH System

K. Kajiwara,[a] C.B. Baxi, J. Lohr, Y.A. Gorelov, M.T. Green, D. Ponce, and R.W. Callis

General Atomics, P.O. Box 85608, San Diego, California 92186-5608
[a]ORISE, Oak Ridge, Tennessee.

Abstract. The thermal performance of three different designs for the steerable mirrors on the ECH launchers installed in the DIII-D tokamak has been evaluated theoretically and experimentally. In each case the disruption forces must be minimized while providing a low loss reflecting surface. One design uses all Glidcop® material, but shaped so that the center is appreciably thicker than the edge. A second design is graphite with a molybdenum surface brazed to the graphite. The latest design is laminated copper/stainless steel construction with a thin copper reflecting surface. All three mirrors employ passive radiative cooling. The mirror temperatures are measured by resistance temperature devices (RTDs) which are attached at the back surfaces of the mirrors. The temperature increases are moderate for the laminated mirror, which has the best overall performance.

INTRODUCTION

Electron cyclotron (EC) waves can be used for plasma heating and current drive in tokamak devices. One important consideration is that the EC wave can be propagated in free space. Therefore, in a reactor the launcher can be located far from the plasma to reduce the neutron damage and employ miter bends which eliminate neutron leakage through the wave guide. Another advantage is that, because of the high electron cyclotron resonance frequency, steering of the rf beam using simple movable quasi-optical mirrors is possible. In free space propagation, the thermal performance of the mirrors becomes important, since the rf beam naturally assumes a Gaussian power profile with very high central power density. With the movable launcher mirrors located in the vacuum vessel, water cooling is difficult and results in serious problems in case of leaks. For non-CW rf operation with pulse lengths up to about ten seconds, enhancement of the radiation cooling from the back of the mirrors makes it possible to operate without active cooling. In DIII-D, five designs for passively cooled mirrors have been used. In this paper, we discuss the three most successful.

DIII-D ECH SYSTEM

The electron cyclotron heating (ECH) system consists of six 110 GHz gyrotrons and three launchers which include two wave guides each [1]. Three of the gyrotrons were made by Communications and Power Industries (CPI) and the others were made by Gycom. The CPI gyrotrons have chemical vapor deposition (CVD) diamond output windows which supports 1 MW-10 s operation for Gaussian output rf beams [2]. The Gycom gyrotrons have boron nitride windows, which limit the pulse lengths for these tubes to 2 s with ~1 MW rf generation. The rf is transported by ~100 m of 31.75 mm diameter evacuated corrugated waveguide carrying the $HE_{1,1}$ mode. Each wave guide has a rf pair of grooved polarizers which can produce arbitrary elliptical polarization of the wave.

Each launcher has two sets of waveguides and mirrors. Each mirror set consists of a weakly focusing fixed mirror and a steerable flat mirror (Fig. 1). The steerable mirrors can direct the rf beams in the toroidal and poloidal directions. Each mirror has two RTDs at the back surface and each waveguide has one RTD to monitor temperature increases during the rf pulses. Each launcher has two Langmuir probes and a camera port for observation of possible arcing at the mirrors or waveguides.

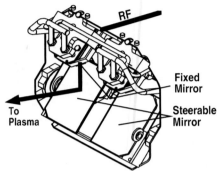

FIGURE 1. The launchers have weakly focusing fixed mirrors and flat steering mirrors with both poloidal and toroidal scan capability. Each launcher assembly accommodates the rf beams from two gyrotrons.

MIRROR DESIGNS

Poloidal and toroidal steering is provided using movable mirrors of different designs to direct the rf beams. Eddy current induced forces arising during disruptions are particularly problematic for the actuator assemblies on the movable mirrors, which have limited ability to react the forces. The first attempt at designing launcher mirrors for DIII-D used graphite covered by evaporated molybdenum and thin copper coatings to create a low loss reflecting surface. The design was excellent thermally and had very low eddy current induced forces during disruptions. However, surface arcing on the mirrors during rf operation at relatively low energy levels generated gas and damaged the coatings.

An improved mirror design was made from Glidcop®. This mirror had a thin copper reflecting surface with a thicker center, or boss, to provide thermal inertia where the power density of the Gaussian beam was greatest. The back surface of this mirror was grooved and blackened to increase radiative cooling [Fig. 2(a)] [3]. This mirror was calculated to withstand a 10 s and 800 kW rf beam without excessive temperature increase or long term ratcheting from repeated pulses. However, the design suffers from relatively large disruptive eddy currents because of the volume of low resistivity Glidcop®. A robust actuator assembly was required to react forces with this mirror.

A modification of the graphite/molybdenum approach used a brazed molybdenum reflecting surface, which eliminated the arc damage, but had a relatively high resistivity and the highest surface temperature of the tested designs [Fig. 2(b)]. This modification did produce a mirror capable of 5 s 800 kW operation.

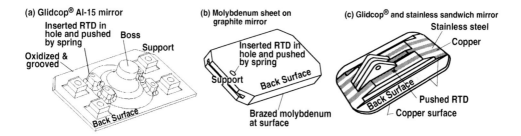

FIGURE 2. The thermal performance was studied for three different mirror designs: (a) Glidcop® mirror with a machined boss to increase the thermal mass with relatively low magnetic torques during disruptions, (b) graphite mirror with brazed molybdenum reflecting surface, (c) stainless steel and Glidcop® laminated structure with a thin copper reflecting surface supported by the sandwich.

The most recent mirror, with overall best performance, is called the "butcher block" mirror, which has a sandwich structure of Glidcop® and stainless steel [Fig. 2(c)]. The reflecting surface on this mirror is a thin copper layer supported by the sandwich. This structure can withstand modest disruption forces while providing a quality reflecting surface. The thermal performance allows rf pulses of 800 kW for 10 s. Heat is conducted from the reflecting surface by the Glidcop® and radiated to the surrounding structure. This design has the best overall performance of the three designs in the present study and meets the power, pulse length and duty cycle requirements for DIII-D experiments.

EXPERIMENTAL RESULTS

Measurements of the mirror temperatures during rf pulses into plasma discharges permit the three designs to be compared. The time dependencies of the RTD measurements are presented in Fig. 3(a) for the Glidcop®, moly/graphite and butcher block mirrors. The rf power was 450–600 kW and the pulse width was 800 ms. Data for the Glidcop® mirror during the plasma discharge and for the first few seconds after are not available. The highest temperature was recorded for the moly/graphite mirror design because of high resistivity of the molybdenum surface. The butcher block mirror has two temperature peaks. The RTD is mounted on the copper part of the sandwich on the mirror back surface. The first peak results from conduction through the copper and the second peak is due to slower conduction through the stainless steel. The lowest peak temperature is observed for the Glidcop® mirror as expected, but the compromise here is that eddy current forces are higher for this mirror than for the others.

Disruptions result in additional mirror heating due to induced currents and increased plasma radiation impinging on the mirrors. Figure 3(b) shows the effect of a disruption on the moly/graphite launcher mirror. The curves are the time evolution of the RTD signals in the cases of normal plasma termination and disruptive termination of a 0.3 MA plasma. The plasma parameters for these two shots are similar, with central electron temperature of 2.7 keV and central density of $2.5 \times 10^{19} m^{-3}$. The

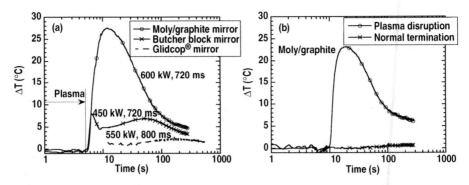

FIGURE 3. Time evolution of the temperature increase measured by the RTD for the different mirror designs: (a) Glidcop®, moly/graphite and butcher block mirror with rf power of 450–600 kW and pulse width 720–800 ms (data for the Glidcop® mirror during plasma discharges and the first few seconds after a plasma discharge are not available); and (b) the moly/graphite mirror in the case of disruptive and normal plasma terminations.

temperature of the mirror is clearly increased by the disruption and the time dependence of the heating response is similar to the case of normal rf injection. The input power from the plasma to the mirror during the disruption is estimated at 20 kJ by comparing with calculations for rf injection.

The peak temperature is a function of the pulse length, as shown in Fig. 4. The peak temperature of the moly/graphite mirror is about five times higher than the butcher block. The butcher block mirror temperature is only slightly higher than the Gliddcop® mirror, showing that the butcher block construction has good thermal performance in spite of the low eddy current design.

DISCUSSION

The solid lines in Fig. 4 show the peak temperature of the Glidcop® mirror calculated by finite element code COSMOS [4] at the RTD compared with measurements. The theoretical prediction is higher than the experimental result, by up to a factor of three. The discrepancy is understood to arise from poor thermal contact between the mirror and the RTD. In the vacuum situation, the RTD is mechanically pushed against the mirror, but without any heat conducting filler to improve the thermal contact. The thermal impedance for this installation is estimated by a simple experimental calibration. By holding an ice cube

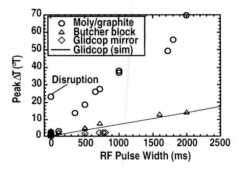

FIGURE 4. The peak temperature as a function of the rf pulse length. The solid line shows the simulation result for the Glidcop® mirror.

against the mirror surface with a thin plastic barrier sheet for 60 s, a known temperature reservoir can be applied to the mirror. The time evolution for this situation can be calculated by COSMOS as a function of the heat transfer between the ice cube and mirror. The heat transfer coefficient between the ice and mirror is estimated by using additional temporary RTDs attached to the mirror using thermal grease to produce a good contact. These RTDs are attached at the center of mirror surface and the bottom near the position of the ice cube. Figure 5 shows the results. The time evolutions of measurements by the additional RTDs are well explained by the code, assuming a heat transfer coefficient of 1350 W/m^{2}K. By comparing the heat transfer coefficient inferred from the plasma measurements with this experimentally derived value, a thermal resistance of 2×10^{-4} m^{2}K/W for the case without grease is determined. Figure 6 shows the result of recalculation of the original heating measurements assuming this thermal resistance of 2×0^{-4} m^{2}K/W with good agreement between the COSMOS model and the measurements.

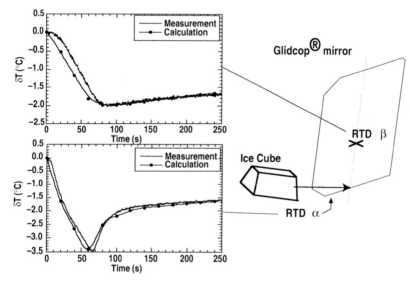

FIGURE 5. Experimental arrangement for the ice cube calibration in which the ice cube was held on the mirror surface through thin plastic for 60 s. The calculation assumes a heat transfer coefficient of 1350 W/m^{2}K between the ice cube and mirror surface.

CONCLUSION

The temperature increases of three mirrors, a Glidcop® mirror, a mirror which was made from graphite with a brazed molybdenum surface, and a mirror which was made from a laminate of copper and stainless steel, are observed during rf pulses using RTDs which are attached at the back surfaces of the mirrors. The highest temperature increase was observed for the moly/graphite mirror and the lowest temperature was for the Glidcop® mirror, which has the best thermal performance. The temperature increase of the laminated mirror was moderately higher than the Glidcop® mirror, but lower than the molybdenum/graphite mirror. Because of the low eddy currents for the laminate design and the acceptable thermal performance, the design has been accepted for the DIII-D launcher systems.

ACKNOWLEDGMENT

Work supported by the U.S. Department of Energy under Contracts DE-AC03-99ER54463, DE-AC02-76CH03073, and DE-AC05-76OR00033. The author (K.K.) acknowledges R.A. Ellis at PPPL for providing the launchers.

REFERENCES

[1] Lohr, J., et al., Proc. of 12th Joint Workshop on Electron Cyclotron Emission and Electron Cyclotron Resonance Heating (2002).
[2] Gorelov, Y.A., et al., Proc. of 27th Int. Conf. on Infrared and Millimeter waves (2002).
[3] Baxi, C.B., et al., Proc. of 19th IEEE/NPSS Symp. on Fusion Engineering (2002).
[4] COSMOS, A Finite Element Analysis Code, Structural Research, Santa Monica, California, USA.

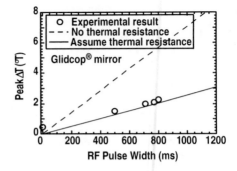

FIGURE 6. Peak temperature as a function of rf pulse width for the Glidcop® mirror. The dashed line is the previous calculation that is the same data in Fig. 4 solid line but with a recalculation made assuming the thermal resistance of 2×10^{-4} $m^2 K/W$ as determined in the ice cube experiments.

Absorption Of X-Wave At The Second Harmonic In HSX

K.M.Likin, J.N.Talmadge, A.F.Almagri, D.T.Anderson, F.S.Anderson, C.Deng, S.P.Gerhardt, K.Zhai

University of Wisconsin, Madison, WI, USA

Abstract. Second harmonic extraordinary mode ECH is used in the HSX stellarator at 0.5 T to break down and heat the plasma. To measure the absorbed power a set of absolutely calibrated microwave diodes have been installed inside the machine. In the QHS and Mirror configuration, the absorption efficiency is high (about 0.9) and drops (to about 0.5) in the anti-Mirror mode. A comparison with ray tracing predictions is made.

INTRODUCTION

On the HSX stellarator, microwave power (up to 100 kW at 28 GHz) is used for neutral gas breakdown followed by heating of the plasma at the second harmonic of the electron cyclotron frequency. The linear polarized beam with $E \perp B$ is launched from the low magnetic field side and is focused on the plasma center with a spot size of 4 cm. The main HSX magnetic configuration with quasi-helical symmetry (QHS) has an axis of symmetry in $|B|$ and very low neoclassical transport [1]. In the Mirror configuration, a mirror term is added to the main field at the location of the launching antenna; in the anti-Mirror configuration, it is opposite to the main field. In both configurations the neoclassical transport is greatly increased.

Most ray tracing codes only calculate the absorption for the first pass of the microwave power through the plasma column [2]. One can also use the ray tracing code to calculate the multi-pass absorption for that power which is not absorbed in the first pass [3]. In an experiment, the absorption efficiency and the localization of microwave energy deposition along the machine can be determined with a set of microwave antennas [4].

MICROWAVE POWER ABSORPTION

Ray tracing calculations

In the HSX ray tracing code the standard geometric optics equations and Altar-Appleton-Hartree formula for the cold plasma refractive index are used to calculate the ray trajectories in 3-D geometry [5]. To get the local absorption coefficient the

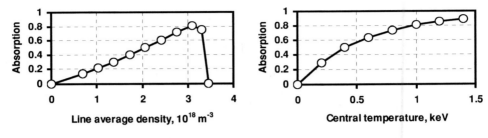

FIGURE 1. Single-pass absorption versus the plasma density and the electron temperature at 0.5 T.

finite temperature should be taken into consideration [6]. The power in each ray is weighted in accordance with a Gaussian profile of the incident wave.

The Biot-Savart method is used to calculate the magnetic field and a linear interpolation over 50*16*51 data points is made to find the effective plasma radius. In typical runs of the code, the line average plasma density is $2 \cdot 10^{18}$ m^{-3}, the central temperature is 0.4 keV and 65 parallel rays represent the incident beam.

The single-pass absorption coefficients versus the line average plasma density and versus the central electron temperature are shown in Fig.1. For the second pass through the plasma we use a wide incident beam (12 cm) with 20 degrees of divergence and it was found that the second-pass absorption efficiency is about 40% at $2 \cdot 10^{18}$ m^{-3} line-average density and the total absorption in two passes reaches 70% while the absorbed power profile does not significantly broaden. Thus, if 50% of launched power is absorbed in the first pass one can expect almost total absorption in a few passes.

Microwave antenna set-up

The location of the microwave antennas around HSX is presented in Table 1. Antennas 2 and 5, 3 and 6 are installed symmetrically with respect to the ECH launch port. Quartz barrier windows are mounted on 20 cm extensions outward from the vacuum vessel wall. The estimated power reflection coefficient from this window (3.3 mm in thickness) is 0.15. The receiving antennas are mounted just behind the windows. Each antenna consists of an open-ended K-band waveguide, attenuator and microwave detector.

TABLE 1. Location Of Microwave Antennas.

Antenna Name	Toroidal Angle With Respect To ECH Port, degs.	Distance Along The Plasma Axis, m	Area inside the vacuum chamber, m^2	Attenuation, dB
Detector #1	6.3	0.181	0.12	50.2
Detector #4	36.0	0.942	0.1188	48.6
Detector #2	69.3	1.675	0.1297	49.2
Detector #5	-69.3	1.675	0.1297	49.5
Detector #3	103.5	2.642	0.1264	49.2
Detector #6	-103.5	2.642	0.1264	49.6

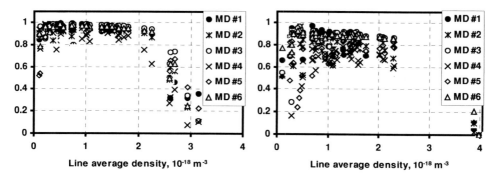

FIGURE 2. Multi-pass absorption efficiency in QHS (left) and Mirror (right) magnetic configurations.

All antennas have been calibrated with a 100 mW source and these calibration curves are used in interpolating the experimental data to get the absolute power. The E-plane of the rectangular waveguide antennas is oriented perpendicular to the magnetic field so that mostly X-wave power is detected. The antennas can be rotated in order to measure the cross-polarized waves as well. The signals from the microwave detectors are amplified with a gain of 30 and collected by the data acquisition system. After a few reflections from the wall the microwave power distribution is assumed to be isotropic and the non-absorbed power can be estimated from the following expression:

$$P_{non} = \frac{U_i \cdot S_i \cdot S_v}{G_i \cdot Wg \cdot Att \cdot trs_w}$$

where U_i – acquired signal; G_i – amplifier gain; S_i – diode sensitivity; Att – attenuation coefficient; trs_w – transparency of quartz window; Wg - effective area of waveguide antenna; S_V – area inside the vacuum chamber at diode location. We define the multi-pass absorption coefficient as $\eta = 1 - P_{non}/P_o$, where P_{non} – measured non-absorbed power and P_o – the reference power in a "cold" plasma discharge at plasma density above the cut-off.

Measurements

The multi-pass absorption efficiency versus the plasma density in QHS and Mirror configurations is shown in Fig.2. One can see that the absorption is localized near the launch antenna, i.e. most power is absorbed in a few passes through the plasma column. At low plasma density ($< 0.5 \cdot 10^{18}$ cm^{-3}), the absorption drops in Mirror mode while it remains high in QHS. At plasma densities greater than the X-mode cut-off, the power level at antennas further from the ECH mirror is low (about 10 kW) compared to the launched power (40 kW). At these densities, the diamagnetic loop does not indicate any significant plasma stored energy (< 2 J). In the anti-Mirror configuration the stored energy is low (5-7 J) and the multi-pass absorption efficiency does not exceed 0.5. In this configuration, the heating occurs on electrons deeply trapped in the local magnetic well and they leave the confinement volume in a short time because they are on so-called direct loss orbit [7].

Rotating antennas around their axis checks the absorption of waves with different polarization. It was found that all waves damp along the machine in the same fashion. Waves with ordinary polarization are not absorbed at the second harmonic and only the cascade-conversion of O-wave into X-wave in during reflection from the wall can lead to high power absorption.

Another method to determine the absorbed power is based on an evaluation of the slope of the diamagnetic loop signal before and just after the heating power turn-off. In all regimes of HSX operation, the value obtained with this technique is 50% of the launched power or less. One possible reason for the discrepancy between the diode measurement and the diamagnetic loop is the high losses due to radiation, ionization, and some other process.

SUMMARY

Power of the extraordinary wave at the second harmonic is absorbed with high efficiency (0.9) in the HSX plasma. At high plasma density, the results of the measurements made with microwave antennas are in good agreement with predictions of ray tracing calculations. However, at low plasma density there is a discrepancy between the measured and theoretical values. At low plasma density the microwave heating power can generate high energy electrons on which the microwave power is well absorbed, and the total absorption is higher than the ray tracing calculations predict. A fraction of the absorbed power (at high plasma density up to 50%) can go to ionization, dissociation, radiation, recombination and charge exchange, and it may explain the difference in absorbed power obtained with the microwave antennas and from the slope of the diamagnetic loop signal.

ACKNOWLEDGMENTS

The work is supported by DOE grant #DE-F602-93ER54222.

REFERENCES

1. Anderson, F.S.B., Almagri, A.F., Anderson, D.T., et.al., *Fusion Technology*, **27**, p.273 (1995).
2. Likin, K.M. and Ochirov, B.D., *Sov. J. Plasma Phys.* **18**, pp. 42-46 (1991).
3. Goldfinger, R.C. and Batchelor, B.D., *Nuclear Fusion* **29**, pp. 1745-1749 (1989).
4. Batanov, G.M., Kolik, L.V., Likin, K.M., et.al. *Sov. J. Plasma Phys.* **18**, pp. 33-38 (1991).
5. Stix, T.H., *Waves In Plasmas,* New York: Springer-Verlag New York, Inc., 1992, pp. 69-87.
6. Alikaev, V.V., Litvak, A.G., Suvorov, E.V., Friman, A.A., *High-frequency plasma heating*, edited by Litvak, A. G., New York: American Institute of Physics, 1992, pp. 3-14.
7. Gerhardt, S.P., Abdou, A., Almagri, A.F., et.al., IAEA-CN-94/EX/P3-22, 19th Intl. Conf. On Plasma Phys. and Contr. Nucl. Fusion, Lyon, France, (2002).

The 110 GHz Microwave Heating System on the DIII-D Tokamak

J. Lohr, R.W. Callis, J.L. Doane, R.A. Ellis,[a] Y.A. Gorelov, K. Kajiwara,[b] D. Ponce, and R. Prater

General Atomics, P.O. Box 85608, San Diego, California 92186-5608
[a]*Princeton Plasma Physics Laboratory, Princeton, New Jersey.*
[b]*ORISE, Oak Ridge, Tennessee.*

Abstract. Six 110 GHz gyrotrons in the 1 MW class are operational on DIII-D. Source power is >4.0 MW for pulse lengths ≤2.1 s and ~2.8 MW for 5.0 s. The rf beams can be steered poloidally across the tokamak upper half plane at off-perpendicular injection angles in the toroidal direction up to ±20°. Measured transmission line loss is about -1 dB for the longest line, which is 92 m long with 11 miter bends. Coupling efficiency into the waveguide is ~93% for the Gaussian rf beams. The transmission lines are evacuated and windowless except for the gyrotron output window and include flexible control of the elliptical polarization of the injected rf beam with remote controlled grooved mirrors in two of the miter bends on each line. The injected power can be modulated according to a predetermined program or controlled by the DIII-D plasma control system using real time feedback based on diagnostic signals obtained during the plasma pulse. Three gyrotrons have operated at 1.0 MW output power for 5.0 s. Peak central temperatures of the artificially grown diamond gyrotron output windows are <180°C at equilibrium.

INTRODUCTION

The last of the six 110 GHz gyrotrons in the DIII-D complex has successfully completed acceptance testing at full operational parameters of 80 kV and 40 A. The tube generated 1.0 MW rf output for 5.0 s. pulses. The completion of these tests marks the end of the present phase of system expansion and the beginning of a period of experiments in which additional operational flexibility will be added without increasing the number of gyrotrons installed. Three Gycom gyrotrons, with pulse lengths limited by heating of their boron nitride output windows, have been generating about 750 kW each for pulse lengths up to 2.1 s. and now three CPI gyrotrons equipped with low loss artificially grown diamond output windows have all been tested to 1.0 MW at 5.0 s pulse length.

All the Gycom gyrotrons were designed intentionally to spread the output rf beam over the 10 cm diameter of the BN windows to limit the peak power loading. Because the DIII-D transmission line is circular corrugated waveguide carrying the low loss $HE_{1,1}$ mode, a Gaussian rf beam had to be reformed using phase correcting mirrors effectively to excite the $HE_{1,1}$ mode. Although this could be done so that a good quality Gaussian waist was formed at the waveguide input, the efficiency of this recovery process was never much higher than about 87% and sometimes was lower.

With the advent of artificially grown diamond gyrotron windows capable of handling the full power Gaussian beam without damage, it became possible to couple to waveguide simply using a single ellipsoidal mirror with ≈93% overall efficiency.

HARDWARE

The six gyrotrons are connected to three dual articulating launchers, which can direct the rf beams poloidally over the tokamak upper half plane and ±20° toroidally to the angles for peak current drive efficiency in both the co- and counter-current drive directions. Following damage to one launcher sustained due to rf-driven arcing, these launchers are now being monitored by video cameras and Langmuir probes in addition to the normal temperature measurements on the mirrors using resistance temperature devices (RTDs). The launchers have also been made more robust thermally by replacement of the aluminum launcher waveguides with stainless steel guides having silver plating on the inner bores. Photographs of one of the launchers on the bench and installed in the tokamak are shown in Fig. 1. The video view of the launcher from the tokamak flange seen in D_α light during a plasma discharge is shown in Fig. 2.

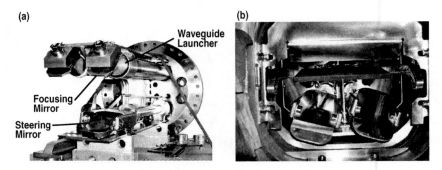

FIGURE 1. View of one of the articulating launchers on the bench. The mirrors can rotate to scan toroidally and tilt for the poloidal scan, although the two motions are not orthogonal.

One consistent challenge for the DIII-D installation has been the estimation of the power injected into the plasma from each of the launchers. Although calorimetric measurements of the power loading of the gyrotron components and the transmission line are made on every shot, and these can be related to the injected power, the accuracy of this procedure when compared with diagnostic measurements of plasma electron heating has been poor. A prototype design for a power monitor to be located near the tokamak is shown in Fig. 3. This vacuum compatible device simply consists of a small gap in the transmission line surrounded by a cylindrical volume into which a small fraction of the transmitted power is radiated. The cylinder has an annular ring of TiO_2 at the location of the waveguide gap, which absorbs most of the leakage rf power. A pair of RTD sensors differentially measures the temperatures of the cylinder wall at the

TiO$_2$ strip and the end plate, giving a time dependent signal the peak value of which is proportional to the integral of the rf power during the pulse. In preliminary tests, a 250 ms pulse at ≈500 kW gave a 2°C peak temperature difference between the RTDs.

FIGURE 2. Video view of a launcher assembly seen in plasma light. The lower tube is the mirror actuator and the upper is the launcher waveguide. The second launcher is on the right border.

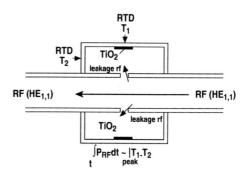

FIGURE 3. Schematic of a prototype rf power monitor. The vacuum compatible unit measures heating from leakage rf power.

GYROTRON CONDITIONING

In general, gyrotrons are delivered to DIII-D from the manufacturers following a period of testing at the factory. In the case of Communications and Power Industries (CPI) production, the tubes are factory tested at about 550 kW for 10 s pulses and for short pulses several ms in duration at 1.0 MW rf output. The available test stand at CPI is limited to 25 A for long pulse operation, limiting the maximum performance testing.

Recently a gyrotron was delivered to DIII-D after major replacement of sections of the collector but without any factory testing. The *ab initio* conditioning performed at DIII-D required about 2.5 months of around the clock operation to achieve full parameter operation at 1.0 MW rf output. In Fig. 4, the progression of pulse length extension at different gyrotron electron beam currents is plotted as a function of calendar day. Initial operation was at low beam currents, but the current was rapidly increased to 25 A, after which pulse extension to 1.0 s required about 4 weeks. After an additional week, 2.5 s pulses were achieved, but an increase in beam current to 30 A restarted the process, with about a week required to return to 1.0 s pulses at the higher current. The current was increased to the full value, 40 A, whereupon several days were required to achieve 2.0 s and after about a week 5.0 s pulses had been fired.

The conditioning process was begun at each step by reducing the magnetic field from the upper main coil until rf generation was lost, establishing the low field operating limit. By only changing the upper main coil, the field geometry near the gun is barely affected. The magnetic field was increased by ≈2% so that operation was reliable, albeit at the expense of output power. Once full length pulses were obtained, the magnetic field was carefully decreased until the peak in the output power was observed. Conditioning was interrupted twice for repairs to the high voltage power supply.

The output power was measured calorimetrically using several algorithms. A typical subset of calorimetry traces is presented in Fig. 5. Analysis of these traces was performed both by using digital oscilloscopes and by an automatic system. Fluctuating baselines were problematic for both methods. The generated rf power estimates rely on measured window loss, on the relationship between cavity rf power loading and generated power, on the efficiency at peak tuning and on the rf power actually absorbed in the dummy loads. The power/heating relationships were measured in the CPI test stand, where a very simple power collection setup routinely achieves total power accountability >97% overall. In the DIII-D system, coupling to waveguide, rf beam conditioning and focusing, a relatively long line to the dummy loads, several miter bends and a complex shared cooling water system combine to decrease the accuracy of the calorimetric measurements. Unmonitored rf power of at least 100 kW has been measured using an infrared camera viewing the transmission line components when the rf beam is directed into the dummy loads. If this power is taken into account, the dummy load measurements still have a 10% discrepancy compared with the cavity measurements. In Fig. 6, the results of the automatic analysis are presented.

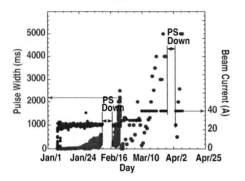

FIGURE 4. Summary of the progression of *ab initio* gyrotron conditioning. The gyrotron was conditioned to 1.0 MW, 5.0 s operation at 80 kV and 40 A in about 2.5 months of around the clock operation.

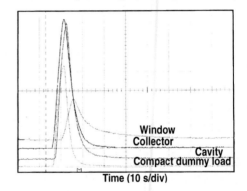

FIGURE 5. Calorimetric measurements of rf heating of the diamond output window, the collector, cavity and compact dummy load. The vertical scales are arbitrary.

The figure plots measurements of the generated rf power inferred from several calorimetric data sets taken separately at full gyrotron parameters, 80 kV, 40 A and peak tuning of the magnetic fields as a function of pulse length. The data fall into bands at constant power, indicating systematic errors in the measurement using each data set. The highest power estimate comes from subtracting the collector power from the input electrical power and assuming an efficiency of 34%. This method typically gives ≈1.3 MW for the generated rf power estimate. Possibly the most accurate power measurement is obtained from the cooling for the gyrotron cavity. This loading was accurately measured at CPI and should be 2.32% of the generated power. Using this measurement, the automatic system indicates peak generated power of 1 MW at 5.0 s pulses. The power absorbed in the gyrotron output window is only about 0.2%, which gives calorimetric temperature differences of less than a degree C for most operations. The window cooling water also cools the window support and exit waveguide. This makes it difficult to infer the window transmitted power to sufficient accuracy for plasma experiments. In most cases the automatic calorimetry system underestimates the output power but when the same data are analyzed manually with a digital oscilloscope transmitted powers in agreement with the cavity measurements are obtained.

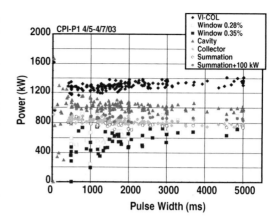

FIGURE 6. Summary of calorimetric data as a function of pulse length. The data for each of the cooling circuits are multiplied by predetermined scale factors to give a measurement of the generated rf power. The data reveal systematic patterns of disagreement at the ±20% indicating that additional corrections must be applied.

SUMMARY

The DIII-D 110 GHz ECH system has been expanded to a six gyrotron system with flexible launch capability. The system generates about 5 MW and about 4 MW reaches the tokamak for experiments.

ACKNOWLEDGMENT

Work supported by the U.S. Department of Energy under Contracts DE-AC03-99ER54463, DE-AC02-76CH03073, and DE-AC05-76OR00033.

Start-up of Spherical Tokamak by ECH on LATE

T. Maekawa, H. Tanaka, M. Uchida, H. Igami, T. Yoshinaga, and K. HIGAKI

Graduate School of Energy Science, Kyoto University, Kyoto 606-8502, Japan

Abstract. (1) Physical process of initial formation of closed flux surface has been studied under the condition of steady external vertical fields by using a 2.45 GHz magnetron (5kW). After the injection of microwave power the plasma current increases gradually until up to 0.4 kA. Once the current reaches this value it quickly increases up to 1 kA within 2-3 msec. The magnetic measurements suggests that a small initial closed flux is formed at Ip=0.4kA and expands quickly to full closed flux in contact with the inboard limiter in accordance with the quick increase of the plasma current. (2) Plasma current up to 2 kA has been started-up by injecting microwaves in the range of 20-50 kW from a 2 GHz klystron and adjusting the external vertical field temporally. Magnetic measurement shows that last closed flux surface with a minor radius of a=15 cm and a major radius of R =25 cm (R/a=1.7) is formed. (3) A new ECH system using a 5 GHz klystron (200kW,100 msec) is under construction. With this system, better coupling to Bernstein waves from EM waves compared with the 2.0 and 2.45 GHz cases may be obtained.

EXPERIMENTAL APPARATUS

Main objective of the LATE (Low Aspect ratio Torus Experiment, Fig.1) device is to realize start-up and formation of spherical tokamak (ST) plasmas by electron cyclotron heating and current drive (ECH /ECCD) alone without Ohmic heating power. Vacuum vessel of LATE is a cylinder with an inner diameter of 100 cm and a height of 100 cm. The center stack is a stainless steel cylinder with an outer diameter of 11.4 cm and encloses 60 turns of toroidal field coil. The coil current is kept constant at It=1kA (60 kAT) during the microwave pulses. The external vertical field is generated by combination of three sets of vertical field coils driven by a pre-programmed power supply. The working gas is hydrogen. The magnetic measurement is performed with 13 flux loops including 5 loops inside the center stack.

EXPERIMENTAL RESULTS

Physical process of initial formation of closed flux surface has been studied under the condition of steady external vertical fields by using a 2.45 GHz magnetron (5kW). After the injection of microwave power, the gas pressure decreases quickly and then, keeps a low value, while the plasma current increases gradually until up to 0.4 kA as shown in Fig.2. Once the current reaches this value it quickly increases up to 1 kA within 2-3 msec. The magnetic measurements suggests that a small initial closed flux is formed at Ip=0.4kA and expands quickly to full closed flux in contact with the inboard limiter in accordance with the quick increase of the plasma current (Fig.2). Such a formation of full closed flux surface can be obtained only with the external vertical field with the decay index of $n > 1.18$ at the major radius of R=20 cm. While the formation of initial small closed magnetic surface may be ascribed to the current generation in the open vertical field with the appropriate mirror ratio [1], The mechanism for quick increase of the current in accordance with the quick expanding of the closed flux surface is unknown.

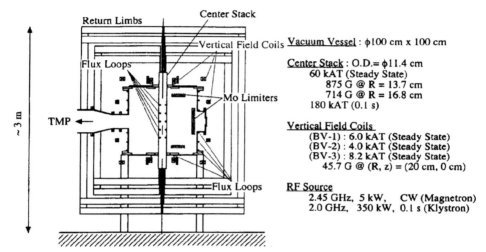

FIGURE 1. Low Aspect ratio Torus Experiment (Side View).

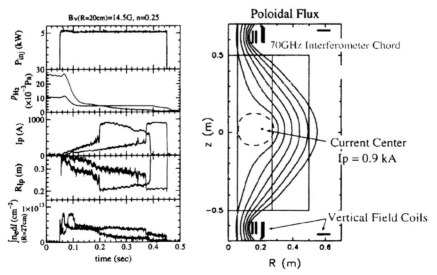

FIGURE 2. Formation of initial closed flux surface

Plasma current up to 2 kA has been started-up by injecting microwaves in the range of 20-50 kW from a 2 GHz klystron and adjusting the external vertical field temporally as shown in Fig.3. The plasma current ramps up with appropriate ramp up rate of the external vertical field. Magnetic measurement shows that last closed flux surface with a minor radius of a=15 cm and a major radius of R =25 cm (R/a=1.7) is formed in the last phase of the discharge. The line-averaged electron density is about 1.7 times of the plasma cutoff density for 2.0 GHz (5.0×10^{10} cm^{-3}).

Discharges with much higher electron density are obtained with slightly lower filling gas pressure and lower external magnetic field as show in Fig.4. The electron density reaches 6 times of the cutoff density, but the plasma current is low (Ip=0.6 kA).

These results suggest that ECH/ECCD by mode-converted Bernstein waves takes place for the present case, in contrast to the previous experiment where the current start-up was achieved by the extraordinary mode in the very low-density regime compared with the cutoff density [2]. Effectiveness of Bernstein wave for ECH/ECCD is crucial for the start-up and formation of ST plasmas since the electromagnetic modes cannot penetrate into over dense plasmas. A new ECH system using a 5 GHz klystron (200kW,100 msec) is under construction. With this system, better coupling to Bernstein waves from externally injected electromagnetic waves compared with the 2.0 and 2.45 GHz cases may be obtained since the wave length is sufficiently shorter than the plasma size. Furthermore, plasma performance may be improved with the stronger toroidal field for 5GHz ECH.

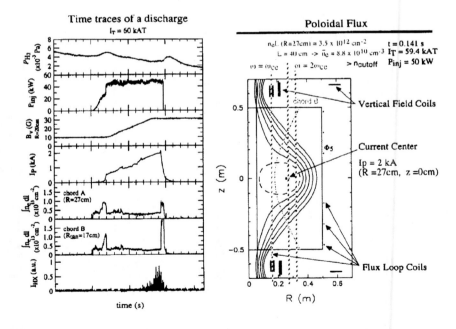

FIGURE 3. Discharge with relatively large plasma current and low density.

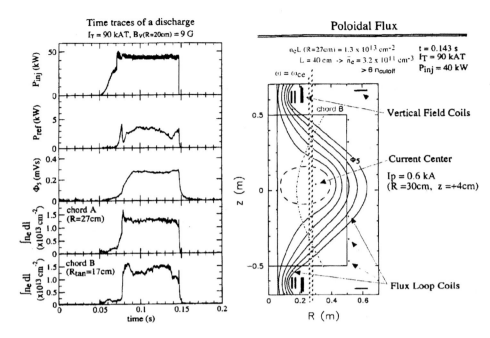

FIGURE 4. Discharge with relatively high density and low plasma current.

REFERENCES

1. C.B.Forest et al., Physical Review Letters Vol.68 (1992) 3559.
2. S.Tanaka et al., Nuclear Fusion, Vol.33 (1993) 505.

Fokker-Planck simulations of X3 EC wave absorption experiments in the TCV tokamak

P. Nikkola*, S. Alberti*, S. Coda*, T. P. Goodman*, R. W. Harvey[†], E. Nelson-Melby* and O. Sauter*

*Centre de Recherches en Physique des Plasmas, Association Euratom-Confédération Suisse, Ecole Polytechnique Fédérale de Lausanne, CRPP - EPFL CH-1015 Lausanne
[†] CompX, P.O. Box 2672, Del Mar, CA 92014-5672, USA

Abstract. Fokker-Planck modeling of the absorption of third harmonic electron cyclotron waves is presented. In the TCV tokamak an advanced electron cyclotron heating system is used to both heat the plasma (ECRH) and drive current (ECCD). Two frequencies are available for heating, 3 MW at 82.7 GHz in X-mode matching the second harmonic of the cyclotron frequency(X2), and 1.5 MW at 117.7GHz which couples to the third harmonic (X3). The X3 waves have a smaller absorption coefficient as the optical thickness of the plasma strongly decreases with the harmonic number. In order to maximize the damping, waves are launched vertically from the top of the vacuum vessel, following the cold X3 resonance. This method alone has been measured to lead to 66% absorption. The Fokker-Planck (F-P) modeling is in agreement with the linear ray tracing code TORAY-GA and both satisfactory reproduce the experimental results.

A second method is based on nonlinear enhancement of the absorption coefficient. A suprathermal electron population with a temperature of about 5 times the bulk temperature is created with the X2 waves. The X3 waves are efficiently damped by this electron population, leading to the measured full absorption of the X3 waves. In this scenario, nonlinear modeling is required, and the F-P calculations are in qualitative agreement with the experiments. However, the simulations suffer from numerical difficulties, and require a large number of mesh points in 3 dimensions (velocity, pitch angle, and radial coordinates). Work is in progress for a better quantitative comparison with the experiment.

INTRODUCTION

One of the main research activities on the TCV tokamak ($R_0 = 0.88$ m, $a = 0.25$ m, $B_0 = 1.44$ T) is the study of Electron Cyclotron Resonance Heating/Current Drive (ECRH/ECCD) in various plasma regimes. ECRH is the only additional heating system with a total of 4.5 MW of Electron Cyclotron (EC) power and the system is constituted of 3.0 MW of 82.7 GHz X-polarized waves at the 2nd EC harmonic (X2), and 1.5 MW of 117.7 GHz waves matching the 3rd harmonic (X3). The power density can be as high as $P_{EC} = 30$ MW/m^3, largest in the world in the medium-large size machines. At this high power level, the nonlinear effects on the ECCD have been shown to be important [1,2]. However, the question of nonlinear effects on the wave absorption is less studied because usually the linear codes predict full absorption of X2 waves or O1 in larger tokamaks like ITER.

The situation is different for the X3 waves since the optical thickness of the plasma is a strongly decreasing function of the harmonic number n, $\alpha \sim (k_B T_e / m_e c^2)^{n-1}$. To increase X3 absorption, the waves are launched vertically approximately parallel to the cold resonance. The absorbed power has been measured to be about 66% in this case [3]. TORAY-GA [4] predicts 43%-75%, depending on the exact launching angle, and similar results are obtained with quasilinear calculations performed by the CQL3D Fokker-Planck (F-P) code [5].

Another way to increase X3 absorption is to create a nonthermal electron population with X2 waves. Since the absorption is effectively proportional to

$$P_{abs} \sim -f \frac{\partial \ln f}{\partial u_\perp} \qquad (1)$$

where f is the electron distribution function and $u_\perp$ is the perpendicular velocity with respect to the magnetic field, increasing f more than decreasing $-\partial \ln f/\partial u_\perp$, as compared to a Maxwellian plasma, leads to nonlinear enhancement of P_{abs}, see equation (2). This nonlinearity has been experimentally shown to lead to 100% absorption of X3 waves [6].

VERTICAL LAUNCH OF THE X3 WAVES, LINEAR ABSORPTION

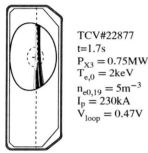

FIGURE 1. Vertical X3 launching geometry and the key plasma parameters.

TCV#22877
t=1.7s
$P_{X3} = 0.75$MW
$T_{e,0} = 2$keV
$n_{e0,19} = 5$m^{-3}
$I_p = 230$kA
$V_{loop} = 0.47$V

The measured X3 absorption in the plasma considered in [3], using the vertical X3 launching geometry, as shown in figure 1, is about 66%. In figure 2a we show the comparison of TORAY-GA calculation with CQL3D results with different velocity meshes. The absorption of CQL3D is in good agreement with the EC wave damping routine of TORAY-GA. In the figure it is also shown that changing the central value of the radial diffusion coefficient, D_0, from 0 to 1 m^2/s or setting the electric field to zero did not affect the damping. The power density profiles of the two codes are almost identical as shown in figure 2b. This calculated absorption of 42-43% is smaller than the observed value. However, the absorbed power is very sensitive to the exact launching angle as shown experimentally and seen with the simulations [3]. In the plasma analyzed in this paper, changing the launcher angle by less than 0.5° increased the TORAY-GA result by a factor of about 2. Thus the difference between the TORAY-GA and CQL3D results and the measured values are within the error bars of the exact launcher aiming and equilibrium reconstruction.

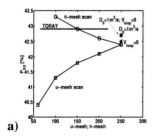
a)

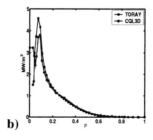
b)

FIGURE 2. a) The CQL3D result is in agreement with the TORAY-GA damping routine. Electric field and radial transport have no effect on the damping. b) Also the power absorption profiles of the two codes agree well.

PREHEATING WITH X2 WAVES, NONLINEAR ABSORPTION

With a finite launching angle between the X2 waves and the normal to the magnetic field, Φ_T^{X2}, a suprathermal electron population is created because the resonance line reaches higher energies than with $\Phi_T^{X2} = 0$ (ECRH). This follows directly from the relativistic EC resonance condition $\omega = \Omega/\gamma + k_\parallel v_\parallel$.

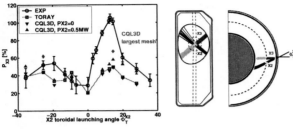

FIGURE 3. The scan of $k_\parallel$ of X2 waves and the measured absorption of X3 waves. The simulation results are discussed in the text. The launching geometry is shown on right.

In figure 3 the measured X3 absorption, using the DML signal with X3 modulation, is shown [6]. 100% absorption of X3 waves is observed with an angle of $\Phi_T^{X2} = 15°$ of the X2 waves. In addition, there is an asymmetry between negative (CNT-CD) and positive (CO-CD) launching angles. The assumed reason for the observed synergy in absorption between X2 and X3 waves is that with X2 waves a suprathermal electron population is created if $\Phi_T^{X2} \neq 0$. This population has larger temperature but also f is larger at a high velocity as compared to f_M. Then, according to equation (1), P_{abs} can be higher if

$$f(\mathbf{u}^*) > \frac{\partial_{u_\perp} \ln f_M|_{\mathbf{u}^*}}{\partial_{u_\perp} \ln f|_{\mathbf{u}^*}} f_M(\mathbf{u}^*) \quad (2)$$

where f_M is the Maxwellian electron distribution function, the X3 wave absorption occurs at the velocity $\mathbf{u}^*$, and $\partial_{u_\perp} \equiv \partial/\partial u_\perp$.

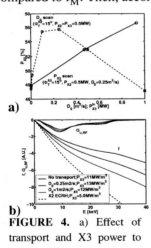

FIGURE 4. a) Effect of transport and X3 power to X3 absorption, b) f and $G_{u,RF}$ with different D_0. Also is shown f with $\Phi_T^{X2} = 0$.

The linear CQL3D results, with zero X2 power (P_{X2}), agree well with TORAY-GA. However, these values are systematically below the observed level. With $P_{X2} = 0.47$ MW, implying a non-Maxwellian distribution function, the F-P calculations show an increase in absorption as compared to the linear TORAY-GA results. The absorbed power increases in some cases by up to 30%. Therefore some synergy is also observed in the simulations. However, in CO-CD with Φ_T^{X2} between $5° - 15°$, the experimental absorption still exceeds the calculated one. In figure 3 also the calculation with the largest possible mesh is shown, giving absorption of about 70%. However, the result was not fully converged as the mesh-size is limited by the 3 GB of memory. Thus in order to simulate the strong synergy, a very large mesh is needed. The reason for the asymmetry in figure 3 is not yet known.

This synergy means that part of the X3 power is absorbed in the non-Maxwellian part of the distribution function. Therefore it is interesting to study the effect of transport in

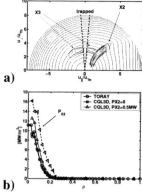

FIGURE 5. a) Quasilinear diffusion coefficients of X2 and X3 waves ($\Phi_T^{X2} = -25°$) and b) comparison of TORAY-GA and CQL3D calculated X3 absorption profiles ($\Phi_T^{X2} = 20°$).

the calculations since with D_0 we can control the suprathermal tail. In figure 4a is shown the absorbed X3 fraction when D_0 is changed from zero (no transport) to $1\,\mathrm{m^2/s}$. There is a clear maximum at $D_0 \approx 0.25\,\mathrm{m^2/s}$. This is in accordance with equation (2) since without transport f is large at high **u** but its logarithmic derivative is reduced as compared to the finite D_0 case. Thus, maximum absorption occurs at a finite D_0. In figure 4b the cuts of f at pitch angle 45° are shown as well as $G_{\mathbf{u}RF} = \mathbf{u}^2 \Gamma_{\mathbf{u},RF}$, where $\Gamma_{\mathbf{u},RF} = \mathbf{D} \cdot \partial_{\mathbf{u}} f$ and **D** is the quasilinear diffusion tensor. Clearly without transport $-\partial_{u_\perp} \ln f$ is smaller than with $D_0 > 0$ where $G_{\mathbf{u},RF} \neq 0$. With $\Phi_T^{X2} = 0$, f is much smaller than with $\Phi_T^{X2} \neq 0$ and the increase in $-\partial_{u_\perp} \ln f$ can not compensate this. The absorption of X3 is also nonlinear in the X3 input power itself, figure 4a, even though this nonlinearity is lower order. The contours of $G_{\mathbf{u},RF}$ and f are shown in figure 5a. The ECRH resonance of X3 waves and the non-symmetric ECCD resonance of X2 waves overlap and the X2 line reaches to higher energies. The comparison of TORAY-GA and CQL3D calculated absorption profiles is shown in figure 5b for $\Phi_T^{X2} = 20°$. The profiles are almost identical, the CQL3D result with $P_{X2} = 0.5\,\mathrm{MW}$ being somewhat broader to account for the difference in total absorbed power. Also the absorption profile of X2 waves is shown.

SUMMARY

The CQL3D Fokker-Planck code is shown to be in accordance with the linear wave damping routine of TORAY-GA for 3rd harmonic heating in very different launching geometries. The enhancement of the absorption due to X2 ECCD is reproduced. The peak absorption is however not fully reproduced by the simulations. This could be due to numerical difficulties, requiring a very large mesh. Work is in progress to improve the convergence.

ACKNOWLEDGMENTS

This work was partly supported by the Swiss National Science Foundation.

REFERENCES

1] R.W. Harvey et al, Phys. Rev. Lett. **88** (2002) 205001.
2] P. Nikkola et al, Submitted to Nuclear Fusion.
3] S. Alberti et al, 29th EPS Conf. on Contr. Fusion and Plasma Phys., Montreux, Switzerland (2002).
4] K. Matsuda, IEEE Trans. Plasma Sci. PS-17 (1989) 6.
5] R.W. Harvey and M.G. McCoy, Proc. IAEA TCM/Advances in Simulation and Modeling in Thermonuclear Plasmas, Montreal (1992).
6] S. Alberti et al, Nucl. Fusion **42** (2002) 42.

Complete Suppression of the m=2/n=1 NTM Using ECCD on DIII–D

C.C. Petty, R.J. La Haye, T.C. Luce, M.E. Austin,[a] R.W. Harvey,[b]
D.A. Humphreys, J. Lohr, R. Prater, and M.R. Wade[c]

General Atomics, P.O. Box 85608, San Diego, California 92186-5608
[a]*University of Texas, Austin, Texas.*
[b]*CompX, Del Mar, California.*
[c]*Oak Ridge National Laboratory, Oak Ridge, Tennessee*

Abstract. Complete suppression of the $m=2/n=1$ neoclassical tearing mode (NTM) is reported for the first time using electron cyclotron current drive (ECCD) to noninductively generate current at the radius of the island O-point. Experiments on the DIII-D tokamak show that the maximum shrinkage of the $m=2/n=1$ island amplitude occurs when the ECCD location coincides with the $q=2$ surface. Estimates of the ECCD radial profile width from the island shrinkage are consistent with ray tracing calculations but may allow for a factor-of-1.5 broadening from electron radial transport.

INTRODUCTION

Neoclassical tearing modes (NTMs) are magnetic islands that grow from a finite amplitude initial condition owing to a helical deficit in the bootstrap current that is resonant with the spatial structure of the local magnetic field [1]. The onset of NTMs represent a significant limit to the plasma performance at higher poloidal beta. For example, high beta values often destabilize the $m=2/n=1$ mode that can lock to the wall, grow rapidly and disrupt the plasma [2]. Here m is the poloidal mode number and n is the toroidal mode number for tearing modes resonant at safety factor $q=m/n$. Several tokamaks have demonstrated the suppression of the $m=3/n=2$ NTM using ECCD positioned at the $q = 1.5$ location [3–5].

This paper reports the first use of ECCD to suppress the important $m=2/n=1$ neoclassical tearing mode on DIII-D. Up to five gyrotrons operating at 110 GHz are used in these experiments, with a maximum combined power of 2.7 MW absorbed in the plasma. The poloidal cross-section of the DIII-D plasma, the electron cyclotron wave trajectories and the electron cyclotron resonances are shown in Fig. 1. The electron cyclotron waves first pass through the third harmonic of the electron cyclotron frequency ($3f_{ce}$) where less than 5% of the power is calculated to be damped before propagating on to the second harmonic resonance ($2f_{ce}$) where the remaining power is absorbed. To drive current at the same location as the $m=2/n=1$ NTM, the launching antennas are steered so that the electron cyclotron waves are absorbed near the $q=2$ surface, as indicated in Fig. 1.

COMPLETE SUPPRESSION OF THE m=2/n=1 NTM BY ECCD

Complete suppression of the $m=2/n=1$ NTM has been achieved for the first time using ECCD to replace the "missing" bootstrap current at the $q=2$ surface. A comparison of two discharges, one with complete suppression of the $m=2/n=1$ NTM and one with only partial suppression, is given in Fig. 2. The energy content is regulated using closed loop feedback of the neutral beam injection (NBI) power in these discharges. The $m=2/n=1$ NTM is triggered by temporarily raising the feedback value of normalized beta (β_N) until it reaches the ideal limit for a plasma without a conducting wall ($\approx 4\,\ell_i$, where ℓ_i is the internal inductance); the feedback value of β_N is then reduced to avoid driving the tearing mode to a large amplitude. Soon after the tearing mode starts to grow, 2.7 MW of ECCD is injected into the plasma near the island location, driving 40 kA of current according to the CQL3D quasilinear Fokker-Planck code [6]. For the complete suppression case in Fig. 2 (solid lines), the plasma control system (PCS) is put into a "search and suppress" mode to make small changes in B_T to find and hold the optimum ECCD position for island stabilization. Figure 2 shows that the PCS makes one adjustment to B_T of ≈ 0.01 T (equi-valent to moving the ECCD location by 0.9 cm along the midplane), after which the $m=2/n=1$ NTM is completely suppressed.

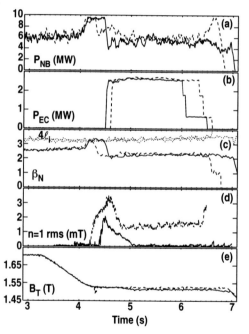

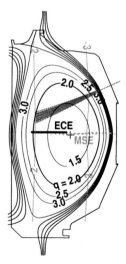

Figure 1. Configuration for NTM suppression using ECCD showing the q surfaces, 2nd and 3rd harmonic resonance locations, and projection of the electron cyclotron wave trajectories. The locations of the ECE and MSE measurements are also indicated.

Figure 2. Time history of complete suppression (solid lines) and partial suppression (dashed lines) discharges showing (a) NBI power, (b) ECCD power, (c) normalized beta and ideal no-wall stability limit (dotted lines), (d) rms amplitude of $n=1$ tearing mode measured at the wall, and (e) toroidal magnetic field strength.

For the partial suppression case (dashed lines), the B_T value and thus the ECCD location is not optimized.

SENSITIVITY TO ECCD LOCATION AND WIDTH

Experiments on DIII-D show that the suppression of the $m=2/n=1$ NTM is sensitive to the location of the ECCD with regard to the $q=2$ surface. This is shown in Fig. 3, where the normalized radius of the $q=2$ surface ($\rho_{q=2}$) determined from equilibrium reconstruction and the normalized radius of the ECCD location (ρ_{EC}) determined from ray tracing codes [7,8] are plotted during the course of a B_T scan. The amplitude of the $m=2/n=1$ NTM measured by Mirnov coils outside the plasma reaches a minimum when the ECCD passes through the $q=2$ surface.

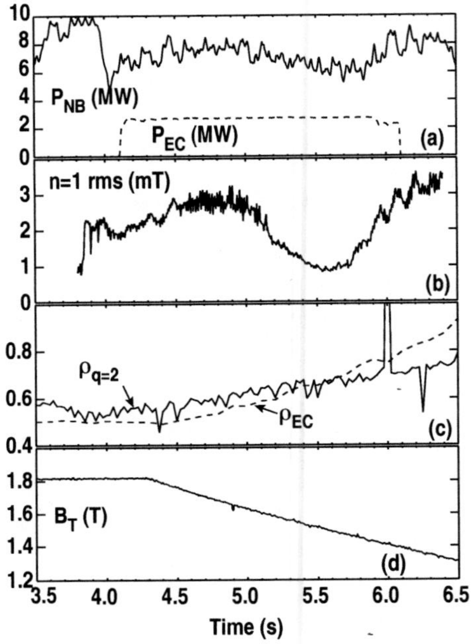

Finally, modeling the island shrinkage due to localized ECCD allows an upper limit to be placed on profile broadening caused by radial transport of the current carrying electrons. The evolution of the radial width (w) of the tearing mode is modeled by the modified Rutherford equation [5],

Figure 3. Time history B_T sweep showing (a) NBI (solid) and ECCD (dashed) powers, (b) rms amplitude of the n=1 tearing mode measured at the wall, (c) normalized radius of $q=2$ surface (solid) and ECCD location (dashed), and (d) toroidal magnetic field strength.

$$\frac{\tau_R}{r}\frac{dw}{dt} = \Delta'r + \varepsilon^{1/2}\left(\frac{L_q}{L_p}\right)\beta_{pol}\left[\frac{rw}{w^2+w_d^2} - \frac{rw_{pol}^2}{w^3} - \frac{8qr\delta_{EC}}{\pi^2 w^2}\left(\frac{\eta J_{EC}}{J_{BS}}\right)\right], \quad (1)$$

where J_{EC}/J_{BS} is the peak ECCD current density normalized to the local bootstrap current density, $\eta = 0.4/(1 + 2\,\delta_{EC}^2/w^2)$ for non-modulated ECCD with perfect alignment between the peak ECCD and the island O-point, δ_{EC} is the FWHM of the ECCD radial profile, and the other terms are defined in Ref. [5]. Radial transport of the current carrying electrons enters into Eq. (1) in two ways: δ_{EC} increases and J_{EC}/J_{BS} decreases. Assuming that the total driven current is independent of δ_{EC}, the net effect of electron radial transport is unfavorable for mode suppression. Figure 4 shows the calculated $m=2/n=1$ island width from Eq. (1) as a function of δ_{EC} for the discharge shown in Fig. 3 at time 5.6 s when the mode has reached its minimum amplitude and $dw/dt = 0$. The TORAY-GA code calculates a peak ECCD current density that is 2.5 times the local bootstrap current density for $\delta_{EC} = 2.6$ cm in the absence of electron radial transport. For fixed ECCD magnitude, $J_{EC}/J_{BS} \propto \delta_{EC}^{-1}$. Figure 4 shows that the predicted minimum width of the $m=2/n=1$ island during the B_T ramp increases with increasing δ_{EC}, with $\delta_{EC} = 2.6$ cm being very close to achieving complete elimination of the island according to Eq. (1). The measured island width of 8 cm is also indicated in Fig. 4. The theoretical and experimental island widths agree for an ECCD profile width of 3.3 cm with a lower limit of 2.6 cm and an upper limit of 3.9 cm. Modeling this amount of ECCD profile broadening using CQL3D places an upper limit of 0.7 m²/s on the fast electron transport.

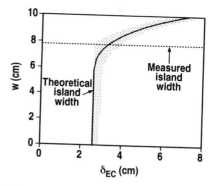

Figure 4. Predicted minimum width from Eq. (1) of m=2/n=1 island (solid line) during B_T sweep; the error bars are due to the 10% uncertainty in the gyrotron power. The measured minimum island width is also indicated.

SUMMARY

For the first time, the important $m=2/n=1$ NTM has been completely suppressed using co-ECCD on the DIII-D tokamak. By noninductively driving a small fraction of the total plasma current (typically 3%) within the island located at the $q=2$ surface, the non-modulated ECCD is able to replace the "missing" bootstrap current and the $m=2/n=1$ NTM can be stabilized. Modeling of the island shrinkage using the modified Rutherford equation indicates that the width of the ECCD radial profile is consistent with ray tracing calculations but may also allow for a factor-of-1.5 broadening due to radial transport of the current carrying electrons.

ACKNOWLEDGMENTS

This work is supported by the U.S. Department of Energy under Contract Nos. DE-AC03-99ER54463 and DE-AC05-00OR22725, and Grant Nos. DE-FG03-97ER4415 and DE-FG03-99ER54463.

REFERENCES

1. Chang, Z, *et al.*, *Phys. Rev. Lett.* **74**, 4663 (1995).
2. La Haye, R.J., Lao, L.L., Strait, E.J., Taylor, T.S., *Nucl. Fusion* **37**, 397 (1997).
3. Gantenbein, G., *et al.*, *Phys. Rev. Lett.* **85**, 1242 (2000).
4. Isayama, A., *et al.*, *Plasma Phys. Control. Fusion* **42**, L37 (2000).
5. La Haye, R.J., *et al.*, *Phys. Plasmas* **9**, 2051 (2002).
6. Harvey, R.W., McCoy, M.G., "The CQL3D Fokker-Planck Code," in *Proc. IAEA Technical Committee Meeting, Montreal, 1992* (International Atomic Energy Agency, Vienna, 1993) p. 498.
7. Cohen, R.H., *Phys. Fluids* **30**, 2442 (1987).
8. Matsuda, K., *IEEE Trans. Plasma Sci.* **17**, 6 (1989).

Assessment of Electron Banana Width Effect in ECCD Experiments

K. L. Wong, V. S. Chan* and C.C. Petty*

Plasma Physics Laboratory, Princeton University, Princeton, New Jersey 08543
*General Atomics, P.O. Box 85608, San Diego, CA 92186

Abstract. It is shown that the energetic trapped electron dynamics can be described by the Smoluchowski equation, and a simple criterion is derived to assess its significance in electron cyclotron current drive (ECCD) experiments. In small tokamak experiments similar to the WT-3 device where the density is low and the inductive electric field can change rapidly, the finite electron banana width can cause the energetic electrons to pile up near the trapped-passing boundary. In the DIII-D experiment, this effect becomes negligibly small.

Results from electron cyclotron current drive experiments on the DIII-D tokamak[1] have been analyzed using the CQL3D code [2,3] – a 3D Fokker-Planck code with two dimensions in velocity space and one dimension in configuration space. Since the code assumes zero electron banana width, it is important to check the validity of this assumption. The banana width for thermal electrons is usually very small; it is the more energetic electrons produced by the electron cyclotron waves that deserve careful consideration. In this note, we present a simple model to assess the finite banana width effect in ECCD experiments.

Absorption of electron cyclotron wave raises the electron perpendicular energy and enhance the trapped electron population. Trapped electrons can become passing electrons via pitch angle scattering. If there is Ohmic current driven by an inductive electric field, this electric field tends to increase the parallel velocity of those electrons moving in the direction opposite to the Ohmic current. Although this is a very small effect on the trapped electrons due to their short bounce period, in the very weak collision regime where the average change in electron parallel velocity from pitch angle scattering is much smaller than that due to the inductive electric field in half of the bounce period, most of the trapped electrons will eventually emerge carrying a co-current. However, in today's tokamak experiments, the opposite is true, i.e., the average change in electron parallel velocity from pitch angle scattering is much larger than that due to the inductive electric field, and therefore only half of the trapped electrons will eventually emerge carrying a co-current.

Figure 1 shows the trapping region in velocity space for the electrons at the outboard side of the magnetic surface where the magnetic field is at its minimum value. As explained in Ref. [4], electrons resonant with cyclotron waves diffuse along nearly constant energy contours in velocity space. Steady state solutions of the Fokker-Planck equation show that the electron flow in velocity space forms a vortex pattern [5]. The cyclotron waves have no net effect on the non-resonant electrons. Their dynamics are governed by Coulomb collisions with the bulk plasma.

For simplicity, let us consider trapped electrons with constant velocity v. The change in $v_\parallel$ due to pitch angle scattering can be approximated by a random walk process. Let x denote the value of $v_\parallel$ when the trapped electron is on the midplane at the maximum major radius position moving in the direction opposite to the plasma current. It is a positive definite label for the trapped electron; its value can be anywhere inside the domain $(0, x_b)$ where x_b is the value of $v_\parallel$ at the trapped/passing boundary. Let x_0 denote the initial parallel velocity. The probability for the electron to have $v_\parallel = x$ at time t is $y(x, x_0, t)$ which is governed by the diffusion equation

$$\frac{\partial y}{\partial t} = D\frac{\partial^2 y}{\partial x^2} \qquad [1(a)]$$

with the initial condition

$$y(x, x_0, 0) = \delta(x - x_0). \qquad [1(b)]$$

Equation [1(a)] has the well known solution [6]

$$y(x, x_0, t) = \left(\frac{1}{4\pi Dt}\right)^{1/2} \exp\left[\frac{-(x - x_0)^2}{4Dt}\right] \qquad [1(c)]$$

where D is the diffusion coefficient in velocity space related to the 90 deg. pitch angle scattering coefficient υ by $D \sim (\pi/2)^2 v^2 \upsilon$. When the effects from plasma drag, inductive electric field and non-uniform magnetic field are taken into account, the equation of parallel motion for the electron becomes:

$$\frac{dv_\parallel}{dt} = -v_s v_\parallel - \frac{e}{m} E_\parallel - \frac{\mu}{m} \frac{\vec{B} \cdot \nabla B}{B} \qquad (2)$$

where v_s is the inverse slowing time. Averaging Eq. (2) over the banana orbit, and putting $\langle v_\parallel \rangle = \alpha x$, Eq. (2) becomes

$$\frac{dx}{dt} = -\langle v_s \rangle x - \frac{e}{m}\frac{1}{\alpha}\langle E_\parallel \rangle \quad . \qquad (3)$$

Obviously, $\alpha < 1$, but its magnitude is of the order of unity. The last term in Eq. (3) is due to the finite banana width; the trapped electron sees a bounce-averaged parallel electric field which is proportional to the banana width Δ_b and the spatial gradient of $E_\parallel$, i.e., $\langle E_\parallel \rangle \sim (\Delta_b/a)(V_L/2\pi R)$ where V_L is the tokamak loop voltage. During the startup phase of a tokamak, $E_\parallel$ is higher at the plasma edge. Such a $E_\parallel$ profile tends to raise x towards x_b, i.e., to change the trapped electron into a passing electron. For typical experimental parameters, dx/dt from Eq. (3) is much smaller than that due to pitch angle scattering and therefore its effect is negligibly small. However, if $E_\parallel$ changes in time, it can have noticeable effect on the trapped electrons as explained in the following.

In pulsed tokamak experiments, the inductive current density profile evolves in time, so does the inductive electric field because they are related to each other through Ohm's law. Take time derivative of Eq. (3) to obtain

$$\frac{d^2 x}{dt^2} = -\langle v_s \rangle \frac{dx}{dt} - \frac{e}{m}\frac{1}{\alpha}\left\langle \frac{dE_\parallel}{dt} \right\rangle \quad . \qquad (4)$$

When this is combined with the pitch angle scattering effect, the velocity space dynamics of the trapped electron can be described by the kinetic equation

$$\frac{\partial y}{\partial t} = D \frac{\partial^2 y}{\partial x^2} + C \frac{\partial y}{\partial x} \quad [5(a)]$$

where

$$C = \frac{e}{m} \frac{1}{\alpha} \frac{\langle dE_\parallel/dt \rangle}{\langle v_s \rangle} . \quad [5(b)]$$

This is the well known Smoluchowski equation [7] usually applied to problems associated with Brownian motion under a constant force. A rigorous derivation of Eq. [5(a)] is given in Ref. [7]. The second term on the right hand side of Eq.(4) represents a "force" in velocity space. Depending on the values of the coefficients, there may be more terms on the RHS of Eq. [5(a)]. Only the leading term is kept here.

During ECCD, electron cyclotron waves raise the perpendicular velocity of those electrons with $v_\parallel$ that can satisfy the resonance condition. When effects of plasma drag and pitch angle scattering are included, Fokker-Planck code simulations show a vortex pattern [5] in the 2-D velocity space as schematically shown in Fig. 1: electrons leaving the trapped region flow around the vortex and re-enter through a different point in the trapped/passing boundary. Here we can only analyze the problem in one dimension velocity space. In order to mimic this electron cyclotron wave effect on the energetic electrons, we can assume total reflection at the trapped/passing boundary, i.e., electrons leaving the trapped region re-enter at the same point so that there is no net particle flux at the trapped/passing boundary. This is equivalent to solving Eq. (5) with the boundary condition:

$$-D \frac{dy}{dx} - Cy = 0 . \quad (6)$$

at $x = x_b$. The solution that satisfies the initial condition specified in Eq. [1(b)] is

$$y(x, x_0, t) = \frac{1}{2\sqrt{\pi Dt}} \left\{ \exp\left[\frac{-(x_b - x - x_0)^2}{4Dt}\right] + \exp\left[\frac{-(x_b - x + x_0)^2}{4Dt}\right] \right.$$

$$\left. \exp\left[\frac{C}{2D}(x_b - x - x_0) - \frac{C^2 t}{4D}\right] - \frac{C}{D\sqrt{\pi}} \exp\left[\frac{C}{D}(x_b - x)\right] \int_{\frac{x_b - x - x_0 + Ct}{2\sqrt{Dt}}}^{\infty} dz \exp(-z^2) \right] . \quad (7)$$

This is very similar to the "sedimentation problem" [5] solved decades ago. In the limit of t approaching infinity, only the last term survives. The steady state solution has the simple form:

$$\lim_{t \to \infty} y(x, x_0, t) = -\frac{C}{D} \exp\left[\frac{C}{D}(x_b - x)\right] . \qquad (8)$$

Obviously the steady state solution depends on the ratio C/D but not on x_0. It is depicted on Fig. 2 for the plasma parameters $n_e = 2 \times 10^{12}$ cm^{-3}, $Z_{eff} = 2$, $R = 65$ cm, $a = 15$ cm, $V_L = 2$ volts which changes in a time scale $\tau = 30$ ms. The steady state solution tilts towards the trapped/passing boundary due to the finite banana width effect; this is because the second term in Eq. [5(a)] produces a convective flow towards the trapped/passing boundary. The piling up of particles at x_b may be noticeable only in small tokamak low-density experiments[8]. When we put the DIII-D parameters into Eq. (7), $Cx_b/D \ll 1$ and $y(x,x_0,t)$ becomes insensitive to x at large values of t as expected because of the domination of the diffusion term. Therefore, in large tokamak experiments, this effect is negligibly small, and the assumption of zero banana width is well justified. Finally, we would like to stress the fact that our analysis is based on a simple model good only for qualitative assessment. While the essential physics is retained, the equations are simplified to the extent that they become analytically tractable. More quantitative analysis can be done with a particle simulation code.

Acknowledgments

This work is supported by the US DOE under Contract No. DE-AC02-76CH03073 and No. DE-AC03-99ER54463. The authors are grateful to Dr. R. Prater for many helpful discussions on ECCD experimental results.

References

[1] C.C. Petty et al., Nucl. Fusion 41, 551 (2001).
[2] G.D. Kerbel and M.G. McCoy, Phys. Fluids 28, 3629 (1985).
[3] R.W. Harvey et al., Phys. Plasmas 7, 4590 (2000).
[4] N.J. Fisch, Rev. Mod. Phys. 59, 175 (1987).
[5] C.F.F. Karney and N.J. Fisch, Phys. Fluids 22, 1817 (1979); also Nucl. Fusion 21, 1549 (1981).
[6] A. Einstein, Ann. d. Physik 17, 549 (1905).
[7] S. Chandrasekhar, Rev. Mod. Phys. 15, 1 (1943).
[8] M. Asakawa et al., Fusion Engineering and Design 53, 237 (2001).

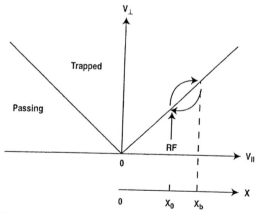

Fig. 1. Schematic representation of the trapping region in velocity space and the particle flow during ECCD.

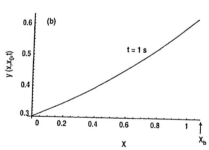

Fig. 2. Analytic solution of Eq.(5) with the boundary condition given by Eq. (9) for small tokamak parameters showing the pile-up at trapped/passing boundary.

ELECTRON BERNSTEIN WAVE HEATING AND CURRENT DRIVE

Electron Bernstein Wave Studies on COMPASS-D and MAST

V. Shevchenko, E. Arends, Y. Baranov, M. O'Brien, G. Cunningham, M. Gryaznevich, B. Lloyd, A. Sykes, F. Volpe

EURATOM/UKAEA Fusion Association, Culham Science Centre, Abingdon, Oxon, OX14 3DB, UK.

A.D.Piliya, A.N.Saveliev and E.N.Tregubova

Ioffe Institute, Politekhnicheskaya 26, 194021 St.Petersburg, Russia.

The present paper reports on electron Bernstein wave (EBW) emission, heating and current drive experiments conducted on COMPASS-D and MAST. Plasma heating and current drive experiments have been conducted in COMPASS-D, where EBWs were excited from the high field side of the tokamak at different toroidal launch angles. It has been found experimentally that X-mode injection perpendicular to the magnetic field provides maximum heating efficiency. Non-inductive currents of up to 100 kA were found to be driven by the EBW-mode in the counter direction. This result is consistent with EBW beam-tracing and quasilinear Fokker-Planck simulations. EBW emission measurements are a valuable tool for optimisation of heating scenarios. Both X-mode and O-mode polarised emission were observed from overdense plasmas at angles perpendicular to the magnetic field and at angles close to the optimum for Bernstein-extraordinary-ordinary (B-X-O) mode conversion. EBW spectra from high electron cyclotron harmonics are presented and discussed. A steerable antenna has been recently installed in MAST. Preliminary results on ECRH breakdown and EBW plasma heating in MAST are discussed.

1. INTRODUCTION

Electron Bernstein waves (EBW) experience very localised damping on electrons at the electron cyclotron resonance but, unlike the extraordinary (X) mode and the ordinary (O) mode, the damping remains strong even at high harmonics of the electron cyclotron (EC) frequency ω_{ce}. EBW do not have any density cut-offs inside the plasma and can, therefore, access plasmas of arbitrary densities for frequencies above ω_{ce}. These features of EBW present the possibility of efficient means for electron cyclotron resonance heating (ECRH) and current drive in high beta plasmas, particularly in spherical tokamaks, where the X-mode and O-mode propagation into the plasma can only be assured at high ω_{ce} harmonics leading to weak damping of these modes inside the plasma.

Plasma heating with the use of EBW excitation via the O-X-B mode conversion mechanism has been successfully demonstrated on the W7-AS stellarator [1]. EBW heating experiments with direct mode conversion of the X-mode, launched from the high field side (HFS) were carried out on the WT-3 tokamak [2]. First EBW heating and current drive (CD) experiments with HFS excitation have been conducted on the COMPASS-D tokamak [3] and reported in [4]. Recently, EBW heating experiments were started on MAST [5] with an array of steerable antennas, which has been installed in the MAST vessel in September 2002. It is intended to excite the EBW-mode in the plasma mainly via the O-X-B mode conversion mechanism using the existing 1.4 MW 60 GHz gyrotron system. In the present work we present the extended scope of EBW experiments and modelling on COMPASS-D and MAST.

2. EBWH AND EBW CD EXPERIMENTS ON COMPASS-D

The COMPASS-D tokamak is equipped with five HFS launch mirror antennas, each adjustable to any toroidal launch angle by rotation about the vertical axis of the input waveguide. It presents the possibility for EBW excitation via direct mode conversion of the slow X-mode launched from the HFS, to the EBW-mode at the upper hybrid resonance (UHR), allowing the study of EBW plasma heating and current drive. According to theory, the X-mode is subject to very weak damping at the W_{ce} resonance for launch angles nearly perpendicular to the magnetic field. Hence, the slow X-mode propagates to the UHR where it is always totally converted to the EBW-mode, with subsequent absorption near the W_{ce} resonance (see Fig. 1). At launch angles far from perpendicular the X-mode absorption becomes stronger, resulting in predominant X-mode plasma heating. By varying the toroidal launch angle, one can estimate the relative heating and CD efficiencies of the X-mode and the EBW-mode (see Fig. 2).

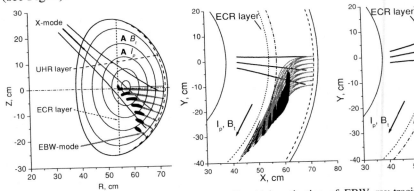

Fig. 1. Poloidal projection of EBW ray-tracing results for the X mode launched perpendicularly to B_t. Black areas indicate strong damping of EBWs.

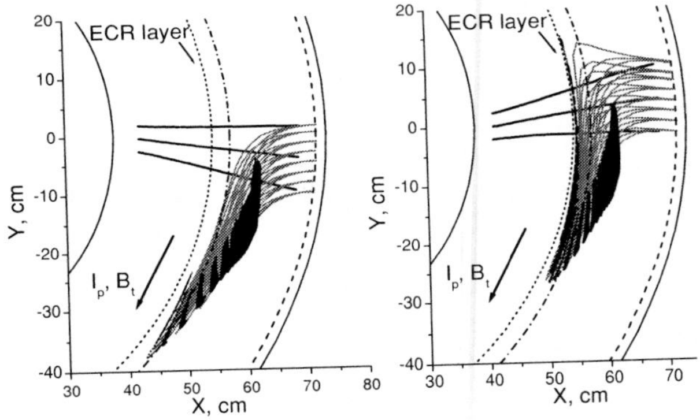

Fig. 2. Toroidal projection of EBW ray-tracing results for toroidal launch angles of −16.7° (left) and +16.7° (right). EBWs completely damp within black areas, where $k_\parallel > 0$ and $w > W_{ce}$. Note sign(I_{CD}) = sign[$k_\parallel$/(W_{ce}-w)].

A highly reproducible plasma scenario has been chosen for these experiments. The line averaged density and plasma current were sustained at $1.8 \cdot 10^{19}$ m^{-3}, 150 kA respectively. Experiments have been conducted at two different values of central toroidal field, 2.05 T and 1.9 T. ECRH power of 600 kW at a frequency of 60 GHz, with a pulse duration of 100 ms, was injected into the plasma at the beginning of the flattop (100 ms) of plasma current and density. The toroidal launch angle was changed shot by shot in the range ±32.6° from perpendicular, with a step of ~8.4°. The power was injected with polarisation perpendicular to the magnetic field. Electron temperature T_e and density n_e profiles were measured with multipoint Thomson scattering (TS) every 50 ms.

During ECRH the electron temperature appears to be noticeably higher for launch angles close to perpendicular to the magnetic field, while the electron temperature profiles do not show any significant transformation, such as peaking or flattening, over the whole range of launch angles. Changes in the surface loop voltage, V_{loop}, required to maintain constant plasma current during ECRH do not have any obvious dependence on the launch angle. Typically the plasma was sustained at a central electron temperature of 1.5 keV and a loop voltage of about 0.9 V before ECRH injection. During ECRH the loop voltage usually dropped down to the value of 0.5 V with slight variations within 0.1 V over the range of launch angles. Because of the temperature changes one would expect much greater reductions in the loop voltage ($S_p \sim Z_{eff}^{-1} T_e^{3/2}$), especially at launch angles close to perpendicular when the electron temperature reaches 3.5 keV.

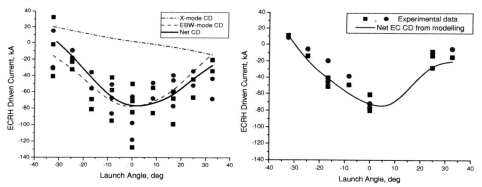

Fig. 3. Noninductive current driven in the plasma estimated from experimental data (points) and ray-tracing modelling (lines) at the central toroidal field of 2.05 T.

Fig. 4. As in Fig. 3 at the central toroidal magnetic field of 1.9 T. Circles and squares correspond to different time slices during ECRH.

The measured electron temperature and loop voltage are consistent with the presence of a significant non-inductive current in the plasma [4]. Because T_e and n_e profile shapes are measured not to change during the ECRH pulse, and Z_{eff} changes are estimated to be relatively small, the non-inductive current driven in the plasma can be estimated from the experimental V_{loop} and T_{e0} with the use of the simple relation [4]

$$I_{CD} = I_p - I_{BS}^{ECRH} - (I_p - I_{BS}^{OH}) \frac{V_{loop}^{ECRH}}{V_{loop}^{OH}} \left(\frac{T_{e0}^{ECRH}}{T_{e0}^{OH}} \right)^{3/2}, \qquad (1)$$

where the plasma current I_p is fixed (150 kA), and the estimated bootstrap current I_{BS}^{OH} before ECRH is about 12 kA, while during ECRH it varies from 15 to 20 kA over the range of launch angles. The experimental and modelling results for two values of toroidal field are summarised in Fig. 3 and Fig. 4. The angular dependence of the driven current appears to be less symmetric with respect to perpendicular launch for the toroidal field of 1.9 T. It is clear that quite large currents ≤100 kA must be driven in a direction counter to the plasma current in order to maintain almost constant loop voltage over the range of electron temperatures during ECRH. The maximum of the driven current corresponds to launch angles close to perpendicular to the magnetic field. At perpendicular launch only a small fraction of the RF power (≤10%) is absorbed in the plasma as the X-mode, while the main part of the power is converted into the EBW-mode and then absorbed near the w_{ce} resonance. This indicates that the EBW-mode is responsible for the non-inductive current driven in both cases of central toroidal field.

A new EBW ray-tracing code [6] has been developed for EBWH and CD modelling in COMPASS-D. The plasma equilibrium obtained from EFIT and the electron density and temperature profiles measured with TS were used as input parameters for the EBW ray-tracing code. This code allows estimation of power deposition profiles with relevant wave vector k profiles for both the X-mode and the EBW-mode in a full tokamak equilibrium. These results have been used as input data for a quasi-linear relativistic Fokker-Planck code [7], modified for EBW applications, in order to simulate currents driven by the X-mode and by the EBW–mode alone. The results of EBW CD modelling are shown in Fig. 1 - Fig. 4. The net ECRH driven current is predominantly defined by the EBW fraction and is in good agreement with experimental results. The normalised EBW current drive efficiency, estimated from the experimental data, presents the value of $h_{20CD} = n_{e20}RI_{CD}/P_{ECRH} \approx 0.035$ (10^{20} A/W m^2), which is higher than the typical figure for O/X-mode EC CD but less efficient than for lower hybrid CD.

3. EBW EMISSION MEASUREMENTS

Two mode conversion processes, namely, the X-B tunnelling and the O-X-B conversion allow EBW excitation with the RF power launched from the low field side (LFS). These processes are very sensitive to the launch geometry and edge plasma parameters. Consequently, monitoring of the reciprocal EBW emission processes is extremely useful for optimisation of EBW heating applications.

On COMPASS-D the plasma emission was measured in the frequency range of 53.0-66.5 GHz with a 14 channel heterodyne radiometer [8]. The receiving antenna was located in the midplane of the torus and directed perpendicularly to the magnetic field. Experimentally, enhanced X-mode EC emission from ω_{ce} and $2 \cdot \omega_{ce}$ resonances in overdense plasma has been observed only in H-mode [9]. The B-X mode conversion efficiency, estimated from these measurements, could reach up to 15%, while more often observed values were in the range of 1-5%.

On MAST the EBW emission is studied with a frequency scanning heterodyne radiometer [10], which covers three frequency bands 16-26 GHz, 26-40 GHz and 40-60 GHz. The full frequency scan, which takes 0.32 ms with a fast sampling or 0.64 ms with a standard sampling rate, is arranged in 32 frequency steps for each frequency band. As a result, the plasma emission spectrum is measured over 96 frequencies during the shot. The majority of spectra have been measured with the antenna viewing optimised for the B-X-O conversion at one of the frequencies within the radiometer range. However, the B-X converted emission was also studied. The main features of the EBW emission can be illustrated in a number of experimentally measured spectra. As soon as the plasma density reaches the value required for B-X-O conversion, the EBW signals from different ω_{ce} harmonics appear in the spectrum (see fig. 5). Here the emission from the first three harmonics is clearly seen during the initial stage of the discharge. The dark gaps in the spectrum always separate neighbouring harmonics. These gaps correspond to the plasma layers, where cyclotron harmonics are coincident with the UHR [11]. The spectral maxima of harmonics move up in frequency during this phase of the discharge because the plasma current ramps up, while the toroidal field remains constant. At 88 ms the emission from the third harmonic has a well-pronounced jump down in frequency and simultaneous increase in intensity. At this time the plasma enters H-mode. The associated increase in edge density gradient enhances the mode conversion efficiency resulting in the increase of the emission intensity at all harmonics. The L-H transition also shifts the mode conversion layer from the bulk plasma outwards to the LCFS.

However, if the plasma is sustained in a high density H-mode longer than 50 ms the behaviour of EBW emission changes dramatically. After 0.14 s in Fig. 6 the ELMs modulate the EBW emission intensity in an unusual way. During ELM-free periods $2\omega_{ce}$ emission is suppressed, while the emission from ω_{ce} and $3\omega_{ce}$ harmonics is enhanced. Conversely, the $2\omega_{ce}$ emission recovers during ELMs, when other harmonics are suppressed. At higher plasma densities similar unusual behaviour has been observed with the emission from the $3\omega_{ce}$ harmonic. Another important feature is that the EBW emission intensity gradually decays if the plasma is sustained in the H-mode quite a long time. The EBW emission decay can be quite substantial especially at high plasma current (~1 MA) and low TF (see Fig. 5). It was never observed in low current (<0.5 MA) plasmas at high TF. In some discharges the emission intensity drops down by a few times and the decrease is more pronounced at lower harmonics than at higher. It should be noted that after a giant ELM or internal reconnection the emission behaviour

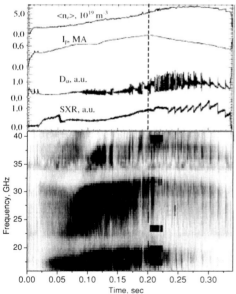

Fig.5 Shot #7680 EBW emission spectrogram measured in a plasma with I_p ramp down. Darker areas correspond to higher intensity.

comes back to normal. Typically, the decay occurs simultaneously at all frequencies with the time scale of 20-30 ms, which is about half of the energy confinement time in these discharges. At the same time plasma temperature and density profiles, as measured by multi-point TS, do not exhibit any significant changes in the gradient zone. Full wave modelling of the O-X-B mode coupling, based on the experimentally measured profiles and plasma equilibrium, shows that no degradation of the coupling efficiency occurs during the H-mode. In fact, the coupling efficiency must be close to 100% in a wider angular range.

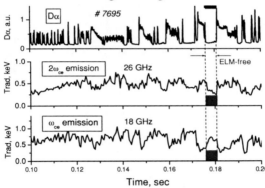

Fig.6 Inverted EBW emission modulation with ELMs in a high density plasma at low TF(I_{TF}=77 kA).

Apparently, both these effects can be attributed to the plasma current redistribution initiated by the H-mode. Indeed, the plasma equilibrium is quite different in the L and H mode. In the H-mode, at least, a higher bootstrap fraction must be driven in the gradient zone close to plasma layers, where the mode conversion occurs. EBW ray-tracing simulations show that even small local perturbations in the edge poloidal magnetic field can change ray trajectories dramatically. In general, such perturbations result in bending of the EBW ray trajectories. It means that EBW emission must originate closer to the periphery where the plasma is cooler.

In support of this hypothesis it was found experimentally that EBW emission is very sensitive to plasma current ramps. As the plasma current ramps up no emission decay was observed, even at low TF and high plasma current, but as the current ramps down the EBW emission intensity degrades very quickly (Fig. 5). A higher rate of I_p reduction typically provokes faster emission suppression. So, in EBWH experiments I_p ramp down must be avoided especially at low TF.

4. FIRST EBW HEATING EXPERIMENTS ON MAST

ECRH breakdown experiments have been conducted on MAST as a first test of the steering capabilities of the antenna. The gyrotrons were fired in pairs during ECRH breakdown experiments in MAST. RF beams (150 kW) have been launched into the vessel tangentially to the ω_{ce} resonance surface. The pulse duration was 20 ms for each beam with a time delay of 10 ms between gyrotrons.

First RF breakdown has been achieved in a pure TF (I_{TF} = 91 kA) with the X-mode and the O-mode (see Fig. 7). In both cases the breakdown appeared at the fundamental resonance surface, indicated by the bright light visible on the CCD image. The breakdown appeared only when at least two gyrotrons were fired. The brightness of the breakdown area was much higher for the O-mode launch at similar parameters. A small tiny ring of plasma has been regularly observed during the O-mode breakdown. The ring is located about ~35 cm above the midplane and its brightness and size increase with increasing power. The upward shift of the bright area corresponds to the direction of electron drift. The toroidal structure of the ring suggests that some current must be driven within it. Estimates from diamagnetic signals give the value of driven current about 3 kA. The current has two maxima, one close to the top of the plasma cylinder and another close to the bottom, with almost zero value near the midplane. Apparently, this current can be attributed to a pressure

 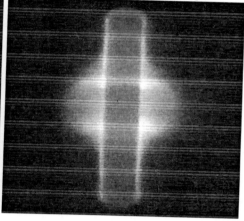

Fig. 7 CCD image of the ECRH breakdown phase at ω_{ce} in a pure toroidal magnetic field.

Fig. 8 CCD image of the ECRH breakdown in toroidal and poloidal (5 mT) magnetic fields.

driven current because it does not change sign when the sign of $k_{\parallel}$ is changed. When a vertical field of 5 mT was applied, the pressure driven current becomes more obvious (see Fig. 8). The plasma current estimated from magnetic measurements was about 5 kA with the maximum at the midplane.

The ECRH timing was varied and power levels up to 0.8 MW were injected. For initial EBW heating studies, high density H-mode target plasmas ($<n_e> \sim 6 \cdot 10^{19}$ m^{-3}) at $I_p \sim 0.8$-0.9 MA and maximum toroidal field were employed. This plasma scenario is optimised for efficient O-X-B coupling but it is not optimised for EBW penetration into the plasma core. In such a configuration one can expect only very peripheral power deposition.

EBW emission is typically strongly modulated by ELMs: longer and deeper ELM-free periods give rise to longer and brighter emission bursts. The radiative temperature (up to 3 keV) measured in the ELM-free phase during the EBWH pulse is higher than the electron temperature (1.5 keV) in the plasma centre, while before EBWH it is usually in good agreement with TS measurements (see Fig. 9). It is reasonable to conclude that during relatively long ELM-free periods the efficiency of O-X-B coupling is high and a significant amount of RF power is deposited into the plasma. This power produces a 'hot' non-Maxwellian tail in the electron distribution function, which quickly dissipates during ELMs or in the L-mode phase. The TS measurements at chords corresponding to the plasma periphery show significant distortion of the scattered spectra consistent with the presence of fast electrons during the ECRH pulse.

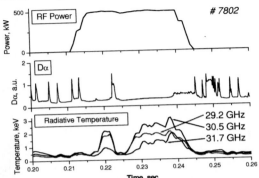

Fig. 9 EBW emission during ECRH pulse in the ELM-free phase.

The ray tracing simulations for the frequencies corresponding to the maximum in the emission spectrum at $2\omega_{ce}$ confirm that the power is emitted from the plasma layers where the ECRH power was deposited at $5\omega_{ce}$. Similar emission spectra were observed also at the fundamental and $3\omega_{ce}$ harmonics. The maximum emission level has been measured when the RF power was launched at the angles close to the optimal for O-X-B conversion. This is a strong evidence of positive interaction between the O-X-B converted EBW-mode and the plasma.

In summary, we have shown experimentally that the X-mode launched from the HFS provides more efficient plasma heating with the excitation of EBWs than with the X-mode alone. EBWs can provide current drive efficiency higher than the typical figure for O/X current drive. The extensive studies of EBW emission from overdense plasma have been conducted on COMPASS-D and MAST, which are very useful for optimisation of EBW heating applications. First EBW heating experiments have been conducted on MAST indicating positive interaction between EBWs and the plasma.

ACKNOWLEDGEMENTS

This work was funded jointly by United Kingdom Engineering and Physical Sciences Research Council and by EURATOM.

REFERENCES

[1] Laqua H.P., *et al.*, Phys. Rev. Lett. **78**, 3467 (1997).
[2] Maekawa T., *et al.*, Phys. Rev. Lett. **86**, 3783 (2001).
[3] Hayward R.T. *et al.*, *Proc. of the 13th IEEE Symposium*, Knoxville, 2-6 October 1989, (New York, IEEE, 1990), p. 854.
[4] Shevchenko V., *et al.*, Phys. Rev. Lett., **89**, 265005-1 (2002).
[5] Darke A.C., *et al.*, *Proc. of the 18th Symposium on Fusion Technology*, Karlsruhe, Germany, 22-26 August 1994 (Elsevier, Amsterdam, 1995), Vol. 1, p. 799.
[6] Tregubova E., *et al.*, *Proc. 30th EPS*, St. Petersburg, Russia, *to be published*.
[7] Shoucri M. and Shkarovsky I., Comp.Phys.Comm. **82**, 287 (1994).
[8] D. Chenna Reddy and T. Edlington, 1996 Rev. Sci. Instrum. **67** (2), p. 462
[9] V. Shevchenko, *Proc. 28th EPS*, Funchal, Madeira, Portugal, 2001, paper P3.101.
[10] V. Shevchenko, *Proc. 27th EPS*, Budapest, Hungary, 2000, paper P3.120.
[11] V. Shevchenko, 2000 Plasma Phys. Reports, Vol. **26**, No 12, p.1000

Electron Bernstein Wave Experiments in the MST Reversed Field Pinch

J.K. Anderson*, M. Cengher*, P.K. Chattopadhyay*, C.B. Forest*,
M. Carter[†], R.W. Harvey[‡], R.I. Pinsker[§], and A.P. Smirnov[¶]

*Department of Physics, University of Wisconsin, Madison WI 53706, USA
[†]Oak Ridge National Laboratory, Oak Ridge, Tennessee, USA
[‡]CompX, Del Mar, California 92186, USA
[§]General Atomics, San Diego, California 92014, USA
[¶]Moscow State University, Moscow, RUSSIA

Abstract. A system to heat electrons in the Madison Symmetric Torus through the electron Bernstein wave is currently being developed. This is an attractive heating scheme for the overdense reversed field pinch plasma, where electron cyclotron heating and current drive are inaccessible. Low power experiments (~ 1 watt) have shown that a significant fraction of launched electromagnetic power successfully couples to the electron Bernstein wave. Furthermore, these experiments have found an optimized launch with finite $n_\perp$. Initial results from experiments at moderate power (~ 150 kW for several milliseconds, driven by a pair of S-band traveling wave tube amplifiers) are presented.

INTRODUCTION

Electron Bernstein waves (EBWs) show promise for providing localized heating and current drive in over-dense plasmas, found for example in the reversed field pinch (RFP) or spherical torus. [Forest et al., 2000, Ram and Schultz, 2000]. Current profile control may be necessary for stabilization of MHD activity in these configurations and is difficult by other means. A major technical difficulty is developing a robust technique for coupling high levels of power to the EBW from external antennas. Here we report on the status of experimental EBW research on the Madison Symmetric Torus RFP on two primary fronts: low power (~ 1 W) coupling studies and moderate power (~ 100 kW) heating experiments.

Two scenarios have been proposed for converting external electromagnetic waves to the EBW. Preinhalter and Kopecky suggested heating based on OXB conversion: [Preinhaelter and Kopecký, 1973] an O-mode launched at the plasma edge is converted to the slow X-mode (at $\omega = \omega_{pe}$) which propagates to the upper hybrid resonance and is fully converted to the Bernstein mode. This in turn propagates inward to the cyclotron resonance where the power is strongly absorbed. The magnetic field strength in MST places the first-harmonic cyclotron frequency in

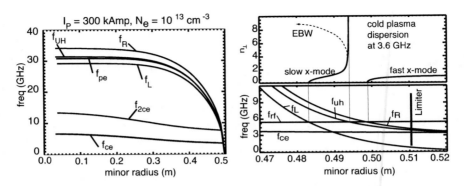

Figure 1: Cutoff and resonance frequencies in ECRF for $I_P = 300$ kA, n_e(central)= 10^{19} m^{-3} MST plasmas. The equilibrium is a solution of the Grad-Shafranov equation fit to the available magnetic data.

the range of about 3 to 8 GHz depending on plasma current and minor radius, and as such characteristic wavelengths are comparable to existing ports in the machine. The important frequencies versus minor radius in the MST are shown in Fig. 1 along with an expanded view in the mode conversion layer of the edge region.

Previous heating experiments have utilized this scheme[Laqua et al., 1998], although direct launch of the X-mode also appears attractive for the RFP where the density gradient scale length is less than the wavelength. [Ram and Schultz, 2000] A perpendicularly launched X-mode tunnels through a cutoff region to the slow branch of the X-mode, and then converts to the EBW at the upper hybrid resonance. Work by Pinsker and Carter [Pinsker et al., 2001] shows a finite $n_\perp$ optimizes coupling of the EBW; both XB and OXB require a steerable antenna for peak efficiency.

Implementation of a steerable optical antenna is difficult in the MST due to the close fitting conducting shell. The scheme under consideration is to use a phased array of waveguides for steering the launched power. This near field grill coupling scheme is not unlike that used for launching lower hybrid power in tokamaks. The primary goals are to test the existing coupling theories and to demonstrate heating in the MST.

EBW Coupling Studies

Thermal levels of emission in the ECRF have been observed from the overdense core of MST.[Chattopadhyay et al., 2002] These measurements confirm mode conversion from the EBW to electromagnetic waves at the edge, and by reciprocity establish the viability of an EBW heating scheme. An effort to experimentally understand the mode conversion process in the RFP is underway.

Coupling studies are performed by varying $n_\perp$ of the launched wave (controlled by introducing a relative phase difference between two adjacent waveguides of the antenna shown in Fig. 2. A 3.6 GHz oscillator drives a traveling wave tube (TWT) amplifier with a CW output on the order of 5 W. The power is divided into two

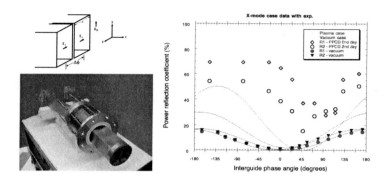

Figure 2: (Upper left: Geometry for two waveguide grill used to test direct launch of EBWs in X-mode polarization. Lower left: photograph of antenna. Right: Experimentally measured reflection coefficients in each waveguide and predictions from from the coupling theory.

arms, one of which has an adjustable (mechanical) phase shifter. The forward and reflected power are measured from the samples of bidirectional couplers in each arm. Figure 2 includes comparisons between data and theory for coupling in vacuum and in plasma for a complete interguide phase scan of the launched X mode. The experiment well matches the theory for the vacuum case, and data for plasma coupling reproduce an important feature from the theory: there is an optimum interguide phasing ($n_\perp$ of the launched wave) where the total reflection reaches a minimum. Further coupling experiments are planned including direct measurements of the edge density profile and the ability to vary the polarization of the launched wave continuously from X-mode to O-mode.

EBW Heating Studies

Initial experiments at about 100 kW through the twin waveguide antenna are underway. A pair of S band traveling wave tube amplifiers, each with a maximum output of about 75 kW, is pulsed for about 4 milliseconds. The power is delivered to the plasma through a very simple transmission line: two lengths of flexible waveguide envelop an isolator and a bidirectional coupler in each arm. A quartz window separates the antenna (under high vacuum) and the rest of the transmission line (at atmospheric pressure). Figure 3 shows a case with good coupling at $n_\perp \sim$

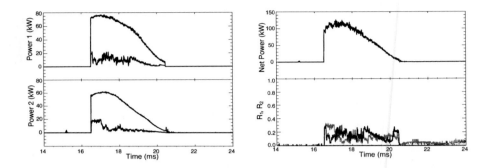

Figure 3: High power experiment. The left hand plots show the forward and reflected power in each arm of the antenna, and the top right plot shows the total net power delivered to the plasma boundary. The lower right plot shows the reflection coefficients for each arm; this particular test (at $n_\perp \sim 0.5$) shows good coupling: $R < 25\%$.

0.5. The forward and reflected power in each arm are shown along with the net power (forward - reflected). The primary reportable result is successful delivery of high power (over 100 kW) to the plasma boundary. Any heating effects that may have occurred during the preliminary studies went unobserved due to insufficient diagnostics. A thorough assessment of diagnostic needs and an extension to 360 kW (6 tubes) for a flat 10 millisecond pulse are planned for the immediate future.

Summary

Electron Bernstein waves make accessible heating and current drive in the ECRF for the overdense reversed field pinch plasma. Experiments aimed at measuring coupling properties of the EBW (low power) and initial heating experiments (moderate power) are underway on the Madison Symmetric Torus. Coupling experiments have found an optimum launch angle, and 100+ kW are available for heating tests. This work is supported by USDOE.

References Cited

P.K. Chattopadhyay, J.K. Anderson, T. Biewer, D. Craig, C.B. Forest, R.W. Harvey, and A.P. Smirnov. Electron bernstein wave emission from an over-dense reversed field pinch plasma. *Phys. Plasmas*, 9:752, 2002.

C. B. Forest, P. K. Chattopadhyay, R. W. Harvey, and A. P. Smirnov. Off-midplane launch of electron bernstein waves for current driven in overdense plasmas. *Phys. Plasmas*, 7:1352, 2000.

H. P. Laqua, H. J. Hartfuß, and W7-AS Team. Electron Bernstein wave emission from an overdense plasma at the W7–AS stellarator. *Phys. Rev. Lett.*, 81:2060, 1998.

R.I. Pinsker, M.D. Carter, C.B. Forest, and P.K. Chattopadhyay. Coupling to the electron Bernstein wave with waveguide antennas. In *Bull. of American Phys. Soc.*, volume 46, page 302, Long Beach, CA, 2001.

J. Preinhaelter and V. Kopecký. Penetration of high-frequency waves into a weakly inhomogeneous magnetized plasma at oblique incidence and their transformation to Bernstein modes. *J. Plasma Physics*, 1973.

A.K. Ram and S. D. Schultz. Excitation, propagation, and damping of electron Bernstein waves in tokmaks. *Phys. Plasmas*, 7:4024, 2000.

Electron Bernstein Wave Applications for Stellarators

M. D. Carter*, D. A. Rasmussen*, G. L. Bell*, T. S. Bigelow* and J. B. Wilgen*

Oak Ridge National Laboratory, Oak Ridge, Tennessee

Abstract.
Electron Bernstein Waves (EBW) provide a mechanism to heat electrons in over-dense plasma using microwave sources. The mode conversion between ordinary or extra-ordinary waves and EBW depends on density gradients in the plasma edge, the angle of incidence, and the polarization of the incoming waves. General considerations for the efficiency of mode conversion to EBW in stellarator geometries are discussed. For cold, dense edge-plasma conditions, Coulomb collisions can cause parasitic absorption in the initial propagation region of the EBW, but collisional damping should not be a problem for the sharp edge gradients and temperatures $\gtrsim 50$ eV in the Quasi-Poloidal Stellarator (QPS). Scenarios for using fixed launch angles with polarization optimization to maximize transmission during the density evolution in QPS are possible.

INTRODUCTION

The Quasi-Poloidal Stellarator (QPS) [1] is a compact, 2 field period, optimized stellarator configuration as shown in Fig. 1. The magnetic field variation for a "bagel-like" cut through the device is shown in the right hand graph of Fig. 1. The QPS plasma parameters are expected to be over-dense, preventing the use of traditional microwave coupling, and purpose of this paper is to analyze possible schemes for heating these over-dense plasmas by mode-conversion processes to electron Bernstein waves (EBW) near the upper hybrid resonance in the plasma edge.

For linear polarizations, two physically distinct mode-conversion processes are possible. The first, "direct X-B", converts incoming X-mode microwaves into EBW via a tunneling process. The second, "O-X-B", converts incoming O-mode microwaves into outgoing X-modes that are converted to inwardly propagating EBW upon reflection at the upper hybrid resonance. For "O-X-B" conversion, the optimum polarization for incoming waves is actually mixed; hence, the desired polarization for the incoming waves is actually "quasi-O". Typically, one or the other of "direct X-B" or "O-X-B" provides the highest mode conversion efficiency depending on the plasma parameters. For low frequency, very over-dense applications, the direct "X-B" scheme is preferred. However, for the frequency ($\gtrsim$ 28GHz) and density parameters expected for QPS, the "quasi-O-X-B" mode conversion mechanism provides the most efficient scheme, and conversion efficiencies can approach 100% if the polarization and launch angle are properly adjusted. In this paper, we consider mode-conversion in the edge, and assume the launch location can be chosen to provide EBW ray trajectories to heat the core-plasma.

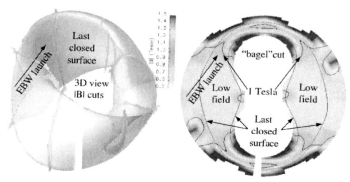

FIGURE 1. Shading maps the magnetic field strength for toroidal cuts in 3D, and a "bagel-like" cut through the equatorial plane of QPS. Contours of the last closed flux surface and the 1 Tesla locations are shown along with the intended scheme for electron Bernstein launch.

COLLISIONAL EDGE DAMPING OF EBW

The GLOSI warm plasma code[2] is used to study the mode conversion layer at the upper-hybrid resonance and the initial propagation of the EBW. Collisional damping of the short wavelength EBW after mode conversion is a concern when considering the antenna and launch scenario, as shown in Fig. 2. In some devices, limiting structures near the beam entry to the plasma are needed to provide the required density gradients for effective mode conversion, and these limiting structures can lead to recycled gas in the region of the launcher. This gas can lead to damping of the EBW in the edge region after mode conversion as shown in the left graph of Fig. 2. However, the vacuum vessel and field line structure for QPS is expected to provide short density gradient scale-lengths without any limiting structures. Thus, for QPS, it is expected that microwaves can be beamed in from a launcher located in a vacuum region having very low neutral density, and Coulomb collisions will provide the only significant source of EBW damping in the edge. The graph on the right in Fig. 2 shows that edge temperatures need to rise above ~ 50 eV as the EBW enter the plasma to prevent significant collisional damping in the edge of QPS. However, temperature profiles in QPS are such that collisional damping of the EBW in the edge is not expected be an issue so long as the upper hybrid resonance for mode conversion of the microwave beam does not occur in a cold ergodic region.

OPTIMIZED TRANSMISSION

At a plasma/vacuum interface at $x = 0$ with plasma on the left and vacuum on the right, the wave field components can be written as

$$\vec{E}^{mn}(x) = \vec{E}_l^{mn} e^{-ikx} + \vec{E}_r^{mn} e^{+ikx} \quad (1)$$

for left (subscript l) and right-going (subscript r) plane waves in vacuum, and similarly for the RF magnetic field, $\vec{B}$.

Only the impedance relationship between tangential components at $x = 0$ is needed to determine the reflection and transmission properties of the plasma. We again use

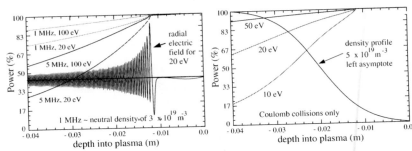

FIGURE 2. Curves show the power remaining as EBW propagate into the plasma from the right for various neutral (left) and Coulomb (right) collisional parameters. Low electron temperatures lead to strong damping because of short EBW wavelengths and also higher Coulomb collision rates. The calculations were performed using a hyperbolic tangent profile as show in the graph on the right.

the GLOSI[2] code to calculate this 2×2 plasma surface impedance for each Fourier mode, $\widetilde{Z}^{mn}$, such that $\vec{E}_t^{mn}(x=0) = \widetilde{Z}^{mn} \cdot \vec{B}_t^{mn}(x=0)$, where the t subscript denotes the tangential components. Each m and n Fourier component represents a launch angle relative to the plane of the interface with n denoting wave-numbers parallel to the static magnetic field, and m denoting wave-numbers perpendicular to both the static magnetic field and the plasma gradients. Using $\widetilde{Z}^{mn}$ with the vacuum dispersion relation for left and right going waves, allows directional impedances to be generated

$$\vec{E}_{tl}^{mn} = \widetilde{Z}_{El}^{mn} \cdot \vec{B}_{tl}^{mn} \quad ; \quad \vec{E}_{tr}^{mn} = \widetilde{Z}_{Er}^{mn} \cdot \vec{B}_{tr}^{mn} \quad ; \quad \vec{B}_{tr}^{mn} = \widetilde{Z}_{Br}^{mn} \cdot \vec{B}_{tl}^{mn}. \tag{2}$$

From these directional impedance relationships, the transmission, T, can be calculated for any polarization and launch angle from Poynting's theorem on the vacuum side of $x=0$,

$$T = 1 - \frac{\text{Real}(\vec{E}_{tr}^{mn} \times \vec{B}_{tr}^{mn*})}{\text{Real}(\vec{E}_{tl}^{mn} \times \vec{B}_{tl}^{mn*})} \tag{3}$$

To find the optimum polarization at each launch angle, the OPTPOL code is used to scan both polarization angles (relative amplitudes of incoming tangential components and their phase), calculate T from Eq. 3 for each polarization, and record the polarization producing maximum transmission. Results showing the maximum mode-conversion efficiency for various density values above X and O-mode cutoff are shown in Fig. 3 for 28 GHz launch. Fig. 4 shows the polarization angle required to obtain the optimum transmission shown in Fig. 3 for a left-hand-density value of $5 \times 10^{19} \text{m}^{-3}$. The polarization angles are described in terms of the microwave magnetic field components such that $B_\parallel = \cos(\alpha)$ and $B_\perp = \sin(\alpha) e^{i\beta}$, and the tangential index of refraction used for axes in the figures can be converted to a launch angle for the incoming microwave beam.

CONCLUSIONS

EBW can be used to heat over-dense plasma in QPS with polarization control if the density gradient scale-lengths are 0.01 m, and the core density is $\gtrsim 3$ times cutoff ($\gtrsim 1.5 \times 10^{19} \text{m}^{-3}$ for 28 GHz). A crude estimate for 53 GHz mode conversion based

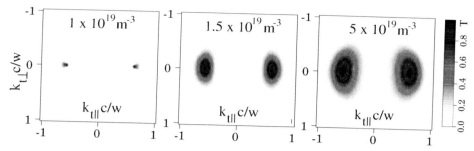

FIGURE 3. Transmission for varying plasma densities shows that efficient conversion can be obtained if the density is high enough on the plasma-core side. The horizontal axis is the index of refraction parallel to the static field in vacuum. The vertical axis in the vacuum index of refraction in the direction perpendicular to both the static magnetic field and the plasma density gradient.

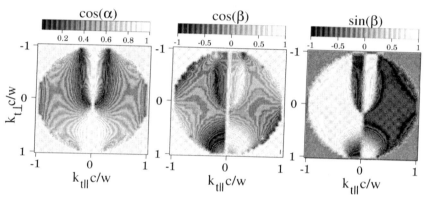

FIGURE 4. The polarization required for optimum transmission at any launch angle is somewhat circularly polarized for "quasi O-X-B" mode conversion with core densities well above cut-off. The left-hand-density value was $5 \times 10^{19} m^{-3}$ for this calculation.

on these results indicate that core plasma densities of $\gtrsim 2 \times 10^{19} m^{-3}$ will be needed for EBW launch. Thus, a density gap exists between the cut-off for cold-plasma microwaves and that required to efficiently mode convert to EBW, and microwave heating is expected to perform poorly in this gap. Methods to "leap-frog" over the density gap using multi-frequency microwave sources or ohmic heating are possible for QPS. Collisional damping of the EBW in the edge after mode conversion is not expected to be a problem for QPS for edge electron temperatures above $\gtrsim 50$ eV.

ACKNOWLEDGMENTS

Research was sponsored in part by Oak Ridge National Laboratory, managed by UT-Battelle, LLC, for the U.S. Dept. of Energy under contract DE-AC05-00OR22725.

REFERENCES

1. D. A. Spong, et al., *Nuclear Fusion* **41**, 711 (2001)
2. C. Y. Wang, et al., *Phys. Plasmas* **2**, 2760 (1995)

General Properties of Scattering Matrix for Mode Conversion Process between B Waves and External EM Waves and Their Consequence to Experiments

T. Maekawa, H. Tanaka, M. Uchida and H. Igami

Graduate School of Energy Science, Kyoto University, Kyoto 606-8502, Japan

Abstract. General properties of scattering matrix, which governs the mode conversion process between electron Bernstein (B) waves and external electromagnetic (EM) waves in the presence of steep density gradient, are theoretically analyzed. Based on the analysis, polarization adjustment of incident EM waves for optimal mode conversion to B waves is possible and effective for a range of density gradient near the upper hybrid resonance, which are not covered by the previously proposed schemes of perpendicular injection of X mode [1] and oblique injection of O mode [2]. Furthermore, the analysis shows that the polarization of the externally emitted EM waves from B waves is uniquely related to the optimized polarization of incident EM waves for B wave heating and that the mode conversion rate is the same for the both processes of emission and the injection with the optimized polarization.

CONFIGURATION AND GENERAL ANALYSIS

We consider a plane slab geometry where the plasma is inhomogeneous only along the x direction with the surface at x=0 as shown in Fig.1. The electron density increases with x, and the external magnetic field, B0, is along the z direction. When a plane EM wave is injected to the plasma, a part of the wave power reaches the UHR layer, is mode-converted to a B wave and leaves the UHR layer, and the rest power is reflected back toward the vacuum free space. In another process of a plane B wave coming to the UHR layer, a part of the wave power is reflected as a B wave and the rest power reaches the plasma surface and is emitted to the free space as an EM wave. In both cases, the parallel component of wave number to the magnetic field, $k_{//}$, is conserved, since the system is homogeneous along z direction. The y component, k_y, is also conserved, and here we consider the case of $k_y=0$ for simplicity. We use the convention of $\exp(-i\omega)$ for the time variation.

Using the wave number of the free space, $k_0 = \omega/c$, we define two linearly polarized modes as $e1(z') = \exp(ik_0 z' - i\omega t)\hat{x}'$, and $e2(z') = \exp(ik_0 z' - i\omega t)\hat{y}'$ for incoming EM waves to the UHR layer (see Fig.1). We also define two linearly polarized modes as $f1(z'') = \exp(ik_0 z'' - i\omega t)\hat{x}''$, and $f2(z'') = \exp(ik_0 z'' - i\omega t)\hat{y}''$ for out going EM waves from the UHR layer. Any incoming EM wave is expressed as a linear combination of e1 and e2 as $E = a1 e1 + a2 e2$, and its time averaged power flux is given by the inner product, $EE^* = a1 a1^* + a2 a2^*$, except for a constant multiplier, where a1 and a2 are the complex amplitudes of e1 and e2, respectively, and asterisk denotes complex conjugate. Here, e1 and e2 constitute a power orthogonal base ($e1 e2^* = 0$, $e1 e1^* = e2 e2^* = 1$). This is the same for f1 and f2 for outgoing EM waves. We can introduce basic modes for B waves as well and denote e3 for the incoming mode to the UHR layer and f3 for the outgoing mode. When incoming waves towards the

UHR layer are $a1e1+a2e2$ for EM waves and $a3e3$ for B waves, outgoing waves are written as $b1f1+b2f2$ for EM waves and $b3f3$ for B waves and following linear relationship,

$$\begin{pmatrix} b1 \\ b2 \\ b3 \end{pmatrix} = \begin{pmatrix} S11 & S12 & S13 \\ S21 & S22 & S23 \\ S31 & S32 & S33 \end{pmatrix} \begin{pmatrix} a1 \\ a2 \\ a3 \end{pmatrix} \tag{1}$$

or $b=Sa$ in matrix notation, holds. Since total power flux is always conserved for any combinations of a and b, that is, $aa^*=bb^*$, S is an unitary matrix ($S^*S^t = I$, where superscript t denotes transposed matrix and I is the unit matrix).

If we denote the wave solution for Eq.(1) by $E(r,t)$ and $B(r,t)$, their time reversal field, $E(r,t)=E(r,-t)$ and $B(r,t)=-B(r,-t)$, is another wave solution for the same plasma system with reversed external magnetic field of $B0=-B0$. By rotating the total system including new wave field by 180 degree around the x axis in Fig.1, the plasma system becomes the same as the original one including the direction of external magnetic field, and therefore, we can use S in Eq.(1) for the rotated $E(r,t)$ field. We apply above procedure to the original outgoing EM waves, $b1f1$ and $b2f2$. First they become, $b1\exp(ik0z''+i\omega t)\hat{x}''$ and $b2\exp(ik0z''+i\omega t)\hat{y}''$ by time reversal, and then, $b1\exp(-ik0z'+i\omega t)\hat{x}'$ and $-b2\exp(-ik0z'+i\omega t)\hat{y}'$ by the spatial rotation, and finally, $b1^*\exp(ik0z'-i\omega t)\hat{x}'$ and $-b2^*\exp(ik0z'-i\omega t)\hat{y}'$ by taking complex conjugate to return the time variation back to the original convention. By doing the same procedure to the others, it turns out that $b1^*e1-b2^*e2$ and $b3^*e3$ constitute incoming waves, and $a1^*f1-a2^*f2$ and $a3^*f3$ outgoing waves for the new wave field as,

$$\begin{pmatrix} a1^* \\ -a2^* \\ a3^* \end{pmatrix} = \begin{pmatrix} S11 & S12 & S13 \\ S21 & S22 & S23 \\ S31 & S32 & S33 \end{pmatrix} \begin{pmatrix} b1^* \\ -b2^* \\ b3^* \end{pmatrix} \tag{2}$$

This is rewritten as $a=S_{tr}b$, where

$$S_{tr} = \begin{pmatrix} S11^* & -S12^* & S13^* \\ -S21^* & S22^* & -S23^* \\ S31^* & -S32^* & S33^* \end{pmatrix} \tag{3}$$

and we know that S_{tr} is another inverse matrix of S, that is, $S_{tr}=S^{-1}=S^{*t}$ from Eq.(1). Therefore, the following relations hold,

$$S21=-S12, S32=-S23 \text{ and } S31=S13.$$

Outgoing EM wave generated from the incoming B wave of e3 is given by Eq.(1) as,

$$E_{EBE} = S13f1 + S23f2 \tag{4}$$

If we write as $S13=r1\exp(i\theta 1)$, $S23=r2\exp(i\theta 2)$, power mode-conversion rate to EM wave is $T_{EBE}=E_{EBE}E_{EBE}^*=r1^2+r2^2$. On the other hand, if we suppose an incident EM wave of which power flux is unity, $E_{inc}=a1e1+a2e2$ ($a1a1^*+a2a2^*=1$), then B wave generated from E_{inc} is written as $E_B = (a1S31+a2S32)f3 = (a1S13-a2S23)f3$. E_B attains its maximum for the following combination of $a1=r1/(r1^2+r2^2)^{1/2}$, $a2=-r2\exp\{i(\theta 1-\theta 2)\}/(r1^2+r2^2)^{1/2}$, that is, when

$$E_{inc} = (S13^*e1 - S23^*e2)\exp(i\theta 1)/(r1^2+r2^2)^{1/2} \tag{5}$$

and the optimal mode-conversion rate is given by $T = T_{max} = r1^2 + r2^2 = T_{EBE}$, exactly the same value as EBE mode-conversion rate.

From the unitary condition ($SS^{*t}=I$), the absolute values and relative phase of the matrix element S13 and S23 can be written in terms of the matrix elements concerning the EM waves alone, S(EM)=(S11,S12,S21,S22), as follows,

$$r1 = sqrt\,(1 - S11S11^* - S12S12^*)$$
$$r2 = sqrt\,(1 - S21S21^* - S22S22^*) \qquad (6)$$
$$exp\{i(\theta 1 - \theta 2)\} = -(S11S21^* + S12S22^*)\,/\,r1r2$$

Thus, we can obtain T_{EBE} and also optimize the polarization of incident EM wave by using S(EM) alone.

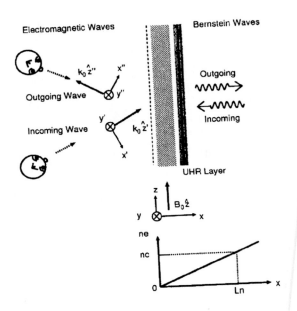

FIGURE 1. Plasma slab and coordinate

AN EXAMPLE OF NUMERICAL RESULTS

For numerical analysis we employ a simple plasma profile where external magnetic field is homogeneous with $B=B0z$ and the electron density linearly increases and reaches a flat region far beyond the plasma cutoff point at $x=Ln$ as shown in Fig.1. Here, Ln denotes the scale length of the density gradient. For these plasmas we have obtained the matrix S(EM) by using the cold plasma resonance absorption model for EM waves [3].

Figure 2(a) shows the emission rate T_{EBE} as a function of parallel refractive index to the external magnetic field, $N//=k///k0$ for the case of a density gradient of $Ln/\lambda 0=0.25$. For reference each fraction of the linearly polarized components of f1 and f2 is also plotted. For this steep density gradient case, T_{EBE} is high for a wide range of $N//=0-0.8$. The polarization of EBE wave at $N//=0.475$, where $T_{EBE} =1$, is shown in Fig.2(b). It is plotted on the x''y'' plane viewed from the vacuum side as shown in Fig.1. That for the reverse emission case of $N//=-0.475$ is shown in Fig.2(d), which is the mirror image of Fig.2(b) with a flat mirror on the xy plane. It is noted that in both cases, the rotation direction of wave electric field is left-hand to the external magnetic field. Polarization of the optimized wave injected with $N_{//}=0.475$ is plotted in Fig.2(c), which is plotted on the x'y' plane view from the vacuum side. Two polarization in Figs.2(c) and (d) have the same elongation and rotation direction, but their ellipses are mirror images each other with a mirror on the xz plane.

All above relationship between each polarization can be shown analytically from Eqs.(4)-(6) regardless the values of $N_{//}$ and Ω/ω, and therefore, holds always independently on individual profiles of the density and magnetic field along x. Of particular importance is that between Figs2.(c) and (d). This result implies that we can adjust the polarization of incident waves for optimal mode-conversion to B waves for the heating experiments by measuring the EBE wave polarization at the same angle of injection of emission. Especially when $T_{EBE}=Tmax=1$ we can explain this relationship using the principles of time reversal invariance and symmetry of mirror reflection for the physical systems. Namely, we suppose that the optimized incident wave is fully mode-converted to B wave and the reflection wave can be neglected. Then, its time-reversed wave solution represents that incoming B wave is fully mood-converted to the EBE wave with the same polarization as that in Fig.2(c) with only its rotation direction reversed. Since the external magnetic field is reversed for time reversal solution, we turn back the field direction by taking mirror reflection for the total system including the EBE wave with a mirror on the xz surface. Thus, the resulted polarization of EBE wave becomes exactly the same one as in Fig.2(d).

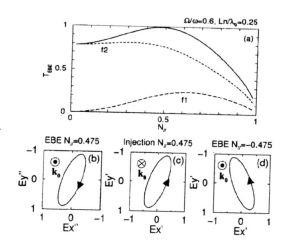

FIGURE 2. Mode conversion rate (transparency) of EBE, and fractions of linearly polarized waves in x'' direction (f1) and in y'' direction (f2), as functions of $N_{//}$. (b)-(d): Polarizations (loci of tip of wave electric field vector as time goes) for (b) EBE wave, (c) incident EM wave optimized for mode conversion and (d) EBE wave in reversed direction.

REFERENCES

1. Ram, A.K. and Schults, S.D., Phys. Plasmas **7**, 4084 –4094 (2000)
2. Cairns, R.A. and Lashmore-Davies, C.N., Phys. Plasmas **7**, 4126-4134 (2000).
3. Stix, T.H. "Waves in Plasmas", American Institute of Physics, New York, 1992, Chapter 13.

Study of electron Bernstein wave absorption using VORPAL

Chet Nieter*, John R. Cary*, Robert W. Harvey[†], R. Dominguez[†] and A.P. Smirov[†]

*University of Colorado, Boulder CO 80309
[†]CompX, Del Mar CA

Abstract. In certain toroidal devices, the electron cyclotron frequency may smaller than the electron plasma frequency. Since electron cyclotron waves can not propagate at frequencies below the plasma frequency, electron-Bernstein waves provide an alternative method of RF heating at the electron cyclotron frequency. We have started preliminary runs of electron Bernstein wave absorption using the versatile plasma simulation code VORPAL. VORPAL allows the study of wave absorption from first principles using a PIC model for the plasma electrons and a full finite difference Maxwell solver for the electromagnetic fields. We will present simulations for parameters typical for spherical tokamaks such as NSTX. Possible additions to code which will expand the parameter space that can be studied and allow for study of current drive will be discussed.

THE VORPAL FRAMEWORK

The VORPAL code framework make strong use of object oriented design to provide a variety of features included the ability to set the dimension of the simulation at run time, a general domain decomposition, and the the availability of multiple models for both the electromagnetic fields and the plasma. A full description of the code and the methods used in its development can be found in reference [1]. Online documentation for the code can be found at [2].

Multi-Dimensional

By templating most of the classes in the VORPAL libraries over dimension and floattype we are able to support simulations of one, two, and three dimensions and of either float or double precision with one code base. One of the principal challenges in writing a multi-dimensional code is the field updates. Normally the fields would be stored as multi-dimensional arrays and the updates would be done with nested loops. This can not be done for an arbitrary dimensional code since the dimension and hence the number of nested loops is not known at the outset.

To solve the problem of field updates, a combination of recursion and template metaprogramming is used. The field data is stored internally in a one dimensional array. A generalization of an iterator is used which understands how the one dimensional index of the field data relates to the corresponding position in the multi-dimensional grid.

These iterators have methods to bump their position on the grid in any direction. Updater classes are then created which consist of a collection of iterators, usually representing a stencil for a finite difference approximation of the relevant differential equation. These updater classes are then "walked" across the grid by what are referred to as walker classes. The walker classes are templated over dimension and they call the next lowest dimension recursively. The walker class for the lowest dimension is then specialized to actually perform the update.

Domain Decomposition

VORPAL is designed to run as both a serial code on stand alone workstations and as a parallel code on systems that run MPI. As part of our general philosophy of flexibility, we have developed a general three dimensional domain decomposition for VORPAL. Any decomposition that can be constructed from a collection of bricks is possible. With this general domain decomposition static load balancing can be done and dynamic load balancing is in the final stages of development. To minimize overhead, VORPAL overlaps communication with computation as much as possible.

To achieve this general domain decomposition, we use the idea of a slab. A slab is an object whose sides are straight lines and whose corners are all right angles. So in one dimension a slab is a straight line, in two dimensions a slab is a rectangle and in three dimensions a slab is a brick. The intersection of any two slabs of the same dimension is another slab. Each domain is described by two slabs, one which is referred to as the physical region is all the cells for which the processor is responsible for updating. The other, which is referred to as the extended region is the physical region plus one layer of guard cells in each direction. The cells that a domain needs to send to a neighboring processor is simply the slab that is the intersection of the sending processors physical region with the receiving processors extended region.

Available Models

VORPAL flexible object oriented framework makes it possible to have multiple models for both the electromagnetic fields and the plasma in the same code. By interacting with the rest of the code through standard interfaces, different models can be used depending on the relevant physics being studied. Hybrid simulations can be done by using multiple models to represent different regions of the plasma or different species.

VORPAL has a finite difference finite time domain Maxwell solver for the electromagnetic fields based off the Yee mesh. Externally applied fields can be modeled using classes that produced profiles that vary in time or space according to some functional description. The Yee field can be combined with any number of externally applied fields. An electrostatic solver is currently under development.

The plasma particles can currently be modeled with either a cold fluid model or a particle-in-cell (PIC) model. The cold fluid model directly advects the fluid velocity rather than using flux transport to find the fluid momentum. This allows for regions of zero plasma density. The PIC model uses policy classes to provide for a variety of different particle dynamics, including a full relativistic Boris push, a non-relativistic electrostatic push, and free streaming particles.

SIMULATION OF ELECTRON BERNSTEIN WAVES

We have begun looking at the possibility of using VORPAL to study electron Bernstein (EB) wave generation and absorption in fusion devices. Much of the current computational work in this area uses ray tracing codes that use linear or quasi-linear approximations to model the EB waves. When the EB waves are absorbed at the cyclotron resonance, their group velocity goes to zero resulting in a build up of field energy at the resonance layer. Direct numerical simulations of this process using VORPAL would explore such non-linear regimes. Since VORPAL uses a full Maxwell solver for the electromagnetic fields, it could also simulate the generation of EB waves from mode conversion at the plasma interface.

Our ultimate goal is to carry out fully 3D simulations that will allow us to calculate coupling and heating efficiency for electron Bernstein modes in toroidal devices, such as NSTX and MST. Typical parameters would be minor radius of 0.6 meters, major radius of 1 meter, resonant magnetic field of 0.3 T, and so electron cyclotron frequency of 8.4 GHz. The electron temperature might vary from tens of eV at the edge to a few KeV at resonance. The density in the regime noted above would be such that the electron plasma frequency is some 2-3 times the resonant electron cyclotron frequency. Our estimates of the electric field from proposed MST experiments are in the range of a few 100 KeV/m.

We demonstrate VORPAL's ability to address these situations by performing an exploratory simulation of electron Bernstein wave generation and propagation. We consider the situation of a wave guide or antenna at the edge of a toroidal device generating radio frequency radiation that propagates into the plasma. We used slab geometry with x being the direction into the plasma, y being the toroidal direction (the direction of the magnetic field), and z being the poloidal direction, the direction of symmetry.

To reduce the computational requirements further, we simulated only the outer 18 cm of plasma. We simulated 37.5 cm of the toroidal direction. The magnetic field was set to a constant 0.12 T, corresponding to an electron cyclotron frequency of 3.4 GHz. Then 4 KeV electrons have a thermal gyro-radius of 1.76 mm. In keeping with earlier statements, the central plasma density was chosen so the electron plasma frequency is three times the electron cyclotron. This gives $1.9 \times 10^{18} \ m^{-3}$. We ramped up the plasma from zero density out to 2 cm from the edge to full density over the next 1 cm. To resolve the wave length of the EB wave and the antenna we use 400 cells in the x-direction and 120 in the y-direction. The Courant condition gives us a time step of 1.4 ps and we run for 2250 time steps so the waves can propagate the length of the simulation. This simulation takes about two hours on a Linux work station, so larger regions and 3D simulations are certainly possible on parallel machines. In Fig. 1 shows the x-component of the electric field after 2150 time steps. We see waves propagating into the plasma with a wavelength on the order of the electron gyro-radius and a group velocity on the order of the electron thermal velocity, which agrees with the dispersion relation for EB waves.

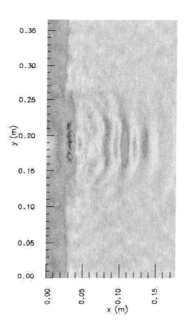

FIGURE 1. Contour plot of the x-component of the electric field of the electron Bernstein wave simulation.

CONCLUSIONS AND OUTLOOK

We have simulated electron Bernstein wave generation using VORPAL. However we had to use field strengths which are several orders of magnitude higher that would normally be used in experiments. This was done to over come the noise generated by using a PIC model for the plasma. We are planning on adding a warm fluid model to VORPAL to allow us to simulate the generation and propagation of EB. To study the resonant absorption of the waves, we are looking at combining the warm fluid model with the PIC model to perform δ-f simulations of the plasma. These methods should be ideal for the problem since one expects the distribution to be Maxwellian with some small deviation due to heating.

REFERENCES

1. Nieter, C., and Cary, J. R., Vorpal: a versatile plasma simulation code, to be published in J. Comput. Phys. (2003).
2. Cary, J. R., and Nieter, C., http://www-beams.colorado.edu/vorpal/, accessed May 2003 (2002).

Modeling of EBW Coupling with Waveguide Launchers for NSTX

R.I. Pinsker, G. Taylor,[a] and P.C. Efthimion[a]

General Atomics, P.O. Box 85608, San Diego, California 92186-5608
[a] *Princeton Plasma Physics Laboratory, Princeton, NJ 08543*

Abstract. The theory originally developed for the MST EBW waveguide coupling experiments [1] is used to model coupling to the EBW in the National Spherical Torus eXperiment (NSTX).

Numerous recent papers have described the coupling between electron Bernstein waves (EBWs) in the plasma interior and waves that can propagate in vacuum (O- and X-modes). This topic is of interest both for electron cyclotron emission diagnostics in overdense plasmas and for possible high-power heating methods appropriate to such overdense plasmas as are characteristic of reversed-field pinches and spherical torii. In this paper, we apply the theory developed for the Madison Symmetric Torus (MST) EBW waveguide coupling experiments [1,2] to the case of NSTX, using parameters for NSTX taken from Ref. [3].

The coupling theory developed in Refs. [1] and [2] is an application of the Brambilla "grill" theory in the lower hybrid range of frequencies [4] to the range of electron cyclotron harmonics. A weakly collisional cold-plasma model is used. Just as in the theory of Ram, et al. [5], the resulting reflection coefficients are nearly identical to those obtained with a full kinetic plasma model, in which the EBW appears explicitly. The difference is that the energy that is absorbed near the upper hybrid resonance (UHR) layer by the weak collisions in the cold plasma model is instead converted to inward-propagating EBWs in the full kinetic description. This has been checked by comparing the surface admittances obtained using this cold plasma model with those computed by the GLOSI code [6], which includes the lowest order EBW in the model. Excellent agreement is found in all cases, verifying the basic assumption of the model.

The program embodying this model has two parts: 1) the calculation of the surface admittances as functions of n_y and n_z, and 2) the calculation of all of the integrals convolving these admittances with the appropriate spectral functions for a phased array of identical rectangular waveguides operating in the fundamental mode. The results of those integrals then yield the reflection coefficients and the amplitudes of the first few evanescent modes in each of the waveguides in the array for an arbitrary excitation (phasing). First, we check the analytic theory results given by Ram for the 'straight-in' case ($n_y=n_z=0$) using only the first part of the program.

We take f = 11.6 GHz, and we assume a purely toroidal field with B(x) = (4 kG)(85 cm)/[85 cm + 67 cm − x (cm)]. We assume the following form of the density profile in the extreme edge region: $n(x) = (3 \times 10^{12} \text{ cm}^{-3}) \exp[(x-g)/L_n]$ for $0 < x < x_{end}$, and in the nominal case, we take $L_n = 0.7$ cm, $g = 1.5$ cm, $x_{end} = 1.7$ cm. The density profile and the locations of the cutoffs and UHR are shown in Fig. 1. The solid curve is the nominal ($L_n = 0.7$ cm) density profile, while the nearly horizontal dashed lines represent the densities at which R=0 (right-hand cutoff), S=0 (UHR), and L=0 (left-hand cutoff).

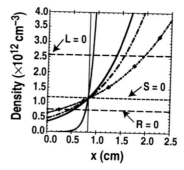

Figure 1. Density profiles for scan of gradient at UHR for NSTX.

To keep all of the other physics constant as the gradient scale-length L_n is varied, the "gap" g is adjusted to keep the position of the UHR fixed at x = 0.83 cm, and x_{end} is adjusted so that the density at x_{end} is constant. The other curves in the figure show a few of the resulting family of density profiles with different values of L_n, one with much shorter L_n and two with shallower gradients.

For the straight-in case, the translation of the surface admittance to a reflection coefficient $|\rho|^2$ is straightforward. In the case where the calculation ends at a density higher than the L=0 cutoff, there is no energy propagation to larger x possible in the cold plasma model, so that the power mode-conversion coefficient is just $1 - |\rho|^2$. The mode-conversion coefficient thus computed as a function of the gradient scale length is shown in Fig. 2 as the solid curve. Also, we compute the same quantity for these profiles, except x_{end} is chosen such that the slow X-mode is still propagating at the high density boundary of the calculation zone. The resulting $1 - |\rho|^2$ is shown as the dashed curve with superimposed × symbols. This case corresponds to the classic Budden problem, with the right-hand cutoff and the UHR, as opposed to the "triplet resonator" case. In the Budden problem, $1 - |\rho|^2$ does not correspond to the power mode converted at the UHR, since energy can propagate out of the calculational zone on the slow X-branch. In the limit of extremely short L_n, there is no reflected wave, and all of the power is transmitted to the slow X-mode branch. Therefore, the quantity $1 - |\rho|^2$ goes to 1 as L_n goes to zero, but the mode-conversion coefficient also goes to 0.

Also in Fig. 2, the numerical full-wave solutions are compared with the analytic theories given by Ram for those two cases, as well as to the experimental data given in Ref. [3]. The analytic estimate which is compared with the experimental data in Ref. [3] does not appear to be reproduced by the numerical solution. The numerical result for the transmission coefficient does not exceed about 80% even at the optimum value of L_n, and there is no sign of oscillations due to the interference between the forward and reflected waves between the UHR and L=0 layers. The simple formula for the quantity $1 - |\rho|^2$ in the Budden case (solid line with no symbols) reproduces the numeric results rather well, and in the range of L_n so far accessed experimentally there is not much difference between the Budden case and the full triplet case. The optimum gradient is significantly steeper than the setup of that preliminary experiment permitted.

All of these calculations so far have considered only the single plane wave with $n_y=n_z=0$. But the work in Refs.[1] and [2] has shown that there is a very strong effect of non-zero n_y in these parameter ranges. This effect is the same one that, in a much lower frequency range, causes the asymmetry in FW antenna loading between co- and counter-current phasing [8]. Mathematically, the effect is a result of the terms in the wave equation involving the products of n_y and spatial derivatives of the off-diagonal element of the dielectric tensor. These terms are dropped in the WKB approximation, and they are first order in n_y and in the gradient of the off-diagonal tensor element, K_{xy}, or D in Stix's notation. This means that the effect depends on the sign of n_y and on the sign of D. These non-WKB terms are most important exactly when the gradient is steep compared to the X-mode wavelength, which in this case means for L shorter than a few centimeters. Even for values of L_n much longer than this, these terms are still important as long as an UHR layer is present in the coupling zone, since these terms all diverge as S vanishes.

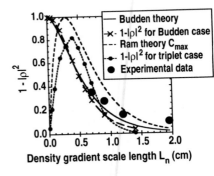

Figure 2. Comparison of numerical full wave solutions for "straight-in" propagation with approximate analytic formulas.

For a discontinuous ("step") density profile, a unidirectional surface wave [2] propagates in the y-direction in the plane of the discontinuity. When a continuous density profile is assumed, what was a true surface wave in the step profile case becomes a very slowly penetrating wave. As the density gradient decreases, then the distance from the right-hand cutoff to the upper hybrid layer increases and the amplitude of the fields to excite the UHR becomes low. Hence, the distance that the slowly penetrating wave must propagate along the y-direction before the power is absorbed (again, really mode-converted) at the upper hybrid becomes large.

We next consider the simplest possible n_y-specific launcher and apply our code to compute the reflection coefficients for NSTX-like parameters. We take the same rf and plasma parameters as in the single plane wave calculations, and we take the launcher to consist of a pair of vacuum waveguides stacked poloidally, of opening dimensions 2.2 cm along the static magnetic field and 1.1 cm vertically. Note that the Brambilla-type models do not model the situation where a waveguide is protruding some distance above a ground plane; in that situation, currents on the top and bottom surfaces of the waveguide tend to make the guide more directive than with a ground plane.

First, the phasing between the forward waves in the two guides is set at zero, to most nearly approximate the single "straight-in" plane wave case, and we compute the reflection coefficient in the guides as a function of the assumed gradient scale length. The resulting transmission coefficient is shown as the solid line in Fig. 3. The transmission coefficient has much less variation with L_n on the shallow gradient side than in the "straight-in" plane wave case. This is a result of the rather efficient excitation of the surface-wave-like mode, propagating nearly poloidally in one direction. The maximum transmission coefficient is obtained by phasing the guides so that the peak of the excited spectrum coincides with the maximal excitation of that mode. The dashed line with the superimposed × symbols shows the transmission coefficient obtained at the optimum poloidal phasing angle, and the remaining dashed curve with the filled circles shows the poloidal phase angle at which that optimum is achieved. The transmission coefficient remains above 80% for all gradient scale lengths greater than about 2 mm. However, the poloidal localization of the start of the EBW propagation would be expected to get less well defined the larger the value of L_n.

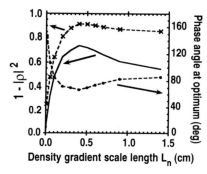

Figure 3. Transmission coefficient for two waveguide array coupler for in-phase operation, operation at the optimal y-phasing, and the phase angle for which that optimum is obtained for NSTX conditions.

The combination of local limiters to reduce the value of L_n with some poloidal phasing would probably result in such a low reflection coefficient that the poloidal phasing could be fixed at a value of 75 or 80 degrees, instead of the relatively complicated system necessary to make that phasing adjustable. The limiters would not be expected to strongly interact with the launched waves, since the launched wave would primarily go 'up' or 'down' depending on the sign of the toroidal field and the corresponding phasing, while the limiters are separated in the toroidal (horizontal) direction.

ACKNOWLEDGMENT

Work supported by the U.S. Department of Energy under Grant DE-FG03-99ER54522 and Contract DE-AC02-76CH03073.

REFERENCES

[1] Pinsker, R.I., et al., "Calculation of Direct Coupling to the Electron Bernstein Wave With a Waveguide Antenna," *Proc. 14th Top. Conf. on RF Power in Plasmas*, edited by T.K. Mau and J.S. deGrassie, AIP Conference Proceedings, Oxnard, 2001, p. 350.
[2] Pinsker, R.I., et al., "Calculation of Coupling to the Electron Bernstein Wave With a Phased Waveguide Array," submitted to *Plasma Phys. Control. Fusion* (2003).
[3] Taylor, G., *Phys. Plasmas* **10**, 1395 (2003).
[4] Brambilla, M.., *Nucl. Fusion* **16**, 47 (1976).
[5] Ram, A.K. and Schultz, S., *Phys. Plasmas* **7**, 4084 (2000).
[6] Wang, C.Y., et al., *Phys. Plasmas* **2**, 2760 (1995).
[7] Ram, A.K. et al., *Phys. Plasmas* **3**, 1976 (1996).
[8] Jaeger, E.F., *Nucl. Fusion* **38**, 1 (1998).

ECE in MAST: Theory and Experiment

J. Preinhaelter[1], V. Shevchenko[2], M. Valovic[2], P. Pavlo[1], L. Vahala[3], G. Vahala[4], and the MAST team[2]

1) *EURATOM/IPP.CR Association, Institute of Plasma Physics, 182 21 Prague, Czech Republic*
2) *EURATOM/UKAEA Fusion Association, Culham Science Centre, Abingdon, OX14 3DB, UK*
3) *Old Dominion University, Norfolk, VA 23529, USA*
4) *College of William & Mary, Williamsburg, VA 23185, USA*

Abstract. A 3D model of Electron Bernstein Waves propagation, absorption and X- and O-mode conversion has been developed and used to determine ECE emission from the MAST plasma.

Introduction. Extensive ECE data from 16 to 60 GHz are available in MAST [1]. The characteristic low magnetic field and high plasma density of a spherical tokamak do not permit the typical radiation of O and X modes from the first five electron cyclotron harmonics. Thus only electron Bernstein modes, (modes not subject to a density limit), which mode convert into electro-magnetic waves in the upper hybrid resonance region, can be responsible for the measured radiation [2].

3D plasma model. To interpret the experimental results we develop a realistic 3D model of the MAST plasma. The instantaneous magnetic field and its spatial derivatives are reconstructed from a 2D splining of two potentials determined by an EFIT equilibrium reconstruction code, assuming toroidal symmetry. The plasma density and temperature profiles (Fig. 1) in the whole *RZ* cross-section of the plasma are obtained from mapping the high spatial resolution Thomson scattering measurements on magnetic surfaces and then interpolating between the low and high field sides values.

Antenna position. The intersection of the antenna pattern with the separatrix determines both the spot position (at which the antenna is aimed) as well as the components of the wave vector of the outgoing waves (see Fig. 2). Only linearly polarized waves are detected and the plane of polarization can be selected by the orientation of the polarization rotator [1]. Since we wish to detect the emission of oblique O-modes, the ECE antenna

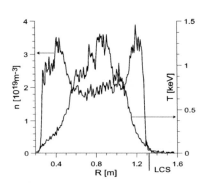

FIGURE 1. Plasma density and temperature profiles measured by Thomson scattering, MAST, shot #4958, *t*=120ms.

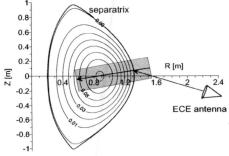

FIGURE 2. Antenna position. The direction of view is inclined about 16° from the equatorial plane upward and the angle between its projection to the equatorial plane and the vertical plane going through the tokamak axis and the antenna position is about 20°.

is oriented so that at the plasma boundary the wave vector and the electric field of outgoing waves are lying in the plane spanned by the edge density gradient and the magnetic field.

Conversion efficiency. In Fig. 2 we also show the auxiliary plane-stratified plasma slab geometry, which is used to determine the mode conversion efficiency by a numerical full wave solution of the wave propagation. Because the conversion region is only several centimeters thick the plane-stratified model is adequate. From a cold plasma model [3] we obtain the values of the global conversion efficiency $C_{EBW-X-O}$. It represents the conversion efficiency of both processes: the direct EBW-X conversion as well as the process in which an obliquely incident EBW is first converted to an X mode which subsequently mode converts to an O-mode [4] at the plasma resonance.

Ray tracing. To determine the radiative temperature we must study the propagation of EBW in 3D. For this purpose we adopt standard ray tracing [5]. The antenna beam is supposed to be Gaussian and is replaced by a set of rays. An angular divergence of the rays is determined from the radius of the antenna aperture (0.05m), the antenna to the Last Closed Surface (LCS) distance (~1m) and the radius of the circular area visible to the antenna (it contains 90% of detected power) placed at the LCS. This radius slightly decreases with frequency: from 0.18m at f=16GHz to 0.14m at f=40GHz. Because we use an electrostatic approximation for EBW we start the EBW rays at the upper hybrid resonance and the short distance between the spot of the vacuum rays on the LCS and the UHR is connected by a straight line in the direction of the density gradient. In performing the calculations we also use mutual interchangeability between emission and absorption.

Rays starting from the layer parallel to equatorial plane are focused to a very small region (Fig. 3). Rays above and below the equatorial plane are strongly damped near the plasma surface because the strong increase of their $N_\parallel$ along the ray [6]. The ray equations describe the motion of the EBW wave packet. To obtain power absorption we integrate the time evolution equation for power simultaneously with the ray [7],

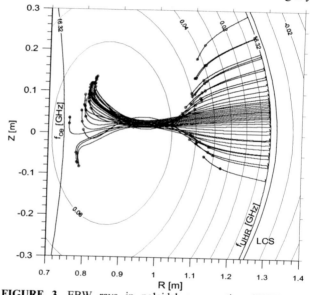

FIGURE 3. EBW rays in poloidal cross-section (#4958, t=120ms, f=16.32GHz). The first dot on the rays denotes the beginning of collisionless damping, the second dot on the ray denotes the position where the wave is fully absorbed.

$$dP/dt = -2\gamma(t)P, \quad \gamma(t) = \mathrm{Im}\{D(\omega,k(t),x(t))\}/d\mathrm{Re}\{D(\omega,k(t),x(t))\}/d\omega,$$

where the time Landau decrement is given in the usual way.

Radiative temperature. Absorption along the ray is non-local (Fig. 4) with reabsorption of the radiation playing an important role. To determine the radiative temperature we must solve the radiative transfer equation [8]

$$dP/dt = \eta - \alpha P$$

simultaneously with the ray evolution equation. Here the absorption coefficient $\alpha = 2\gamma(t)$ and the emissivity η is given by Kirchoff's law which states that the emission is proportional to the absorption multiplied by the Rayleigh-Jeans black body radiation intensity, so $\eta = \alpha \omega^2 T(t)$. Integrating the radiative equation from $t=+\infty$ to 0 we find that the emitted power can be expressed by the Rayleigh-Jeans law with T_{rad} instead of the local temperature T. So

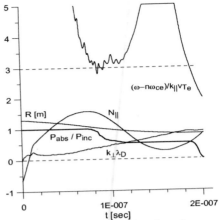

FIGURE 4. Power absorption along the ray. #4958, $f=16.32$, ray 12.

$$P \sim \omega^2 T_{rad}, \text{ where } T_{rad} = \int_0^\infty 2\gamma(t')T(t')\exp\{-\int_0^{t'} 2\gamma(t'')dt''\}dt'.$$

Simulated ECE power detected by antenna. The intensity of ECE detected by the antenna can be expressed as

$$I_{ECE} = const \times \sum_{rays} W_{Gauss} C_{EBW-X-O} \omega^2 T_{rad} \Delta S$$

where $W_{Gauss} = \exp\{-2d^2/(D/2)^2\}$, D is the diameter of the visible area to the antenna at the LCS, d is the distance of the ray from the central ray and ΔS is the segment of surface of visible area corresponding to the ray. Using this procedure (see Figs. 5a,b) we are able to determine the ECE power incident on our antenna (Figs. 6, 7).

Slightly worse results were obtained for shot #7685 at $t=240$ms (Fig. 7). It is an example of a well developed H-mode with high density in the SOL. ECE interpretation is difficult since the EBW–X-O conversion region is shifted past the LCS, making the detected ECE weak and noisy.

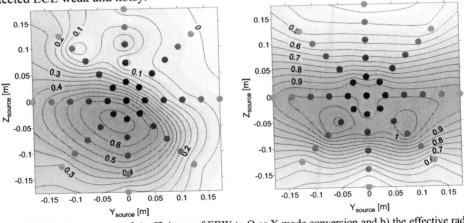

FIGURE 5. Contour map of a) efficiency of EBW to O or X mode conversion and b) the effective radiative temperature [keV], for $f=16.32$GHz. Y_{source} and Z_{source} are coordinates in a visible area circle placed at the LCS. (about 0.96m from the antenna for #4958 and $t=120$ms). Dots correspond to individual rays.

Conclusions. Agreement between the experimental data for ECE and our estimates is good for shots where the mode conversion region lies within the transport barrier inside the LCS. For shots where the conversion region is situated in the SOL, where there are large density fluctuations, the ECE is noisy and we obtain only qualitative agreement.

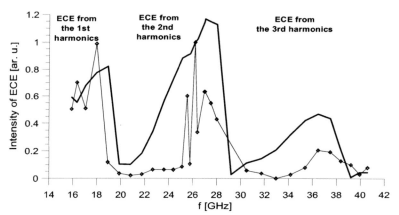

FIGURE 6. ECE from MAST, shot #4958, $t=120$ms. Black diamonds are experiment data and the thick full line is numerical simulation of ECE power incident on antenna. Reference frequency $f=26.16$GHz, for which $I_{ECE}=1$.

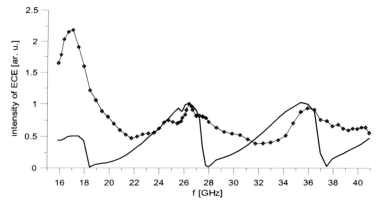

FIGURE 7. ECE from MAST, shot #7686, $t=240$ms. Reference frequency $f=26.38$.

Acknowledgement. Part of this work was funded jointly by UK Engineering and Phys. Sci. Council and by EURATOM.

References
[1] Shevchenko V., *27th EPS, Budapest*, 2000, ECA Vol. **24B**, paper P3.120.
[2] Laqua H.P., et al., *29th EPS, Montreux*, 2002, ECA Vol. **26B**, D-5.010 (2002)
[3] Irzak M.A., et al., *Plasma Physics Reports* **25**, 601 (1999).
[4] Preinhaelter J., Kopecky V., *J. Plasma Phys.* **10**, 1 (1973), part 1.
[5] Pavlo P., Krlin L., Tluchor Z., *Nucl. Fusion* **31**, 711 (1991).
[6] Forest C.B., et al., *Physics of plasmas* **7**, 1352 (2000).
[7] Bonoli P., et al., *Phys. Fluids* **29**, 2937 (1986).
[8] Bornatici M., et al., *Nucl. Fusion* **23**, 1153 (1983).

Relativistic Effects in Heating and Current Drive by Electron Bernstein Waves

A. K. Ram, J. Decker, and A. Bers

Plasma Science & Fusion Center, M.I.T, Cambridge, MA 02139, U.S.A.

R. A. Cairns

University of St. Andrews, Fife KY16 9SS, Scotland

C. N. Lashmore-Davies

Culham Science Centre, UKAEA, Abingdon, Oxon OX14 3DB, U.K.

ABSTRACT

The high-β magnetically confined plasmas in spherical tori (ST), like NSTX and MAST, provide a unique opportunity for a wide variety of applications of electron Bernstein waves (EBW). These applications range from heating of the ST plasma to modifying and controlling its current profile. Using the fully relativistic dielectric tensor for a Maxwellian distribution function, this paper presents initial results illustrating the effect of relativity on the dispersion characteristics of EBWs. It is found that, even at temperatures relevant to present STs, the relativistic dispersion properties of EBWs are significantly different from their non-relativistic counterpart.

INTRODUCTION

The EBWs, excited at the edge of the ST plasmas by mode conversion of an externally launched extraordinary or ordinary mode [1, 2], propagate into the overdense plasma and damp on electrons at the Doppler-shifted electron cyclotron resonance (or its harmonics). Depending on the poloidal location of the EBW excitation, the magnitude of the parallel (to the magnetic field) wavenumber can be either less than one (for equatorial excitation) or greater than one (for excitation away from the equatorial plane) [1]. Furthermore, for EBWs $k_\perp \rho_e \gtrsim 1$ ($k_\perp \rho_e$ is the electron Larmor radius normalized to the perpendicular wavelength) so that small Larmor radius expansions are not valid. Previous studies have found relativistic effects to be important for conventional electron cyclotron waves [3].

RELATIVISTIC DIELECTRIC TENSOR

There are two numerically useful representations for obtaining the relativistic conductivity tensor $\bar{\bar{\sigma}}$ for a relativistic Maxwellian distribution function [4]. The first form of the conductivity tensor is:

$$\bar{\bar{\sigma}} = \frac{1}{4\pi} \frac{\omega_p^2}{\omega_c} \frac{c^4}{v_t^4} \frac{1}{K_2\left(\frac{c^2}{v_t^2}\right)} \int_0^\infty d\xi \left\{ \frac{K_2\left(R^{1/2}\right)}{R} \bar{\bar{T}}_1 - \frac{K_3\left(R^{1/2}\right)}{R^{3/2}} \bar{\bar{T}}_2 \right\} \quad (1)$$

where ω_p, ω_c, v_t are the rest mass electron plasma frequency, cyclotron frequency, and the thermal velocity, respectively, K_ν is the modified Bessel function of the second kind of order ν, and

$$\bar{\bar{T}}_1 = \begin{pmatrix} \cos\xi & -\sin\xi & 0 \\ \sin\xi & \cos\xi & 0 \\ 0 & 0 & 1 \end{pmatrix} \quad (2)$$

$$\bar{\bar{T}}_2 = \frac{c^2}{\omega_c^2} \begin{pmatrix} k_\perp^2 \sin^2\xi & -k_\perp^2 \sin\xi(1-\cos\xi) & k_\perp k_\parallel \xi \sin\xi \\ k_\perp^2 \sin\xi(1-\cos\xi) & -k_\perp^2(1-\cos\xi)^2 & k_\perp k_\parallel \xi(1-\cos\xi) \\ k_\perp k_\parallel \xi \sin\xi & -k_\perp k_\parallel \xi(1-\cos\xi) & k_\parallel^2 \xi^2 \end{pmatrix} \quad (3)$$

$$R = \left(\frac{c^2}{v_t^2} - i\xi\frac{\omega}{\omega_c}\right)^2 + 2\left(\frac{k_\perp c}{\omega_c}\right)^2 (1-\cos\xi) + \frac{k_\parallel^2 c^2 \xi^2}{\omega_c^2} \quad (4)$$

For any equilibrium distribution function $f_0(p_\perp, p_\parallel)$, the second form of the conductivity tensor is:

$$\bar{\bar{\sigma}} = -\frac{i}{2}\frac{\omega_p^2}{\omega_c}\left\langle \sum_{n=-\infty}^{\infty} \frac{1}{n-\bar{\omega}}\left(\frac{1}{\kappa T}\frac{p_\perp}{m\gamma}\right) \bar{\bar{\sigma}}_N f_0(p_\perp, p_\parallel) \right\rangle \quad (5)$$

where

$$\bar{\bar{\sigma}}_N = \begin{pmatrix} \frac{n^2}{\zeta^2} p_\perp J_n^2 & -i\frac{n}{\zeta} p_\perp J_n J_n' & \frac{n}{\zeta} p_\parallel J_n^2 \\ i\frac{n}{\zeta} p_\perp J_n J_n' & p_\perp J_n'^2 & i p_\parallel J_n J_n' \\ \frac{n}{\zeta} p_\parallel J_n^2 & -i p_\parallel J_n J_n' & \frac{p_\parallel^2}{p_\perp} J_n^2 \end{pmatrix} \quad (6)$$

$$\zeta = \frac{k_\perp p_\perp}{m\omega_c}, \bar{\omega} = \frac{1}{\omega_c}\left(\omega\gamma - k_\parallel \frac{p_\parallel}{m}\right), \omega_c = \frac{eB_0}{m}, \langle ..... \rangle = \int_0^\infty dp_\perp p_\perp \int_{-\infty}^\infty dp_\parallel \quad (7)$$

NUMERICAL RESULTS

We have developed two computational codes based on the two equivalent relativistic representations discussed above. For a variety of cases we find that the two codes lead to identical results. In the mode conversion region near the edge of NSTX-type plasmas [1, 2], we find that the results obtained from the relativistic description are the same as those from the non-relativistic description. Thus, the mode conversion formalism developed in [1, 2] is not modified by relativistic effects. In Fig. 1 we show the dispersion characteristics of EBWs as a function of ω/ω_c for a uniform plasma with electron temperature of 3 keV, $\omega_p/\omega_c = 36$, and $n_\parallel = 0.2$. We note that the relativistic effects broaden the resonance (Fig. 1b) at the second harmonic where the real part of $n_\perp$ becomes small, and narrow down the resonance at the fundamental electron cyclotron frequency where $n_\perp$ becomes large. In Fig. 2 we show the dispersion characteristics of EBWs as a function of $n_\parallel$ for a uniform plasma with electron temperature of 3 keV, $\omega_p/\omega_c = 36$, and $\omega/\omega_c = 1.8$. It is now evident that there are significant differences between the relativistic and non-relativistic properties of EBWs. These local differences would accumulate along a ray path in toroidal geometry, leading to a very different spatial location of the damping region for EBWs. This will be examined in the future.

ACKNOWLEDGEMENTS

This work is supported by DoE Grant Numbers DE-FG02-91ER-54109 and DE-FG02-99ER-54521, and partially by the UK Department of Trade and Industry and EURATOM.

References

[1] A. K. Ram and S. D. Schultz, *Phys. Plasmas* **7**, 4084 (2000).

[2] A. K. Ram, A. Bers, and C. N. Lashmore-Davies, *Phys. Plasmas* **9**, 409 (2002).

[3] M. Bornatici, R. Cano, O. De Barbieri, and F. Engelmann, *Nuclear Fusion* **23**, 1153 (1983), and references therein.

[4] B. A. Trubnikov, in *Plasma Physics and the Problem of Controlled Thermonuclear Reactions*, edited by M. A. Leontovich (Pergamon Press Inc., New York, 1959) Vol. III, p. 122.

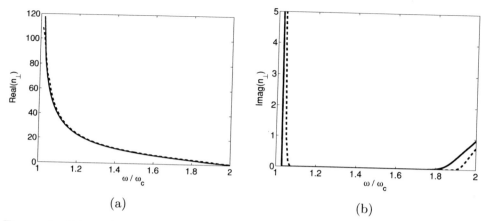

Figure 1: Dispersion characteristics of EBWs as a function of ω/ω_c for $n_\| = ck_\|/\omega = 0.2$, $(\omega_p/\omega_c)^2 = 36$, and $T_e = 3$ keV: (a) real part of $n_\perp = ck_\perp/\omega$; (b) imaginary part of $n_\perp$. The solid line represents the solution of the dispersion relation in the fully relativistic case, and the dashed line is for the non-relativistic case.

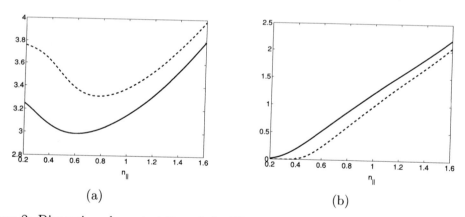

Figure 2: Dispersion characteristics of the EBW as a function of $n_\| = ck_\|/\omega$: (a) real part of $n_\perp$; (b) imaginary part of $n_\perp$. The parameters are as cited in the text. The solid line represents the solution of the dispersion relation in the fully relativistic case, and the dashed line is for the non-relativistic case.

Electron Bernstein Wave Research on NSTX and CDX-U

G. Taylor[a], P.C. Efthimion[a], B.Jones[b], G.L. Bell[c], A. Bers[d],
T.S. Bigelow[c], M.D. Carter[c], R.W. Harvey[e], A.K. Ram[d],
D.A. Rasmussen[c], A.P. Smirnov[f], J.B. Wilgen[c], J.R. Wilson[a]

[a] Princeton University, Princeton, NJ 08543, USA
[b] Sandia Laboratory, Albuquerque, NM 87185, USA
[c] Oak Ridge National Laboratory, Oak Ridge, TN 37831, USA
[d] Plasma Science and Fusion Center, Cambridge, MA 02139, USA
[e] CompX, Del Mar, CA 92014, USA
[f] Moscow State University, Moscow, Russia

Abstract. Studies of thermally emitted electron Bernstein waves (EBWs) on CDX-U and NSTX, via mode conversion (MC) to electromagnetic radiation, support the use of EBWs to measure the T_e profile and provide local electron heating and current drive (CD) in overdense spherical torus plasmas. An X-mode antenna with radially adjustable limiters successfully controlled EBW MC on CDX-U and enhanced MC efficiency to ~ 100%. So far the X-mode MC efficiency on NSTX has been increased by a similar technique to 40-50% and future experiments are focused on achieving ≥ 80% MC. MC efficiencies on both machines agree well with theoretical predictions. Ray tracing and Fokker-Planck modeling for NSTX equilibria are being conducted to support the design of a 3 MW, 15 GHz EBW heating and CD system for NSTX to assist non-inductive plasma startup, current ramp up, and to provide local electron heating and CD in high β NSTX plasmas.

INTRODUCTION

CDX-U [1] and NSTX [2] are high β spherical tori that contain overdense ($\omega_{pe} \gg \omega_{ce}$) plasmas that are not accessible to low harmonic electron cyclotron waves, and hence preclude the use of established technologies such as ECRH and ECCD. Since electron Bernstein waves (EBWs) propagate in overdense plasmas and absorb strongly at electron cyclotron resonances, they may be used for local electron temperature measurements, electron heating (EBWH) and current drive (EBWCD). Coupling to EBWs is possible via mode conversion (MC) of electromagnetic waves in the vicinity of the plasma edge [3,4]. Recent studies of thermally emitted EBWs via MC have evaluated the EBW MC physics both to develop a local electron temperature diagnostic and, as a result of the symmetry of the MC process [5], to support the development of EBWH and EBWCD.

EBW MODE CONVERSION MEASUREMENTS

Optimized EBW MC has been demonstrated on CDX-U, with almost complete conversion of thermally emitted EBWs to X-mode electromagnetic radiation, in agreement with theoretical predications using the measured density scale length (L_n) at the MC layer [6]. A local, radially scanned, limiter surrounding a quad-ridged antenna produced controlled steepening of L_n from > 3 cm to 7 mm in the vicinity of the EBW to X-mode MC layer, resulting in an order of magnitude increase in the MC efficiency of fundamental EBW emission, so that $T_{rad} \sim T_e$, measured by Thomson scattering (Fig. 1(a)).

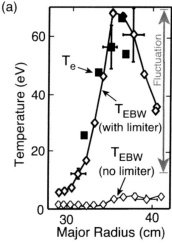

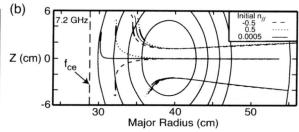

Figure 1. (a) Radiation temperature of MC EBW emission with and without a local limiter on CDX-U. When the limiter shortens L_n to 7 mm at the EBW MC layer, the EBW T_{rad} and T_e profiles are similar. The EBW T_{rad} exhibits large amplitude fluctuations (vertical grey arrows). (b) GENRAY calculation showing significant displacement for EBW rays near the midplane resulting from 1-2 cm vertical shifts in the plasma column causing large fluctuations in measured T_{rad}.

Large fluctuations in EBW MC efficiency were observed on CDX-U (vertical grey arrows in Fig. 1(a)) and, while these were fairly strongly correlated with L_n fluctuations at the MC layer, GENRAY [7] ray tracing calculations (Fig. 1(b)) indicate that vertical plasma oscillations of only 0.3 cm could account for oscillations in T_{rad} of 20-80 eV, or about half the observed T_{rad} fluctuation. Detailed analysis of the CDX-U EBW MC experiments is presented elsewhere [8]. Since the NSTX plasma is much larger and better controlled than CDX-U, fluctuations in EBW emission due to refraction on NSTX are expected to be considerably smaller. Initial studies of EBW emission during NSTX plasmas showed $\Delta T_{rad}/T_{rad} \sim 20\%$, confirming this expectation. These initial EBW emission measurements also indicated MC efficiency < 5% for L-mode discharges and ~ 15% during H-modes, consistent with theoretical calculations using the measured L_n at the MC layer [9].

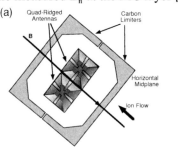

Figure 2. (a) Schematic diagram showing the layout of the X-mode EBW antenna installed on NSTX. Two carbon limiters surround a pair of quad-ridged antennas used for X-mode EBW radiometry and O-mode reflectometry. (b) Photograph of the antenna installed near the mid-plane of NSTX.

Recently, EBW MC experiments on NSTX, that use the HHFW antenna limiters to steepen L_n at the MC layer, have demonstrated 40-50% EBW MC [10]. The maximum MC efficiency was limited by the connection length between the antenna limiters which constrained the shortest achievable L_n at the MC layer to ~ 7 mm. An X-mode

EBW antenna with a local limiter calculated to achieve $L_n \sim 3$ mm has been installed in NSTX (Fig. 2). An O-mode reflectometer integrated into the antenna will measure the local L_n at the MC layer. This antenna is predicted to achieve better than 80% EBW MC.

EBW HEATING AND CURRENT DRIVE

While the EBW MC studies on CDX-U and NSTX were motivated by the need to develop a fast electron temperature profile diagnostic for high β, overdense plasmas, they also validate the MC physics for EBWH and EBWCD [5]. EBWH and EBWCD can help optimize the magnetic equilibrium and suppress deleterious MHD in ST plasmas that might otherwise prevent access to high β operation [11]. Deposition at r/a > 0.8 may be required for MHD suppression in NSTX. Placing the EBW launcher well above or below the mid-plane on the low field side may have several benefits; large uni-directional $n_{//}$ shifts, needed for efficient CD, can result even with an $n_{//} \sim 0$ launch [12], trapped particle effects that significantly reduce the EBWCD efficiency near the mid-plane can be minimized, and the launcher can be located where there is generally less competition for vacuum vessel access.

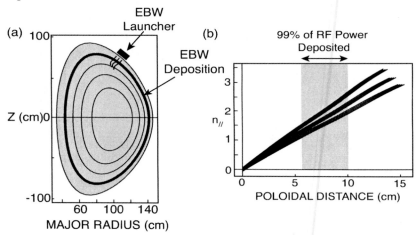

Figure 3. (a) GENRAY ray tracing calculation for 15 GHz EBWs launched from a location at a poloidal angle 85 degrees above the mid-plane of NSTX for a β = 30% plasma equilibrium. EBW rays are projected on to a poloidal cross-section. (b) Plot of $n_{//}$ versus poloidal projected distance along the ray show a significant shift in $n_{//}$ within 5-10 cm of the plasma edge.

Figure 3(a) shows a GENRAY [7] calculation for a bundle of 15 GHz EBW rays launched from 85 degrees above the mid-plane with a range of $n_{//}$ between – 0.1 and + 0.1. Figure 3(b) shows the significant $n_{//}$ upshift that occurs during the first 5-10 cm the rays travel into the plasma. 99% of the EBW power is deposited at the Doppler-shifted second harmonic resonance within 6-10 cm from the plasma edge, shown by the thickened flux surface line in Fig. 3(a) and the grey shaded region in Fig. 3(b).

Modeling of the EBWCD was performed with the CQL3D bounce-averaged Fokker-Planck code [13] for the case shown in Fig. 3. CQL3D results for 1 MW of launched EBW power are shown in Fig. 4. The EBW power is locally deposited near r/a = 0.8 (Fig. 4(a)) and the resulting CD profile is shown in Fig. 4(b). The CD efficiency for this case is 0.06 A/W, assuming 100% MC between the injected RF power and the EBW. Since $T_e = 0.6$ keV and $n_e = 1.6 \times 10^{19}$ m^{-3} at the CD location which has a major radius of 1 m, this corresponds to a dimensionless current drive efficiency, $\zeta_{ec} = 0.53$ [14]. This is about three times the value of ζ_{ec} obtained with the same plasma with CD at r/a = 0.8 near the mid-plane, is similar to the value obtained for EBWCD near the magnetic axis [15] and is about 2-3 times the ζ_{ec} for ECCD in DIII-D [16].

The EBW deposition profile and CD efficiency can be modified by changes in the density and temperature profile, a sensitivity study is being conducted to investigate this and an EBW launcher design is being developed that will allow control of $n_{//}$ and the polarization of the electromagnetic launch wave for optimum coupling to the EBW. A single steerable mirror launcher combined with a rotatable reflective grating polarizer is being considered, since it provides the greatest flexibility for optimizing EBW coupling and control of the EBW power deposition. The launch frequency that provides the widest radial access to NSTX, which typically operates at 0.35 to 0.45 T, is about 15 GHz. Presently, no high power RF sources with pulse durations of about 1 s exist at this frequency. MIT has proposed the development of an 800 kW 15 GHz gyrotron tube, a 3 MW EBW system using this tube is being considered for NSTX to assist non-inductive plasma startup, current ramp up, and to provide local electron heating and CD in high β NSTX plasmas.

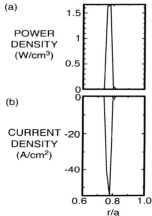

Figure 4. (a) The EBW power deposition profile and (b) EBWCD current density profile for the case in Fig.3 for 1 MW of RF power.

ACKNOWLEDGMENTS

This work is supported by US Department of Energy (DoE) contract nos. DE-AC02-76CH03073, DE-FG02-91ER-54109 and DE-FG02-99ER-54521.

REFERENCES

[1] Menard, J., et al., *Phys. Plasmas*, **6**, 2002-8 (1999).
[2] Ono, M., et al., Proc. 17th IAEA Fus. Energy Conf. (IAEA, Vienna, 1999), **3**, p. 1135.
[3] Preinhaelter, J., and Kopecky, V., *J. Plasma Phys.* **10**, 1-12 (1973).
[4] Ram, A.K., and Schultz, S.D., *Phys. Plasmas* **7**, 4084-4094 (2000).
[5] Ram, A.K., Bers, A., and Lashmore-Davies, C.N., *Phys. Plasmas* **9**, 409-418 (2002).
[6] Jones, B., et al., *Phys. Rev. Lett.*, **90**, 165001 (2003).
[7] Smirnov, A.P., and Harvey, R.W., Bull. Am. Phys. Soc. **40**, 1837 (1995).
[8] Jones, B., PhD. Thesis, Princeton University (2002).
[9] Taylor, G., et al., *Phys. Plasmas*, **9**, 167-170 (2002).
[10] Taylor, G., et al., *Phys. Plasmas*, **10**, 1395-1401 (2003).
[11] Efthimion, P.C., et al., Paper IAEA-CN-94/EX/P2-12, 19th IAEA Fus. Energy Conf., Lyon, France, October 2002.
[12] Forest, C.B., et al., *Phys. Plasmas*, **7**, 1352-1355 (2000).
[13] Harvey, R.W., and McCoy, M.G., *Proc. IAEA. Com. on Advances in Simulation and Modeling of Thermonuclear Plasmas*, Montreal, Quebec (IAEA, Vienna, 1993) p. 489.
[14] Luce, T.C., et al., *Phys. Rev. Lett.*, **83**, 4550-4553 (1999).
[15] Taylor, G., et al., *Proc. 12th Joint Workshop on ECE and ECRH*, edited by G. Giruzzi, World Scientific, Singapore, 2003, pp. 151-160.
[16] Petty, C.C., et al., *Proc. 14th Topical Conference on Radio Frequency Power in Plasmas*, edited by T.K. Mau and J. deGrassie, AIP Conf. Proc. 595, Melville, New York, 2001, pp. 275-281.

RF SOURCES AND SYSTEMS

Archimedes Plasma Mass Filter

Richard Freeman, Steve Agnew, Francois Anderegg, Brian Cluggish,
John Gilleland, Ralph Isler, Andrei Litvak, Robert Miller, Ray O'Neill,
Tihiro Ohkawa, Steve Pronko, Sergei Putvinski, Leigh Sevier,
Andy Sibley, Karl Umstadter, Terry Wade, and David Winslow

Archimedes Technology Group 4810 Eastgate Mall
San Diego, California 92121

Abstract. The Archimedes' Plasma Mass Filter is a novel plasma-based mass separation device. The basic physics of the Filter concept and a description of its primary application for nuclear waste separation at Hanford will be presented along with initial experimental results from a Demo device. The Demo is a 3.89 m long cylindrical device with a plasma radius of 0.4 m and an axial magnetic field up to 1600 Gauss. The plasma is produced by helicon waves launched by two four-strap antennas placed symmetrically either side of a central source region. One strap of each antenna is powered by one of four phase controlled 1 MW transmitters operating in the frequency range from 3.9 – 26 MHz. Each end of the device has ten concentric ring electrodes used to apply an electric field to rotate the plasma. Application of a parabolic voltage profile results in a rigid body rotation. Heavy ions above the cut-off mass number are extracted radially and collected by a heavy ion collector surrounding the source injection region while light ions are collected at the ends of the cylinder. Initial experiments will use noble gas and trace metals to demonstrate separation before attempting to operate with complex waste characteristic of Hanford.

INTRODUCTION

The Archimedes Filter represents a next generation advance in separations technology and is a straightforward application of plasma principles to effect a separation into two fractions: a heavy and a light. The Filter utilizes plasma rotation to achieve the mass separation, so it operates in a manner that is similar but distinct from a plasma centrifuge. While the plasma centrifuge is highly collisional and represents a continuous incremental mass separation, the Filter operating regime is collisionless and its mass separation is discontinuous. A Plasma Mass Filter system employs an axial magnetic field and a radial electric field to produce plasma rotation. Simple radial force balance results in all heavy ions above a "cutoff mass" being radially ejected to the wall whereas the lighter ions below the "cutoff mass" are confined and directed along the axial magnetic field to collectors at each end of the device. For a parabolic voltage profile, the plasma rotates as a rigid body and the mass "cutoff" is not dependent on the ion birth position. The magnetic and electric fields determine the location of the "cutoff mass", A_c, according to the relation:

$$A_c = (M_i/Z)/M_H = eB^2R^2/(8VM_H) \qquad (1)$$

where M_i/Z is the mass to charge ratio of the cutoff atom, M_H is the mass of hydrogen, B is the magnetic field, R is the plasma radius, V is the voltage at the center, and e is the electronic charge.

The Filter has many potential applications in areas where separation of species is otherwise difficult or expensive. In particular, radioactive waste sludges at Hanford have been a particularly difficult issue for pretreatment and immobilization. The present approach at Hanford is to vitrify all of the waste, including 37,000 MT of high level waste (HLW). Figure 1 illustrates the application of the Archimedes Plasma Mass filter to the waste at Hanford where over 75% of the sludge (excluding water, carbon, and nitrogen) has mass less than 59 g/mol while 99.9% of radionuclide activity has mass greater than 89 g/mol. Hence, it is clear that a mass separation of this waste would greatly reduce the quantity of radioactive material to be vitrified. More details concerning the application of the Filter to Hanford high level waste (HLW) can be found in Ref 1.

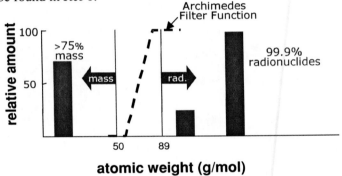

FIGURE 1. Archimedes Filter function in "gap" between Hanford high level waste sludge mass and mass of 99.9% of radionuclides.

Plasma geometry and density distribution determine the Filter function slope or width. Note that masses close to the Filter mass cutoff are not completely removed from the light fraction while the heavier radionuclides are, in principle, completely removed from the light fraction.

Plasma Mass Filter System

A Plasma Mass Filter [2] system consists of four primary subsystems: feed preparation; injection and vaporization; ionization and separation; and collection of light and heavy oxides. Given the large amount of sodium present in the Hanford waste along with the desire to minimize the atom-mols of material to process in the Filter, a natural "solvent" for Hanford high-level waste is molten sodium hydroxide. Following feed preparation via denitration and decarbonation, sodium hydroxide is by far the largest component in Hanford waste. This waste slurry is fed into a plasma calciner (or some other suitable calcination step) that produces a sodium hydroxide melt. The hydroxide melt is fluid in the range 400-800° C and can carry the waste oxide mixture to the filter injectors.

At the filter injectors, the hydroxide mixture will be rapidly heated with a 5 MHz inductively heated thermal plasma. This sodium hydroxide torch is specifically designed to vaporize oxide mixtures carried by molten NaOH. The NaOH melt is nebulized at the injector and rapidly heated by the ~5000 °C plasma in the torch. These inductively-coupled plasma torches operate solely on nebulized NaOH vapor without any added gas providing an extremely effective vaporization method. This high temperature vapor stream from the injector is directed into the Filter plasma at the center of the device. There the vapor is completely ionized by the main plasma, which is heated by helicon excitation at 5 MHz by two antenna assemblies that are integral to the device.

The Archimedes Plasma Mass Filter demonstration device (Demo) has been constructed to demonstrate the basic physics and engineering features using a non-radioactive surrogate waste. The basic Demo parameters are listed in Table 1 and Figure 2 shows a cross-sectional view of the Demo and a photograph of the actual device.

TABLE 1. Demo Parameters

Plasma Radius (m)	0.4
Length (m)	3.89
Maximum Magnetic Field (Gauss)	1500
Maximum Electrode Voltage (V)	700
Maximum RF Power (MW)	4

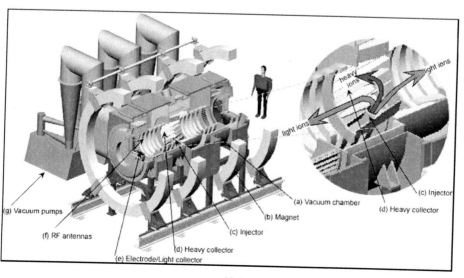

FIGURE 2. Demo Filter a) cross-sectional view and b) as built photo

As described above, ions that exceed the tuned cutoff mass quickly spin out of the plasma and deposit on the heavy collector. Ions with M/Z less than the cutoff mass remain in the plasma and deposit at the light collector/electrode assemblies at each end. Current design of a Commercial Filter unit targets a throughput of around 0.7 MT/day of mixed oxides. This would produce 0.175 MT/day of heavy oxide and 0.525 MT/day of light oxyhydroxide. Oxygen and hydrogen atoms are more difficult to ionize that most other species and therefore will persist as neutral species. As a result, both heavy and light deposits will be fully oxidized and suitable for feed to either IHLW or ILAW processing. The strategies for injection and collection are complimentary. For injection, rapid heating to very high temperature is essential to minimize fractionation or distillation of the materials being vaporized. This rapid heating occurs with a residence time adequate for even the most refractory of materials to completely atomize. For waste collection, correspondingly rapid cooling is likewise needed to help minimize fractionation of plasma components. The material that accumulates at the light collector will be removed from the chamber as a sodium hydroxide melt while the heavy collector oxide will require a once per day cleaning operation. A variety of alternate approaches for cleaning the heavy collector are being explored.

Modeling and Simulation

The basic performance of the Plasma Mass Filter has been explored in various simulations. An orbit code was developed to look at basic single particle orbits, and an ionization model was developed and coupled with a Monte Carlo plasma simulation for selected ions to study collisional effects in a background plasma. The ionization model allows us to use experimental ionization cross-sections and potentials to model the charge state of each species. Once ionized, the trajectory of each ion can likewise be modeled in the magnetic and electric fields of the Filter plasma and this forms the basis of the Monte Carlo simulation. Furthermore, the ion and neutral radiation and conduction losses determine the power needed to sustain the plasma throughput.

Electrode design was aided by a particle-in-cell code, XOOPIC, studied in collaboration with John Verboncoeur of UC Berkeley. Experimental tests of electrode physics and plasma rotation were conducted in collaboration with George Tynan of UCSD and Roger Bengtson of UT Austin.

The initial design of the RF antennas and predictions of plasma heating were developed with the support of John Shearer using the Antena code and with ORNL RF modelers through an Archimedes funded CRADA. The ORNL work was particularly useful for antenna phasing studies and for determining the effects of conducting surfaces outside the plasma region. Ultimately, a full cylindrical code included was developed at Archimedes and used for the final design.

RF System Features

Helicon waves [3-6] are used to produce and heat the plasma in the Archimedes Plasma Mass Filter. The waves are launched with two phase-controlled four-strap m=0 antennas. A photograph of the antenna arrays installed in the Demo is shown in Figure 3.

FIGURE 3. RF Antenna Arrays in Demo

The antenna straps have an inside diameter of 82.6 cm and the four straps are 10.3 cm wide and are separated by 10 cm. The total length of the device is 3.89 m, and the 0.4 m radius Archimedes' plasma is much larger than conventional helicon discharges. The straps are powered by four Thales cw amplifiers capable of producing 1 MW each in a frequency range from 3.9 MHz to 26.1 MHz. These amplifiers were derived from the commercial short wave broadcast systems that use a CQK 650-1 tetrode. The power is delivered to the Demo using Spinner 50 kV 9 3/16" 50 ohm transmission lines with ceramic insulators; Spinner also developed the vacuum feedthroughs based on a modified Asdex design. Forward and reflected power monitors are included in the transmission lines. A variable series and parallel capacitor are connected to each strap to allow limited matching; the capacitors are standard 40 kV 1500 pF high power rf capacitors manufactured by Comet. Archimedes is separately developing a phase control system to actively control the phase between straps in each antenna array.

Initial Experimental Results

In addition to the rf forward and reflected power monitors, diagnostics for the initial experiments included a microwave interferometer to measure line averaged density, multiple cameras located strategically around the device, including a camera recording a borescope view axially along the plasma, vertical spectroscopy arrays located at three axial locations, and a survey spectrometer viewing the plasma along the machine axis were available. Figure 4 shows a time trace of the plasma density, argon ion and argon neutral lines, and the RF forward power.

Initial experiments were conducted primarily using a single array of four bare inconel antenna straps in contact with the plasma. Plasma densities in argon gas of up to 10^{18} m^{-3} were achieved with a single strap at a frequency of 7.15 MHz and a magnetic field of 150 Gauss. The dependence of line averaged plasma density measured with a microwave interferometer versus operating pressure is shown in Figure 5. Capacitive discharges [5] from the bare Inconel straps to each other and to the end walls resulted in substantial sputtering and metal contaminated plasmas. The initial experiments have been repeated with the antenna straps wrapped with ceramic material to reduce capacitive discharges and have substantially reduced the chrome and iron impurities.

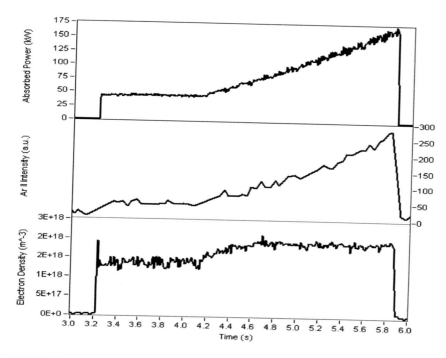

FIGURE 4. Time dependence of a) RF forward power; b) argon ion and neutral lines; and c) line averaged plasma density.

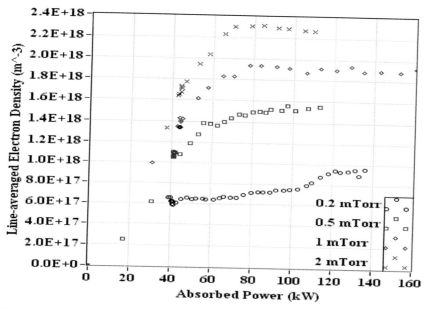

FIGURE 5. Line-averaged plasma density versus forward power for different argon neutral pressures.

Additional reduction of capacitive coupling between straps and to the end of the device was obtained by inserting Mica Mat septa between straps. These measures to reduce the capacitive coupling have improved the uniformity of the plasmas achieved. The design of an enclosed antenna array with a Faraday screen [7] is underway. Rotation experiments are commencing along with experiments to extend the operating parameters to higher magnetic field.

Summary

A novel Archimedes Filter was described and its application to high-level waste at Hanford was outlined. Initial experiments are underway with the objective to demonstrate the physics and engineering viability of the Plasma Mass Filter concept. Initial results with argon demonstrate the viability of producing high density plasmas in a large device with modest powers.

REFERENCES

1. John Gilleland, "Application of Archimedes Filter for Reduction of Hanford HLW" Waste Management Conference, Feb. 2002.
2. "Plasma Mass Filter" - Tihiro Ohkawa - US Patent #6,096,220 (August 1, 2000)
3. R. W. Boswell, Plasma Phys. Controlled Fusion 26, 1147 (1984).
4. F. F. Chen, Plasma Phys. Controlled Fusion 33, 339 (1991).
5. Francis F. Chen and Rod W. Boswell, IEEE Transactions on Plasma Science, Vol. 25, No. 6, 1245 (December 1997)
6. Rod W. Boswell and Francis F. Chen, IEEE Transactions on Plasma Science, Vol. 25, No. 6, 1229 (December 1997)
7. "Shielded RF Antenna" – Richard L. Freeman and Robert L. Miller – US Patent #6,356,025 (March 12, 2002)

Radio-Frequency Sustainment of Laser Initiated, High-Pressure Air Constituent Plasmas

Kamran Akhtar, John E. Scharer, Shane M. Tysk, and Mark Denning

Electrical and Computer Engineering Department
University of Wisconsin-Madison, WI 53706, USA

Abstract. In this paper we investigate the feasibility of creating a high-density ~ 10^{12}-10^{14} cm^{-3}, large volume seed plasma in air constituents by laser (300 mJ, 20(±2) ns) preionization of an organic gas seeded in high-pressure gas mixtures and then sustained by efficient absorption of rf power (1-25 kW pulsed) through inductive coupling of the wave fields. A multi-turn helical antenna is used to couple radio-frequency power through a capacitive matching network. A 105 GHz interferometer is employed to obtain the plasma density in the presence of high collisionality utilizing phase shift and amplitude attenuation data. TMAE Plasma decay mechanisms with and without the background gas are examined.

Introduction

Development of high-density (10^{13} cm^{-3}), large volume (~1000 cc) atmospheric pressure plasma with reduced power budget is of significant interest for variety of applications including material processing, biological decontamination and drag modification at supersonic speed. The theoretical minimum radio-frequency (rf) power required to initiate an air discharge of ~10^{13} cm^{-3} at sea level is 9 kW/cm^3.[1] The model was based on a mono-energetic electron-beam excitation and DC electric field sustainment. Non-equilibrium pulsed electron heating experiments suggest reduction of time-averaged rf power budget required to initiate and sustain the high pressure discharges. In this paper we present the possibility of initiating a discharge by laser ionization of an organic gas tetrakis (dimethyl-amino) ethylene (TMAE) seeded in high-pressure gas that can be efficiently sustained by substantially reduced rf power.

Laser Plasma Source

The general schematic of the laser-initiated and rf sustained plasma source is shown in Fig. 1(a). An excimer laser that runs in ArF mode (6.42 eV) produces a 193 nm uniform intensity ultra-violet (UV) beam with a 20(±2) ns pulse of 300 mJ. The laser output cross-section of 2.8×1.2 cm^2 is increased to 2.8×2.8 cm^2 using a lens system of fused quartz plano-convex and plano-concave lenses. The laser beam enters a 5 cm plasma chamber of length 150 cm through a 2.8 cm diameter Suprasil quartz window. Gas mass flow controllers and a swirl gas injection system are also located at this end. A turbo-molecular pump is used to pump down the plasma chamber to a base pressure of 10^{-6} Torr. Radiofrequency source is a 10 kW unit at 13.56 MHz with a very fast turn on/off time.

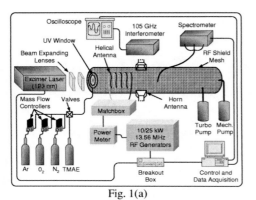 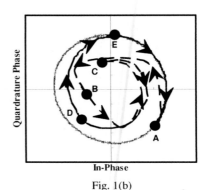

Fig. 1(a)   Fig. 1(b)

Fig.1. (a) Schematic of the experimental arrangement of the laser-initiated and RF sustained plasma and (b) interferometer trace showing the phase and amplitude variation for 35 mTorr TMAE plasma.

A 105 GHz (QBY-1A10UW, Quinstar Technology) quadrature-phase, millimeter wave interferometer is used to characterize the temporal development of plasma following the application of the 20 ns laser pulse and the radiofrequency power. We employ a measurement technique and a model we developed where both phase and amplitude change data are used to determine both the plasma density and the effective collision frequency.[2] An interferometer trace showing the phase and amplitude variation for a 35 mTorr TMAE plasma is shown in Fig.1(b). The outside circle represents the phase variation for the vacuum condition.

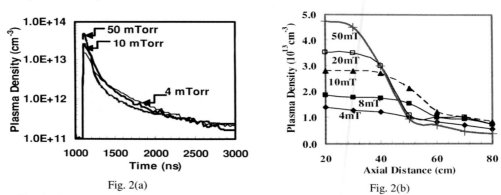

Fig. 2(a)   Fig. 2(b)

Fig. 2. The temporal (a) and axial (b) plots of TMAE plasma density for various TMAE vapor pressures.

A high-density (10^{13} cm^{-3}), large volume (~500 cc) TMAE plasma is created by laser photoionization. The plasma formation follows the path A→B→C shown in Fig. 1(b) as measured 20 cm from the Suprasil window. The plasma is formed by the application of the 20 ns laser pulse and during this interval the line average plasma density reaches its maximum value of 4×10^{13} cm^{-3}. After the application of the laser pulse, the TMAE plasma decays temporally along the path C→D→E→A. Temporal and axial plots of TMAE plasma density are shown in Figs. 2(a) and 2(b), respectively. It is observed that the plasma density increases with increasing TMAE vapor pressure.

However, the increase in peak plasma density is small above 40 mTorr. In addition the axial plasma density decay is also more rapid at higher vapor pressure. This is due to the strong laser attenuation near the window at higher pressure.

Since this experiment is performed to study the effect of background gases on the TMAE plasma formation and decay characteristics, we present a temporal plot of plasma density for different background gases at 760 Torr in Fig. 3(a). A TMAE vapor pressure of 16 mTorr is maintained. It is evident that a high density TMAE plasma can be created in air and it is possible that this plasma can be sustained by an efficient coupling of radiofrequency power. The continuity equation for TMAE plasma decay, including the two-body (electron-ion) recombination coefficient, α_r, the three-body recombination coefficient, β_r ($\sim\beta_g+\beta_e$), involving either a neutral atom (β_g) or an electron (β_e) as the third particle, and electron attachment term, κ_a, is given by the following equation,

$$\frac{\partial n_e}{\partial t} = -\alpha_r n_e^2 - \beta_r n_e^2 - \kappa_a n_g^2 n_e \qquad (1).$$

Since the TMAE molecule is a strong electron donor, here κ_a represents the electron attachment process involving the background oxygen gas of density n_g.

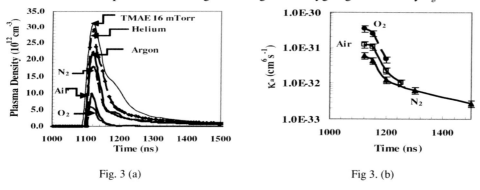

Fig. 3 (a)   Fig 3. (b)

Fig. 3 (a)Temporal plot of TMAE plasma density for different background gas at 760 Torr and (b) the electron attachment coefficient.

In pure TMAE plasma at pressures < 50 mTorr, two-body electron-ion recombination is the main loss mechanism. There is an increase in the recombination coefficient with time exhibiting the process of delayed ionization.[3] It varies from 4.5×10^{-7} cm^{-3}s^{-1} at t =1120 ns after the application of the 20 ns laser pulse at 1100 ns to 4.6×10^{-6} cm^{-3}s^{-1} at t=2500 ns. However in the presence of background gas at atmospheric pressure, three-body electron-ion recombination with neutrals as the third particle contributes to electron loss. It remains almost constant during the plasma decay as it depends on the neutral particle density. From the density decay plot a measure of the three-body decay rate is obtained. It varies from $\beta_r/n_g \sim 4\times10^{-26}$ cm^{6}s^{-1} for helium to $\beta_r/n_g \sim 10^{-25}$ cm^{6}s^{-1} for argon with the background gas at 760 Torr. With air constituents as the background gas, the dominant loss mechanism is the process of electron attachment to oxygen. The electron attachment rate coefficient decreases with a temporal decrease in the plasma density as shown in Fig. 3(b). The experiment shows that the life-time of the laser produced plasma is long enough such that rf power can be coupled efficiently through inductive wave coupling.

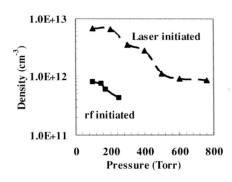 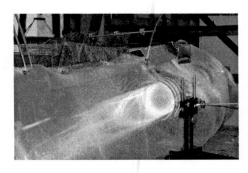

Fig. 4 Comparison of laser initiated-rf sustained and rf initiated-rf sustained argon plasma discharge in a 10 cm diameter tube at modest rf power of 2.0 kW

A 5-turn helical antenna of quarter-inch copper tube of axial coil length 13 cm and internal diameter 12 cm is used to couple rf power through a capacitive matching network to the laser initiated discharge in a 10 cm diameter chamber. Here we present an initial result of laser-initiated and rf sustained experiment with argon as the background gas. A program written in LabView is used to control the sequence of operation and data acquisition. A discharge was successfully initiated by laser preionization of 10 mTorr of TMAE seeded in argon at up to 760 Torr and then sustained at modest rf power of 2 kW. Large volume (~ 2500 cm^3) plasma of line average density of 10^{12}cm^{-3} is obtained. This is a significant improvement over the radiofrequency initiation and sustainment that was limited only up to 250 Torr at the same power level.[4]

Summary

The feasibility of sustaining a laser initiated seeded plasma with radio-frequency power is demonstrated. Future research will examine emission spectra of the rf sustained seed gas-air mixture together with the interferometry data. We will also examine high-power pulse discharges to reduce time-average power budget.

Acknowledgement

This work is primarily supported by Air Force Office of Scientific Research Grant F49620-00-1-0181 and in part by NSF Grant ECS-9905948.

References

1. Vidmar, R.J., and Stalder, K.R., AIAA 2003- 1189, January (2003).
2. Akhtar K, Scharer J, Tysk S., and Kho E., Rev. Sci. Instrum., **74**, 996 (2003).
3. Ding, G., Scharer, J., Kelly, K., Phys. Plasmas, 334 (2001).
4. Kelly K., Scharer J., Paller E., and Ding G., J. App. Phys., **92**,698 (2002).

RF Frequency Oscillations in the Early Stages of Vacuum Arc Collapse *

Steven T. Griffin[†] and Y.C. Francis Thio[¶]

[†]*Department of Electrical and Computer Engineering, University of Memphis, EN-206 Memphis, TN, 38152, USA*

[¶] *SC-55/Germantown Building, U.S. Department of Energy MD, 1000 Independence Ave. S.W., Washington, D.C., 20585-1290, USA*

* Work supported in part by the 2002 NASA Faculty Fellowship Program at Marshall Space Flight Center and conducted at the Propulsion Research Laboratory, Marshall Space Flight Center

Abstract. RF frequency oscillations may be produced in a typical capacitive charging / discharging pulsed power system. These oscillations may be benign, parasitic, destructive or crucial to energy deposition. In some applications, proper damping of oscillations may be critical to proper plasma formation. Because the energy deposited into the plasma is a function of plasma and circuit conditions, the entire plasma / circuit system needs to be considered as a unit. To accomplish this, the initiation of plasma is modeled as a time-varying, non-linear element in a circuit analysis model. The predicted spectra are compared to empirical power density spectra including those obtained from vacuum arcs.

PLASMA ACCELERATING GUN

Coaxial plasma gun development utilizing classic capacitive charging / discharging techniques for pulsed power generation [1-2] has been done at the Marshall Space Flight Center. For these guns, it is important to advance the understanding of the behavior of the pulse power systems specific to this application and a number of related applications such as plasma thrusters based on a Field Reversed Configuration (FRC) and other techniques [3].

In a typical capacitor discharge pulsed power system, a bank of low inductance storage capacitors is used to drive a pulse-shaping network through a switch. The intention is to trickle charge the capacitor bank over a relatively long period of time. The energy contained in the bank is transferred at a moderate rate from the capacitor bank to the pulse-shaping network. The pulse-shaping network then transfers the energy rapidly to the load. As a result the majority of the energy is time compressed into higher instantaneous powers at the load.

For the coaxial plasma guns developed at MSFC, several of these functions are combined into a single structure giving a simpler device. As seen in Fig. 1, a stainless steel rod is positioned along the central axis of a stainless steel tube. A Teflon spacer is inserted over the anode and inside the cathode as depicted in the Fig.1. A capacitor bank (two parallel 330 μF capacitors with 30 nH of stray inductance) is connected through low inductance bus work to the breach of the plasma gun. In this configuration, the vacuum gap between the anode and cathode is used to stand off the capacitor voltage – replacing the switch. The coaxial gun configuration serves as a pulse-shaping network. The plasma in gun serves as a load.

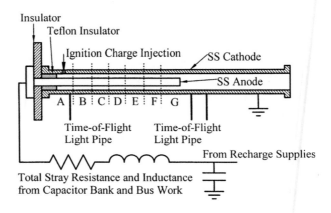

FIGURE 1. Diagram of MSFC Coaxial Plasma Gun showing Seven Arc Regions A.- G.

The capacitor bank is charged and the gun triggered by injecting charge into the breach at the location indicated in Fig. 1. Since the mean-free-path of the electrons in the ambient gas (He at $P_{ambient}$ = 10mTorr) exceed the transverse dimensions of the coaxial line, a vacuum arc forms and is driven by magnetic pressure towards the muzzle of the gun. Vacuum arcs depend on wall materials liberated by spatially localized "hot spots" to provide material to sustain the plasma [4]. In this version of the gun, initial formation of the vacuum arc is aided by the evolution of material from the Teflon insulating spacer in the breach. The resulting generation of lower ionization potential species and the ionization of the entire media bootstraps the trigger charge injection into a vacuum arc.

Data was taken from the system shown in Fig. 1. This included Atomic Emission Spectroscopy (AES), fiber optic based measurement of the time-of-flight, and the breach voltage and current. Of interest here is the RF frequency oscillations associated with the breach voltage and current during the firing of the capacitor bank into the plasma gun filled with 10 mTorr of He. A rich power spectral density was observed. A typical result from the voltage data is shown in Fig. 2 b). A non-linear model of the circuit performance of the system was constructed and the principal phenomena of the power spectral density isolated by comparison to AES, time-of-flight and circuit computer simulation.

The circuit model included the main storage capacitors, stray bus inductance and resistance, a lumped element model of the plasma gun transmission line properties, and temporally switched reactive components and components to model lossy medium. The latter were used to model the non-linear, time variant nature of the evolving plasma. Approximate values for the plasma modeling elements were computed based on independently determined plasma conditions. The resulting modeling system was validated against the previously mentioned data.

RESULTS

Only typical results of the model developed are presented in Fig. 2 below. In Fig. 2 b) is actual power spectral data obtained from the plasma gun of Fig. 1. This is a portion of the total power spectral density, which extended to 500 MHz. In the region shown in Fig. 2 b) a number of resonant peaks are observed. Those less than 1 MHz are generally associated with stray inductance interacting with various circuit elements. The resonant peaks between 1 – 10 MHz are the circuit manifestations of the ionic plasma frequencies associated with various species contained in the plasma. Using a line ratio technique analogous to the classical techniques in optical hydrogen spectroscopy, makes it possible to remove the dependence on plasma conditions such as ionic density, and to identify the species responsible for the various resonant peaks. In addition, it is possible to de-convolve the circuit's system response from the data – sharpening the overall spectra. The species represented here include He^+, He_2^+, C^+, CF_3^+, $(CF_2)_2\text{-}CF^+$, etc. These have been verified by standard atomic emission spectroscopic techniques for the atomic species and by vibrational band emissions for the molecular species with existing spectral databases. The species observed are consistent with prior work [5 - 6] for the decomposition of Teflon in the presence of energetic media such as He plasma. For comparison the pulse power circuit was modeled and the resulting spectral density generated as in Fig. 2 a). The results of a purely circuit model can be compared to the actual spectra to aide in the identification of plasma-generated versus circuit-generated resonant peaks. It is also possible to introduce the resonant peaks coupled in from the plasma into the circuit simulation via non-linear switching of reactive elements into and out of the circuit. Details of this are presented under separate cover.

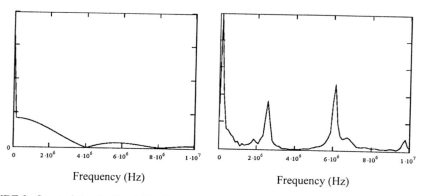

FIGURE 2. Spectral density for a) Non-linear circuit simulation and b) Vacuum arc data. Vertical re-scale in arbitrary units for direct comparison

CONCLUSION

A technique for analyzing the impedance of a plasma gun is presented. This has been applied to He plasma formation in the presence of a Teflon surface in the breech of an existing gun. The capacitive discharge pulsed power system has been modeled, the power spectral density predicted and the simulation results compared to the actual. This approach is useful to isolate the plasma related resonance in this coupled system. It is possible to model the terms coupled in from the plasma via the non-linear introduction of stray circuit components.

ACKNOWLEDGMENTS

The authors would like to thank James W. Smith, Richard Eskridge, Jeff Richeson and Jason Cassibry for their technical contributions to this work. Special thanks to Chuck Karr, Susan Dunnavant and Jeanelle Bland for their role in providing administrative support for the presentation. Thanks to the 2002 NASA Faculty Fellowship Program at Marshall Space Flight Center (MSFC) for financial support for the summer activities and travel support. Finally we would like to thank the Propulsion Research Laboratory at MSFC for making extensive capital equipment available for the summers of 2001 and 2002.

REFERENCES

1. Thio, Y.C.F., Eskridge, R., Martin, A., Smith, J., Lee, M. and Griffin, S., "Plasma Accelerator Development for Dynamic formation of Plasma Liners: A Status Report" in *Second International Magnetized Target Fusion Workshop Proceedings*, Reno, NV: University of Nevada, 2001, p.37.
2. VanDevender,. J.P. et. al., *Proceedings of the IEEE* **80**, 931-994 (1992).
3. Grabowski, C., Cavazos, T., Gale, D., Sommars, W., Degnan, J.H., Intrator, T.P., Siemon, R.E., Taccetti, J.M., Waganaar, B. and Wurden, G.A., "Crowbar Switch for Field Reversed Configuration Formation Discharges" in *Second International Magnetized Target Fusion Workshop Proceedings*, edited by K. Levin, et. al., Department of Physics, Reno, NV: University of Nevada, 2001, p.30.
4. Lafferty, J.M., *Vacuum Arcs Theory and Applications*, New York: Academic Press, 1980.
5. Vaslie, C. and Seymour, R.B., *Handbook of Polyolefins*, New York: Marcel Dekker Inc., 1993.
6. Balmain, K.G. and Dubois, G.R., *IEEE Transactions on Nuclear Science* **NS-26**, 5146-5151 (1979).

PSPICE MODEL OF LIGHTNING STRIKE TO A STEEL REINFORCED STRUCTURE

Neil Koone and Brian Condren

BWXT Pantex L.L.C.

Abstract. Surges and arcs from lightning can pose hazards to personnel and sensitive equipment, and processes. Steel reinforcement in structures can act as a Faraday cage mitigating lightning effects. Knowing a structure's response to a lightning strike allows hazards associated with lightning to be analyzed. A model of lightning's response in a steel reinforced structure has been developed using PSpice (a commercial circuit simulation). Segments of rebar are modeled as inductors and resistors in series. A program has been written to take architectural information of a steel reinforced structure and "build" a circuit network that is analogous to the network of reinforcement in a facility. A severe current waveform (simulating a 99th percentile lightning strike), modeled as a current source, is introduced in the circuit network, and potential differences within the structure are determined using PSpice. A visual three-dimensional model of the facility displays the voltage distribution across the structure using color to indicate the potential difference relative to the floor. Clear air arcing distances can be calculated from the voltage distribution using a conservative value for the dielectric breakdown strength of air. Potential validation tests for the model will be presented.

Introduction

Knowing a structure's response to a lightning strike allows hazards associated with lightning to be analyzed. This paper outlines a novel approach for evaluating a facility's Faraday cage protective qualities by constructing a simulated circuit, which is analogous to the facility's structural steel. This simulated circuit is then injected with a current waveform simulating lightning at a circuit node that would correspond to a likely attachment point on the structure. PSpice, a commercial circuit solver, is used to solve the circuit. Voltage, current, and power information about each element of the circuit is captured for the duration of the simulation in a database. Three dimensional models and animations of voltage distribution can then be analyzed to evaluate the structure's response to a lightning strike.

Model

The Equivalent Circuit Lightning Model (ECLM) consists of three sets of programs: an equivalent circuit generator which creates a PSpice netlist that is analogous to a facility's conductive reinforcement, a database parser that stores and sorts information about the solved equivalent circuit, and visualization tools which allow voltage profiles for the three dimensional structure to be displayed and evaluated.

In order to obtain the maximum amount of data from a lightning strike, detailed simulations of mock facilities were created. Architectural information about a steel reinforced structure is input into a program that creates a PSpice netlist. Steel reinforced structures typically contain horizontal and vertical intersections or meshes of carbon steel (rebar). The grid formed by the rebar is broken down into individual circuit elements where each intersection forms a node and the segment of rebar between nodes forms an electrical circuit. Each segment of steel reinforcement is modeled as an inductor and resistor in series. Rebar diameter and spacing are used to calculate inductance and resistance of sections of rebar between junctures. The lightning current waveform used for the models is the 99th percentile worst-case lightning strike as indicated in equation below.

$$I(t) = I_o \left(e^{-\alpha t} - e^{-\beta t} \right) \quad (1)$$

where I is current, I_o = 204.8 kA, α = 1/(144.6 µs), and β = 1/(0.510 µs).

PSpice is used to solve the electrical circuit. The netlist circuit element PSpice input files generated for this conference typically contain 37,000 circuit elements with some having up to 327,000 circuit elements (for 4 inch rebar spacing models). Voltage, current, and power are output by PSpice in a common simulation data format (CSDF) file. The values for voltage at each node intersection between horizontal and vertical rebar are extracted using a program written in Delphi to parse the data and save the data into a database. Data is obtained for every node at several time steps for the duration of the simulation resulting in data files ranging from 20MB to 200 MB in size.

Once data is parsed from the PSpice CSDF file and converted to a database format, node voltage information can be displayed using 3-D computer rendering. A 3-D model was written using Delphi and OpenGL to display node voltage values. A color array scale is used to display the intensity of the voltage with respect to ground (the floor). When creating the netlist, Cartesian coordinate values were placed in the names of the circuit elements. These coordinates are extracted to determine the node location in 3-D space.

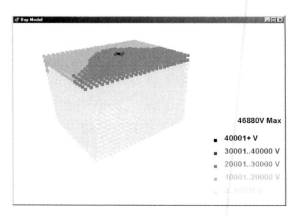

FIGURE 1. Lightning strike to center of roof of a facility constructed of #6 rebar with a 12" spacing.

Results

The models simulated to produce the data for this report are of fictitious facilities. The facility models contain rebar of variable size and spacing. The facility simulations demonstrate the maximum voltage values obtained for a facility of dimensions 20ft x 25ft x 15ft (L x W x H). Two layers of rebar "grids" are incorporated in the model with 10% random connections between inner and outer layers. The spacing between rebar, both vertical and horizontal, was varied for each size of rebar simulated. A blast relief roof system was modeled into the facility, which consists of a rebar reinforced double layer roof with a connection along only one wall (hinge). The purpose of a hinged roof is to relieve pressure in the event of an explosion rather than contain the explosion. The hinge roof design is used around the world where artillery and other volatile material are stored and processed.

Figure 2 shows the maximum facility voltage for simulations constructed of rebar having 0.500" to 1.125" diameter rebar.

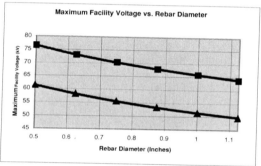

FIGURE 2. Maximum facility voltage for facilities having 9" and 12" spacing and diameters ranging from 0.500" to 1.125".

Figure 3 shows the maximum facility voltage for simulations constructed of rebar having 0.500", 0.750" and 1.125" rebar with 8", 9", 10" and 12" spacing.

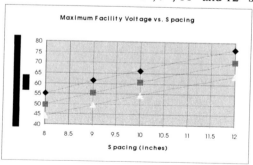

FIGURE 3. Maximum facility voltage for facilities composed of #4, #6, and #9 having 8" to 12" spacing.

Validation

The Equivalent Circuit Lightning Model makes the following assumptions: rebar segments can be modeled as inductors and resistors in series, a lightning strike can be represented as a current waveform in PSpice, and voltage and current distributions in the PSpice circuit are equivalent to the distributions that would occur in the modeled facility due to a lightning strike. In order to test these assumptions the following validation method is being proposed.

A known current pulse will be injected into small scale, physical models constructed of wires or small rods that have known materials properties. Test measurement equipment will be attached to various junctures (nodes) within the small scale, physical model. Potential differences within the physical model resulting from the current pulse at these locations will be recorded. The small scale, physical model will then be simulated using ECLM method. And a comparison of results of the validation test to the simulations will be made.

Figure 4 depicts experimental setup for validation testing.

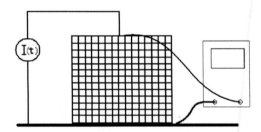

FIGURE 4. Experimental setup for validation testing.

Conclusions

The Equivalent Circuit Lightning Model method may be used to simulate voltage distributions in steel reinforced facilities. ECLM simulations have shown that increasing rebar diameter lowers the maximum facility by a small factor. ECLM Simulations also show that reducing rebar spacing reduces the maximum facility voltage a more significant factor that is roughly proportional to the reduction in rebar spacing.

Acknowledgments

This research was funded as a Plant Directed Research and Development project for BWXT Pantex L.L.C.

Investigation of a Light Gas Helicon Plasma Source for the VASIMR Space Propulsion System

Jared P. Squire[*], Franklin R. Chang-Diaz[*], Verlin T. Jacobson[*], Tim W. Glover[*], F. Wally Baity[†], Mark D. Carter[†], Richard H. Goulding[†], Roger D. Bengtson[¶], and Edgar A. Bering, III[‡]

[*]*Advanced Space Propulsion Laboratory, NASA Johnson Space Center, Houston, TX 77059*
[†]*Oak Ridge National Laboratory, Oak Ridge, TN 37831*
[¶]*University of Texas at Austin, Austin, TX 78712*
[‡]*University of Houston, Houston, TX 77204*

Abstract. An efficient plasma source producing a high-density ($\sim 10^{19}$ m^{-3}) light gas (e.g. H, D, or He) flowing plasma with a high degree of ionization is a critical component of the Variable Specific Impulse Magnetoplasma Rocket (VASIMR) concept. The high degree of ionization and a low neutral background pressure are important to eliminate the problem of radial loss and axial drag due to charge exchange. We have performed parametric (e.g. gas flow, power (0.5 – 3 kW), and magnetic field studies of a helicon operating with gas (D$_2$ or He) injected at one end, with a high magnetic mirror downstream of the antenna. The downstream mirror field has little effect on the exhaust flux up to a mirror ratio of 10. We have explored operation with a cusp and a mirror field upstream. The application of a cusp increases the plasma flux in the exhaust by a factor of two. Plasma flows into a large (5 m^3) vacuum (< 10^{-4} torr) chamber at velocities higher than the ion sound speed. High densities ($\sim 10^{19}$ m^{-3}) have been achieved at the location where ICRF will be applied, just downstream of the magnetic mirror.

INTRODUCTION

The Variable Specific Impulse Magnetoplasma Rocket (VASIMR) is a high power rf driven magnetized plasma rocket, capable of high exhaust velocities, > 100 km/s.[1] Research focuses on three major areas: helicon[2,3] plasma production, ion cyclotron resonant frequency (ICRF) acceleration[4] and plasma expansion in a magnetic nozzle. The present VASIMR experiment (VX-10) performs research to demonstrate the thruster concept at a total rf power on the order of 10 kW. The helicon must produce a plasma stream from light gases (H$_2$, D$_2$, and He) with a high degree of ionization that flows into a high magnetic field, ~ 0.5 T, where rf power at about 2 MHz can be converted to particle energy at the ICRF fundamental resonance. The high degree of ionization is critical to reduce charge exchange losses. This paper focuses on recent experimental results in developing such a light gas helicon plasma source.

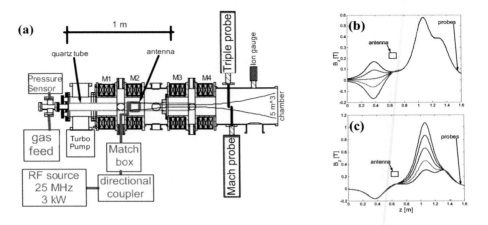

FIGURE 1. a) Schematic of the VX-10. b) Axial magnetic profile with an upstream field scan; c) magnetic profile with peak field scan.

EXPERIMENT

Figure 1(a) shows a schematic of the VX-10 experimental system. The system consists of four cryogenic electromagnet coils integrated into the vacuum chamber. Figure 1(b) shows typical magnetic field profiles for helium discharges. A quartz tube passes down the axis inside the vacuum and is sealed against the port cover on the upstream end. A gas choke located near the magnetic field maximum greatly reduces the neutral gas flow from the helicon to the main chamber. A 25 MHz, 3.3 kW source drives a water-cooled Boswell type double-saddle antenna (10 cm long) located inside the vacuum around the quartz tube and roughly centered under the M2 coil.

Gas (He or D_2) flows into the upstream end of the quartz tube. The discharges are pulsed for approximately 3 seconds to keep the pressure in the exhaust region below 10^{-4} torr. Data presented here are measured downstream of the peak magnetic field in the exhaust. We use two reciprocating probes, Langmuir triple probe and Mach probe.[5] The Langmuir probe density profiles, with helium, were normalized (factor 1.4) with the line averaged density measurement using an interferometer. The Mach number is calculated as the average of two models[5,6] with the error bars representing the difference between models. We calculate the total plasma flux from the probe radial profiles. Uncertainties in Mach probe models dominate errors, with shot-to-shot variations of less than 5%.

RESULTS

Others[7,8,9] have reported that helicons operate most efficiently near the lower hybrid frequency, $\omega_{rf} \approx \omega_{LH} \approx (\omega_{ce}\omega_{ci})^{1/2}$. We also find this for VX-10. Densities measured at the probes are approximately 10^{18} m^{-3} and are flowing at or greater than the ion sound speed. Field line mapping indicates densities near 10^{19} m^{-3} at the location of the ICRF antenna. Measurements nearer to the helicon antenna[10] indicate slower flow velocities and higher densities.

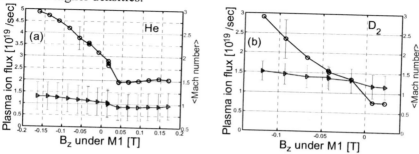

FIGURE 2. Plasma flux (O) and Mach number (▷) with a B_z scan at M1, z = 0.38 m. a) He and b) D_2.

Other experiments have reported increased plasma density when a cusp is placed in the vicinity of the helicon antenna.[11] We have observed a similar increase in the plasma flux with application of a cusp upstream of the antenna. As Fig. 1(b) shows, we have performed a scan of the field upstream of the antenna while keeping the field strength at the antenna location approximately constant. The measured plasma flux with a strong reversed field is more than double the case without the cusp, for both He and D_2, as shown in Fig. 2. For deuterium, positive upstream field gave rise to difficulty with plasma startup.

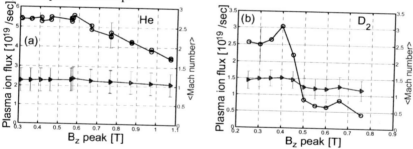

FIGURE 3. Plasma flux (O) and Mach number (▷) for a scan of the peak B_z. a) He and b) D_2.

We scanned the peak field downstream of the antenna while keeping the field strengths at the antenna and probes constant, as shown in Fig. 1(c). We do not find a strong effect on the plasma flux due to the mirror field, up to a mirror ratio of about 10, as shown in Fig. 3. For deuterium, we see a mode change at about a mirror ratio of 10, indicated by the flux reduction. We note, however, that the stray field from M3 changes the gradient at the antenna for the high mirror ratios.

Finally, we find that with the installation of a gas choke that there is evidence that we achieve a plasma stream with a high degree of ionization (~100%) as desired for ICRF experiments. We previously have reported these results with helium[12] and show the similar results for deuterium here. When we decrease the injected gas rate below a critical value, the plasma flux decreases proportionally and the electron temperature (T_e) begins to rise up to nearly double, shown in Fig. 4(a). The rise in T_e is corroborated by a gridded energy analyze, but the indicated value is likely higher than the actual value. Much above the critical input gas rate, the output flux degrades substantially. Also, a helicon power scan, with the gas flow near the critical value, shows plasma flux proportional to power up to a value that starts to saturate, as shown in Fig 4(b).

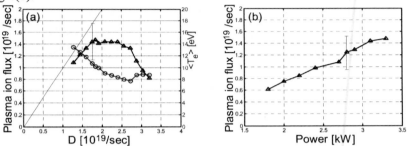

FIGURE 4. a) Plasma flux ($\triangle$) and T_e ($\bigcirc$) with a scan of the input neutral flux. b) Plasma flux with a power scan.

SUMMARY

We have explored the plasma flux from a flowing helicon discharge with helium and deuterium when a magnetic cusp is applied upstream of the antenna and a magnetic mirror downstream. We find that the cusp increases the plasma flux measured in the exhaust by more than a factor of two. The mirror field has little effect on the flux below mirror ratios of 10. Lastly, we achieve a plasma stream with a high degree of ionization that will make a good target for future ICRF experiments.

REFERENCES

1. Chang Díaz, F. R., "An Overview of the VASIMR Engine: High Power Space Propulsion with RF Plasma Generation and Heating", RADIO FREQUENCY POWER IN PLASMAS, 14th Topical Conference, Oxnard, CA, AIP Conference Proceedings 595, 3 (2001).
2. Boswell, R. W. and Chen, F. F., *IEEE. Transactions of Plasma Science,* **25,** 1229-1244, (1997).
3. Chen, F. F. and Boswell, R.W., *IEEE. Transactions of Plasma Science,* **25,** 1245-1257, (1997).
4. Breizman, B. N. and Arefiev A. V., *Phys. Plasmas* **8,** 907 (2001).
5. Hutchinson, I.H., *Phys. Rev. A*, **37,** 4358 (1987)
6. Stangeby P. C., *Phys. Fluids,* **27,** 2699 (1984)
7. Zhu, P. and Boswell, R. W., Phys. Rev. Lett. **63,** 2805 (1989).
8. Yun, S.-M, Kim., J.-H., and Chang, H.-Y., *J. Vac. Sci. Technol. A* **15,** 673 (1997).
9. Carter, M. D., et al., P*hys. Plasmas*, **9,** 5097 (2002).
10. Squire, J. P, Chang Díaz., F. R., Glover, T. W., Jacobson, V. T., Chavers, D. G., Bengtson, R. D., Bering III, E. A., Boswell, R. W., Goulding, R. H. and Light, M., "Progress in Experimental Research of the VASIMR Engine" *Transactions of Fusion Science and Technology*, **43,** 111 (2002).
11. Guo, X. M., Scharer, J., Mouzouris, Y., and Louis, L., *Phys. Plasmas* **6,** 3400 (1999).
12. Squire, J. P., et al., "Experimental Research Progress Toward the VASIMR Engine", 28th International Electric Propulsion Conference, Toulouse, France, Conference Proceedings (2003).

Experimental Measurements and Modeling of a Helicon Plasma Source with Large Axial Density Gradients*

S.M. Tysk, C.M. Denning, J.E. Scharer, B.O. White, and M.K. Akhtar

Department of Electrical and Computer Engineering
University of Wisconsin – Madison 53706

Abstract. An investigation of wave magnetic field, density and temperature profiles, electron distribution function and wave-correlated Argon optical emission in a helicon plasma source is carried out. Diagnostics include Langmuir and wave magnetic field probes, interferometer, monochromator, and retarding field energy analyzer. The UW helicon experimental facility operates with argon pressures in the range of 1-300 mTorr. A variable capacitor matching network is used to couple up to 1.3 kW of pulsed RF power to a half-turn double-helix antenna. A uniform axial magnetic field of 200-1000 G is applied. Densities in the range of 10^{11}-3×10^{13} cm^{-3} are obtained. Wave-correlated optical emission and modulation is externally measured at 443 nm corresponding to an upper state 35 eV above the neutral ground state and phase correlated with the 13.56 MHz antenna current. MAXEB, AntenaII, and nonlinear ionization codes are used to model the conditions present in the system and to provide a comprehensive picture of wave field behavior and fast and thermal electron ionization processes in a helicon source with strong axial density gradients.

INTRODUCTION

Helicon plasma sources are high efficiency, high density sources that operate over a wide range of densities (10^{11}-10^{13} cm^{-3}). Helicons are typically cylindrically symmetric with an axial magnetic field and operate at a radio frequency well above the ion cyclotron frequency. These sources have a wide range of applications including argon lasers[1], materials processing, and space based thruster systems[2].

An important question for helicon sources is the cause of the highly efficient ionization. The role of a non-Maxwellian fast electron component of the electron distribution function[3] in the helicon ionization process is currently an important issue. The existence of such a population of fast electrons is investigated using an optical diagnostic similar to that used by Ellingboe et al.[4] and Scharer et al.[5] on the WOMBAT helicon experimental facility. An important difference in this research is that the optical probe is located outside the Pyrex plasma chamber so that the plasma perturbations of the probe are negligible. Optical space-time measurements of the Ar II line emitted from a state 35 eV above the neutral ground state at 443 nm are made. Emission at 443 nm is indicative of the presence of a population of fast electrons which are required to populate the Ar II state through electron collisions.

Experimental Facility

A schematic of the radio-frequency helicon plasma source utilized for these experiments is shown in Figure 1. The plasma chamber is a 10 cm diameter Pyrex tube. The plasma chamber is pumped to a base pressure of $<10^{-6}$ Torr and 3 mT of argon is flowed through the antenna and chamber whereas the gas feed on WOMBAT is downstream. RF Power in the range of 600-800 W is coupled to the plasma with a half-turn double-helix antenna (dominant m = +1 mode) via a capacitive matching network.

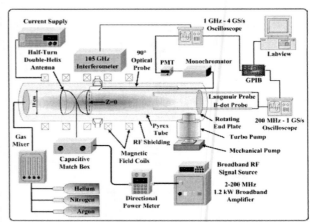

Figure 1: Schematic of UW Helicon Facility

Less than 0.75 % of the RF power is reflected over the entire operating range. A set of 5 electromagnets is used to create a uniform axial magnetic field of 200-1000 Gauss. External optical probe and internal Langmuir, B-dot, and energy analyzer diagnostics are utilized. Internal diagnostics are retractable dog-leg systems capable of axial and radial scans that minimize plasma perturbations. Data is acquired using oscilloscopes and a GPIB interface and analyzed with Labview programs.

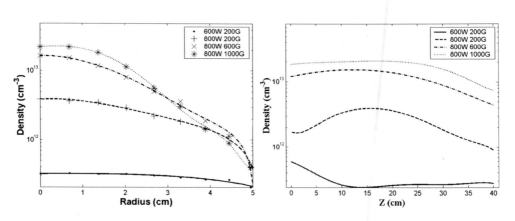

Figure 2: Radial Density Profiles (Z=15 cm)   **Figure 3:** Axial Density Profiles (r=0 cm)

There are three distinct modes of operation for the aforementioned range of powers and axial (B_0) fields. The low density inductive mode exists at low power and low B_0 (<700W, <200G). The moderate density transition mode exists at lower power and lower B_0 (~700-800W, <400G). The high density helicon "blue" mode exists at higher power and higher B_0 (>800W, > 400G). These modes can be characterized by their axial and radial density profiles as shown in Figures 2 and 3. The inductive mode has a relatively uniform radial profile and an axial profile that peaks at the antenna and decays away. The transition mode has a broad radial profile and a highly downstream peaked axial profile. The helicon "blue" mode has a highly centrally peaked radial profile and moderate downstream peaked axial profile. Optical emission at 443 nm does not occur in the inductive mode begins in the transition mode and increases as power and magnetic field is increased.

Experimental Results

In this experiment the RF power delivered to the plasma is pulsed with a 10% duty cycle and a 6 ms pulse length. In the transition mode (800W, 200G) an initial high density transient (6×10^{12} cm^{-3}) occurs for the first 0.5 ms of the pulse and levels out for the following 2.0 ms (4×10^{12} cm^{-3}) and drops to a lower density (2×10^{12} cm^{-3}) for the final 3.5 ms of the pulse. Binning measurements are taken at 1.5 ms into the pulse with a 6 μs window that spans 85 RF periods. In the transition mode a traveling emission peak is observed as shown in Figure 4. The peak bin travels with a resonant electron phase velocity ($v_{\varphi 443} = \omega(\Delta z/\Delta\varphi)$) 2.4×10^6 m/s corresponding to a resonant energy of 16.4 eV. Figure 4b shows the modulation depth for the transition mode increasing away from the antenna. This is indicative of an increasing contribution to the emission from wave correlated excitation. The modulated emission present in the helicon "blue" mode is close to the random level and would be the result of emission from "thermal" random background collisions. The dotted line in Figure 4b represents the measured average noise modulation resulting from a random signal input into the optical binning diagnostic.

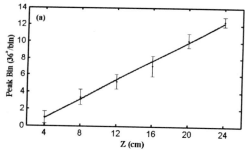

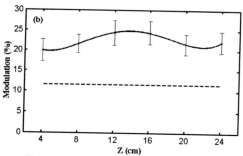

Figure 4: Transition Mode (800W, 200G) (a) Wave Correlated Emission (b) Modulation

The B_z wave magnetic field is measured with a small Pyrex enclosed five-turn loop B-dot probe. The electrostatic signal is rejected using a hybrid combiner with a common mode rejection ratio of 38 dB. The AntenaII code is then used to model the E_z field in accordance with lab data using an axial averaged density. Figure 5 plots the wave phase velocity from the B-dot probe lab data and from the result of the AntenaII run along with the resonant electron phase velocity from the optical binning experiment. The phase velocity for all three cases show good agreement with an average phase velocity

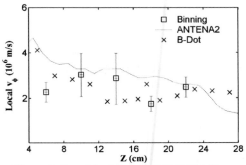

Figure 5: Transition Mode (800W, 200G) Phase Velocity Comparison

of 2.5×10^6 m/s corresponding to an average resonant electron energy of 18 eV. The wave phase velocity increases after this point and the emission drops off to a negligible value after z = 25 cm.

CONCLUSION

Ar II optical emission measured in the transition mode of the UW helicon experimental facility indicates a population of fast electron moving at the helicon wave phase velocity. Modeling results indicate a contribution to ionization of the argon case by non-Maxwellian components of the electron distribution function. This mode agrees well with the results of the WOMBAT experiment conducted at similar densities but at higher power and lower axial field magnetic field. The high density blue mode preliminary results indicate significantly reduced modulation indicating ionization primarily due to thermal processes. Energy analyzer data and further detailed modeling will describe the character of the helicon source in the transition and helicon "blue" modes.

ACKNOWLEDGEMENTS

*Research supported by NSF and in part by AFOSR.

REFERENCES

[1] P. Zhu and R. Boswell, Physics of Fluids B 3, 869 (1991)
[2] F. R. Chang Diaz, J. P. Squire, R. D. Bengston, B. N. Breizmann, F. W. Baity, and M. D. Carter, *Proc. of the 36th AIAA/ASME/SAE/ASEE Joint Propulsion Conference (Huntsville, AL 2000)*, No. 2000-3756
[3] A.W. Molvik, T.D. Rognlein, J.A. Byers, R.H. Cohen, A.R. Ellingboe, E.B. Hooper, H.S. McLean, B.W. Stallard, and P.A. Vitello, J. Vac. Sci. Techno. A 14 (3), 984 (1996).
[4] A. Ellingboe, R. Boswell, J. Booth, and N. Sadeghi, Physics of Plasmas 2 (6), 1807 (1995)
[5] J. Scharer, A. Degeling, G. Borg, and R. Boswell, Physics of Plasmas 9 (9), 3734 (2002).

RF-Plasma Coupling Schemes for the SNS Ion Source

R. F. Welton, M. P. Stockli, S. Shukla, Y. Kang

Accelerator Systems Division, Spallation Neutron Source*
Oak Ridge National Laboratory, Oak Ridge, TN, 37830-6473

R. Keller and J. Staples

Lawrence Berkeley National Laboratory, Berkeley, CA 94720

Abstract. The ion source for the Spallation Neutron Source (SNS) is a radio-frequency, multicusp source designed to deliver beams of 45 mA of H⁻ with a normalized rms emittance of 0.2 π mm mrad to the SNS accelerator. RF power with a frequency of 2 MHz is delivered to the ion source by an 80-kW pulsed power supply generating nominal pulses 1 ms in width with a 60-Hz repetition rate. The ion source—designed, constructed and commissioned at Lawrence Berkeley National Laboratory (LBNL)—satisfies the basic requirements of commissioning and early operation of the SNS accelerator. To improve reliability of the ion source and consequently the availability of the SNS accelerator and to accommodate facility upgrade plans, we are developing optimal RF–plasma-coupling systems at Oak Ridge National Laboratory (ORNL). To date, our efforts have focused on design and development of internal and external ion source antennas having long operational lifetimes and the development and characterization of efficient RF matching networks.

INTRODUCTION

The Spallation Neutron Source (SNS) is a second-generation pulsed neutron source dedicated to the study of the dynamics and structure of materials by neutron scattering and is currently under construction at Oak Ridge National Laboratory (ORNL). Neutrons will be produced by bombarding a liquid Hg target with a 1.4-MW, 1-GeV proton beam produced using a series of linear accelerators and an accumulator ring [1,2]. SNS performance upgrades are being discussed where 3- to 5-MW of beam power would be required. To meet the 1.4-MW baseline requirement, the ion source must deliver ~45 mA of H⁻ within a 1-ms pulse (60 Hz) into a normalized rms emittance of 0.2 π mm mrad.

---

*SNS is a collaboration of six US National Laboratories: Argonne National Laboratory (ANL), Brookhaven National Laboratory (BNL), Thomas Jefferson National Accelerator Facility (TJNAF), Los Alamos National Laboratory (LANL), Lawrence Berkeley National Laboratory (LBNL), and Oak Ridge National Laboratory (ORNL). SNS is managed by UT-Battelle, LLC, under contract DE-AC05-00OR22725 for the U.S. Department of Energy.

CP694, *Radio Frequency Power in Plasmas: 15th Topical Conference on Radio Frequency Power in Plasmas*, edited by Cary B. Forest
© 2003 American Institute of Physics 0-7354-0158-6/03/$20.00

This report discusses collaborative ion source development efforts undertaken between ORNL and LBNL, focusing specifically on the ion source RF–plasma coupling systems: antenna and matching network.

THE H⁻ MULTICUSP ION SOURCE

A schematic diagram of the H⁻ ion source is shown in Fig. 1. The source plasma is confined by a multicusp magnetic field created by a total of 20 samarium-cobalt magnets lining the cylindrical chamber wall and 4 rows of magnets lining the back plate. RF power (2 MHz, ~45 kW) is applied to the antenna shown in the figure through a transformer-based impedance-matching network. A magnetic dipole (150-300 Gauss) filter separates the main plasma from a smaller H⁻ production region where low-energy electrons facilitate the production of large amounts of negative ions. A heated collar, equipped with eight cesium dispensers, each containing ~5 mg of Cs_2CrO_4, surrounds this H⁻ production volume. The RF antenna is made from copper tubing that is water cooled and coiled to 2 1/2 turns. A porcelain enamel layer insulates the plasma from the oscillating antenna potentials. More details of this source design can be found in reference 3.

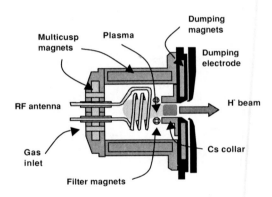

FIGURE 1. Schematic diagram of the SNS ion source.

RF ANTENNA DEVELOPMENT

The multicusp, RF-driven, positive/negative ion sources developed at LBNL have performed quite well in a wide variety of applications [4]. Most of these applications have involved low duty-factor, pulsed operation in which thinly coated (100-200 um) porcelain enamel coatings were sufficient to guarantee long operational lifetimes. The SNS source, on the other hand, requires long pulses (1 ms), high repetition rates (60 Hz) and high RF peak powers (~45 kW) which tends to severely shorten antenna lifetime, in some cases, to only several hours. In a collaboration between LBNL and ORNL, we began improving the lifetime of this ion source component. Detailed accounts of this work can be found in an earlier report [5], and we will only summarize here.

Electromagnetic modeling has shown that a 1- to 2-kV RF potential develops across the length of the antenna because of its inductance. Large RF electric fields can develop between different parts of the antenna and between the antenna and the plasma chamber, since one leg of the antenna is grounded to the plasma chamber through a resistor. If the antenna coating is not sufficiently thick or has too large a dielectric constant, the majority of this electric field will exist within the plasma sheath as opposed to within the insulating coating. Large electric fields in the plasma sheath accelerate charged plasma particles into the coated antenna, causing sputter ejection of the coating materials as well as vaporization of the coating material from localized heating. Eventually, a thin spot in the coating develops that enhances the electric field, driving the process until a hole is burnt through the coating. Once bare conductor of the antenna is exposed to the plasma, the plasma itself can conduct a significant portion of the RF current normally carried by the antenna and thereby greatly reduce the inductive power coupling to the plasma. This entire process can be greatly accelerated if a manufacturing defect exists in the coating.

Quantitative models have been developed and applied to fusion plasmas to determine the fraction of a given electric field that exists within the plasma sheath, versus the field inside a dielectric wall material [6]. Using the plasma parameters of the SNS source and material properties of the porcelain enamel coating, it has been shown that essentially all of the electric field within the plasma sheath can be eliminated provided the following two conditions are met [5]. First, the coating is sufficiently thick, greater than 0.5 mm, several times the thickness of the original coating. Second, the dielectric constant of the porcelain is reduced by removal of the TiO_2 (K=86) component from the porcelain mixture.

Based on these calculations, a local company [7] developed a multilayer coating technique to achieve the specified coating thickness and composition. Since TiO_2 is added to porcelain enamel mixtures purely as a color pigment with no structural importance, it was easily eliminated from the mixture resulting in a coating that appears clear. This approach has yielded coatings as thick as 1 mm, which was achieved through the successive application of ~10 enamel layers. Multi-layer coatings also significantly reduce the chance that single-layer defects will penetrate the entire coating thickness.

To date, antennas fabricated in this fashion have allowed the successful commissioning of the front end of the SNS at LBNL (~500 hours of operation) and again at ORNL (~1000 hours of operation) [8]. In addition to these accelerator-commissioning activities, which have generally required low duty-factor operation, we have also performed several high duty-factor lifetime tests. On one occasion, the source operated continuously for 107 hours with ~25 kW of applied RF power at the full 6% duty factor with no visible damage to the antenna observed. During the last few hours of the test, a ~25-mA beam was extracted from the source. Another such test also occurred at LBNL on the front-end system where the source was operated for 125 hours at 2 to 3% duty factor, again producing ~25 mA with no antenna damage visible.

To ensure an adequate ion source lifetime at the nominal SNS beam current and duty factor, in the absence of specific lifetime tests, we have developed a contingency strategy for coupling RF power into the source: use of an external antenna. This approach is similar to that employed at DESY for their very low duty factor application [9]. To date, thermal and mechanical finite-element analysis has been performed on an Al_2O_3 plasma chamber to determine an optimal design. One such design, shown in Fig. 2, was developed in conjunction with designers at ISI Corporation. The design features an alternative ion source backflange where the helical antenna is submerged in de-ionized water surrounding an Al_2O_3 plasma chamber. This system is currently under consideration for development.

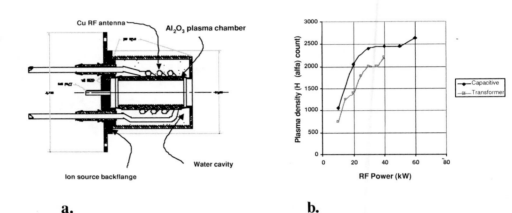

FIGURE 2 (a) Initial design of the SNS external RF antenna and (b) comparison of the optimal RF coupling that could be achieved by each matching network (see text).

RF MATCHING NETWORKS

Efficient power transfer between the RF generator and ion source antenna is extremely important since the extracted H⁻ current is, to first order, directly proportional to the RF power coupled to the plasma [5]. We have investigated two alternative schemes to accomplish this coupling: (i) a transformer-based matching network and (ii) a simple capacitive voltage divider circuit. Both matching networks were developed at LBNL and are described in detail in reference 10 but have not yet been critically evaluated for plasma-generation efficiency at power levels required by the SNS ion source.

Briefly, the transformer-based network (i) directly excites the primary windings of a ferrite transformer (6 turn). The secondary winding (1 turn) is attached to a series LC circuit that includes the ion source antenna (inductance) and a variable vacuum capacitor (50 to 2300 pF) that is adjusted to allow the circuit to resonate at 2 MHz. The impedance matching is accomplished by varying the turn number of the ferrite transformer. The other matching network (ii) consists of a parallel LC circuit where a vacuum capacitor (1100 to 2500 pF) is resonated with an inductance, which includes the ion source antenna. The impedance of the RF generator is matched to that of the parallel LC circuit by a second vacuum variable capacitor (50 to 2300 pF), which is connected between the generator and LC circuit.

Figure 2b shows a comparison between the relative plasma densities that could be achieved using each individually optimized matching network. Each network was electrically tuned for optimum power transfer, and the H_2 flow was held at 15 SCCMs as required for optimal H^- production. Water cooling was added to several components of the capacitive voltage divider matching network to allow operation at high power levels. The relative plasma density was monitored by observing the Balmer-α line emitted from the plasma by using an optical spectrometer. Fig. 2b clearly shows that the capacitive network is capable of more efficient coupling to the source and is also stable at higher total RF powers. At a nominal operating power of 30 kW, the capacitive network produced a ~20% denser plasma than the transformer-based network currently employed. Based on these findings we will consider the benefit of ultimately switching to the lower-loss capacitive system.

REFERENCES

1. Holtkamp, N., et al., "The SNS Linac and Storage Ring: Challenges and Progress Towards Meeting Them," *Proceedings of the European Particle Accelerator Conference*, EPAC'02, Paris, France, ID: 191 - TUZGB002.
2. White, M., "The Spallation Neutron Source (SNS)," *Proceedings of the Linear Accelerator Conference*, LINAC '02, Gyeongju, Korea, ID: MO101.
3. Keller, R., et al., "Ion-source and LEBT Issues with the Front-End Systems for the Spallation Neutron Source," *Rev. Sci. Intrum.* 73, 914 (2002).
4. Leung, K. N., "The Application and Status of the Radio Frequency Multicusp Source," *Rev. Sci. Intrum.* 71, 1064 (2000).
5. Welton, R.F., et al., "Ion Source Antenna Development for the Spallation Neutron Source," *Rev. Sci. Intrum.* 73, 1008 (2002).
6. J.R. Myra et al., J. Nucl. Mater. 249 (1997) 190.
7. Cherokee Porcelain Enamel Corporation, 2717 Independence Lane, Knoxville, TN 37914.
8. A. Aleksandrov, "Commissioning of SNS Front End Systems at ORNL," PAC 2003.
9. J. Peters, "Internal versus External RF Coupling into a Volume Source," EPAC'02, Paris, France, ID: 710 – THPRI025.
10. J. Staples, "High-Efficiency Matching Network for RF-Driven Ion Sources," *Proceedings of the Particle Accelerator Conference*, PAC '01, Chicago, USA, ID: WPAH014.

THEORY AND MODELING

Effects of Non-Maxwellian Plasma Species on ICRF Propagation and Absorption in Toroidal Magnetic Confinement Devices[1]

R.J. Dumont*[†], C.K. Phillips[†] and D.N. Smithe**

*Association Euratom-CEA sur la Fusion Contrôlée, F-13108 St Paul lez Durance, France
[†]Princeton Plasma Physics Laboratory, P.O. Box 451, Princeton, NJ 08543, U.S.A.
**Mission Research Corp., Newington, VA 22122, U.S.A.

Abstract. Auxiliary heating supplied by externally launched electromagnetic waves is commonly used in toroidal magnetically confined fusion experiments for profile control via localized heating, current drive and perhaps flow shear. In these experiments, the confined plasma is often characterized by the presence of a significant population of non-thermal species arising from neutral beam injection, from acceleration of the particles by the applied waves, or from copious fusion reactions in future devices. Such non-thermal species may alter the wave propagation as well as the wave absorption dynamics in the plasma. Previous studies have treated the corresponding velocity distributions as either equivalent Maxwellians, or else have included realistic distributions only in the finite Larmor radius limit. In this work, the hot plasma dielectric response of the plasma has been generalized to treat arbitrary distribution functions in the non-relativistic limit. The generalized dielectric tensor has been incorporated into a one-dimensional full wave all-orders kinetic field code. Initial comparative studies of ion cyclotron range of frequency wave propagation and heating in plasmas with nonthermal species, represented by realistic distribution functions or by appropriately defined equivalent Maxwellians, have been completed for some specific experiments and are presented.

INTRODUCTION

The physics underlying the interaction between a fusion plasma and waves belonging to the Ion Cyclotron Resonance Frequency (ICRF) range can be fairly complicated, and highly dependent on the scenario under consideration. Due to the range of wavelengths involved in this interaction, it is generally needed to resort to full-wave codes, solving directly the wave equation. Recently, various refinements have been added to these descriptions. For example, in some situations such as High Harmonic Fast Wave electron heating (HHFW) or Ion Bernstein Wave heating (IBW), the ion Larmor radius (ρ_i) can become comparable to, or even larger than the wavelength and a second order expansion in $k_\perp \rho_i$, $k_\perp$ being the perpendicular wavenumber, may not be sufficient. This is why the so-called all-orders full wave codes have been developed and have proven successful in providing a realistic theoretical description of the wave-plasma interaction.

One such code is the one-dimensional (1-D) all-orders full wave code METS[1][2].

---
[1] This work was supported at PPPL by U.S. DOE contract #DE-AC02-76CH03073

Owing to its low requirements in terms of computational power, it provides a useful tool to describe experiments not possibly simulated by a second-order code, as well as a benchmark for ray-tracing calculations, whose range of application is determined by the validity of the WKB approximation. Thanks to the recent progresses of computing resources, hitherto unexplored situations have been studied and new physical elements have been unveiled. All-orders two-dimensional (2-D) codes, such as AORSA-2D, have been developed and used to show the 2-D nature of the physics associated with some ICRH-based schemes, such as mode-converted wave heating[3]. However, the computation requirements are fairly high and so far, these codes have been limited to Maxwellian populations. In this work, the influence of the non-thermal populations commonly met in fusion experiments is investigated. To this aim, METS has been extended to handle the corresponding non-Maxwellian distributions, by resorting to massively parallel computers to perform the large number of additional velocity integrals.

In this paper, the theoretical background will be shortly introduced, and the results obtained for three different scenarios (1) mode-converted IBW with simultaneous Tritium Neutral Beam Injection (NBI) in TFTR; (2) HHFW in the presence of a Deuterium Beam on NSTX and (3) ^{3}He minority heating with alpha particles on ITER, are presented.

LOCAL FULL-WAVE ANALYSIS

The key element of the problem is the wave equation, which can be written as

$$\nabla \times \nabla \times \mathbf{E} - \frac{\omega^2}{c^2}\left(\mathbf{E} + \frac{i}{\omega\varepsilon_0}\mathbf{J}_p\right) = i\omega\mu_0\mathbf{J}_s \qquad (1)$$

In this equation, $\mathbf{E}$ and $\mathbf{J}_p$ are the electric field and the plasma current. $\mathbf{J}_s$ is the current provided by external sources. METS uses ingoing and outgoing waves boundary conditions, which has the advantage of allowing the evaluation of the per-pass absorption. A slab geometry is employed, the plasma parameters varying along $\hat{\mathbf{e}}_x$, whereas $\hat{\mathbf{e}}_z$ is chosen to be the direction of the toroidal magnetic field. This allows to write the Discrete Fourier Transform of the wave equation under the form

$$\sum_j \exp(ik_j x_i)\left[\mathbf{k}_j \times \mathbf{k}_j \times \overline{\overline{\mathbf{1}}} + \frac{\omega^2}{c^2}\overline{\overline{\mathbf{K}}}(x_i,\mathbf{k}_j)\right] \cdot \mathbf{E}(k_j) = 0 \qquad (2)$$

where x_i is the i-th spatial grid point and $k_j \equiv 2\pi j/L$, where L is the length of the chord along which the computation is performed. $\mathbf{k}_j \equiv k_j\hat{\mathbf{e}}_x + k_y\hat{\mathbf{e}}_y + k_z\hat{\mathbf{e}}_z$. k_y is assumed to be constant and k_z varies in accordance with the toroidal upshift, i.e. $k_z(R) \equiv k_{z,ant}R_{ant}/R$, R being the major radius, R_{ant} the antenna major radius and $k_{z,ant}$ representing the launched spectrum.

$\overline{\overline{\mathbf{K}}}$ is the dielectric kernel, which is a generalized version of the usual dielectric tensor. In a weakly inhomogeneous plasma (i.e. in a plasma such that $\rho_i \ll L_B$, L_B being the equilibrium scale length), it is obtained by differentiating the absorption kernel $\overline{\overline{W}}$

$$\overline{\overline{K}}(x,k_x) = \overline{\overline{1}} + \left(1 - i\frac{\partial}{\partial k_{2,x}} \cdot \frac{\partial}{\partial x}\right)\overline{\overline{W}}(x,k_{1,x},k_{2,x})\bigg|_{k_{1,x}=k_{2,x}=k_x} \tag{3}$$

$\overline{\overline{W}}$ is needed to obtain both $\overline{\overline{K}}$ and W_{abs}, the spatial absorption on each species and its computation is the main task of METS.

When the particular case of a Maxwellian distribution function is considered, $\overline{\overline{W}}$ can be expressed analytically, possibly including the magnetic field parallel gradient effects[4]. In this work, however, the distribution function is supposed to be arbitrary. This implies that the parallel gradient effect must be neglected (the extensive discussion of this feature is beyond the scope of this paper and shall be presented in a forthcoming article) and that both $v_\parallel$ and $v_\perp$ integrations have to be performed numerically, which is a computationally intensive task. This makes it necessary to use massively parallel computers. This version of the code has been thoroughly benchmarked versus the original version, by comparing the results obtained using the analytical expressions for $\overline{\overline{W}}$ and the results obtained with a numerical Maxwellian, where the velocity-space integrations are actually performed. Provided an adequate velocity space grid is used, it has been found that the results are generally indistinguishable[5].

MODE-CONVERTED ION BERNSTEIN WAVE IN TFTR

ICRF waves have been used on TFTR ($R_0 = 2.84$m, $a_0 = 0.98$m), as a tool to heat electrons, through the absorption of a mode-converted Ion Bernstein wave[6]. A Deuterium-Tritium (D-T) shot is considered, with the following parameters: $B_0 = 4.7$T, $n_{e0} = 4.7 \times 10^{19}$m^{-3}, $T_{e0} = 6.8$keV. In this discharge, the following thermal ion species are taken into account: Deuterium, Hydrogen and Carbon, all with central temperature $T_{i0} = 31$keV. The frequency of the Fast Wave launched from the low field side antenna is $f_{FW} = 80$MHz with parallel wavenumber $k_\parallel^{ant} = 7$m^{-1}. In addition, a Tritium beam was injected in the plasma, with $E_b = 80$keV. The Tritium ions concentration, obtained from spectroscopic measurement is $\eta_T \equiv n_T(0)/n_e(0) = 0.42$. To simulate these beam ions, an isotropic slowing-down distribution function is first employed[7].

In order to treat a non-Maxwellian population with a Maxwellian code, it is usual to compute the equivalent Maxwellian distribution for this population. This is done by evaluating the equivalent temperature and density profiles to ensure that both the number of particles and the energy content are the same for the two distributions. METS can be used to investigate the validity of this assumption, by comparing the results obtained in both cases.

The power deposited on each species is computed. In this discharge, only electrons and Tritium ions are found to absorb a significant fraction of the wave power. On Fig. 1(a) (resp. (b)), the power deposited on electrons (resp. Tritium ions) is shown as a function of major radius.

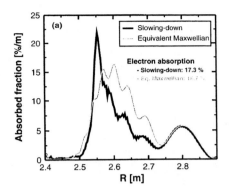 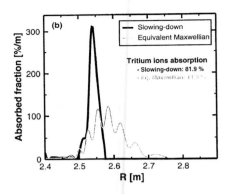

FIGURE 1. Profile of the power deposited on electrons *(a)*; on Tritium ions *(b)* for a slowing-down distribution function *(thick line)* and for its equivalent Maxwellian *(thin line)*.

It can be observed that the electron power deposition profile exhibit similar shapes. The net absorption obtained for both distribution functions are in agreement within 2%. On the other hand, the profiles of the Tritium ions absorption look very different, even though again, the net absorption agree. With the slowing-down distribution, the power deposition is much more peaked and shifted towards high field side. Clearly, the conclusion is that the use of an equivalent Maxwellian is inappropriate to describe such a mode conversion scenario.

FAST WAVE ELECTRON HEATING IN NSTX

High Harmonic Fast Wave heating is commonly employed on the spherical torus NSTX to heat electrons and to drive non inductive toroidal current[8]. When the wave is injected in a plasma with simultaneous Deuterium Neutral Beam injection, it is reasonable to expect a modification of the propagation and absorption properties of the wave under the influence of the non-Maxwellian fast ion population, which can divert a fraction of the wave from being absorbed by the electrons.

We simulate a typical HHFW+NBI shot with $B_0 = 0.45T$, $n_{e0} = 2.75 \times 10^{19} m^{-3}$ and $T_{e0} = 1keV$. The wave frequency is $f_{FW} = 30MHz$ and the launched parallel refractive index is $k_\parallel^{ant} = 14m^{-1}$. Three thermal ion species with $T_{i0} = 1keV$ are taken into account: Deuterium, Hydrogen and Carbon. In this case, the absorption is expected to take place approximately between the 8-th and the 12-th harmonic of the hydrogen cyclotron resonance, which makes the use of an all-orders code necessary. A fast Deuterium component, induced by the 80keV beam, is considered, such as $n_{D,fast}(0)/n_e(0) = 0.15$. Simulating the beam ions with an isotropic slowing-down distribution, as in the previous section, it was shown[5] that the net absorption reaches 94%, to be compared with 70% in the absence of beam. The main point is that when the beam ions are considered, the wave power gets split among electrons and fast Deuterium ions. In fact, the electron absorption drops from 70% to 24% whereas the beam ions absorb 70%

of the power. It appears that this situation is quite well reproduced when an equivalent Maxwellian is substituted to the isotropic slowing-down distribution, as was shown in a previous work[5]. Similar tests have been performed with different parameters and a good agreement is obtained in each case. This legitimates the use of an equivalent Maxwellian to model such a scenario in NSTX, which has the advantage of largely reducing the computation time.

In order to improve the description of the beam ions, the isotropic slowing-down distribution used above can be replaced by using a slowing-down model allowing to take into account the anisotropic nature of a beam distribution[7]. The beam ions source is assumed to be Gaussian-shaped $K(\theta) = A \cdot \exp(-(\theta - \theta_0)^2/\Delta\theta^2)$ where A is a constant and the beam characteristics, θ_0 and $\Delta\theta$, respectively the injection angle with respect to the magnetic field and the beam angular width, are considered as given. In the simulations presented here, $\Delta\theta = 30°$ is used.

The effect of the distribution function anisotropy on the wave absorption can be examined. On Fig. 3(a) and (b) are presented the electron and Fast Deuterium power deposition profiles obtained when the beam ions are described by an isotropic slowing-down distribution function or when a tangential Gaussian-shaped source is used ($\theta_0 = 0°$).

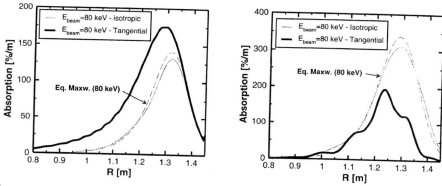

FIGURE 2. Power deposited on electrons *(a)*; on the beam ions *(b)* when an isotropic slowing-down distribution is used to simulate the fast ions *(thin solid line)* or when a Gaussian, tangential beam injection is considered *(thick solid line)*. Also shown is the equivalent Maxwellian result *(dashed line)*.

Without beam, the power is found to be completely absorbed by the electrons, the single-pass absorption being 70% (see ref. [5]). When the beam ions are taken into account by means of an isotropic distribution function, the fast ion absorption reaches 70% whereas the electron absorption drops to 24%. If an anisotropic distribution function is used, for tangential injection, the electron absorption increases to 44% and the fast ion absorption is reduced by a factor two (35%). This is a consequence from the fact that the dynamics of the wave-ions interaction is essentially taking place in the perpendicular direction. Meanwhile, the anisotropic distribution function obtained for tangential injection being essentially peaked in the parallel direction, the ions tend to absorb less power, more being thus available to the electrons. Another observation is that if the equivalent Maxwellian is suited to approximate the isotropic case, it fails to describe the anisotropic result accurately, overestimating the fast ion absorption.

Finally, the effect of the distribution function anisotropy on the wave absorption is investigated. On Fig. 3(a) and (b) are shown the electron and Fast Deuterium power deposition profiles obtained when the beam injection angle is varied between 0 and 30°

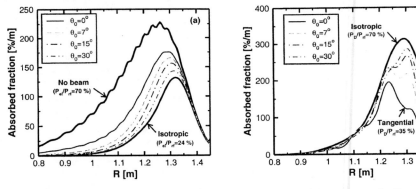

FIGURE 3. Power deposited on electrons *(a)*; on the beam ions *(b)* for an anisotropic, Gaussian beam, simulated by a slowing-down distribution function when the angle of injection is varied between 0 and 30°. Also shown are the deposition profiles obtained when no beam is taken into account *(a)*, as well as when an isotropic slowing-down distribution is used.

It appears that even a moderate modification of the injection angle has significant consequences on the absorption by fast ions (and consequently by electrons as well). For instance, the results obtained when $\theta_0 = 30°$ are close to the isotropic deposition profiles. Obviously, this point has to be considered in Fast Wave Electron Heating experiments since even a slight misalignment of the beam, or a geometrical imperfection, can have significant consequences, by increasing the fast ion absorption to the detriment of the electrons.

MINORITY HEATING IN ITER

Of the various possible ion heating schemes proposed for ITER, fundamental Helium-3 minority heating has received a lot of interest[9]. By using the same type of analysis with an isotropic slowing-down distribution to model the alpha population, it was shown that the fusion ashes could divert a significant fraction of the power initially aimed at heating the minority ions[5].

A further step can be achieved by using METS to describe not only the non-Maxwellian nature of the alpha particles, but introducing the formation of a RF-induced tail for the minority ion. To this aim, a fairly simple model is used, where the corresponding distribution is assumed to be Maxwellian in the parallel direction with temperature T_{i0}, whereas a tail of temperature $T_{t,min}$ develops in the perpendicular direction, with a turnover velocity v_*, above which the pitch-angle scattering become weak. f is then assumed to have the form $f(\mathbf{u}) = F(u_\perp) \cdot f_{max}(u_\parallel)$ (**u** being the velocity normalized to $T_{i0}^{1/2}$) and

$$F(u_\perp) = \begin{cases} A\exp(-u_\perp^2) & \text{if } u_\perp < u_* \\ A\exp\left[-u_*^2\left(1 - \frac{T_{i0}}{T_{t,min}}\right)\right] \cdot \exp\left(-u_\perp^2 \frac{T_{i0}}{T_{t,min}}\right) & \text{else} \end{cases} \quad (4)$$

To simulate this type of discharge in ITER, the following parameters are used: $R_0 = 6.3$m, $a_0 = 2.0$m, $B_0 = 5.2$T, $n_{e0} = 1 \times 10^{20}$m^{-3}, $T_{e0} = 25$keV. For the wave $f_{FW} = 53$MHz and $k_\parallel^{ant} = 10$m^{-1}. D, T, and ^{6}C ions are all taken to be thermal species with $T_{i0} = 31$keV. Both Deuterium and Tritium are assumed to have the same concentration ($\eta_D = \eta_T = 42\%$) and $\eta_{He-3} = 3\%$. The fusion born Helium-4 ions are modeled by means of a slowing-down distribution characterized by $E_b = 3.5$MeV and $\eta_{He-4} = 1\%$, with a peaked density profile. For these parameters and assuming the power density to be $p_{rf} = 1$MW/m^3, the tail temperature, as predicted with a simple Stix model, is found to be $T_{t,min}(0) \approx 130$keV and the turnover velocity is roughly twice the thermal velocity.

On Fig. 4(a), the absorption profiles are presented when the alpha population is considered or neglected (in the latter case, the ^{6}C density is used to ensure electro-neutrality). The minority ion population is supposed to be Maxwellian. On Fig.4(b), the alphas are considered, with the minority species exhibiting a RF tail

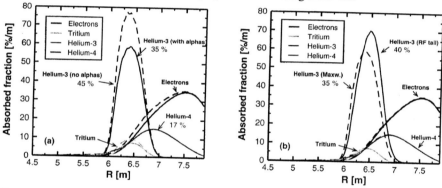

FIGURE 4. Comparison of the power deposition profiles (a) when the effects of the fusion ashes are considered (solid) or neglected (dashed); (b) when the alphas are considered and the minority ion is described by a Maxwellian distribution function (dashed) or when a RF tail is considered (solid).

In the presence of alpha particles, a significant fraction of the wave power is unable to reach the minority ion. The situation is slightly improved when a RF tail is considered owing to the shift of the minority absorption towards the low field side of the plasma. It should be emphasized, however, that a definite conclusion on this subject requires further investigations, especially to improve the model for the RF tail. Indeed, one can envision that the enhanced absorption of the RF power extends the tail further, which in turn increases the wave absorption, to the detriment of the alpha population. Such a simulation requires several iterations between a non-Maxwellian wave code and a Fokker-Planck code. Work is still in progress in this area.

CONCLUSIONS

METS, a 1-D all-orders Full-Wave with the capability to handle non-Maxwellian distribution functions has been presented and employed to simulate various ICRF heating schemes. On TFTR, in a mode-converted IBW electron heating scenario, the results obtained when the Tritium beam ions are described by an isotropic slowing-down distribution function or by its equivalent Maxwellian are compared. It is found that even though the net absorption figures agree well, the fast ion absorption profiles are quite different. The same type of comparison applied to a HHFW electron heating scenario on NSTX, in the presence of NBI fast Deuterium ions is performed. First, the code indicates that these beam ions absorb a significant part of the power, to the detriment of the electrons. Secondly and contrary to the TFTR shot simulation, the equivalent Maxwellian gives a result in fair agreement with the isotropic slowing-down result. However, when an anisotropy of the distribution function is introduced to improve the modeling of a tangential beam injection, the absorbed power predicted in the equivalent Maxwellian case is overestimated for the fast ions, and underestimated for the electrons. Furthermore, it is found that this absorption balance is a sensitive function of the beam injection angle. Finally, the code is employed to simulate a ^{3}He minority heating scenario in ITER, in the presence of alpha particles. It is obtained that if the latter reduce the minority absorption, this deleterious effect is minimized when a RF tail model is used to describe the minority ion heating. In such a case, however, further simulations are clearly needed, to incorporate the deformation of the fast ion distribution function and the resulting modification of the plasma-wave interaction in a self-consistent way.

ACKNOWLEDGMENTS

Fruitful discussions with D.B. Batchelor, L.A. Berry, P.T. Bonoli, M.D. Carter, E. D'Azevedo, D.A. D'Ippolito, R.W. Harvey, E.F. Jaeger, J.R. Myra, A.L. Rosenberg and J.C. Wright are gratefully acknowledged.

REFERENCES

1. D.N. Smithe, Plasma Phys. Controlled Fusion **31**, 1105 (1989).
2. D.N. Smithe, C.K. Phillips, J.C. Hosea, R.P. Majeski, and J.R. Wilson, in *Proceedings of the 12th Topical Conference on Radio Frequency Power in Plasmas*, 1997, Savannah, GA, edited by P.M. Ryan and T. Intrator (American Institute of Physics, New York, 1997), p. 367.
3. E.F. Jaeger, L.A. Berry, E. D'Azevedo *et al.*, Phys. Plasmas **8** (2001) 1573.
4. D.N. Smithe, P.L. Colestock, R.J. Kashuba, and T. Kammash, Nucl. Fusion **27**, 1319 (1987).
5. R.J. Dumont, C.K. Phillips and D.N. Smithe, in *Controlled Fusion and Plasma Physics*, 26B, paper P-5.051 (29th EPS Conference, Montreux, Jun. 17-21, 2002).
6. C.K. Phillips, M.G. Bell, S. Bernabei *et al.*, Nucl. Fusion **40**, 461 (2000).
7. M. Cox and D.F.H. Start, Nucl. Fusion **24**, 399 (1984).
8. J.C. Hosea, R.E. Bell, M. Bitter *et al.*, in *Proceedings of the 28th European Physical Society Conference on Controlled Fusion and Plasma Physics*, 2001, Funchal-Madeira, Portugal, edited by C. Silva, C. Varandas and D. Campbell (European Physical Society, 2001), vol. 25A, p. 1165.
9. D. Van Eester, F. Louche, and R. Koch, Nucl. Fusion **42** (2002) 310.

ECCD for Advanced Tokamak Operations Fisch-Boozer versus Ohkawa Methods [1]

Joan Decker

Plasma Science and Fusion Center, MIT, Cambridge MA 02139

Abstract. Current can be driven using electron cyclotron waves (ECW) by optimizing either the Fisch-Boozer mechanism (ECCD) or the Ohkawa mechanism (OKCD). In ECCD, perpendicular heating due to ECW creates an asymmetric resistivity. In OKCD, current is generated by ECW-induced asymmetric electron trapping. OKCD is a good candidate for off-axis CD where the ECCD effectiveness is reduced due to trapped electrons. The two mechanisms are described using the kinetic, bounce-averaged, Fokker-Planck code DKE with a quasilinear ECW operator. Currents and CD efficiencies for the two methods are calculated and compared in different regions of an advanced tokamak plasma. Numerical results confirm the experimental observations that ECCD is best for central CD but becomes ineffective beyond a certain radial distance from the plasma center. On the low-field side (LFS) of this outboard region, OKCD can very effectively generate localized currents. As it is optimized on the LFS, OKCD requires lower wave frequencies than ECCD - an advantage when considering ECW sources.

INTRODUCTION

Electron cyclotron current drive (ECCD) has been successfully used for full current drive [1], current profile control [2], and the stabilization of MHD instabilities, particularly neoclassical tearing modes (NTM) [3][4][5]. In accordance with the Fisch-Boozer method [6], ECW are used to transfer perpendicular energy to the resonant electrons. This creates an asymmetric resistivity, because more energetic electrons are less collisional. This asymmetry in the resistivity generates a electron flow in the same parallel direction as the resonant electron velocity. However, it has been found experimentally [7][1] that the ECCD efficiency decreases as current is driven further off-axis. This decrease in the ECCD efficiency has been understood to be due to the effect of trapped particles. Because ECCD increases the perpendicular energy of electrons, it also diffuses them to a region of velocity space that is closer to the trapped region. A fraction of these electrons are pitch-angle scattered into the trapped region. Since the bounce period of trapped electrons is much shorter than the collisional detrapping time, half of the electron detrapping will occur through the opposite side of the trapped region, thus creating a counter current. The resulting CD efficiency can thus be strongly reduced.

The Ohkawa method [8] for current drive (OKCD) makes use of electron trapping to generate current. The ECW are launched in the opposite direction from ECCD and the wave parameters are chosen so that the wave-particle interaction induces trapping. This trapping creates a depletion of electrons on the resonant side of the velocity space. The

---

[1] Work carried out in collaboration with Abraham Bers and Abhay K. Ram, *PSFC, MIT*, and Yves Peysson *CEA-Cadarache, France*

fast bounce motion of trapped electrons leads to a symmetric detrapping of electrons. The resulting effect of asymmetric trapping and symmetric detrapping creates a current in the parallel direction opposite from that of resonant electrons.

In the original OKCD description [8] several simplifying and non-self-consistent assumptions were made. Recently, we have shown [9][10] in calculations based upon a complete and self-consistent, neoclassical, 2-D momentum-space Fokker-Planck description of collisions, and quasilinear diffusion due to ECW, that OKCD, with properly chosen ECW frequencies and parallel indices, can effectively generate appreciable currents.

In the following, using our kinetic formulation and code [9][11], we compare ECCD and OKCD in a toroidal geometry. The objective is to determine if the Ohkawa method could be a valuable alternative for far-off axis CD using ECW, where ECCD was found to be ineffective.

KINETIC MODEL

Bounce-Averaged Fokker-Planck Equation

We consider the simplest relevant toroidal geometry in this study of ECCD and OKCD. Toroidal axisymmetry is assumed, and the magnetic flux-surfaces are taken to be circular and concentric.

The steady-state gyro-averaged kinetic equation is given by

$$\mathbf{v}_{gc} \cdot \nabla f = \mathcal{C}(f) + \mathcal{Q}(f) \tag{1}$$

where f is the guiding center distribution function and depends on the 4-D phase-space $f = f(r, \theta, p_{\|}, p_{\perp})$; r is the radial location and θ is the poloidal angle taken from the outboard horizontal mid-plane. The electron momentum is decomposed into its components $p_{\|}$ and $p_{\perp}$ respectively parallel and perpendicular to the magnetic field. $\mathcal{C}$ and $\mathcal{Q}$ are the collisional and RF quasilinear operators respectively. The guiding center velocity $\mathbf{v}_{gc}$ can be decomposed into the parallel motion along the field lines, and a drift velocity: $\mathbf{v}_{gc} = \left(\mathbf{v} \cdot \widehat{b}\right)\widehat{b} + \mathbf{v}_d$, where $\widehat{b}$ is the unit vector in the direction of the magnetic field. Usually, in tokamaks, the characteristic times for the parallel motion, collisions and quasilinear diffusion are much shorter than the drift time. Therefore, in first approximation, the drifts can be neglected and electrons are assumed to remain on a given flux-surface. Consequently, the drift-kinetic equation (1) reduces to the Fokker-Planck equation

$$\frac{v_{\|}}{r} \frac{B_\theta}{B} \frac{\partial f}{\partial \theta} = \mathcal{C}(f) + \mathcal{Q}(f) \tag{2}$$

which gives the 3-D distribution function $f = f(\theta, p_{\|}, p_{\perp})$, and can be solved separately on each flux-surface.

Most tokamaks with reasonably high temperature operate in the low-collisionality regime, or banana regime. In this case, the bounce time τ_b is much shorter that the collisional detrapping time τ_{dt}, so that trapped electrons can perform many bounce periods before being detrapped. Banana orbits are then well-defined and, to the lowest order in the collisionality parameter, f is constant along the field lines and symmetric in the trapped region. Applying the bounce-averaging operator

$$\{\mathcal{A}\} \equiv \frac{1}{\tau_b} \left[\frac{1}{2} \sum_\sigma\right]_T \int_{-\theta_c}^{\theta_c} \frac{d\theta}{2\pi} \frac{r}{|v_\||} \frac{B}{B_\theta} \mathcal{A} \tag{3}$$

the first term in (2) is annihilated. In (3), θ_c is the turning angle for trapped particles, and the sum over $\sigma = v_\|/|v_\||$ applies only to trapped electrons, for which the average must be performed over both the forward and backward motions. As a result, we obtain the bounce-averaged, Fokker-Planck equation

$$\{\mathcal{C}(f)\} + \{\mathcal{Q}(f)\} = 0 \tag{4}$$

which must be solved numerically in the 2D momentum space.

Numerical Code

Equation (4) is solved using the code DKE [11]. This code uses the fully relativistic collisional operator developed by Braams and Karney [12] and the relativistic RF quasilinear operator proposed by Lerche [13]. Because the distribution function is symmetric in the trapped region, only one half of the trapped region is to be considered in the FP calculations [14]. However, this requires a particular treatment of the particle fluxes in momentum space at the trapped/passing boundary, since electrons that are barely trapped can be detrapped either on the co- or counter-passing side, given that the bounce period is much shorter than the collisional detrapping time. With this scheme, the bounce-averaged dynamics include trapping effects implicitly, leading to very fast computer calculations. In addition, non-uniform grids are used both in momentum and pitch-angle coordinates, allowing for finer calculations in the most important regions of momentum space for ECCD and OKCD, particularly near the trapped/passing boundary.

The calculations presented in this paper were carried out for a typical DIII-D plasma, with major radius $R = 1.67$ m, minor radius $a = 0.67$ m, and magnetic field on axis $B_t = 2.1$ T. The temperature and density profiles were taken to be parabolic, with respective core and edge temperatures $T_{e0} = 4.0$ keV and $T_{ea} = 0.0$ keV, and densities $n_{e0} = 3.0 \times 10^{19}$ m^{-3} and $n_{ea} = 0.4 \times 10^{19}$ m^{-3}. The effective ion charge was taken uniformly $Z_{\text{eff}} = 2$.

The EC wave was considered to be a Gaussian beam of width 2 cm, polarized in the quasi-X mode. The wave-particle interaction occured near the second harmonic, $\omega \simeq 2\Omega_{ce}$.

Moments of the distribution function f calculated from (4) give the flux-surface averaged current density J and the flux-surface averaged absorbed power density p_d. A normalized intrinsic CD figure of merit is defined by

$$\eta = \frac{J/(en_e v_{Te})}{p_d/(n_e \nu_e m_e v_{Te}^2)} \tag{5}$$

where ν_e is the electron collisional frequency $\nu_e = (e^2 n_e \ln \Lambda)/(4\pi\varepsilon_0^2 m_e^2 v_{Te}^3)$.

CALCULATIONS ON A SINGLE FLUX-SURFACE

In order to describe the mechanisms of ECCD and OKCD, we first consider a given flux-surface at a normalized radius $\rho \equiv r/a = 0.6$. The EC beam crosses the flux-surface at a poloidal location θ_b. The flux-surface averaged energy flow density incident on the flux-surface is $\langle S_{\text{inc}} \rangle = 230$ kW/m². The calculation of ECCD and OKCD depends primarily on the location of the resonance curve in momentum space. This location is given by the resonance condition equation, which can be written as

$$\gamma - N_\| \frac{p_\|}{mc} - \frac{2\Omega_{ce}}{\omega} = 0 \qquad (6)$$

The two relevant wave parameters are therefore the toroidal refractive index $N_\|$ and the ratio $2\Omega_{ce}/\omega$. Along the ray path, $N_\|$ usually remains relatively constant across the resonance region. Here, we take $N_\| = 0.3$ for ECCD and $N_\| = -0.3$ for OKCD. However, the variations of $2\Omega_{ce}/\omega$ across the resonance region greatly affect the location of the resonance curve and therefore, the driven current. Considering an EC beam propagating horizontally from the low-field side and crossing the cyclotron resonance near $\rho = 0.6$ with $N_\| = \pm 0.3$, we find that the peak of power absorption occurs slightly before the actual resonance, at a value $2\Omega_{ce}/\omega \simeq 0.98$, regardless of the poloidal location θ_b. Therefore, we choose this value, which sets the EC wave frequency.

Two different scenarios are considered. ECCD on the high-field side (HFS) at $\theta_b = 180°$, where trapped particle effects are expected to be minimal, and OKCD on the low-field side (LFS) at $\theta_b = 0°$. The respective 2-D distribution functions are shown on Fig. 1, as well as the "parallel distributions", which are obtained by integration over the perpendicular momentum:

$$F_\| = 2\pi \int_0^\infty dp_\perp \, p_\perp \, (f_0 - f_M) \qquad (7)$$

ECCD

ECCD is affected by trapped particles even if, on the HFS, no trapped particle directly interacts with the EC wave. In fact, the resonant electrons rapidly move to the LFS where they can exchange momentum with trapped particles. After half a bounce period, which is very short compared to the collisional time scale, these trapped particles can transfer momentum to the counter-passing electrons. This explains the momentum increase of electrons with negative $p_\|$, visible on graph (a). It clearly appears as an opposite current on the parallel distribution on graph (c). This counter current reduces the efficiency of ECCD.

The driven current density is $J^{\text{EC}} = 1.7$ MA/m² and the density of power absorbed is $p_d^{\text{EC}} = 3.2$ MW/m³. The figure of merit (5) is then $\eta^{\text{EC}} = 1.8$. For comparison, the current density and figure of merit calculated without including the effect of trapped particles are, respectively, $J_{\text{NT}}^{\text{EC}} = 4.6$ MA/m² and $\eta_{\text{NT}}^{\text{EC}} = 3.1$. Thus, even when ECCD is located on the HFS, the trapped particles effect strongly reduces both the driven current density and the ECCD efficiency.

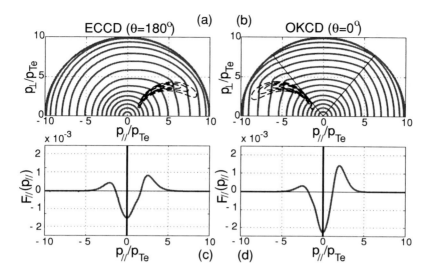

Figure 1: 2D distribution (a)-(b) and corresponding parallel distribution (c)-(d) for the cases of ECCD and OKCD, respectively. The dashed lines on graphs (a)-(b) are a contour plot of the diffusion coefficient. Note that for a Maxwellian, the contours would be equidistant concentric circles.

OKCD

On the LFS, there is a large fraction of trapped particles, so that, in the present case, the resonant region in velocity space -in dashed contours on graph (b)- is located right under the trapped passing boundary. As a consequence, barely counter-passing electrons are trapped under the action of the EC wave, which increases their perpendicular momentum. Because of the fast bounce motion, the distribution function is rapidly symmetrized in the trapped region, and these electrons can be detrapped either on the co- or counter-passing side.

The consequence of an asymmetric trapping due to the wave - which creates a sink of electrons on the counter-passing side -, and the symmetric detrapping, is an accumulation of electrons on the co-passing side. This accumulation is visible on the parallel distribution, shown on graph (d), and generates the Ohkawa current.

The OKCD driven current density is $J^{OK} = 3.4$ MA/m^2 and the density of power absorbed is $p_d^{OK} = 6.0$ MW/m^3. The figure of merit (5) is then $\eta^{OK} = 1.9$. The current density driven by OKCD is sensibly larger than ECCD; however, the density of power absorbed is also larger, so that the figures of merit are comparable. This comparison does not predict how much total current would be driven by a EC beam using the EC or the OK method; however, the larger power absorbed and current densities suggest that OKCD may lead to narrower current profiles, which is important for the accuracy of current profile control and NTM stabilization.

GLOBAL CALCULATION

The distinction between the Fisch-Boozer and Ohkawa effects, which determines the driven current, depends primarily on the location of the wave-particle interaction in momentum space, which varies along the ECW propagation path. Therefore, in order to compare ECCD and OKCD, it is necessary to integrate the total driven current and power absorbed along this path.

OKCD profiles

We consider a ECW launched horizontally, in the mid-plane, from the LFS. Three different OKCD current density profiles are shown on Fig. 2. The radial locations of current deposition were set to $\rho = 0.5/0.6/0.7$, by appropriately choosing the ECW frequency. On the profile centered around $\rho = 0.5$, some negative current is generated first along the ray path, which means that the Fisch-Boozer effect dominates there, because the EC diffusion region in momentum space is far from the trapped region. Closer to the resonance, the Ohkawa effect start to dominate and positive current is driven, because the EC diffusion region in momentum space is near the trapped region. The Fisch-Boozer, negative current decreases if current is driven further off-axis ($\rho = 0.6$) where the

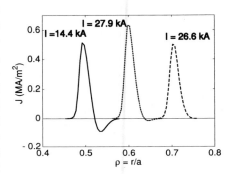

Figure 2: Current density profiles for OKCD at three different radial locations. The total driven current is also indicated. For each case, a beam input power of $P_{\text{inc}} = 1.5$ MW was completely absorbed.

fraction of trapped electrons increases; at some point, it becomes negligible ($\rho = 0.7$). This leads to a significant increase in the OKCD current, which is almost twice higher at $\rho = 0.6$ than at $\rho = 0.5$.

ECCD versus OKCD

In order to compare ECCD and OKCD, we compare the following cases: ECCD on the HFS ($\theta = 180°$), ECCD above the magnetic axis ($\theta = 90°$), and OKCD on the LFS ($\theta = 0°$). The radial location of deposition is modified by varying the EC wave frequency (for $\theta = 0°$ and $\theta = 180°$) or the vertical launching position (for $\theta = 90°$). The ECW is launched with an input power $P_{\text{inc}} = 1.5$ MW and a toroidal launching angle chosen such that $N_{\|} = 0.3$ (ECCD) or $N_{\|} = -0.3$ (OKCD) at the location of deposition. The total driven current is plotted on Fig 3 as a function of the normalized radius.

We can see that the ECCD current decreases steadily with the normalized radius. The decrease is faster for ECCD above the magnetic axis ($\theta = 90°$), which becomes impossible beyond a certain radius ($\rho > 0.6$), in accordance with experimental observations [7]. OKCD current can be driven when the number of trapped particles becomes sufficient ($\rho > 0.4$) and becomes rapidly larger than ECCD at $\theta = 90°$. Beyond some point ($\rho > 0.6$), it becomes even larger than ECCD at $\theta = 180°$. It is interesting to note that, in the present case, OKCD would drive as much current at $\rho = 0.6$ or 0.7 as ECCD at $\theta = 90°$ would drive at $\rho = 0.4$; therefore, OKCD can drive appreciable currents far off-axis, where ECCD cannot. When compared to ECCD at $\theta = 180°$, OKCD gives slightly higher currents for $\rho \geq 0.6$.

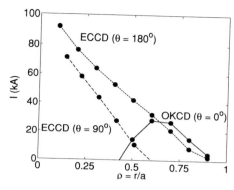

Figure 3: Total current generated by a EC beam of $P_{inc} = 1.5$ MW launched horizontally from the LFS. Two ECCD cases are considered, with deposition at $\theta = 180°$ or $\theta = 90°$, and a OKCD case at $\theta = 0°$.

Varying the Radial Location of Deposition in OKCD

In ECCD and OKCD, the current is deposited near the intersection of the ray path with the cyclotron resonance layer. Experimentally, the radial location of deposition may have to be controlled and modified during the operation. This has been done by changing the location of the resonance layer, by moving the plasma, or changing the magnetic field [4][3]; it has also been done by steering launching mirrors, in order to modify the ray path [5].

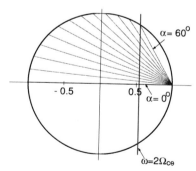

Figure 4: Simulation of OKCD with varying poloidal launching angle. The resonance layer is maintained fixed.

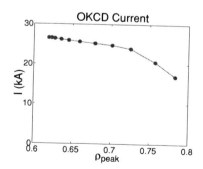

Figure 5: OKCD total current as a function of the current profile peak ρ

The effect of steering the poloidal launching angle is investigated by calculating OKCD with launching from the LFS at an angle α with respect to the horizontal plane.

The ECW frequency, and therefore the resonance layer, is fixed, as shown on Fig. 4. The EC beam input power is $P_{\text{inc}} = 1.5$ MW, and the toroidal launching angle is $\phi = -15°$, so that the parallel refractive index at the location of deposition varies from $N_\parallel \simeq -0.3$ for $\alpha = 0$ to $N_\parallel \simeq -0.2$ for $\alpha = 50°$.

The total driven OKCD current is calculated, and shown on Fig. 5 as a function of the radial location of the current profile peak. The current decreases with ρ, but not as fast as in the case where the radial location is modified by changing the location of the resonance layer (Fig. 3). In fact, the driven current does not vary much over a radial range larger than 10% of the plasma, making OKCD compatible with a steering manipulation for controlling the current deposition location.

CONCLUSION

Experiments have shown that ECCD, effective close to the core, becomes very ineffective or even impossible when driven far off-axis. In this paper, we have shown that in this region, of interest to advanced tokamak operation scenarios, OKCD can offer a valuable alternative, as it drives current with reasonable figures of merit over a large range of far off-axis locations.

ACKNOWLEDGMENTS

Work supported by U.S. DoE Grants DE-FG02-91ER-54109 and DE-FG02-99ER-54521, and Cooperative Grant No. DE-FC02-99ER54512.

References

[1] Sauter, O., et al., *Phys. Rev. Lett.*, **84**, 3322-3325 (2000).

[2] Murakami, M., et al., *Phys. Plasmas*, **10**, 1691-1697 (2003).

[3] Gantenbein, G., et al., *Phys. Rev. Lett.*, **85**, 1242-1245 (2000).

[4] La Haye, R.J., et al., *Phys. Plasmas*, **9**, 2051-2060 (2002).

[5] Isayama, A., et al., *Plasmas Phys. and Cont. Fusion*, **42**, L37 (2001).

[6] Fisch, N.J., and Boozer, A.H., *Phys. Rev. Lett.*, **45**, 720-722 (1980).

[7] Petty, C.C., et al., *Nuclear Fusion*, **41**, 551-565 (2001).

[8] Ohkawa, T., *General Atomic Report* no. 4356.007.001 (1976).

[9] Decker, J., Peysson, Y., Bers, A., and Ram, A.K., "Self-consistent ECCD calculations with bootstrap current", in *Proc. EC-12 Conference*, Aix-en-Provence, France (2002).

[10] Decker, J., Peysson, Y., Bers, A., and Ram, A.K., "On Synergism between Boostrap and Radio-Frequency Driven Currents", in *29th EPS Conference on Plasma Phys. and Cont. Fusion*, Montreux, Switzerland (2002).

[11] Peysson, Y., Decker, J., and Harvey, R.W., "Advanced 3-D Electron Fokker-Planck Transport Calculations", in *AIP Proc. RF- 15 Conference*, Moran, WY (2003).

[12] Braams, B.J., and Karney, C.F.F, *Phys. Fluids B*, **1**, 1355-1368 (1989).

[13] Lerche, I., *Phys. Fluids*, **11**, 1720-1726 (1968).

[14] Killeen, J., et al., *Computational Methods for Kinetic Models of Magnetically Confined Plasmas*, Springer-Verlag, 1986.

Towards predictive scenario simulations combining LH, ICRH and ECRH heating

V. Basiuk, J.F. Artaud, A. Bécoulet, L.G. Eriksson, G.T. Hoang, G. Huysmans, F. Imbeaux, X. Litaudon, D. Mazon, C. Passeron, Y. Peysson

Association Euratom-CEA, 13108 St Paul-Lez-Durance, France

INTRODUCTION

Reliable predictive simulations, combining current, heat and matter transport equation with a 2D equilibrium allowing diagnostic reconstruction such as Faraday angle and MSE angle are of great interest for existing and future tokamak. The CRONOS code [1] with its various power deposition codes (DELPHINE [2], REMA [3], PION [4]) is a powerful tool to prepare such scenario in a reasonable CPU time (few hours, for a one minute plasma discharge). An example of such advanced scenario, with a negative seed of current at the center of the discharge is shown in this paper.

It allows also testing new concept of feedback control, which will be directly implemented on the new real-time network of Tore-Supra. In this concept, the algorithm as to find itself the best and safe way to reach enhance performance (i.e. best plasma fusion power D-D) using different actuators (injected power, ...). On this paper, we will focus on a simple example where the initial and final states are known and we will show why a steady state tokamak allowing long pulse operation is necessary for such control

TOOLS FOR PREDICTIVE SCENARIO

CRONOS code

The heat transport model used in all the simulation is the Bohm/Gyrobohm model, with coefficient adapted to the Tore Supra plasma [5].

For the simulations concerning the feedback control studies, the ion diffusive coefficient X_i, linked to the electron one, X_e, is supposed to stay constant during purely electron additional heating, which is the case in Tore-Supra when using the Lower Hybrid system (LHCD). Then, the ion channel is dominated by the equipartition term.

The electron density are simulated using the following equation (1) :

$$\frac{\partial}{\partial t}(V'n_e)+\frac{\partial}{\partial \rho}\left(V'\langle|\nabla\rho|^2\rangle\gamma_e\right)=V'S_{ne}, \text{ with } \gamma_e=-D_e\frac{\partial n_e}{\partial \rho}-n_e V_e \quad (1)$$

where D_e the electron diffusivity ($D_e = 0.1$ m^2 s^{-1}) and V_e the electron convection speed (V_e=0.6 m s^{-1}) are deduced from experiment in the case of gas injection and active pumping.

Feedback control

In this control, the algorithm will have to improve the plasma fusion power using two actuators: the Lower Hybrid heating system, in the current drive mode to insure a zero loop voltage (steady state) and the gas injection/pumping system to modify the electron density. The figure 1 shows the algorithm which is implemented inside CRONOS. The feedback control has to keep in memory the last state the plasma reached before modifying one actuator. After the change, as the plasma equilibrium state has changed, the control has to wait a given time (few second), reaching the new equilibrium, before comparing the new state to the older

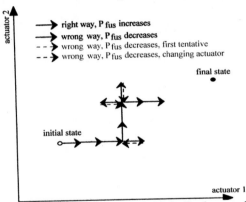

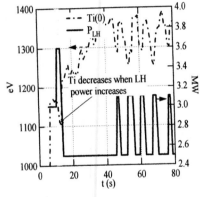

Figure 1: Schematic view of the feedback control. After each change of one actuator (i.e. P_{LH} increases of 300 kW), the algorithm memorized the initial value (t_0) of the plasma fusion and compared it to the new value after a given time ($t_1=t_0+\Delta t$)

Figure 2: Effect of the LH power on the central ion temperature (gray area) and oscillation of the algorithm (after 40 s) due to the Greenwald constraint

Different constraints can be inserted in the algorithm. The main one is that the density must not be higher than the Greenwald density which is evaluated before each change of the actuator. This evaluation is made using a given efficiency for the LHCD (6.5 10^{19} AW^{-1}m^{-3}) and then estimated the plasma current target of the new plasma equilibrium. If this limit is reached, the algorithm not allowed the actuator change and try the next actuator.

For the electron density a combine proportional (coefficient g_{prop}), integral (coefficient g_{int}) and derivative (coefficient g_{der}) control (PID) is used to minimize oscillation around the target value (g_{prop}=1, g_{int}=1 and g_{der}=0.2).

The first feedback control simulation shows (figure 2) that the Bohm-Gyrobohm model with a purely electron additional heating and for low value of line density (n_{bar}) generates a decrease of the central ion temperature (and so of the fusion power) when the LH power increases. This is due to the fact that increasing LH power increases the electron temperature and so the X_e. This fact not allows the algorithm to reach the estimated final state (high value of LH power associated to high value of electron density), when the Greenwald constraint is turned on.

For the moment, no experimental evidence of such an effect is seen on Tore-Supra. To avoid this problem, X_i will be kept constant with its value calculated in the initial state of the simulation (ohmic part of the simulation). The LH power is limited to 7 MW and the density to 10^{20} m^{-3} in line density (n_{bar}). The figure 3 shows the results of 3 simulations where the step time Δt, during which the change of the actuator occurred, increase for 0.5s to 2 s. The $\Delta t>2s$ is the case where we allow the plasma to almost reach its new equilibrium. On the contrary $\Delta t>0.5s$ is a case where the algorithm does not wait the steady state, which is only reach on the last step, when the maximum value of the injected power is obtained at a density near of the Greenwald value. The path in the plasma current versus line density plot is quite different in the three simulations with a similar final result for the plasma fusion power (Figure 3a and 3b). This observation is also observed on the time evolution of the q profile which is different in the three case.

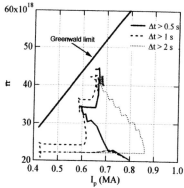

Figure 3a: Path finding of the simulation in the graph (plasma current, line density). The Greenwald limit is also indicated.

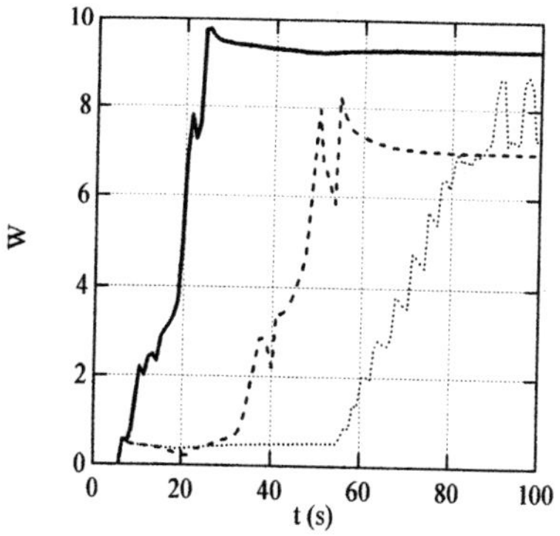

Figure 3b: Time evolution of the plasma fusion for the three simulations

ECRH COMBINED TO LH

The ECRH system will allow to control the plasma current from the center towards the edge of the plasma, using the 118 GHz frequency and modifying the injection toroidal angle (θ). In the simulation, ECRH is combined to LH power to reach a zero loop voltage plasma, with one antenna (P_1) coupling its power in the center ($\theta = -24°$), the two other (P_2 and P_3) at x=0.42 ($\theta = +25°$), where x is the normalized coordinate associated to the toroidal flux. Time slice of the injected powers and the various current are shown in figure 4. A simple proportional feedback control (already implemented on Tore Supra) on the LH power, using a gain of 30 W A^{-1}, is used to

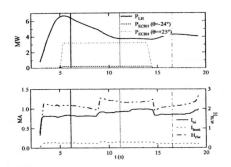

Figure 4: time evolution of the injected power, of the non inductive current, of the bootstrap current and of the enhance confinement compare to the Rebut Lallia Watkins law.

insure a constant flux at the edge of the plasma. The shear effect is added on the heat transport model [6]. A Reversed shear plasma can be obtained, as shown in figure 5 (current and q profile at t = 14s) and 6 (time evolution of q), with a relatively low level of ECRH power (200 kW for the launcher whose toroidal angle is −24 degrees). Reconstruction of the Motional Stark Effect measurement by CRONOS shows that such a very central inversed shear can be seen by the diagnostic.

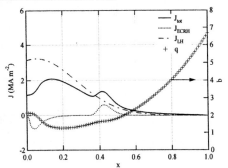

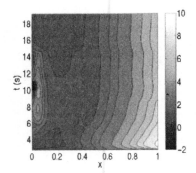

Figure 5: Current power deposition at t=11 s, for ECRH, LHCD and total current and safety factor profile

Figure 6: Contour plot of the q profile. ECRH power begins at 4s and lasts at 14 s.

CONCLUSIONS

The CRONOS code allows preparation and test of every kinds of feedback control, which can be easily installed on the real time system of Tore Supra. It is also a great tools to prepare advanced scenario, to study the effect of different plasma parameters (such as location of the ECRH current seed), to design and optimize diagnostic measurements (Motional Stark effect, Polarimetry).

REFERENCES

1. V. Basiuk, J.F. Artaud et al., Nuclear Fusion to be published in summer 2003
2. F. Imbeaux, Report EUR-CEA-FC-1679 (1999)
3. Krivenski et al., Nucl. Fusion 25 (1985) 127
4. L.G. Eriksson and al, Nuclear Fusion 33 (1993) 1037
5. M. Erba et al, Nuclear Fusion, Vol. 38, N° 7, pp 1013-1028, 1998
6. X. Litaudon et al, Plasma Physics and Controlled Fusion, vol.41 (1999) p.A-733-A746 (1999)

Nonlinear Interaction between RF-Heated High-Energy Ions and MHD-Modes

T. Bergkvist*, T. Hellsten*, T. Johnson* and M. Laxåback*

*Alfvén Laboratory, Association EURATOM-VR, SE-100 44 Stockholm, Sweden

Abstract. Excitation of global Alfvén eigenmodes by fast ions during ICRH is frequently observed in tokamaks. The importance of the phasing of the ICRH antennae for the excitation of these modes have been seen in experiments. The Alfvén eigenmodes will drive the distribution function of the fast ions towards a state where the gradient in phase space is reduced. In general, the fast ions are displaced outwards, which can have a significant effect on the ICRH power deposition and lead to reduced heating efficiency. To calculate the effect on the heating profiles by the excitation of Alfvén eigenmodes and the effect on the resonating ions the Monte Carlo code FIDO, used for ICRH, has been upgraded to include particle interactions with MHD-waves. This allows self-consistent calculations of the mode amplitude and the distribution function during RF heating.

INTRODUCTION

During intense ion cyclotron resonance heating, ICRH, the particle distribution develops a high energy non-Maxwellian tail, consisting mainly of particles with wide trapped or non-standard drift orbits. These high energy particles can under specific conditions excite global Alfvén eigenmodes. First, the eigenmode, which is determined by the equilibrium, has to have a frequency, a radial location and mode numbers corresponding to a location in phase space where there is a considerable amount of resonant high energy particles. Second, the gradient of the particle distribution in phase space has to be such that there is more energy going into the mode from particles decelerated by the interactions, than there is energy going from the mode to the accelerated particles. During the excitation of the mode the gradient of the resonant high energy particle distribution in phase space will be reduced and the density of fast ions in real space will be less peaked, resulting in a broadening of the ICRH heating profiles.

Using a phasing of the ICRH antennae for which the wave is propagating parallel to the plasma current we obtain a high energy particle distribution consisting mainly of trapped and non-standard orbits with turning points close to the midplane. An opposite phasing will drive the high energy particles into banana orbits for which the turning points are driven away from the midplane [1].

To investigate how the Alfvén eigenmodes are affecting the ion cyclotron heating efficiency and also how the phasing of the ICRH antennae influence the conditions for MHD-wave excitation the Monte Carlo code FIDO [2], used to model the distribution function during ICRH, has been upgraded to include interactions with MHD-waves. A flow chart of the FIDO code and the present updates are shown in Fig. 1.

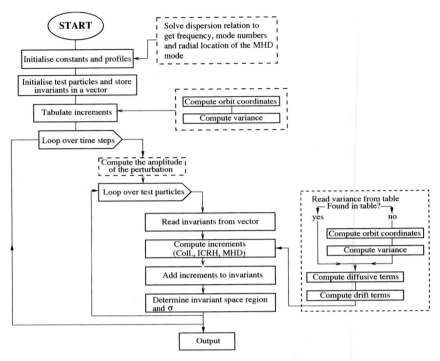

FIGURE 1. Flow chart of the FIDO code including MHD interactions

MHD INTERACTIONS

We assume the MHD-mode to be well localized near the resonant flux surface, leading to a 1-D diffusion in phase space. Here we introduce a decorrelation time for the MHD-particle interactions and assume the initial phase between the particles and the perturbation to be stochastic. The evolution of the distribution function is solved with a Monte Carlo method where the increments are obtained from the change in normalized particle energy due to a plasma displacement [3]

$$\frac{d\tilde{E}}{dt} = \frac{eZ}{m}\mathbf{E}_1 \cdot \mathbf{v}_{d0} + \mu \frac{\partial B_{1\parallel}}{\partial t}. \tag{1}$$

Here $\mathbf{v}_{d0}$ is the zeroth order drift velocity due to gradients and curvature of the zeroth order magnetic field. The electric field from the MHD-mode is given by $\mathbf{E}_1 = -\xi_1 \times \mathbf{B}_0$ where ξ_1 is a first order plasma displacement. The plasma displacement is either obtained from a simple model [4] or from the solution of the linearized MHD equation, $-\rho_0 \omega^2 \xi_1 = \mathbf{F}(\xi_1)$, e.g. using the LION [3, 5] code, which is also used for calculating the wave field for ICRH. Neglecting the parallel electric field, Eq. (1) can be written to first order as

$$\frac{d\tilde{E}}{dt} = \left(\frac{v_\perp^2}{2} + v_\parallel^2\right)\frac{(\mathbf{B}_0 \times \nabla B_0) \cdot \mathbf{E}_1}{B_0^3} - \frac{\mu}{m}(\nabla \times \mathbf{E}_1)_\parallel, \tag{2}$$

where the velocities are of zeroth order. By integrating Eq. (1) over a decorrelation time the expectation value and variance of the normalized energy change is obtained. The diffusive terms in the Monte Carlo operator, which consist of the covariances of the three invariants $\tilde{E} = v^2/2$, $\tilde{P}_\phi = Rv_\phi + \psi q/m_i$ and $\Lambda = B_0 v_\perp^2/Bv^2$, spanning the phase space, are derived from the relations $\Delta \tilde{P}_\phi = n\Delta \tilde{E}/\omega$ [6] and $\Delta \Lambda = -\Lambda \Delta \tilde{E}/\tilde{E}$, where in the last expression the conservation of magnetic moment during MHD interaction has been used. The particle drift is then given by the expectation values of the invariant increments and the gradient of the diffusion coefficient along the characteristic in phase space. Due to the stochastic initial phase between the particle and the mode the integration gives an expectation value of zero, and the deterministic drift is only determined by the gradient of the variance in phase space. To reduce the computational work a resonance condition is imposed stating that the total phase shift during one decorrelation time shall not exceed 2π. In general $\tau_D \gg \tau_b$, where τ_D is the decorrelation time and τ_b is the bounce period, requiring that higher harmonics of the resonance condition have to be considered, leading to $\tau_D(n\Delta_b\phi - m\Delta_b\theta - \omega\tau_b \pm N2\pi)/\tau_b < 2\pi$, where N is the harmonic of the resonance and Δ_b is the change during a bounce period.

DISCUSSIONS

In order to illustrate the effects of the MHD interactions, an approximation for the mode structure, $\xi_1 = \mathbf{A}(r)e^{i(n\phi - m\theta - \omega t)}$, is used. The radial structure, $\mathbf{A}(r)$, is derived from the potential function from Ref. [4]. The shear Alfvén frequency with corresponding mode numbers is obtained from the dispersion relation, which is solved using the LION code. Comparing two ICRH preheated distributions with different phasings, counter current propagating waves ($-90°$) and co current propagating waves ($+90°$), we can deduce, from a scan in mode frequency, that the region of possible mode excitation is strongly affected by the different antenna phasings. We have used parameters corresponding to typical ones for JET when global Alfvén waves are excited. In Fig. 2 a negative particle drift corresponds to excitation of the mode while a positive particle drift corresponds to damping of the mode. In the absence of ICRH, Coulomb collisions and damping the MHD mode amplitude will saturate, 3(a). This occurs when the gradient along the characteristic in phase space vanish, corresponding to a less peaked radial profile 3(b). The collisions will introduce new particles into the resonant regime resulting in a weak contribution to the saturation level, whereas the RF heating more effectively introduces resonant fast ions and hence lead to a continous growth of the mode amplitude.

REFERENCES

1. T. Hellsten, J. Carlsson, and L.-G. Eriksson, *Physical Review Letters*, **74** 3612–3615 (1995).
2. J. Carlsson, L.-G. Eriksson, and T. Hellsten, In *Theory of Fusion Plasmas*, 351, (1994).
3. L. Villard, S. Brunner, and J. Vaclavik, *Nuclear Fusion*, **35** 1173 (1995).
4. H. L. Berk, B. N. Breizman, and M. S. Pekker, *Nuclear Fusion*, **35** 1713–1720 (1995).
5. L. Villard, et al., *Computer Physics Reports*, **4** 95 (1986).
6. L.-G. Eriksson, et al., *Physics of Plasmas*, **6** 513–18 (1999).
7. J. Hedin, et al., *Nuclear Fusion*, **42** 527–540 (2002).

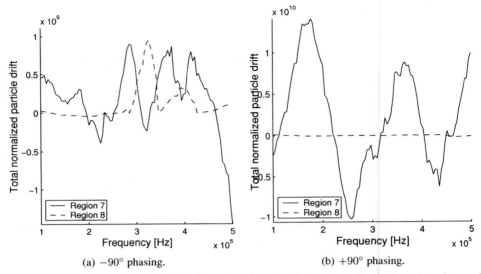

FIGURE 2. Total normalized particle drift as a function of mode frequency. Region 7 consists of trapped banana orbits and region 8 consists of co-current passing orbits on the low field side of the magnetic axis [7]

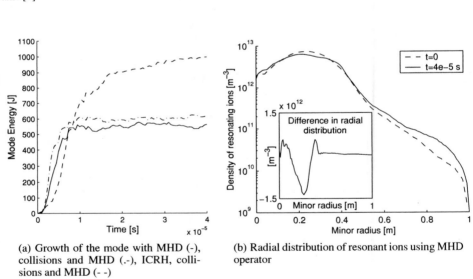

(a) Growth of the mode with MHD (-), collisions and MHD (.-), ICRH, collisions and MHD (- -)

(b) Radial distribution of resonant ions using MHD operator

FIGURE 3. Preheated distribution using a $+90°$ phasing of the ICRH antennae. TAE mode located at minor radius $r = 0.4m$ with frequency $250kHz$ and mode number $m = 4, n = 3$.

Local Mode Analysis of 2D ICRF Wave Solutions

D.A. D'Ippolito,[1] J.R. Myra,[1] E.F. Jaeger,[2] L.A. Berry,[2] D.B. Batchelor[2]

[1]*Lodestar Research Corporation, Boulder, Colorado;* [2]*ORNL, OakRidge, Tennessee*

Abstract. Numerical techniques (windowed Fourier transforms and wavelets) are developed for carrying out a local mode analysis $k(x)$ of ICRF field solutions. It is shown that simultaneous resolution of long- and short-wavelength waves in typical rf mode conversion scenarios requires the use of a modified wavelet transform. The ability to get quantitative information by this technique is assessed and visualization techniques for wave polarization information are illustrated.

INTRODUCTION

With the growing capability of rf simulations,[1-3] there is a strong motivation to develop appropriate post-processing tools for extracting physical information from the numerical solutions. Full-wave ICRF codes yield complicated rf field patterns, and the challenge is to understand these patterns by appealing to the intuitive, but approximate, physics-based notion of local plasma modes (global eigenmodes, transmitted and reflected waves, and mode conversions between different types of waves). Quantitative information on the local wavevectors, amplitudes and wave polarizations is required for a basic understanding of the plasma heating and the ICRF-driven currents and flows. As part of the rf SciDAC project, numerical techniques (windowed Fourier transforms and wavelets) have been investigated for the local mode analysis of ICRF field solutions. Here, we illustrate these techniques for a DIII-D D(H) mode conversion reference case[4] computed by the AORSA-1D code, which includes a model of the 2D poloidal magnetic field. Work is in progress to apply the techniques described in this paper to full 2D rf wave solutions.

TRANSFORM METHODS

We illustrate these methods by considering the 1D case where the function $f(x)$ is represented by its values $f_i = f(x_i)$ on a grid of N spatial points x_j. The transform specifies the mapping $f(x) \to F(k)$ where $k = k_x$ is also represented on a grid of N points. Here, we analyze $E(x)$ obtained from a fast wave (FW) to ion Bernstein wave (IBW) or ion cyclotron wave (ICW) mode-conversion solution with $x = R - R_0$ and $\mathbf{k} = k_x \mathbf{e}_x + (n/R) \mathbf{e}_z$, where R is the major radius of the tokamak and n is the toroidal mode number. Our tests of various transform methods are described below.

The simplest approach is to use the *discrete Fourier transform* to resolve the wave propagation data into global k modes. This identifies all the relevant physical modes in the spatial domain of interest but does not yield any information as to their spatial location, nor does it yield insight into relationships among modes such as mode conversion. To resolve this difficulty, we considered the *Windowed Fourier Transform* (WFT) technique, in which the function E(x) is multiplied by a window w(x) before carrying out the transform. The best results are obtained using a smooth window function such as the Gaussian $w(x) = \text{Exp}[-(x-x_0)^2/(2x_w^2)]$, where x_0 and x_w are the location and width of the window. The WFT method *with a constant window width* works well for a single wave or for the case of multiple waves with similar wavelengths, but it fails for the case of multiple waves with very different wavelengths, such as occurs in mode conversion. For example, a large window is needed to resolve the wavelength (or k) of the FW but does a poor job in giving the spatial location of the IBW; a small window does a good job in localizing the IBW in space but does a poor job in resolving the FW wavelength.

An analysis of this problem shows that it stems from the need to minimize two conflicting types of errors: (1) the "Heisenberg" error $\Delta k_1 = C\pi/\Delta x$, where C is a constant of order unity and Δx is the size of the region in which the transform is carried out (here, the window width), and (2) the "non-local" or "gradient" error $\Delta k_2 \approx (\partial k/\partial x) \Delta x$, where k(x) is the local (eikonal) wavenumber. Note that Δk_1 vanishes in for a large window, whereas Δk_2 is reduced by a small window. The competition between the two effects produces an "optimal" window width which depends on the wavelength, i.e. $x_w = x_w(k)$. Thus, we must generalize the WFT technique to have a window width that scales with k.

A transform involving basis functions that are translated and scaled is called a "wavelet." The *Morlet wavelet* is essentially a WFT with a scaled Gaussian window, $x_w = c_0/k$, where c_0 is a constant that has an optimal value ($c_0 = 5$) for minimizing the error. This wavelet is an improvement over the WFT, but the scaling breaks down at k = 0. The infinite window width at k = 0 leads to large gradient error and false peaks in the spectrum. However, it is essential in treating the FW to IBW mode conversion problem that one resolve k's of both signs and therefore handle the behavior at k = 0.

We have developed a simple modification to the Morlet wavelet that keeps its useful features while still resolving the k = 0 region. We scale the window width as $x_w = c_0/(k^2 + k_0^2)^{1/2}$ so that $x_w \rightarrow c_0/k_0$ as $k \rightarrow 0$, where k_0 is a constant. While $k_0 \neq 0$ spoils the pure wavelet scaling in a small region, it permits us to obtain a physical answer over the whole x-k plane. We call this approach the "*k-wavelet*" method. It should be emphasized that we do not need the "pure" wavelet scaling for our application. We are simply using the wavelet transform for visualization and for extracting the dispersion function k(x). We will show that the k-wavelet method provides a useful tool for graphically obtaining this information.

The k-wavelet transform is implemented in *Mathematica* using fast Fourier transform (FFT) techniques as follows. The wavelet transform $f(x) \rightarrow W(x_0, k)$ involves a spatial convolution of the functions f(x) and a mother wavelet $\Psi(x)$. Using the convolution theorem to cast this into k-space reduces the number of computations from N^2 to $N \ln N$. Thus, we define the k-wavelet transform $W(x_0, k)$ as

$$W(x_0, k) = \frac{1}{N} \Im^{-1}(\Im[f(x)] \Im[\Psi^*(k, x-x_0)]) , \qquad (1)$$

where $\Im[f(x)]$ denotes the forward FFT of $f(x)$, $\Im^{-1}$ denotes the inverse FFT, and Ψ is the k-wavelet function defined by

$$\Psi[k, x-x_0] = \exp[ik(x-x_0)] \exp\left[-(x-x_0)^2 / (2 x_w^2)\right] , \qquad (2)$$

with $x_w(k) = c_0/(k^2 + k_0^2)^{1/2}$ and c_0, k_0 are constants. In the limit $k_0 \to 0$, $\Psi \to \Psi[k(x-x_0)]$ and the exact wavelet scaling is recovered.

APPLICATIONS AND CONCLUSIONS

We first consider a DIII-D D(H) mode conversion reference case[4] computed by the AORSA-1D code, which includes a model of the 2D poloidal magnetic field B_p. We define the k-wavelet spectral power density $P_\alpha(x_0, k) = |W_\alpha(x_0, k)|^2$, where W_α is the k-wavelet transform of $E_\alpha(x)$ defined in Eq. (1). In Fig. 1 we show the contours of $P_\perp(x0, k) = P_x(x_0, k) + P_y(x_0, k)$, obtained by analyzing the AORSA-1D code solutions for two values of B_p corresponding to horizontal slices of the 2D equilibrium. The wavelet analysis illustrates the important result[5,2] that the mode conversion process is sensitive to the poloidal magnetic field. Above the midplane [Fig. 1(a)] the incoming FW converts to a backward-propagating IBW, whereas below the midplane [Fig. 1(b)] the FW converts to a forward-propagating ICW.

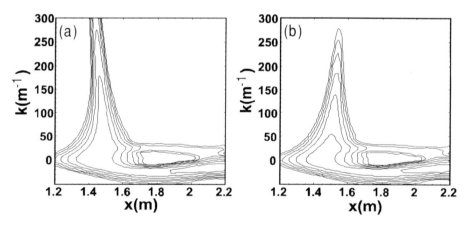

FIGURE 1. k-wavelet transform spectral power density $P_\perp(x_0, k)$ of the rf electric field with $c_0 = 5$ and $k_0 = 40$ m^{-1} for horizontal slices (a) above the midplane, and (b) below the midplane.

A comparison of the contours in Fig. 1(a) with the analytic 1D hot-plasma dispersion relation for the same parameters shows good agreement for the k contours of the incident and reflected FW and the IBW and for the location of the mode conversion surface (near x = 1.5 m). Thus, the k-wavelet transform is useful for obtaining quantitative information about the spatial dependence of the wavenumber.

We have also investigated the use of the wavelet analysis for simultaneous visualization of dispersion, amplitude and wave polarization information in 3D plots by the use of appropriately defined color palettes. In Fig. 2, we show a grayscale print of a 3D plot for the case of Fig. 1(a) with palette chosen to reflect the linear wave polarization, E_x/E_y. A color version of this figure can be viewed in the version of this paper posted on our website.[6]

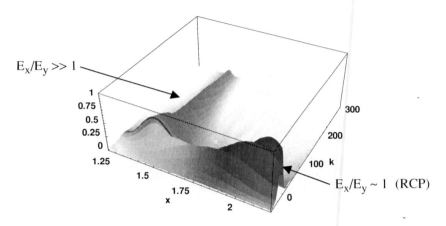

FIGURE 2. k-wavelet transform of $P_\perp(x_0, k)$ of the rf electric field using the same parameters as Fig. 1(a); the color palette indicates the *linear* wave polarization E_x/E_y.

We have shown that the 'k -wavelet" transform provides a useful diagnostic for wave properties in complex situations such as mode conversion where multiple waves with vastly different wavelengths are present simultaneously. The methods used here can be generalized to obtain the wavevector $\mathbf{k}_\perp = (k_x, k_y)$ for 2D rf field solutions, $\mathbf{E} = \mathbf{E}(x,y)$. However, the 1D analysis illustrated here is qualitatively valid when $k_y \ll k_x$, which is satisfied for the case shown here.

ACKNOWLEDGMENTS

This work was supported by US DOE grants DE-FC02-01ER54650, DE-FG03-97ER54392 and contract DE-AC05-00OR22725.

REFERENCES

1. E.F. Jaeger, L.A. Berry, E. D'Azevedo, D. B. Batchelor *et al.*, Phys. Plasmas **9**, 1873 (2002).
2. E.F. Jaeger, L.A. Berry, J. R. Myra, D. B. Batchelor *et al.*, Phys. Rev. Lett. **90**, 195001 (2003).
3. P.T. Bonoli, J.C. Wright, Y. Lin, M. Porkolab *et al.*, Bull. APS **47**, 141 (2002), paper GO1 13.
4. E.F. Jaeger, L. A. Berry, E. D'Azevedo *et al.*, Phys. Plasmas **8**, 1573 (2001).
5. E. Nelson-Melby, M. Porkolab, P. T. Bonoli, *et al.*, Phys. Rev. Lett. **90**, 155004 (2003).
6. D.A. D'Ippolito, J.R. Myra *et al.*, http://www.lodestar.com/LRCreports/kwavelets.pdf .

The Effects of Combined Parallel Gradient and Collisional Decorrelation in the Absorption of RF Waves

Brent Goode*, John R. Cary* and L. A. Berry[†]

Center for Integrated Plasma Studies and Department of Physics, University of Colorado, Boulder, CO 80309-0390
[†]*Oak Ridge National Lab Oak Ridge, TN 37831-8071*

Abstract. We examine the effects on RF wave absorption of two different types of particle wave resonance decorrelations. These two mechanisms are collisions and parallel gradients in the magnetic field. A careful treatment of collisions shows that in the integrand for the plasma dispersion function there is an interaction term which couples both decorrelation mechanisms. In toroidal fusion devices each type of decorrelation dominates in a certain region of space. Of special interest is the region where magnetic field lines intersect the resonant surface with low angle of incidence. In this region neither decorrelation mechanisms dominates over the other, and the combination of the two produces different absorption physics than either mechanism separately.

INTRODUCTION

The plasma response function is at the heart of many calculations of wave driven quantities in plasmas. In order to improve the accuracy of these calculations we wish to add to the physics which is incorporated in the response function. To that end we have calculated a plasma response function which includes the effects of collisions and parallel magnetic field gradients on the same footing. In previous treatments of the problem [1] of plasma response functions with parallel gradients, collisional effects are added after the rest of the problem has been completed. To do this, an estimate has to be made in order to convert from velocity perpendicular to the magnetic field to thermal velocity. By incorporating collisions from the beginning, we do not need to make this estimate, and can show that the estimate previously made was incorrect. The previous treatment also used the assumption that the collision frequency would be small. This causes their result to be valid only at high temperatures. Our analysis was completed without this assumption, which makes it valid at all temperatures. The plasma response function can vary greatly depending on the strength of the parallel gradient. In order to examine this effect we have studied the variation of gradient strength as a function of spatial location within an ideal cylindrical torus, and the varation of the response function over this range of strengths.

THE DERIVATION OF THE RESPONSE FUNCTION

The equation we solve is a driven Fokker-Planck equation given by

$$\frac{\partial f_1}{\partial t} + \vec{v}\cdot\nabla f_1 + \frac{q}{m}(\vec{v}\times\vec{B}(z))\cdot\nabla_{\vec{v}} f_1 - \nu(\nabla_{\vec{v}}\cdot(\vec{v}f_1)) + v_T^2\frac{\partial^2 f_1}{\partial v^2}) = h(f_0,\vec{v},\omega,\vec{k}) \qquad (1)$$

The technique we use is a variation of the characteristic and Green function technique first used by Dougherty [2]. In this technique (1) is separated into a differential equation which defines characteristic trajectories by the matrix b, and a differential equation which generates the Green function involving the matrices a and b. Dougherty used a constant strength magnetic field which made solving the characteristics from the matrix b easy. The use of a non-constant magnetic field makes this more difficult, so we instead start with an equation describing the characteristic position of a particle in a varying magnetic field of the form

$$\vec{B} = \vec{B}_0(1+\frac{z}{L_\parallel}) \qquad (2)$$

with drag like collision. From this we solve for the matrix b and then use this matrix to find the Green function. For out source term we use the most general form, which is a Maxwellian background times a plane wave. After integrating this source with our Green function and integrating the resulting f_1 over velocity we get

$$Z(\zeta) = \int_0^\infty e^{-\frac{1}{\gamma^2}(\gamma x - \frac{3}{2} + 2e^{-\gamma x} - \frac{1}{2}e^{-2\gamma x}) - \frac{1}{\beta^2+\gamma^2}\gamma x - \frac{1}{2\gamma^2}((1-e^{-\gamma x}) - \frac{1}{2}\alpha\gamma x^2)^2 - i\zeta x} dx \qquad (3)$$

where

$$\alpha = \frac{\Omega}{k_\parallel^2 v_T L_\parallel}, \qquad \beta = \frac{\nu}{k_\perp v_T}, \qquad \gamma = \frac{\nu}{k_\parallel v_T},$$

$$\delta = \frac{\Omega}{k_\perp v_T}, \qquad x = k_\parallel v_T t, \text{ and} \qquad \zeta = \frac{(\omega-\Omega)}{k_\parallel v_T}.$$

Very near the midplane the linear variation of field strength is incorrect, and we must use a quadratic field perturbation instead. Starting with a magnetic field described by

$$\vec{B} = \vec{B}_0(1 - \frac{1}{2}\frac{\lambda v_T z^2}{k_\parallel^3}) \qquad (4)$$

the response function becomes

$$Z(\zeta) = \int_0^\infty e^{-\frac{1}{\gamma^2}(\gamma x - \frac{3}{2} + 2e^{-\gamma x} - \frac{1}{2}e^{-2\gamma x}) - \frac{1}{\beta^2+\gamma^2}\gamma x - \frac{(1-e^{-\gamma x})^2}{\gamma^2(1+\frac{1}{3}i\lambda x^3)} - i\zeta x} dx \qquad (5)$$

COMPARISON WITH THE PREVIOUS TREATMENT

To facilitate a qualitative comparison with the response function derived by Smithe[1] we will need to assume that the collision frequency is small. This will allow us to do a

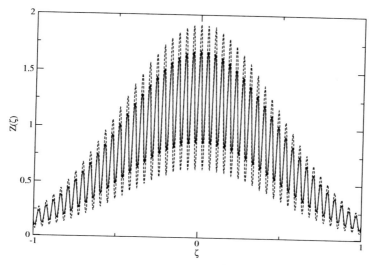

FIGURE 1. Our results (solid line) versus Smithe's results (dashed line)

series expansion of the terms that go like $e^{-\gamma x}$ to first order in γ. The result is

$$Z(\zeta) = \int_0^\infty e^{-\frac{1}{3}\gamma x^3 - \frac{x^2}{2}(1-\frac{1}{2}\alpha x)^2 - i\zeta x} dx \qquad (6)$$

which has the same functional form as Smithe's result. The difference is in the coefficient of the γx^3 term. Smithe found this coefficient to be $\frac{1}{4}$, but here it is $\frac{1}{3}$ here. Since our solution included collisions from the beginning we had no need to make an estimate of the size of $v_\perp$ in terms of v_T as Smithe did. This leads us to the conclusion that Smithe's estimate of the coefficient was incorrect. To see just how much of an effect this correction has we plot the real part of our response function versus the real part of Smithe's response in Figure 1.

THE GEOMETRIC VARIATION OF THE RESPONSE FUNCTION

We expanded the magnetic field as a function of distance along a field line for our solution, but the magnetic field in a toroidal fusion device actually varies as a function of major radius. To find the value of the parameter we call $L_\parallel$ we must expand field strength as a function of major radius. Then we find the conversion between major radial distance and distance along a field line by finding the angle of intersection between the field line and a surface of constant major radius. For an ideal cylindrical torus we find that

$$L_\parallel = \frac{R_0 - a\cos(\Theta)}{\sin(\Theta)\sin(\arctan(\frac{a}{R_0 q(a)}))} \qquad (7)$$

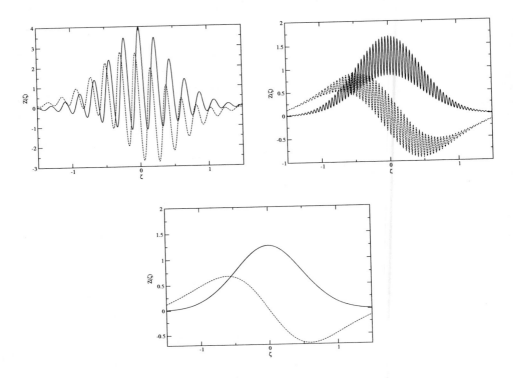

FIGURE 2. The plasma response function for small parallel gradient lengths (top left, $L_{\|} = 40m$), intermediate lengths (top right, $L_{\|} = 200m$), and large parallel gradient lengths (bottom, $L_{\|} = 500m$). The solid line is the real part and the dashed line is the imaginary part.

where a is the minor radial position and Θ is the angle in the poloidal plane away from the midplane. If we assume a major radius of 6m then $L_{\|}$ can vary from 38m to infinity. Over this variation in length the response function shows three distinct types of behavior. For small values of $L_{\|}$ the response function is dominated by parallel gradient effects and is very oscillatory with an average near zero. When $L_{\|}$ is very long collisional effects damp out the parallel gradient effects to produce a response function that is the same as would be produced by thermal effects alone. For intermediate values of $L_{\|}$ both collisional effects and parallel gradient effects interact to produce a response function that shows oscillations around an average of a thermal response function. Examples of each behavior can be seen in Figures 2.

REFERENCES

1. Smithe, D., Colestock, P., Kammash, T., and Kashuba, R., *Phys. Rev. Lett.*, **60**, 801–804 (1988).
2. Dougherty, J. P., *Phys. Fluids*, **7**, 1788–1799 (1964).

Calculation of Fokker-Planck Ion Distributions Resulting from ICRF Full-Wave Code Fields and Collisions

R.W. Harvey, N. Ershov, A.P. Smirnov, P. Bonoli[*], J.C. Wright[*], F. Jaeger[†]
D.B. Batchelor[†], L.A. Berry[†], E. D'Azevedo[†], M.D. Carter[†], D.N. Smithe[‡]

CompX, P.O. Box 2672, Del Mar, CA 92014-5672

Abstract

A fully numerical calculation of ion velocity and spatial diffusion coefficients resulting from full wave code electromagnetic fields in tokamak geometry, has been implemented. EM fields are obtained from the TORIC[1, 2] and AORSA-2D[3] codes. The described coefficient calculator integrates the unapproximated Lorentz equation of motion to obtain the change in velocity after one complete transit of the noncircular tokamak flux surfaces. Averaging velocity changes over initial starting gyro-phase and initial toroidal angle of an ion, gives bounce-averaged diffusion coefficients. Performing this operation for an array of initial parallel and perpendicular velocities and radial locations gives the full set of coefficients. Coupling the coefficients to the CQL3D[3] Fokker-Planck code enables calculation of the equilibrium nonthermal ion distribution function resulting from the EM fields and collisions. This calculation is computationally feasible due to the recent rapid advances of computing hardware. Adequate resolution is obtained in 50 CPU hours. Initial results will be presented, and compared with a Fokker-Planck/ray-tracing calculation.

Full wave solutions for the electromagnetic fields excited in toroidal fusion devices by antennas operating in the ICRF ion cyclotron range of frequencies have been obtained[1, 2, 3]. The important physics of mode conversion, and of wavelength comparable to plasma scale, are included in these calculations.

High power ICRF can produce substantial nonthermal tails on the cyclotron resonant ions, as discussed at this meeting[4] and elsewhere. To increase the range of accuracy of the full wave codes, it is necessary to account for nonthermal ion distributions. This effort reduces to two tasks: (1) Inclusion of nonthermal distributions in the full wave code dielectric tensor [5]; and (2) Calculation of nonthermal distrubutions resulting from RF induced diffusion. The latter effect is the topic of this note.

We obtain that it is reasonably feasible to perform a fully numerical computation of the RF diffusion coefficients by integration of the Lorentz equation of ion motion in the combined equilibrium magnetic and fluctuating full wave EM fields. An ensemble of ion starting conditions is used, and the diffusion coefficients are obtained by suitable averages of the results. This approach includes analytically difficult effects such as realistic toroidal geometry, the overlapping of the ion gyro-orbit with several cyclotron resonant surfaces such as occurs in HHFW for fast ion in

---
[*]PSFC, MIT, Boston, Mass.
[†]ORNL, Oak Ridge, Tenn.
[‡]Mission Research Corp., Newington, Vi.

the NSTX device, and the phase relations between multiple toroidal modes. The resulting diffusion coefficients are used in a Fokker-Planck code[6] to obtain 2D-in-velocity-space distributions on a radial array of flux surface. A fully numerical calculation of electron cyclotron diffusion coefficients was previously reported, using similar methods[7].

A fourth order Rugge-Kutta scheme was found to be sufficient to integrate the ion equations of motion in the geometry and fields depicted in Fig. 1. With step size equal to 1/80 gyro-period, the calculation is adaquatedly converged.

Fig. 2 shows the *change* in velocity u versus time step for four typical ions resonant with the electric field for C-Mod antenna current of 100A. Particle initial conditions differ by the initial gyrophase, here equispaced. The net change in momentum $\Delta u = (u_{final} - u_{initial})$ after one complete transit or bounce motion of the ion in the poloidal plane gives the step size in velocity space. The average over initial azimuthal phases, $\langle \Delta u \rangle$, is zero in the linear case. The average over initial azimuthal phases, $\langle \Delta u^2 \rangle$, is non-zero, and the resulting diffusion is put into a form suitable for the CQL3D bounce-averaged FP code.

The general velocity space diffusion equation, averaged over times which are long compared to the particle gyro-, bounce- and toroidal symmetrizing- times, is[8, 9, 7]

$$\frac{\partial f}{\partial t}|_{RF} = \frac{\partial}{\partial J_i}\left(\overline{D_{ij}} \frac{\partial f}{\partial J_j}\right), \quad \overline{D_{ij}} \equiv \frac{\langle \Delta J_i \Delta J_j \rangle}{2 \Delta t}, \quad i, j = 1:3,$$

where $J_1 = (m_e/Ze)(m_i u_\perp^2/2B)$, $J_2 = \oint v_\parallel dl$, and $J_3 = p_\phi$ are action variables with associated cyclic angle variables, θ_1 = gyrophase, θ_2 = bounce phase, θ_3 = toroidal angle ϕ. The average $\langle \Delta J_i \Delta J_j \rangle$ in the above equation is taken over the cyclic variables. The time Δt is the average time $\Delta t = \tau_b(2\pi/\Delta\phi)$ to return to a particular toroidal angle increment $\Delta\phi$ of the toroidal phase. The bounce time is τ_b. We transform to the coordinates used in CQL3D, that is, pitch angle θ_0 and momentum-per-mass u_0 of the electrons evaluated at the minimum magnetic field (B) point on each flux surface. This gives the equation for f at the minimum B point to be compared with the CQL3D FP form and verifies that the CQL3D momentum diffusion coefficient $D_{u_0 u_0}$ in CQL3D is simply $\langle \Delta u_0^2 \rangle / 2\Delta t$, where the $\langle\rangle$-average is taken as above for the action variables.

Fig. 3 shows the diffusion coefficient from the orbit calculations as outlined above. There is reasonable agreement for the location and shape of the diffusion coefficient with results from ray tracing (with GENRAY[10]) of ICRF waves. The magnitudes of the coefficients are similar, although it is difficult to exactly compare: the WKB method breaks down near the $\omega_{cH^+} = \omega$ layer. Also, the genray-cql3d obtains EM field amplitudes from specified incident power, whereas TORIC EM field amplitudes derive from input antenna current.

Nonthermal distributions are produced at nominal antenna currents, as shown in Fig. 4. Figure 5 compares the specific power density per du, from the fully numerical diffusion calculation and from the genray-cql3d ray tracing. Specific power density is $p_{rf}(u_0) \equiv \int d\theta_0 2\pi \sin\theta_0 u_0^2 (\frac{1}{2}m_i * u^2) \cdot \frac{\partial f}{\partial t}|_{RF}$. The shape or these curves is very similar, giving the preliminary result that the physics obtained with the fully numerical diffusion is correct.

Antenna current was varied from 1A to 1kA, giving $\sim 10^6$ variation in diffusion coefficient. Coefficient shape was constant and amplitude scaled closely to (antenna-current)^2. Computational noise appears not to be an issue. There was evidence of nonlinear broadening of the coefficient at the highest currents. Adaquate accuracy is obtained with 140 minutes of 2.4GHZ Xeon CPU time per flux surface. To achieve multi-processor speed up of the code, it is parallelized using the MPI library.

Refinements of these calculations are underway. This includes improved centering of the ion orbits on each flux surface: that is, ions are assigned to the flux surface bin encomapassing there time-avaraged radial location. Although the Fokker-Planck code has a zero-banana width approximation, the distributions can be correct to first order in (banana width)/(plasma scale length). Attention will be paid to measuring toroidal momentum imparted by the RF to the ions. Comparisons will be made with approximate ion diffusion coefficients which, in the linear, Maxwellian regime, will give ion damping as presently used in the AORSA full wave code[3].

References

[1] M. Brambilla, R. Bilato, P. Bonoli, Radio Frequency Power in Plasmas, 14th Top. Conf., Oxnard, Calif. (2001)

[2] P.T. Bonoli, M. Brambilla, et al., Phys. of Plasmas **7**, 1886 (2000).

[3] E.F. Jaeger, L.A. Berry et al., Phys. of Plasmas **8**, 1573 (2001)

[4] A. Rosenberg, Invited talk I-06, this meeting (2003).

[5] R. Dumont, Invited talk I19, presented by C.K. Phillips (2003).

[6] R.W. Harvey and M.G. McCoy, "The CQL3D Fokker-Planck Code", Proc. of IAEA TCM on Advances in Simulation and Modeling of Thermonuclear Plasmas, Montreal, 1992, p. 489-526, IAEA, Vienna (1993); NTIS document DE93002962.

[7] R.W. Harvey and R. Prater, *ibid.* Oxnard Mtg, p. 298 (2001).

[8] A.N.Kaufman, Physics of Fluids **15**, 1063(1972).

[9] L.-G. Eriksson, P. Helander, Physics of Plasmas **1**, 308 (1994).

[10] A.P. Smirnov and R.W. Harvey, Bull Amer. Phys. Soc. 40, 1837, Abst. 8P35 (1995); "The GENRAY Ray Tracing Code" CompX report CompX-2001-01 (2001).

Acknowledgments

This work is supported by the USDOE RF-SciDAC project, grant DE-FC02-01ER54649.

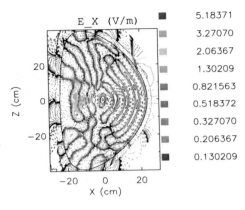

Figure 1: A component of the fluctuating EM field from TORIC, for 1A antenna current.

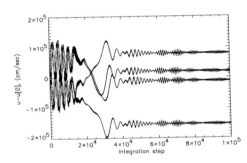

Figure 2: Particle speed changes as a function of step number, starting at $\rho = 0.25a$ on the outer equatorial plane. Initial gyrophases differing by 90 deg. Antenna current is 100 A.

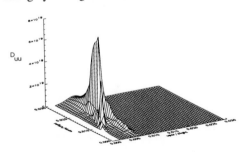

Figure 3: $D_{u_0 u_0}$ coefficient in $u_\parallel, u_\perp$-space, resulting from 1kA on the C-Mod ICRF antenna, $n_\phi = 10$. Maximum value 8.2e19 cgs units.

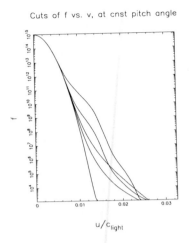

Figure 4: Nonthermal distribution function from diffusion coefficient scaled by 0.1. This gives a low power of 0.1 watts/cm^3.

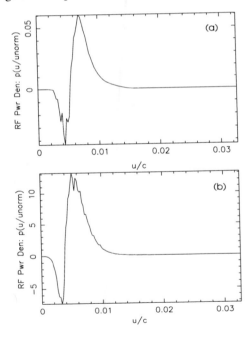

Figure 5: (a) Specific power density per du, from fully numerical diffusion coeffient, and (b) from ray tracing.

Mode Conversion Flow Drive in Tokamaks

E. F. Jaeger,[*] L. A. Berry,[*] J. R. Myra,[†] D. B. Batchelor,[*] E. D'Azevedo[*]

[*]*Oak Ridge National Laboratory, Oak Ridge, Tennessee 37831-8071*
[†]*Lodestar Research Corporation, 2400 Central Avenue P-5, Boulder, Colorado 80301*

Abstract. A two-dimensional integral full-wave model is used to calculate poloidal forces driven by mode conversion in tokamak plasmas. In the presence of a poloidal magnetic field, mode conversion near the ion-ion hybrid resonance is dominated by a transition from the fast magnetosonic wave to the slow ion cyclotron wave. The poloidal field generates strong variations in the parallel wave spectrum that cause wave damping in a narrow layer near the mode conversion surface. The resulting poloidal forces drive sheared poloidal flows comparable to those in direct launch ion Bernstein wave experiments.

INTRODUCTION

Recent experiments on the Tokamak Fusion Test Reactor (TFTR) [1] have shown that directly coupled ion Bernstein waves (IBW) can be used to drive sheared poloidal flow at high cyclotron harmonics in tokamak plasmas. In addition, experiments on the Frascati Tokamak Upgrade (FTU) [2], along with several calculations [3–6], suggest that these externally driven sheared flows can trigger internal transport barriers that reduce micro-turbulence and enhance plasma confinement. Unfortunately, the technology required for direct coupling of IBW to the bulk plasma is extremely difficult and not fully understood. A simpler and perhaps more efficient means of driving sheared flow is to use mode conversion. Mode conversion in radio frequency (rf) heated plasmas has been the subject of both theoretical [7–10] and experimental [11–15] studies. Previous one-dimensional (1-D) simulations of mode conversion [16, 17] near the ion-ion hybrid resonance show a transition from the fast magnetosonic wave to the slow IBW. Without a poloidal magnetic field, the IBW propagates on the high field side of the ion-ion hybrid resonance and is absorbed by electron Landau damping and magnetic pumping. By themselves, these mechanisms do not lead to poloidal flow [4]. However, recent two dimensional (2-D) calculations [10] show that the mode conversion process is significantly modified in the presence of a poloidal field.

In this paper, we apply a 2-D integral full-wave model [10] to calculate the poloidal forces that result from mode conversion near the ion-ion hybrid resonance. Results show that, in the presence of a poloidal magnetic field, mode conversion is dominated by a transition from the fast magnetosonic wave to the slow ion cyclotron wave (ICW).

Unlike the mode-converted ion Bernstein wave, the ICW is a cold plasma wave that usually propagates on the low field side of the mode conversion layer. The poloidal magnetic field generates strong variations in the parallel wave number ($k_\parallel$) that broaden the cyclotron resonance and allow damping of the ICW by both ions and electrons in a narrow layer near the mode conversion surface. The resulting poloidal forces in this layer can drive sheared poloidal flows. In recent experiments [14,15], both ion absorption and poloidal flow have been observed near the mode conversion layer. The possibility of mode conversion to the ICW was first suggested in 1977 by Perkins [7] and has recently been observed both numerically [10] and experimentally [13]. However, the 2-D effect of the poloidal field on the mode-converted ICW and the resulting sheared poloidal flow have not previously been analyzed.

MODE CONVERSION IN TWO DIMENSIONS

Numerical solutions of the integral wave equation are carried out using AORSA [10], a fully spectral, 2-D integral full-wave model. Figure 1 shows a simulation of mode conversion in a He^3–H–D plasma near the ion-ion hybrid layer in Alcator C-Mod [12,13]. Vertical lines indicate the location of the ion-ion hybrid (left) and ion cyclotron resonances (right) for both He^3 and H. The mode conversion region is expanded in Fig. 1(b). Notice that over most of the plasma cross section, the mode-converted waves propagate on the low field side (right) of the ion-ion hybrid resonance, with wavelength decreasing to the right. These are cold plasma, forward waves consistent with ICW. Waves propagating on the high field side of the resonance, in the small region near the magnetic axis, are finite temperature, backward waves consistent with IBW.

The poloidal magnetic field has a strong effect on these results. In Fig. 2, three different values of normalized poloidal field (b_p) are assumed for the He^3–H–D mode conversion scenario. Solid lines show the location of the ion-ion hybrid resonance. For $b_p = 0$, the mode-converted IBW propagates on the high field side (left) of the mode

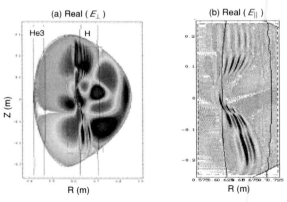

FIGURE 1. He^3–H–D mode conversion in Alcator C-Mod [12,13] with $B_0 = 5.85$ T and $n_\varphi = 10$.

conversion surface in agreement with previous 1-D results without a poloidal field. As the poloidal field is increased in Fig. 2(b), the ICW appears on the low field side of the resonance. For poloidal field magnitudes typical of real tokamaks in Fig. 2(c), the ICW dominates the mode conversion process, except for the region immediately surrounding the magnetic axis.

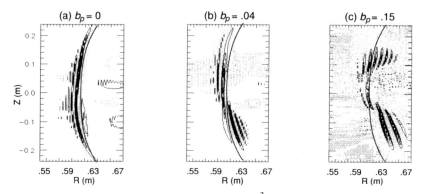

FIGURE 2. Effect of the poloidal magnetic field b_p for He^3–H–D mode conversion with $n_\varphi = 15$.

ENERGY ABSORPTION AND POLOIDAL FLOW DRIVE

Interaction between the plasma and the mode-converted ICW leads to wave damping and poloidal forces that drive sheared flow in the mode conversion layer. The direct force exerted by waves on a plasma can be written as the sum of the wave momentum for each Fourier component ($\mathbf{k}/\omega$) times the power absorbed in that component. This force contains a dissipative part that corresponds to absorption of photon momentum, and a reactive part that reduces to the ponderomotive force in the cold plasma limit [6]. This reactive part can be locally large, but does not drive flux surface average flows because the relevant flux surface averages are zero. Additional forces, due to transport of momentum [4–6], are small in Fig. 1.

Figure 3(a) shows 2-D contour plots of the power absorbed and the dissipative part of the poloidal force for minority ions in Fig. 1. The absorption and poloidal force are skewed toward the lower half of the plasma where the sign of the poloidal field gives upshift in $k_\parallel$. Figure 3(b) shows the flux surface average of the power absorption and poloidal force for electrons and minority ions. The poloidal force on each species follows closely the power absorbed in that species. With 1 MW of total rf power, the magnitude of the peak force on the electrons is about 3 Nt/m^3. This is comparable to the force calculated [5] when 0.36 MW of high harmonic IBW is launched directly in TFTR and absorbed at the fifth harmonic cyclotron resonance [1]. These results are consistent with recent experiments [14,15] and suggest that mode conversion to ICW can be used to drive sheared poloidal flow, thereby creating internal transport barriers and giving access to enhanced confinement regimes.

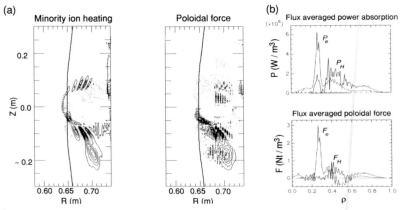

FIGURE 3. (a) Contours of minority ion heating and poloidal force for the example in Fig. 1; (b) flux surface averaged power and poloidal force with 1 MW of rf power absorbed.

ACKNOWLEDGMENTS

The authors wish to thank M. Porkolab, F.W. Perkins, and E. Nelson-Melby for helpful discussions. Research was sponsored by Oak Ridge National Laboratory, managed by UT-Battelle, LLC, for the U.S. Dept. of Energy under contract DE-AC05-00OR22725.

REFERENCES

1. LeBlanc, B. P., et al., *Phys. Rev. Lett.* **82**, 331 (1999).
2. Cesario, R., et al., in *Radio Frequency Power in Plasmas*, edited by S. Bernabei and F. Paoletti, AIP Conf. Proc. 485, New York, 1999, p. 100.
3. Biglari, H., Diamond, P. H., and Terry, P. W., *Phys. Fluids* B **2**, 1 (1990).
4. Berry, L. A., Jaeger, E. F., and Batchelor, D. B., *Phys Rev. Lett.* **82**, 1871 (1999).
5. Jaeger, E. F., Berry, L. A., and Batchelor, D. B., *Phys. Plasmas* **7**, 3319 (2000).
6. Myra, J. R., and D'Ippolito, D. A., *Phys. Plasmas* **7**, 3600 (2000).
7. Perkins, F. W., *Nucl. Fusion* **17**, 1197 (1977).
8. Stix, T. H., *Waves in Plasmas*, American Institute of Physics, New York, 1992, pp. 250–256.
9. Majeski, R., Phillips, C. K., and Wilson, J. R., *Phys. Rev. Lett.*, **73**, 2204 (1994).
10. Jaeger, E. F., et al., *Phys. Plasmas* **8**, 1573 (2001).
11. Majeski, R. et al., *Phys. Rev. Lett.*, **76**, 764 (1996).
12. Bonoli, P. T., et al., *Phys. Plasmas* **7**, 1886 (2000).
13. Nelson-Melby, E., et al., *Phys. Rev. Lett.* **90**, 155004 (2003).
14. Phillips, C. K., et al., *Nucl. Fusion* **40**, 461 (2000).
15. Castaldo, C., et al., in *Proceedings of the 19th IAEA Fusion Energy Conference, Lyon, France, 2002* IAEA, Vienna, 2002.
16. Colestock, P. L., Kashuba, R. J., *Nucl. Fusion* **23**, 763 (1983).
17. Fukuyama, A., et al., *Nucl. Fusion* **23**, 1005 (1983).

Modelling of ICRH induced current and rotation

T. Johnson*, T. Hellsten*, L.-G. Eriksson†, M. Laxåback* and T. Bergkvist*

*KTH, Association Euratom-VR, SE-100 44 Stockholm, Sweden
†Association Euratom-CEA, Cadarache, F-130108, St. Paul lez Durance, France

Abstract. The scenario of ^{3}He(^{4}He) minority ICRH has been studied with the SELFO code, revealing the importance of finite orbit width and orbit topology for ICRH induced currents and torques.

INTRODUCTION

Ion-cyclotron resonance heating (ICRH) provides a tool for profile control in tokamaks through the possibility of localized ion and/or electron heating and through the induced currents and torques. The current and rotation induced by ICRH are small compared to the plasma current and the Alfvén velocity, but they are localized and could therefore provide shear. A number of parameters could be used to optimize rotation and current profiles, e.g. the choice of resonant ion species and their concentrations, the resonance positions and the toroidal mode spectra.

During high power ICRH the distribution functions of the resonant ion species develop anisotropic tails of trapped energetic ions with the turning points close to the cyclotron resonance. Toroidal acceleration induces a radial drift of the turning points of a trapped orbit $\Delta \psi_{tp} = n_\phi/(Ze\omega)\Delta E$, where n_ϕ is the toroidal mode number and ϕ increases in the direction anti-parallel to the plasma current. Thus, asymmetrically propagating waves induces a pinch of resonating ions [1, 2, 3]. For heating with $n_\phi < 0$ turning points moves along the resonance towards the midplane $Z = 0$ where they meet. The orbit is then transformed, or *detrapped*, into a passing one. For energetic ions, or if the resonance is on the low field side of the magnetic axis, the detrapping will exclusively produce co-current passing orbits, *ie* an *asymmetric detrapping* [4].

We shall distinguish between thin *banana* orbits and *potato* orbits with widths comparable to the minor radius [5]. Ions on trapped potato orbits spend most of their time outside the flux surface of the turning points, making the orbits similar to co-current passing orbits. Consequently, ions on trapped potato orbits has a much higher toroidal precession than ions on corresponding banana orbits. The current profile of trapped potato orbits are monopolar, in contrast to the bipolar diamagnetic current from banana orbits.

Momentum absorbed from the wave by the resonant ions is transferred to the bulk plasma through friction and $j \times B$ forces. Also pure perpendicular heating can give rise to friction and $j \times B$ forces, due to orbit broadening.

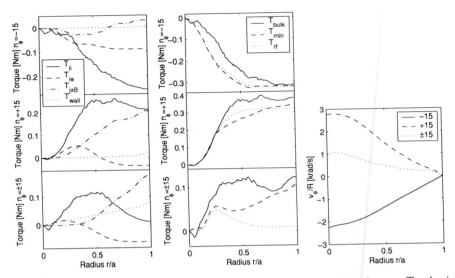

FIGURE 1. To the left are the collisional torque from to the ions T_{ii} and the electrons T_{ie}, the $j \times B$ torque T_{jxB} and the torque lost when ions hit the wall. To the right are the torques to the bulk plasma T_{bulk}, to the minority T_{min} and from the RF wave field T_{rf}. The momentum in the system is conserved if $T_{min}(a) = T_{bulk}(a)$. To the right are the rotation profile.

NUMERICAL ASSESSMENT OF CURRENT AND ROTATION

The ICRH induced current and rotation has been studied with the SELFO code [6]. It calculates the distribution functions of resonant ions using the FIDO code [7], and the magnetosonic wave field using the LION code [8, 9]. The FIDO code is a Monte Carlo code solving a Langevin equation, equivalent to the orbit averaged Fokker-Planck equation with a quasi linear RF-operator [7]. The wave field, which is obtained from the wave equation with the dielectric tensor corresponding to the calculated ion distribution functions, is repeatedly updated as the distribution functions are evolving.

The scenario of on axis minority ^{3}He heating at its fundamental cyclotron frequency in a ^{4}He plasma is analysed for the asymmetric toroidal mode spectra $n_\phi = -15$ and $n_\phi = 15$, and the symmetric spectrum $n_\phi = \pm 15$. The plasma parameters are similar to those of a typical JET plasma: $B_0 = 3.4$T, $I_p = 3$MA, $R_0 = 3$m, $a = 1$m, $P = 5$MW, $n_e = 2.9 \times 10^{19}$m^{-3}, $n_{3He}/n_e = 0.03$, $Z_{eff} = 3$ and $T_e = T_i = 5$keV. This gives a slowing down time of 0.32s and a critical energy $E_c = 210 keV$, where slowing down and energy scattering are comparable [10].

The torques onto the bulk plasma and the minority, as calculated with the SELFO code[1], are shown in figure 1. In the centre the friction and $j \times B$ both contribute to the torque onto the bulk plasma in response to the momentum absorbed from the wave. Further out the two torques partially cancel, as expected for neo-classical transport

---

[1] In the current version of SELFO the friction diagnostics has been improved. It does *not* use corrections by subtracting the torques from a thermal minority population as used in reference [11].

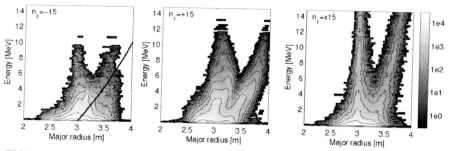

FIGURE 2. Minority ion density in energy and major radius averaged from $Z/a = [-0.3, 0.3]$.

of ions with thin orbits. Although the friction and $j \times B$ forces may be affected by the finite orbit width, the total momentum to the bulk is to lowest order given by the momentum from the wave field, as seen in previous studies [11]. However, for the symmetric spectrum $n_\phi = \pm 15$ the change in the momentum from fast ions lost to the wall is dominating.

The reason for the higher losses to the walls with the symmetric spectrum is illustrated in figure 2, showing the fast ion density in energy and major radius evaluated around the midplane $Z = 0$. As each orbit intersects this plane at two points, the "V" shape represents the orbit width as function of energy. For $n_\phi = 15$ the ions are heated, increasing their orbit width, and pinched outwards by the momentum of the wave until they hit the wall. For $n_\phi = \pm 15$ the volume average of the pinch cancels and the orbit losses occur at higher energies. Although the $n_\phi = -15$ spectrum confines the energetic ions they do not reach as high energy as with the other mode spectra and therefore do not reach the wall. For $n_\phi = -15$ the ions with the highest energy are co-current passing. As the energy increases the Doppler shift of the cyclotron resonances moves along the orbit towards the low field side until they meet in the midplane, see the solid line to the left in Fig. 2 showing the resonance position for $(v_\perp/v)^2 = B/B_0$. This is the highest reachable energy, since further heating would take the ion out of resonance. For the scenario considered here the maximum energy at which the ions are in resonance is less then the energy at which the ions hit the wall.

The ICRH induced rotation is modeled by the steady state of a momentum diffusion equation $\partial_r \left(RnmDr\partial_r v_\phi \right) = rT_{bulk}$ and shown to the right in figure 1. The diffusion coefficient is $D = a^2/(\alpha_m \tau_m)$, where the momentum confinement time is assumed equal to the energy confinement time $\tau_m \sim \tau_e \sim 0.3$s and $\alpha_m = 2$. The rotation profiles for the different spectra shown in figure 1 are consistent with recent experimental results [12].

The minority ion cyclotron current drive including the neoclassical Ohkawa currents [13] is shown in figure 3. All profiles except $n_\phi = -15$ have a bipolar structure due to the diamagnetic currents of trapped and symmetrically detrapped ions. The monopolar current profile for $n_\phi = -15$ is dominated by co-current passing ions on potato orbits in the MeV range, well above E_c. Note how the outward pinch for $n_\phi = 15$ displaces the bipolar structure outwards. From figure 3 it is clear that most of the current with the symmetric spectra $n_\phi = \pm 15$ comes from trapped ions with energies well above E_c. This is contrary to what is predicted for the Fisch minority current drive mechanism [14]. The current due to passing ions gives an illustration of the transition

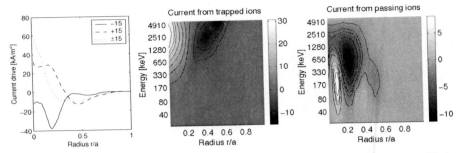

FIGURE 3. To the left is the profile of driven current: solid line $n_\phi = -15$, dashed line $n_\phi = 15$, dotted line $n_\phi = \pm 15$. The current density in logarithmic energy and minor radius ($\delta I = J 2\pi r \, \delta r \, \delta \log(E)$) with $n_\phi = \pm 15$ for trapped ions (in the middle) and passing ions (to the right).

from symmetric to asymmetric detrapping. For energies of a few hundred keV the current has a bipolar structure due to symmetric detrapping [15], while for energies above 1MeV the detrapping is asymmetric exclusively producing co-current passing ions [4].

CONCLUSIONS

The currents and torques induced by on axis ^{3}He minority ICRH have been calculated with the SELFO code. The torques to the bulk plasma are to first order given by the torque absorbed from the wave field. However, for the symmetric spectrum the losses to the wall dominate. For $n_\phi = -15$ it has been found that the losses to the wall are avoided by the production of potato orbits for which the ion energy is limited by the condition for the ions to remain in resonance.

For the scenarios considered here it has been found that the current drive is dominated by ions with energies well above the critical energy, contrary to what is predicted for the Fisch minority current drive mechanism [14]. With $n_\phi = -15$ the current drive is dominated by co-current passing ions on potato orbits, while with $n_\phi = 15$ and $n_\phi = \pm 15$ the current profiles are bipolar from trapped ions.

REFERENCES

1. S. Riyopoulus et al, *Nuclear Fusion*, **26** (1986)
2. L. Chen, J. Vaclavik, and G.W. Hammett, *Nuclear Fusion*, **28** 389 (1988)
3. T. H. Stix, *Waves in Plasmas*, Springer, New York, (1992)
4. J. Carlsson, T. Hellsten, and J. Hedin, *Physics of Plasmas*, **5** 2885 (1998)
5. L.-G. Eriksson and F. Porcelli, *Plasmas Physics and Controlled Nuclear Fusion*, **43** 145 (2001)
6. J. Hedin, T. Hellsten, and L.-G. Eriksson, *Nuclear Fusion*, **40** 1819 (2000)
7. J. Carlsson, L.-G. Eriksson, and T. Hellsten, In *Theory of fusion plasmas*, 351, (1994)
8. L. Villard, et al., *Computer Physics Reports*, **4** 95 (1986)
9. L. Villard, S. Brunner, and J. Vaclavik, *Nuclear Fusion*, **35** 1173 (1995)
10. T.H. Stix, *Nuclear Fusion*, **15** 737 (1975)
11. T. Johnson, et al., In *Theory of fusion plasmas*, (2002)
12. L.-G. Eriksson, In *at this conference*
13. J. W. Connor and J. G. Cordey, *Nuclear Fusion*, **14** 185 (1974)
14. N. J. Fisch, *Nuclear Fusion*, **21** 15–22 (1981)
15. T. Hellsten, J. Carlsson, and L.-G. Eriksson, *Physical Review Letters*, **74** 3612–3615 (1995)

Local Solutions for Generic Multidimensional Resonant Wave Conversion

E. R. Tracy[a] and A. N. Kaufman[b]

[a]*Physics Department, William & Mary, Williamsburg, VA 23185-8795*
[b]*Lawrence Berkeley National Laboratory and Physics Department, UC Berkeley, Berkeley, CA 94720*

Abstract. In more than one spatial dimension, resonant linear conversion from one wave type to another can have a more complex geometry than the familiar 'avoided crossing' of one-dimensional problems. In previous work [1] we have shown that *helical* ray shapes are generic in a mathematical sense. Here we briefly describe how the local field structure can be computed.

INTRODUCTORY COMMENTS

Resonant conversion between various wave types is exploited in RF heating schemes for fusion plasmas. Such conversion can occur in nonuniform plasmas where, for some spatial position x_0 and for a frequency ω_0, two wave types '*a*' and '*b*' can have nearly equal wave vectors $\mathbf{k}_a \sim \mathbf{k}_b \sim \mathbf{k}_0$. The resonance condition implies a matching of the local *phase velocities*, but still allows the two waves to have different *group velocities* and *polarizations*. Thus, the process cannot be reduced to one dimension, even locally, and the possibility of new physics arises. This is discussed more fully in [1-3].

In a separate paper [1], we have considered the question of what ray geometries might be 'generic' in multi-dimensions, and how the WKB connection coefficients can be calculated. We define the meaning of 'generic' conversion just after Eq. (7). This line of research is similar in spirit to the one-dimensional work of Littlejohn and Flynn [4]. In [1] we show that in systems with two or three spatial dimensions (implying that the ray phase space (**x**,**k**) is four- or six-dimensional, respectively), the ray geometry in conversion regions *cannot* be purely hyperbolic. Instead, it is generically a combination of hyperbolic motion in one two-dimensional subspace (analogous to a one-dimensional 'avoided crossing' or 'tunneling' region) and elliptical motion in another. Hence, the combined motion is helical. Such a combination of behaviors is, of course, not possible in the one-dimensional conversion problem. This result was independently derived using a different approach by Colin de Verdiere [5].

In [2] a tutorial introduction is given to ray-based analysis of multi-dimensional conversion, and in [1] we show that generic ray motion in multi-dimensional conversion is helical. Here we sketch the calculation of the wave field for helical conversion. Details will be provided in a longer paper.

STATEMENT OF THE PROBLEM

Consider the linear wave equation (1) for the (three-component) electric field in a non-uniform, time-stationary, plasma:

$$\int d^2x' dt' D_{jk}(\mathbf{x},\mathbf{x}',t-t') E_k(\mathbf{x}',t') = 0, \qquad j,k=1,2,3. \tag{1}$$

Summation over repeated indices is implied. We assume that the plasma has two spatial dimensions for simplicity, but note that the approach can be generalized to higher dimensions. In addition, the WKB connection coefficients do not depend on the number of spatial dimensions, but simply on the fact that only two waves are undergoing conversion.

We assume that the wave equation is conservative and can be derived from an action principle. Using methods described in [2] we convert (1) into the form of a partial differential equation:

$$D_{jk}(\mathbf{x},-i\nabla,i\partial_t) E_k(\mathbf{x},t) = 0, \qquad j,k=1,2,3. \tag{2}$$

Our goal is to solve (2) subject to some specified initial/boundary conditions. A standard tool for this analysis is the WKB method. WKB methods break down in conversion regions and must be augmented by a local treatment which provides an algorithm for calculating the WKB connection coefficients, as described below.

THE LOCAL 2x2 WAVE EQUATION

In the conversion region, the electric field is expanded in the local form

$$\mathbf{E}(\mathbf{x},t) = e^{-i\omega_0 t} e^{i\mathbf{k}_0 \cdot (\mathbf{x}-\mathbf{x}_0)} \left[\psi_\alpha(\mathbf{x}) \hat{\mathbf{e}}_\alpha + \psi_\beta(\mathbf{x}) \hat{\mathbf{e}}_\beta \right] \tag{3}$$

The *uncoupled* polarization vectors $\mathbf{e}_\alpha$ and $\mathbf{e}_\beta$ are (locally) constant and can be constructed using methods sketched in [2]. Inserting the ansatz (3) into (2) gives

$$D_{ij}(\mathbf{x}_0 + (\mathbf{x}-\mathbf{x}_0), \mathbf{k}_0 - i\nabla; \omega_0) \psi_j(\mathbf{x}) = 0, \qquad i,j = \alpha,\beta. \tag{4}$$

where

$$D_{ij} = \mathbf{e}_j^* \cdot \mathbf{D} \cdot \mathbf{e}_k. \tag{5}$$

(Here, and in what follows, the * notation refers to the Hermitian adjoint on vectors and operators.) Suppressing the ω_0-dependence, and Taylor-expanding the wave operator about the conversion point, we have

$$\begin{pmatrix} \hat{D}_{\alpha\alpha} & \hat{D}_{\alpha\beta} \\ \hat{D}_{\alpha\beta}^* & \hat{D}_{\beta\beta} \end{pmatrix} \begin{pmatrix} \psi_\alpha(\mathbf{x}) \\ \psi_\beta(\mathbf{x}) \end{pmatrix} = 0 \tag{6}$$

where

$$\hat{D}_{ij} \equiv D_{ij}(\mathbf{x}_0,\mathbf{k}_0) + \nabla_x D_{ij} \cdot (\mathbf{x}-\mathbf{x}_0) - i\nabla_k D_{ij} \cdot \nabla. \tag{7}$$

Because the polarizations of (3) are the *uncoupled* ones, and because the conversion point lies on the dispersion surface for both uncoupled waves, the constant terms of the diagonal elements $D_{\alpha\alpha}$ and $D_{\beta\beta}$ are zero, while the off-diagonal term is typically a non-zero (complex) coupling constant η. In prior work [2,6], we assumed that the dominant terms in the vicinity of the conversion point were given by the first order corrections to the wave operator along the diagonal. This is, strictly speaking, only correct if the coupling constant is non-zero in the conversion *region* (not just at the conversion *point*). Here we include the more general case as part of the analysis. Using an extension of methods described in [3] and more recent results from [1], it is possible to recast (5) into:

$$\begin{pmatrix} q_1 - i\gamma\partial_1 & q_2 + \Omega\partial_2 \\ q_2 - \Omega\partial_2 & q_1 + i\gamma\partial_1 \end{pmatrix} \begin{pmatrix} \psi_\alpha(q_1,q_2) \\ \psi_\beta(q_1,q_2) \end{pmatrix} = 0. \tag{8}$$

Here q_1 and q_2 are new ray phase space coordinates that are linear combinations of the old **x** *and* **k**, and γ and Ω are constants. Note that Ω is the rate of rotation about the conversion point of the elliptical part of the ray orbit, while γ is the rate of exponentiation of the hyperbolic part [1]. Note also that the diagonal operators commute with the off-diagonal ones. We can now define a *generic* conversion to be one where all terms in the matrix operator of (8) are of equal importance.

SOLUTION OF THE 2x2 WAVE EQUATION

Operating from the left with

$$\begin{pmatrix} q_1 + i\gamma\partial_1 & -(q_2 + \Omega\partial_2) \\ -(q_2 - \Omega\partial_2) & q_1 - i\gamma\partial_1 \end{pmatrix} \tag{9}$$

and defining

$$\hat{D}_1 \equiv q_1 - i\gamma\partial_1, \quad \hat{D}_2 \equiv q_2 + \Omega\partial_2, \quad \hat{D}_3 \equiv q_1 + i\gamma\partial_1, \tag{10}$$

leads to

$$\begin{pmatrix} \hat{D}_3\hat{D}_1 - \hat{D}_2\hat{D}_2^* & 0 \\ 0 & \hat{D}_1\hat{D}_3 - \hat{D}_2^*\hat{D}_2 \end{pmatrix} \begin{pmatrix} \psi_\alpha(q_1,q_2) \\ \psi_\beta(q_1,q_2) \end{pmatrix} = 0. \tag{11}$$

Thus, the α and β subspaces have decoupled. From the form of (11) it is seen that the q_1- and q_2-dependences separate. Further analysis shows that the q_1-dependence involves a parabolic cylinder-like equation (though not self-adjoint), and the q_2-dependence involves a self-adjoint equation like that of a quantum harmonic oscillator. The separation constant plays the role of an effective coupling constant (more precisely, the magnitude squared of the coupling constant). The general solution of (11) is a linear superposition of terms involving products of parabolic cylinder-type functions and harmonic oscillator eigenfunctions, with each term in the series having a different separation constant. The matching to incoming and outgoing WKB waves is done by first computing the expansion coefficients by fitting at large negative values of q_1 to the incoming WKB wave. Then, using the asymptotic behavior of the parabolic cylinder functions at large positive values of q_1, the outgoing WKB wave is calculated.

At the level of the ray picture [2], we find the following result: the incoming WKB wave is a *family* of rays, with an amplitude, phase and polarization assigned to each. The entire family of incoming rays follow helical orbits with helicity $\kappa=\Omega/\gamma$ as they pass through the conversion region and connect smoothly onto the family of outgoing *converted* rays. The *conversion coefficient* provides the amplitude and phase assigned to each of the outgoing converted rays. The *transmitted* family of rays also follows helical orbits. They each are paired with an incoming ray and assigned an amplitude and phase given by multiplication of the data on the incoming ray by the *transmission coefficient*. Except for that small set of rays with effective coupling constant nearly zero (which generates an outgoing Gaussian beam) the transmission and conversion coefficients for each ray are identical to those obtained in [3] since they depend only upon the asymptotics of parabolic cylinder functions. This will be elaborated in a longer paper. We note that similar results were previously obtained by Littlejohn and Flynn [7].

SUMMARY AND CONCLUSIONS

We have briefly described results recently obtained concerning resonant conversion of linear waves in multiple spatial dimensions. Our goal has been to understand *generic* behavior, rather than analyzing a particular physical model. The search for generic results, true for 'typical' representatives of a family of systems, can lead to very general results. However, we have found that magnetized plasmas are typically *not* generic in the sense we use here. This is because the gyro-orbits of magnetized particles exhibit symmetry around the local magnetic field. There are various ways in which genericity might be obtained; for example strong shear flows or fully three-dimensional gyro-orbits associated with complex magnetic field geometry might break the symmetry which makes magnetized plasmas nongeneric. This is work in progress.

ACKNOWLEDGMENT

This work was supported by the USDOE Office of Fusion Energy Sciences.

REFERENCES

1. E. R. Tracy and A. N. Kaufman, "Ray helicity: a geometric invariant for multi-dimensional resonant wave conversion", submitted to Physical Review Letters. Available online at http://arXiv.org/physics/0303086.
2. E. R. Tracy, A. N. Kaufman, and A. J. Brizard, Physics of Plasmas **10**, 2147-2154 (2003).
3. E. R. Tracy and A. N. Kaufman, Phys. Rev. E **48**, 2196-2211 (1993).
4. W. G. Flynn and R. G. Littlejohn, Ann. Phys. **234**, 334-403 (1994).
5. Y. Colin de Verdiere, "The level crossing problem in semi-classical analysis II: the Hermitian case", preprint, March 2003.
6. E. R. Tracy, A. N. Kaufman, and A. Jaun, Phys. Lett. A **290**, 309-316 (2001).
7. R. G. Littlejohn and G. Flynn, personal communication (unpublished).

Momentum Conservation and Nonlinear RF-Induced Flows

J.R. Myra,[1] D.A. D'Ippolito,[1] L.A. Berry,[2] E.F. Jaeger[2] and D.B. Batchelor[2]

[1]Lodestar Research Corporation, 2400 Central Ave., Boulder, CO, 80301
[2]Oak Ridge National Laboratory, Oak Ridge, Tennessee 37831

Abstract. Recent progress on the numerical computation of 2D full-wave field solutions has motivated advances in the nonlinear theory of rf-induced plasma flows for the control of turbulence. Here, an accounting of how momentum is injected into the system by the antenna, and how it can be transferred from waves to plasma flows and to the equilibrium magnetic field coils and walls is given. Equations for both the plasma momentum and the wave momentum are developed. The former is a recapitulation of results from nonlinear flow drive theory. The latter equation yields a generalized form of the Maxwell stress tensor, including plasma dielectric effects. It is shown that momentum is conserved by the plasma-wave-antenna-wall system for poloidal and toroidal flux-surface-averaged flows. In general, however, momentum exchange with the equilibrium magnetic field coils is possible.

INTRODUCTION

Pioneering[1] as well as more recent[2-6] theoretical papers have considered the topic of rf-driven flows in tokamak plasmas. It has been suggested that ICRF waves could be employed both to flexibly control internal transport barriers, and also to enable fundamental physics investigations of nonlinear waves, flows and turbulence.

Recent advances in the numerical computation of rf fields have permitted full-wave solutions of ICRF fast waves undergoing mode conversion in 2D (axisymmetric) tokamak plasmas where the equilibrium poloidal magnetic field plays an important role.[5] In addition to the bipolar sheared-flow layers that were investigated previously, these solutions have illustrated the importance of direct wave-momentum absorption by the plasma leading to net (unipolar) flows. In this paper, we investigate the conservation laws for wave and plasma momentum, and address the question of when forces on the equilibrium magnetic field coils need be considered.

Waves of frequency ω cause plasma motion both on the rapid ω time scale (accounted for in the plasma dielectric) and on the slow ("dc") time scale, treated in the plasma momentum equation below. The external world interacts with the wave-plasma system through the Lorentz force $\mathbf{F} = \rho\mathbf{E} + (1/c)\,\mathbf{J} \times \mathbf{B}$. Each of the quantities, ρ, $\mathbf{E}$, $\mathbf{J}$ and $\mathbf{B}$, have both oscillatory (ω) and dc (slow) contributions. We account for the $\omega^*\omega$ products of external forces in the wave momentum equation and the dc*dc products in the plasma momentum equation in the sections which follow.

PLASMA MOMENTUM

The species-summed plasma momentum equation describes the evolution of plasma flows by the rf (plasma wave) force $\mathbf{F}_{pw}$,

$$\frac{\partial}{\partial t}(nm\mathbf{u}) + \nabla \cdot (nm\mathbf{u}\mathbf{u}) + \nabla p + \nabla \cdot \vec{\Pi}_0 - \nabla \cdot \vec{D}\nabla(nm\mathbf{u}) = \frac{1}{c}\mathbf{J} \times \mathbf{B} + \mathbf{F}_{pw} \qquad (1)$$

Here $\mathbf{u}$ is the dc flow velocity of the plasma, $p = n(T_e + T_i)$ is the equilibrium plasma pressure, $\mathbf{J}$ is the dc plasma current, $\mathbf{B}$ is the equilibrium magnetic field, $\vec{\Pi}_0$ is a viscosity tensor (e.g. due to neoclassical physics) that describes the reaction of the plasma to the driven flows and $\vec{D}$ is a diffusion tensor (e.g. describing turbulent diffusion of toroidal momentum). In Eq. (1) we have neglected $\rho \mathbf{E} \sim \nabla^2\phi\nabla\phi$ relative to ∇p as it is smaller by $(\lambda_d/L)^2 \ll 1$ for a quasineutral plasma. Diffusion describes momentum loss of plasma flows to the wall.

The total nonlinear force of the waves on the plasma is[4,5]

$$\mathbf{F}_{pw} = \left\langle \rho_1 \mathbf{E}_1 + \frac{1}{c}\mathbf{J}_1 \times \mathbf{B}_1 - \nabla \cdot \vec{\Pi}_{ql} \right\rangle \equiv \mathbf{F}'_{pw} - \left\langle \nabla \cdot \vec{\Pi}_{ql} \right\rangle \qquad (2)$$

where $\rho_1 = \Sigma_j\, n_{1j} Z_j e$ and $\mathbf{J}_1$ are the species summed charge density and current [$\sim \exp(-i\omega t)$, i.e. first order in wave fields], $\vec{\Pi}_{ql}$ is the nonlinear stress tensor[3] and $\langle...\rangle$ is a fast (ω) time average. Using only Maxwell's equations and defining $\mathbf{P} = \Sigma_j\, \vec{\chi}_j \cdot \mathbf{E}$ with dielectric susceptibility $\vec{\chi}$

$$\mathbf{F}_{pw} = \left\langle \frac{1}{16\pi}(\nabla \mathbf{E}^*) \cdot \mathbf{P} - \frac{1}{16\pi}\nabla \cdot (\mathbf{P}\mathbf{E} + \vec{\Pi}_{ql}) \right\rangle \qquad (3)$$

After some algebra, in the case where parallel rf forces on ions are negligible,[4,5]

$$\mathbf{F}_{pw} = \mathbf{k}\frac{P_{rf}}{\omega} - \varepsilon^h \nabla U_0 - \nabla_\perp U_1 + \mathbf{b} \times \nabla U_2 \equiv \mathbf{F}_d - \nabla_\perp U_1 + \mathbf{b} \times \nabla U_2 \qquad (4)$$

where P_{rf} is the absorbed rf power density, $\mathbf{k}$ is the wavenumber (in general summed over all modes), ε^h is the Hermitian part of the dielectric tensor and U_0, U_1 and U_2 (whose exact forms are not needed here) are given as bilinear products of the wave electric field. The "direct" term $\mathbf{F}_d$, defined by Eq. (4), has a more general form in terms of sums over modes using the W tensor.[2,4] The U_1 and U_2 terms give rise to a redistribution of the plasma momentum by the waves (e.g. sheared flows with no net momentum input). The $\mathbf{k}P_{rf}/\omega$ term represents momentum exchange between the plasma and the waves. The U_0 term is the standard reactive ponderomotive potential term and does not give rise to flux-surface-averaged poloidal or toroidal flows because it has the form of a gradient in the parallel and toroidal directions. The flux-surface averaged poloidal and toroidal flows come from the direct ($\mathbf{k}P_{rf}/\omega$) and the dissipative stress (U_2) terms.

The terms on the lhs of Eq. (1) are in conservation law form, while those on the rhs contain momentum exchange with the coils and the waves. In general both waves and flows can induce plasma currents that cause the plasma to exchange momentum with the coils. This can be demonstrated by considering the case of flows guided through a turn by a curved magnetic field. Another example is the case of ponderomotive drift currents $\mathbf{J} \sim \mathbf{F}_{pw} \times \mathbf{B}$ induced by rf waves. However, the relevant toroidal and parallel flux-surface-averages of Eq. (1),[6] viz. $<\mathbf{B}\cdot(1)>_\psi$ and $<\text{Re}\zeta\cdot(1)>_\psi$ annihilate $\mathbf{J} \times \mathbf{B}$. Thus, momentum exchange between the plasma and the equilibrium magnetic field coils plays no role in understand *flux-surface-averaged* toroidal and poloidal flows.

FIELD AND WAVE MOMENTUM

The momentum conservation law for the electromagnetic fields is given by[7]

$$\frac{\partial}{\partial t}\left\langle\frac{\mathbf{E}\times\mathbf{B}}{4\pi c}\right\rangle + \nabla\cdot\left\langle\ddot{\mathbf{T}}_{em}\right\rangle = -\left\langle\rho\mathbf{E}+\frac{1}{c}\mathbf{J}\times\mathbf{B}\right\rangle \equiv -\mathbf{F}'_{pw}+\mathbf{F}_{ext} \quad (5)$$

where the external force $\mathbf{F}_{ext}$ is defined in Eq. (10) and the Maxwell stress tensor is

$$\ddot{\mathbf{T}}_{em} = \frac{1}{4\pi}\left[\frac{1}{2}\ddot{\mathbf{I}}(E^2+B^2)-(\mathbf{EE}+\mathbf{BB})\right] \quad (6)$$

with $\ddot{\mathbf{I}}$ the identity tensor. When ρ and $\mathbf{J}$ consist of both plasma charges and currents (ρ_1 and $\mathbf{J}_1$) and "external" (i.e. antenna and wall, ρ_{ext} and $\mathbf{J}_{ext}$) charges and currents, then part of the plasma responses contained in $\mathbf{F}_{pw}'$ can be absorbed into the field momentum terms and Eq. (5) takes the form (after some algebra)

$$\frac{\partial}{\partial t}\left\langle\frac{\mathbf{D}\times\mathbf{B}}{4\pi c}\right\rangle + \nabla\cdot\left\langle\ddot{\mathbf{T}}_w\right\rangle + \left\langle\mathbf{F}'_d\right\rangle = \mathbf{F}_{ext} \quad (7)$$

$$\ddot{\mathbf{T}}_w = \frac{1}{4\pi}\left[\frac{1}{2}\ddot{\mathbf{I}}(\mathbf{D}\cdot\mathbf{E}+B^2)-(\mathbf{DE}+\mathbf{BB})\right] \quad (8)$$

$$\mathbf{F}'_d = -\frac{1}{8\pi}\left[(\nabla\mathbf{D})\cdot\mathbf{E}-(\nabla\mathbf{E})\cdot\mathbf{D}\right] \quad (9)$$

$$\mathbf{F}_{ext} = -\left\langle\rho_{ext}\mathbf{E}_1+\frac{1}{c}\mathbf{J}_{ext}\times\mathbf{B}_1\right\rangle \quad (10)$$

where $\mathbf{D} = \mathbf{E} + \mathbf{P}$. Note that the nonlinear stress tensor $\ddot{\Pi}_{ql}$ is not present in the wave momentum equation, while the vacuum part of the Maxwell stress tensor does not appear in the plasma momentum equation.

The terms in $\mathbf{F}_d'$ are not manifestly conservative, and contain the physics of dc momentum exchange between the waves and the plasma. For example, in the eikonal limit, to zero order in the wave envelope gradient $\mathbf{F}_d' = \mathbf{F}_d = k P_{rf}/\omega$. More generally, the eikonal limit of Eq. (7) yields the momentum moment of the wave-kinetic equation[8]

$$\frac{\partial}{\partial t}(\mathbf{k}N_k)+\nabla\cdot(\mathbf{v}_g\mathbf{k}N_k)+2\gamma\mathbf{k}N_k+\nabla\omega\,N_k = \mathbf{F}_{ext} \quad (11)$$

where $N_k = W_k/\omega$ is the wave action (W_k is the wave energy density), $\mathbf{v}_g$ is the group velocity and γ is the damping rate of the wave. The last two terms on the lhs of Eq. (11) can be derived from $\mathbf{F}_d'$.

In the full wave form [Eqs. (7) – (10)], wave-plasma momentum exchange can occur even for a cold-fluid plasma. Consider for example, the cold-fluid mode conversion from the fast wave to the ion cyclotron wave.[5,9] In the cold-fluid limit the $\mathbf{F}_d'$ term reduces to $-(\nabla\ddot{\varepsilon}):\mathbf{EE}/8\pi$. We can interpret this force as compensating for mode-conversion or reflection-induced changes in the wave momentum flux term $\ddot{\mathbf{T}}_w$. Mode-conversion or reflection alone (without absorption) cannot drive flux-surface-averaged flows. However, this does not rule out radial forces on the plasma which would need to be balanced by corresponding $\mathbf{J}\times\mathbf{B}$ forces on the coils. In a reflection scenario, the wave momentum normal to the reflection surface is not conserved. Also, waves can be guided through a turn by a curved (e.g. poloidal) magnetic field. Again, the $\mathbf{F}_d'$ term compensates for the change in the wave momentum flux and ultimately represents a force that is transmitted to the supporting (coil) structures, analogous to the case of light being guided along a curved path by a fiber optic cable.

Finally, the integrated form of Eq. (5) may be useful for testing momentum conservation of the field solutions from full-wave rf codes,

$$\int d^2 \mathbf{x} \cdot \left\langle \vec{\mathbf{T}}_{em} \right\rangle + \int d^3 \mathbf{x}\, \mathbf{F}'_{pw} = \int d^3 \mathbf{x}\, \mathbf{F}_{ext} \qquad (12)$$

If the integration volume is bounded by a vacuum region in front of the antenna and walls, then $\mathbf{F}_{ext} = 0$ and the $\vec{\mathbf{T}}_{em}$ term describes the input of momentum. If the bounding surface is inside the antenna and walls, then $\vec{\mathbf{T}}_{em} = 0$, and the antenna-wall forces arise from the surface currents and charges by Eq. (10). If the bounding surface is in the vacuum or the walls, then the nonlinear stress term vanishes there (i.e. $\mathbf{F}'_{pw} = \mathbf{F}_{pw}$), and the net force exerted by the antenna and walls equals the net force on the plasma from direct wave momentum absorption.

CONCLUSIONS

We have shown that, as far as flux-surface-averaged toroidal and poloidal plasma flows are concerned, momentum is conserved by a system that consists of the plasma, the wave fields and the antenna-wall boundary conditions on the rf fields. In general, (i.e. not flux-surface averaged, or when considering radial momentum) the system must include the equilibrium magnetic field coils, because of induced $\mathbf{J} \times \mathbf{B}$ forces. These forces can be seen to arise in situations where waves or particles are guided by curved magnetic fields, or where waves undergo reflection or mode conversion. For flux-surface-averaged flows, the plasma and waves exchange momentum by direct absorption of wave momentum and by dissipative stress terms. An integrated form of the field momentum equation including rf forces on the antenna and walls should be useful in testing momentum conservation in full-wave rf codes, and relating it to the net rf force on the plasma that drives flux-surface-averaged flows.

ACKNOWLEDGMENTS

This work was supported by U.S. DOE grants/contracts DE-FC02-01ER54650, DE-FG03-97ER54392 and DE-AC05-00OR22725. The authors gratefully acknowledge stimulating conversations with the RF SciDAC team.

REFERENCES

1. G.G. Craddock and P.H. Diamond, Phys. Rev. Lett. **67**, 1535 (1991).
2. E.F. Jaeger, L.A. Berry and D.B. Batchelor, Phys. Plasmas **7**, 641 (2000); E. F. Jaeger, L. A. Berry, and D. B. Batchelor Phys. Plasmas **7**, 3319 (2000).
3. J.R. Myra and D.A. D'Ippolito, Phys. Plasmas **7**, 3600 (2000).
4. L.A. Berry, et al., this proceedings; J.R. Myra, L.A. Berry, D.A. D'Ippolito, et al., report in preparation (2003).
5. E.F. Jaeger, L.A. Berry, J.R. Myra, D.B. Batchelor, et al., Phys. Rev. Lett. **90**, 195001 (2003).
6. J.R. Myra and D.A. D'Ippolito, Phys. Plasmas **9**, 3867 (2002); and refs. therein.
7. J.D. Jackson, *Classical Electrodynamics* (Wiley, New York, 1962), pp. 189 – 194.
8. T.H. Stix, *Waves in Plasmas* (Springer-Verlag, New York, 1992) p. 91; A. Bers, in *Plasma Physics – Les Houches 1972* (Gordon and Breach, New York, 1975), pp. 126 – 134.
9. E. Nelson-Melby, M. Porkolab, P.T. Bonoli, Y. Lin, et al. Phys. Rev. Lett. **90**, 155004 (2003).

Hybrid Method for RF Heating Problem of Mirror Confinement Machine

B. H. Park, S. S. Kim, J. H. Yeom, K. I. You, J. Y. Kim, M. Kwon and HANBIT team

Korea Basic Science Institute, Daejon, Korea 305-333

Abstract. A RF code named 'hybrid' for heating problem of mirror confinement machine is developed. The meaning of the hybrid is that the heating problem is solved partially by analytic and partially by numerical method without loss of generality. The vacuum region including an antenna current is dealt with analytic technique and usual FDM for the plasma region. In this work we propose proper boundary conditions of electric field on the axis of cylinder. We applied this method to solving the heating problem of HANBIT the mirror confinement machine located at the KBSI site. The dispersion characteristics are studied by calculating antenna radiation resistance for variance of external magnetic field strength and plasma density and also compared with the experimental data. The heating and field profiles are also investigated for given magnetic and density configurations.

I. INTRODUCTION

HANBIT magnetic mirror device has been developed as a joint plasma research facility for basic study of high temperature plasma in Korea. Recent experimental result [2] shows that the slot antenna can heat ions effectively near the ion cyclotron resonance heating (ICRH) condition resulting in abrupt density jumps. However, these density and temperature jumps are not clearly understood yet: one of the possible scenarios is that the ICRH plays a crucial role for this phenomenon.

To investigate the role of RF-wave in plasma production and ion heating in HANBIT, we developed RF code named HYBRID. The key idea of this new algorithm is to find the special boundary condition for electric fields on the temporal imaginary surface located between the antenna and the plasma edge. This boundary condition, relate values on the boundary to the antenna current and values inside the boundary.

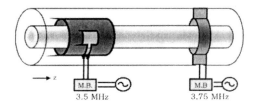

FIGURE 1. Schematic diagram of HANBIT RF system.

II. THEORY OF HYBRID METHOD

Boundary Condition on the Imaginary Surface

Figure 1 shows the schematic diagram of HANBIT RF-system. We separately treat the vacuum region including antenna and the plasma region. Suppose that a surface between the antenna and the plasma. If we have the solutions of source free Maxwell equation in the region enclosed by the imaginary boundary and the chamber wall then the following relation can be derived. [3]:

$$\oint_{\text{imaginary boundary}} (\vec{E} \times \vec{B}^{*\{TE\atop TM\}} + \vec{E}^{*\{TE\atop TM\}} \times \vec{B}) \cdot d\vec{s} = -\frac{4\pi}{c} \int_{\text{vac}} \vec{J}_{\text{ext}} \cdot \vec{E}^{*\{TE\atop TM\}} dv. \quad (1)$$

Here, superscript TE and TM denote each of transverse electric and magnetic modes of source free Maxwell equation in the region defined by the imaginary boundary and chamber wall. The non-scripted E and B are supposed to be a solution of whole RF system. The TE and TM modes in vacuum region mentioned above are

$$E_r^{TE} = \frac{m}{\delta^2 \rho} F_3(\rho)\sin(k_n z), \quad E_\theta^{TE} = \frac{i}{\delta} F_4(\rho)\sin(k_n z), \quad E_z^{TE} = 0,$$

$$B_r^{TE} = -\frac{N}{\delta} F_4(\rho)\cos(k_n z), \quad B_\theta^{TE} = -\frac{imN}{\delta^2 \rho} F_3(\rho)\cos(k_n z), \quad B_z^{TE} = F_3(\rho)\sin(k_n z),$$

$$E_r^{TM} = \frac{N}{\delta} F_2(\rho)\sin(k_n z), \quad E_\theta^{TM} = \frac{imN}{\delta^2 \rho} F_1(\rho)\sin(k_n z), \quad E_z^{TM} = F_1(\rho)\cos(k_n z)$$

$$B_r^{TM} = -\frac{m}{\delta^2 \rho} F_1(\rho)\cos(k_n z), \quad B_\theta^{TM} = -\frac{i}{\delta} F_2(\rho)\cos(k_n z), \quad B_z^{TM} = 0.$$

Here, $\rho = r\omega/c$, $k_n = n\pi/L_c$, $N = ck_n/\omega$ and $\delta^2 = N^2 - 1$ and L_c is chamber length of HANBIT central cell. $F_1, F_2, F_3,$ and F_4 found in the above equations are the abbreviation of following functions. In the after following equations, the subscript a, b and c denote the radial position of the *imaginary surface, antenna,* and *chamber wall.*

$$F_1(\rho) = I_m(\delta\rho)K_m(\delta\rho_c) - K_m(\delta\rho)I_m(\delta\rho_c) \quad F_2(\rho) = I'_m(\delta\rho)K_m(\delta\rho_c) - K'_m(\delta\rho)I_m(\delta\rho_c)$$
$$F_3(\rho) = I_m(\delta\rho)K'_m(\delta\rho_c) - K_m(\delta\rho)I'_m(\delta\rho_c) \quad F_4(\rho) = I'_m(\delta\rho)K'_m(\delta\rho_c) - K'_m(\delta\rho)I'_m(\delta\rho_c)$$

If the region under the imaginary surface including plasma is divided into n_r by n_z meshes then with some manipulations, Eq. (1) for a certain azimuthal mode m becomes

$$-\frac{A_1 \Delta z}{2k_n} \sum_{k=1}^{n_z - 1} (E_\theta^{n_r, k} + E_\theta^{n_r, k+1})(\cos(k_n z_{k+1}) - \cos(k_n z_k)) + \frac{A_2}{\Delta r} \int (r_{n_r} E_\theta^{n_r} - r_{n_r - 1} E_\theta^{n_r - 1})\sin(k_n z) dz$$

$$+\frac{B_1 \Delta z}{2k_n} \sum_{k=1}^{n_z - 1} (E_z^{n_r, k} + E_z^{n_r, k+1})(\sin(k_n z_{k+1}) - \sin(k_n z_k)) + \frac{B_2}{\Delta r} \int (E_z^{n_r} - E_z^{n_r - 1})\cos(k_n z) dz$$

$$= \frac{4\pi}{c} \int J_a \cdot E^{*TE} d^3r \quad (2)$$
$$\frac{D_1 \Delta z}{2k_n} \sum_{k=1}^{n_z-1}(E_z^{n,k} + E_z^{n,k+1})(\sin(k_n z_{k+1}) - \sin(k_n z_k)) + \frac{D_2}{\Delta r} \int (E_z^{n,r} - E_z^{n,r-1}) \cos(k_n z) dz = \frac{4\pi}{c} \int_{vacuum} J_a \cdot E^{*TM} d^3r.$$

Here, super script i and k denotes radial and axial node number and

$$A_1 = 2\pi r_a F_3(\rho_a), \quad A_2 = 2\pi F_4(\rho_a)\frac{c\delta}{\Lambda^{m,n}\omega}, \quad B_1 = -2\pi F_3(\rho_a)\frac{imNc}{\delta^2\omega}, \quad B_2 = -2\pi F_4(\rho_a)\frac{imc^3 k_n}{\Lambda^{m,n}\delta\omega^3}$$

$$D_1 = -2\pi r_a \frac{i}{\delta} F_2(\rho), \quad D_2 = -2\pi r_a F_1(\rho_a)\frac{i}{\Lambda^{m,n}}\frac{c}{\omega}\left\{\frac{m^2}{\delta^2 \rho_a^2} + 1\right\}, \quad \Lambda^{m,n} \equiv \left[1 - N^2 - \frac{m^2}{\rho_a^2}\right].$$

Equation (2) means that the values on the imaginary surface are directly related with the values under the surface and the antenna current. The number of unknown values in Eq. (32) is $2\times(n_z-1)$ then we can find the same number of equations for n_z-1 axial vacuum modes. Therefore, Eq. (2) can be used for boundary condition on the imaginary surface.

Boundary Conditions on the Cylinder Axis

On the cylinder axis, the wave equation is not applicable because of singularity of equation. For a sufficiently thin cylinder including axis, it can be possibly assumed that the plasma and external magnetic field are uniform then we find an analytic expression of electric field in this small cylinder. With the help of this expression we can obtain proper equations of electric field on the axis. For $m = \pm 1$ azimuthal modes,

$$\frac{\partial^2 E^r}{\partial r^2} + \frac{\partial^2 E^r}{\partial z^2} + \frac{\omega^2}{c^2}E^r + \frac{4\pi\omega i}{c^2}(\sigma^{rr} + im\sigma^{r\theta})E^r = 0, \quad E^\theta = imE^r, \quad E_z = 0, \text{ and for } m \neq \pm 1$$

modes, $E_r = E_\theta = E_z = 0$.

III. RESULT AND DISCUSSION

The HYBID code does not solve for vacuum region. Therefore, we can save storage to be apportioned to vacuum region and can treat the antenna current easy way by integrating the current and vacuum mode. In usual case, the antenna current can be modeled in analytic forms so, the source terms in Eq. (2) expressed in algebraic equations.

Figure 2 shows that the solutions well agree with the dispersion relation of magnetized cylindrical plasma. External magnetic field is assumed as uniform to compare with the dispersion relation. There is no cavity resonance mode of m=-1 when $\omega/\Omega_{ci} > 1$. Left graph of Fig. 3 shows that plasma radiation resistance is abruptly increase when the external magnetic field is larger than 2302 kG corresponding to ion cyclotron resonance for 3.5 MHz. Right side of Fig. 3 is an experimental result of density jump for variance of external magnetic field. 107% central magnetic field is correspond to approximately 2302 kG The efficient power coupling may be the possible explanation of the density jump.

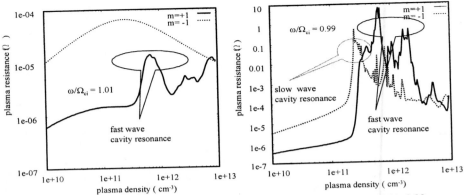

FIGURE 2. Plasma radiation resistance for density variance when $\omega/\Omega_{ci} = 1.01, 0.99$.

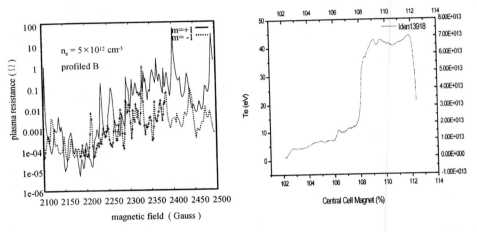

FIGURE 3. Comparison calculated plasma radiation resistance with an experimental result of density jump for variance of external magnetic field.

ACKNOWLEDGMENTS

This work is supported by the Korea Ministry of Science and Technology under the HANBIT Project Contract.

REFERENCES

1. Park, B. H. et al., *Transactions Fusion Technology* **39**, 245 (2001)
2. Kwon, M. et al, *Fusion Science and technology* **43**, 23-31 (2003)
3. Jackson, J. D., "Wave Guides and Resonant Cavities," in *Classical Electrodynamics*, 2nd ed. New York: Wiley, 1975, pp. 370-375.

Advanced 3-D Electron Fokker-Planck Transport Calculations

Y. Peysson, J. Decker[*], R.W. Harvey[†]

*Association Euratom – CEA, CEA/DSM/DRFC
CEA-Cadarache, 13108 St Paul-lez-Durance, France*
[*]*Plasma Science and Fusion Center, Massachusetts Institute of Technology,
Cambridge, MA 02139 U.S.A.*
[†]*CompX, P.O. Box 2672, Del Mar, CA 92014-5672 U.S.A.*

Abstract The first fully implicit 3-D (2-D in momentum and 1-D in configuration spaces respectively) solutions for the bounce-averaged relativistic Fokker-Planck-transport equations based on an incomplete LU matrix factorization is reported. This new method allows a fast and stable convergence towards the steady state solution including the radial transport, a critical point for self-consistent modeling of the current drive problem in presence of radio-frequency wave in tokamak plasmas.

1. INTRODUCTION

Numerical solvers based on a finite difference technique of the linearized electron Fokker-Planck equation are key tools for detailed studies of the non-inductive radio-frequency (RF) current drive towards the achievement of steady state high performance discharges in tokamak plasmas. During the last twenty years, progress in this domain followed regularly those obtained in the field of computers, enabling the description of physical processes with ever increasing realism [1-5]. Up to now, the more detailed kinetic calculations for the current drive problem are based on a three dimensional (3-D) description of the electron dynamics, 2-D in momentum and 1-D in configuration spaces respectively. This allows to take into account accurately of the relativistic and trapped particles effects which both are expected to play a key role for the current generation process in thermonuclear conditions. The radial transport of the fast electron population can be also investigated, a essential point for a fine tuning of the current density profile [6,7].

Simulations of complex scenarios in which current profile shaping is a crucial ingredient is a major challenge from the numerical point of view. In that case, plasma equilibrium, transport properties, RF wave propagation, current drive and current resistive diffusion must all be determined self-consistently [8]. Up to now, as part of a chain of codes, a 3-D Fokker-Planck solver must provide in a quick and precise manner the steady-state electron distribution function at different plasma radii, taking advantage of the strong time ordering between collisional relaxation and resistive diffusion. For this purpose, the implicit or inverse time advancing technique is the most appropriate method, since large time step may be used without onset of numerical instabilities, which considerably speed-up convergence [9]. However, direct application of this method leads to a huge memory consumption for a fine mesh grid since it involves the inversion of a large matrix usually based on a complete LU matrix factorization method. Therefore, existing 3-D bounce-averaged relativistic Fokker-Planck codes rely on an operator splitting technique which strongly reduces memory requirements since at each time step Δt, momentum and spatial dynamics are alternatively described implicitly or explicitly (normal time advancing) [2-5]. The advantage of this method is nevertheless cancelled, because it is restricted to small Δt values in order to avoid convergence towards a spurious bi-stable asymptotic solution, as for the Crank-Nicholson time differencing technique.

In this context, a new fully implicit 3-D relativistic bounce-averaged Fokker-Planck code has been developed for the RF current drive problem. It is based on a incomplete matrix LU factorization that allows to reduce considerably memory requirements while keeping all the advantages of the fully implicit method, namely a fast and stable rate of convergence. It benefits from the highly sparse nature of the matrix to be inverted in the Fokker-Planck problem. In a simplified version, this method "called the 9-P Strongly Implicit Procedure" has been successfully applied for a nine diagonal Fokker-Planck operator [10]. The principle is now generalized to matrices with an arbitrary number of bands. In Sec. 2, main characteristics of the code are described, and in Sec. 3 an example is presented concerning the Lower Hybrid current drive.

2. CODE DESCRIPTION

The starting point of the calculations is the conservative form of the linearized 3-D relativistic bounce-averaged Fokker-Planck solver, $\{\partial f/\partial t\}+\{\nabla_p \cdot S_p\}+\{\nabla_r \cdot S_r\}=0$ where brackets { } denote the bounce average over electron trajectories (trapped and passing particles), S_p and S_r the fluxes in momentum and configuration spaces, and f the electron distribution function. All quantities are determined at the minimum magnetic field value B_{min}. In this code version, a simplified plasma equilibrium is considered where nested magnetic flux surfaces are circular. Neglecting Shafranov shift, it allows analytical evaluation of bounce integrals. Momentum dynamics is described in the spherical coordinate system, while the radial dynamics use the cylindrical one. The Fokker-Planck equation is translated in an algebraic form on to a non-uniform mesh grid which defines a system of cells. The distribution function f is represented by it values at the centers of these cells, while fluxes are required on the edges, so that internal boundary conditions are automatically satisfied. Collisions are described by the Belaiev-Budker operator that is valid from classical to relativistic kinetic energies [5], while RF wave-particle interactions are taken into account within the usual quasi-linear diffusion approximation. In its matrix form the Fokker-Planck equation is given by $\overline{M}X=Y$ where X is the vector whose components correspond to the discrete values of the distribution function and Y is a vector which contains the first order correction of the collision operator describing bulk electron damping on the fast tail, and the the distribution function at the beginning of the time step. The schematic form of $\overline{M}$ is presented in Fig. 1. It is made of n_r diagonally aligned blocks of 15 diagonals, each corresponding to the momentum dynamics at a given plasma radius. Because of the radial dependence of the trapped/passing boundary, their sizes tend to shrink from r/a = 0 to 1. Two bands of coefficients symmetrically placed with respect to the main diagonal represent the momentum dependent fast electron radial transport. The number of matrix coefficients is of the order of $(n_r*n_p*n_\xi)^2$ where n_p and n_ξ are respectively the numbers of momentum and pitch-angle points. Instead of using a complete matrix factorization technique $\overline{M}=\overline{L}\overline{U}$, an incomplete method $\overline{M}=\tilde{L}\tilde{U}$ is employed where both $\tilde{L}$ and $\tilde{U}$ are highly sparse matrices. This factorization is computed in the same (column-oriented) manner as the complete LU factorization except after each column of $\tilde{L}$ and $\tilde{U}$ has been calculated, all entries in that column which are smaller in magnitude than the local drop tolerance are "dropped" from $\tilde{L}$ and $\tilde{U}$. The only exception to this dropping rule is the diagonal of the upper triangular factor $\tilde{U}$ which remains even if it is too small. A standard iterative procedure like bi-conjugate gradient stabilized is then used to solve the set of equations. The numerical efficiency of the method is presented in Fig. 2 for the Lower Hybrid current drive problem at a given plasma radius. By increasing the drop tolerance up to 3×10^{-3}, the memory required to store $\tilde{L}$ and $\tilde{U}$ matrices is reduced by a factor more than 12, while the rate of convergence is improved by 20%, because of the reduced number of numerical operations. Here $\Delta t = 1000$, and the convergence is considered to be achieved when the residual R defined in Ref. [11] reaches 10^{-10}. It is then straightforward to extrapolate the case presented in Fig. 2 to the full three dimensional one, where the distribution function is simultaneously solved at each plasma radius. A memory of the order of 100 Mbytes is needed for studying fast electron transport over 20 radial values. Above a given value, the drop tolerance becomes too high and the information contains by the matrix $\overline{\mathbf{M}}$ is lost in the factorization. The rate of convergence then decreases rapidly, until onset of numerical instabilities.

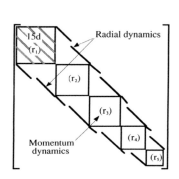

FIGURE 1. Structure of the banded and highly sparse matrix to be factorized in the 3-D fully implicit bounce-averaged Fokker-Planck code. At each radial position, the momentum dynamics is described by a square block characterized by 15 diagonals. The radial dynamics connects blocks between them. The block sizes depend of the radial position of the trapped passing boundary.

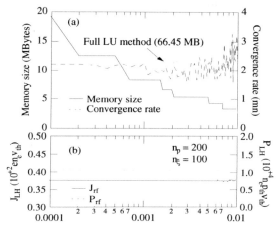

FIGURE 2. (a) Memory size requirement for the $\tilde{L}$ and $\tilde{U}$ matrices storage as well as the rate of convergence for calculating the steady-state solution using the biconjugate gradient stabilized method as a function of the threshold below which coefficients are forced to zero in the incomplete matrix factorization procedure. (b) Both the current and power densities are reported for the Lower Hybrid current drive problem (momentum grid size: 200, pitch-angle grid size: 100, uniform grids).

3. RESULTS

The remarkable performances of this method are illustrated by the example of Lower Hybrid current generation in the presence of a high radial transport. The radial dependence of the quasi-linear diffusion coefficient is given by $D_{LH}=D_{LH}(0)\exp[-(r-r_{max})/\Delta r]$, where $D_{LH}(0) = 1$, $r_{max}/a = 0.4$ and $\Delta r/a = 0.2$, and the resonance domain in momentum space is $[3*v_{th}(0), 5*v_{th}(0)]$. The radial diffusion rate is taken constant over the plasma radius, with a linear dependence with the parallel electron velocity above a threshold velocity of $3*v_{th}(0)$, in order to sketch magnetic turbulence induced transport. All profiles parameters correspond to advanced scenario studies of the Tokamak C-MOD: pure deuterium plasma, $T_e = 7.5\left(1-(r/a)^2\right)^2$ [keV], $T_i = 0.533 T_e$, $n_e = 2.3 \times 10^{+20}\left(1-(r/a)^2\right)^{0.5} + 0.2$ [m^{-3}]. As shown in Fig. 3, the current profile is strongly broaden by the large fast electron radial transport here taken. This result has been obtain after only 5 iterations assuming that $R < 10^{-10}$ at all plasma radii with $\Delta t = 1000$. In Fig. 4 the effect of the fast electron transport is clearly seen on the distribution functions. At the location where the RF power is absorbed at $r/a = 0.275$, the quasilinear domain has no more sharp boundaries, while at $r/a = 0.695$, in the region where the direct quasi-linear contribution is negligible, a significant tail is formed. It is important to notice that the code accurately describe the trapping or detrapping effect of fast particles under the fast electron transport because of the radial dependence of the trapped/passing boundary. The gain in time is of several order of magnitude, since time steps less than $\Delta t = 10$ had to be used with the operator spliting technique to ensure a normal convergence. On a regular workstation, convergence is then achieved in few minutes depending on the grid sizes and the value of the residual R as a convergence criterion.

Aside from the intrinsic performances of the method which takes into account that all computers today have large memory capabilities, this fast fully implicit 3-D code offers also interesting possibilities concerning new physical investigations. Hence, self-consistent calculations of the wave- induced radial transport on RF current drive [12], a problem which may be very important for the case of the Lower Hybrid wave can be now performed very easily, without excessive computer time consumption.

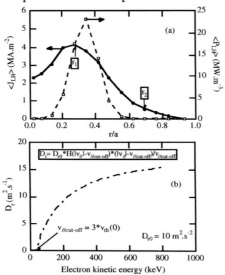

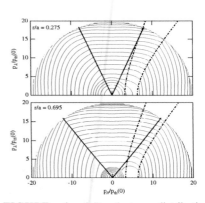

FIGURE 3. (a) Power and current density profiles for Lower Hybrid current drive in presence of fast electron radial transport. (3) Energy dependence of the radial diffusion rate. Arrows r_1 and r_2 indicate radial locations where distribution function are represented in Fig. 4

FIGURE 4. 2-D electron distribution functions at different plasma locations. The dotted lines represent the trapped/passing boundaries, while the dotted-dashed line the domain of resonance for the Lower hybrid wave. Radial position are indicated in Fig. 3.

REFERENCES

1. C.F.F. Karney, Comput. Phys. Rep. **4** (1986) 183.
2. M.R. O'Brien, M. Cox, J.S. McKenzie, Nucl. Fusion **31** (1991) 583.
3. R.W. Harvey, M.G. McCoy, 'USDOC NTIS No DE93002962' report.
4. G. Giruzzi, Plasma Phys. Control. Fusion **35** (1993) A123.
5. I.P. Shkarofsky and M. Shoucri, Bucl. Fusion **37** (1997) 539.
6. Y. Peysson, Plasma Phys. Control. Fusion **35** (1993) B253.
7. B. W. Harvey et al., Phys. Rev. Letters, **20** (2002) 205001.
8. V. Basiuk et al., accpeted in Nucl. Fusion summer (2003).
9. W.H. Press et al., in *Numerical Recipes*, Cambridge Univ. Press, Cambridge (1986).
10. Y. Peysson and M. Shoucri, Comput. Phys. Commun. **109** (1998) 55.
11. J.P.S. Bizarro and P. Rodrigues, Nuclear Fusion, **37** (1997) 1509.
12. K. Kupfer and A. Bers, Phys. Fluids B3 (1991) 2783.

Plasma Dielectric Tensor for Non-Maxwellian Distributions in the FLR Limit

C.K. Phillips[1], A. Pletzer[1], R.J. Dumont[2] and D.N. Smithe[3]

[1] *Princeton Plasma Physics Laboratory, P.O. Box 451, Princeton, NJ 08540, U.S.A.*
[2] *Association Euratom-CEA sur la Fusion Contrôlée, F-13108 St Paul lez Durance, France*
[3] *Mission Research Corp., Newington, VA 22122, U.S.A.*

Abstract. Previous analytical and numerical studies have noted that the presence of fully non-Maxwellian plasma species can significantly alter the dynamics of electromagnetic waves in magnetized plasmas. In this paper, a general form for the hot plasma dielectric tensor for non-Maxwellian distributions is derived that is valid in the finite Larmor radius approximation. This model provides some insight into understanding the limitations on representing non-Maxwellian plasma species with equivalent Maxwellian components in modeling RF wave propagation and absorption.

INTRODUCTION

Laboratory fusion plasmas as well as many space plasmas can be comprised of both thermal and non-thermal species. In collisionless space plasmas, turbulent heating or shock processes can accelerate particles, resulting in velocity-space distributions that are Lorentzian or power-law-like in nature [1-3]. Neutral beam injection and fusion reactions in laboratory fusion plasmas both introduce energetic ions, which follow a slowing-down type distribution in velocity space. Finally, when electromagnetic waves are applied to heat or else to drive noninductive currents in magnetized plasmas, the wave-induced particle acceleration results in velocity-space distributions that feature energetic "tails" or extended "quasilinear plateaus". In all of these situations, the question that arises is whether or not these non-thermal plasma species have a noticeable impact on electromagnetic wave dynamics in these plasmas.

Previous analytical and numerical studies [1-6] have shown that wave dynamics can be affected if a sizeable non-thermal ion population is present in the plasma. Most of these studies have focused on modifications to wave absorption or to instability thresholds. A number of these studies [4,5,7] have noted that power absorption on a non-Maxwellian distribution can be approximated by that on an equivalent Maxwellian, chosen so that the thermal speed of the equivalent Maxwellian is equal to the velocity-space averaged perpendicular speed of the non-thermal distribution.

More recently, a 1D all-orders local full wave, METS [8], has been extended to include the effects of non-thermal species on both wave propagation and absorption. Results from this code indicate that the absorption and wave propagation in plasmas with isotropic, non-Maxwellian species can be reasonably simulated with equivalent

Maxwellian in many regimes. However, the spatial profile of power deposition on short wavelength kinetic waves can be narrower and anisotropic effects can lead to larger discrepancies with models based on equivalent Maxwellians [8].

In this paper, the hot plasma dielectric susceptibility for a fully non-Maxwellian but still gyrotropic particle distribution function is derived that is valid in the finite Larmor radius limit (FLR). This model provides some insight into understanding the limitations on representing non-thermal species as equivalent Maxwellians in modeling RF wave absorption and propagation. It also provides the basis for generalizing the plasma dielectric operators in FLR-based 2D full wave simulation codes.

DERIVATIONS

The hot plasma susceptibility for a given species, "s", described by the gyrotropic particle velocity distribution $f_0(v_\perp, v_{//})$, in a homogeneous, uniformly magnetized plasma can be written in the following form:

$$\tilde{\chi}_s = \frac{\omega_{ps}^2}{\omega} \int_0^\infty 2\pi v_\perp dv_\perp \int_{-\infty}^\infty dv_{//} \, \hat{z}\hat{z} \frac{v_{//}^2}{\omega} \left(\frac{1}{v_{//}} \frac{\partial f_0}{\partial v_{//}} - \frac{1}{v_\perp} \frac{\partial f_0}{\partial v_\perp} \right)_s +$$
$$\frac{\omega_{ps}^2}{\omega} \int_0^\infty 2\pi v_\perp dv_\perp \int_{-\infty}^\infty dv_{//} \sum_{n=-\infty}^{n=\infty} \left[\frac{v_\perp U}{\omega - k_{//}v_{//} - n\Omega} \tilde{T}_n \right]_s \quad , \quad (1)$$

where $\tilde{T}_n$ is given in Equation (10-48) of reference [9] by Stix. In the FLR limit, the expressions in Equation (1) may be simplified by replacing the Bessel functions in $\tilde{T}_n$ by their series expansions, integrating by parts the terms involving $\int_0^\infty 2\pi v_\perp dv_\perp (v_\perp^m) \partial f_0 / \partial v_\perp (\dots)$, and retaining terms through order $\lambda \sim k_\perp^2 w_\perp^2 / 2\Omega_s^2$, where Ω_s is the cyclotron frequency and $w_\perp^2$ is the velocity space average of the perpendicular velocity. The resulting form of the plasma susceptibility is a generalization of the Maxwellian-based FLR susceptibility given by Stix in Equations (59-63) in Chapter 10 in reference [9]. It may be written in similar but generalized form as:

$$\chi_{xx} = \frac{\omega_p^2}{\omega} \left\{ \frac{1}{2}[\tilde{A}_{1,0} + \tilde{A}_{-1,0}] - \frac{\lambda}{2}[\tilde{A}_{1,1} + \tilde{A}_{-1,1}] + \frac{\lambda}{2}[\tilde{A}_{2,1} + \tilde{A}_{-2,1}] \right\} \quad (2)$$

$$\chi_{xy} = i\frac{\omega_p^2}{\omega} \left\{ \frac{1}{2}[\tilde{A}_{1,0} - \tilde{A}_{-1,0}] - \lambda[\tilde{A}_{1,1} - \tilde{A}_{-1,1}] + \frac{\lambda}{2}[\tilde{A}_{2,1} - \tilde{A}_{-2,1}] \right\} \quad (3)$$

$$\chi_{xz} = \frac{\omega_p^2}{\omega}\left(\frac{1}{2}\frac{k_\perp}{\Omega}\right)\left\{[\tilde{B}_{1,0}+\tilde{B}_{-1,0}]-\lambda[\tilde{B}_{1,1}+\tilde{B}_{-1,1}]+\frac{\lambda}{2}[\tilde{B}_{2,1}+\tilde{B}_{-2,1}]\right\} \quad (4)$$

$$\chi_{yy} = \frac{\omega_p^2}{\omega}\left\{2\lambda\tilde{A}_{0,1}+\frac{1}{2}[\tilde{A}_{1,0}+\tilde{A}_{-1,0}]-\frac{3\lambda}{2}[\tilde{A}_{1,1}+\tilde{A}_{-1,1}]+\frac{\lambda}{2}[\tilde{A}_{2,1}+\tilde{A}_{-2,1}]\right\} \quad (5)$$

$$\chi_{yz} = i\frac{\omega_p^2}{\omega}\left(\frac{k_\perp}{\Omega}\right)\{\tilde{B}_{0,0}-\lambda\tilde{B}_{0,1}-\frac{1}{2}[\tilde{B}_{1,0}+\tilde{B}_{-1,0}]$$
$$-\lambda[\tilde{B}_{1,1}+\tilde{B}_{-1,1}]-\frac{\lambda}{4}[\tilde{B}_{2,1}+\tilde{B}_{-2,1}]\} \quad (6)$$

$$\chi_{zz} = \frac{2\omega_p^2}{\omega k_{//}w_\perp^2}\int_{-\infty}^{\infty}dv_{//}\int_0^\infty 2\pi v_\perp dv_\perp \; v_{//}\left[\frac{-1}{v_\perp}\frac{\partial f_0}{\partial v_\perp}\frac{w_\perp^2}{2}\right]$$

$$+\frac{\omega_p^2}{\omega}\left[\frac{2\omega}{k_{//}w_\perp^2}\right]\{(1-\lambda)\tilde{B}_{0,0} \quad (7)$$

$$+\int_{-\infty}^{\infty}dv_{//}\int_0^\infty 2\pi v_\perp dv_\perp \frac{v_{//}}{\omega-k_{//}v_{//}}\left[\frac{1}{2}\frac{w_\perp^2}{v_\perp}\frac{\partial f_0}{\partial v_\perp}+f_0\left(\frac{1-k_{//}v_{//}}{\omega}\right)\right]$$

$$+\frac{1}{2}\frac{\omega_p^2}{\omega}\lambda\left\{\frac{2(\omega-\Omega)}{k_{//}w_\perp^2}\tilde{B}_{1,0}+\frac{2(\omega+\Omega)}{k_{//}w_\perp^2}\tilde{B}_{-1,0}\right\},$$

where, for j=0,1:

$$\tilde{A}_{n,j} = \int_{-\infty}^{\infty}dv_{//}\frac{1}{\omega-k_{//}v_{//}-n\Omega}\int_0^\infty 2\pi v_\perp dv_\perp H_j(v_{//},v_\perp) \quad (8)$$

$$\tilde{B}_{n,j} = \int_{-\infty}^{\infty}dv_{//}\frac{v_{//}}{\omega-k_{//}v_{//}-n\Omega}\int_0^\infty 2\pi v_\perp dv_\perp H_j(v_{//},v_\perp) \quad (9)$$

$$H_0(v_{//},v_\perp) = \frac{1}{2}\frac{k_{//}w_\perp^2}{\omega}\frac{\partial f_0}{\partial v_{//}}-(1-\frac{k_{//}v_{//}}{\omega})f_0(v_{//},v_\perp) \quad (10)$$

$$H_1(v_{//},v_\perp) = \frac{1}{2}\frac{k_{//}w_\perp^2}{\omega}\frac{\partial f_0}{\partial v_{//}}\frac{v_\perp^4}{w_\perp^4}-(1-\frac{k_{//}v_{//}}{\omega})f_0(v_{//},v_\perp)\frac{v_\perp^2}{w_\perp^2}. \quad (11)$$

In the limit that $f_0(v_\perp,v_{//}) = f_{max}(v_\perp) h(v_{//})$, then the generalized susceptibility reduces to that given in reference [9].

DISCUSSION

The local hot plasma susceptibility for a given species, "s", described by the gyrotropic particle velocity distribution $f_0(v_\perp,v_{//})$, that is valid in the FLR approximation has been derived by expanding the full hot plasma susceptibility to first order in $\lambda \sim k_\perp^2 w_\perp^2 /2\Omega_s^2$. In the limit that $k_{//} \Rightarrow 0$, the FLR-based susceptibility for a general distribution differs from that of a Maxwellian distribution only in terms of $O(\lambda)$. The xx, xy, and yy elements will be the same as that of an equivalent Maxwellian in this limit, provided that the thermal speed of the Maxwellian is chosen to equal $w_\perp^2$. Hence, the propagation of waves, such as fast waves or ion Bernstein waves, which depends on these elements, can be simulated exactly using the equivalent Maxwellian. The χ_{xz} and χ_{yz} elements will be approximately equal in this limit, provided the ratio:

$$\frac{I_V}{\langle v_{//} \rangle} = \frac{\int_0^\infty 2\pi v_\perp dv_\perp \int_{-\infty}^\infty dv_{//} \; v_{//} f_0(v_{//},v_\perp) \frac{v_\perp^2}{w_\perp^2}}{\int_0^\infty 2\pi v_\perp dv_\perp \int_{-\infty}^\infty dv_{//} \; v_{//} f_0(v_{//},v_\perp)} \quad (12)$$

is close to unity. More generally, the FLR-based susceptibility elements for a general, gyrotropic distribution can be computed using equivalent Maxwellians, if the (velocity)^n moments for the general distribution are similar to those of the equivalent Maxwellian. Finally, the expressions for the FLR-based susceptibility given in Equations (2)-(11) may be utilized to generalize the dielectric operator in 2D FLR-based full wave codes. Such a generalization is required in order to self-consistently integrate such codes with Fokker-Planck packages.

ACKNOWLEDGMENTS

This work was supported by U.S.D.O.E. contract DE-AC02-76CH03073.

REFERENCES

1. Gedalin, M., Lyubarsky, Y., Balikhin, M. and Russell, C.T., *Phys. Plasmas* **8**, 2934 (2001).
2. Gedalin, M., Strangeway, R.J., and Russell, C.T., *J. Geophys. Res.* **107**, SSH **1**-1,6 (2002).
3. Heilberg, M. and Mace, R.L., *Phys. Plasmas* **9**, 1495 (2002).
4. Koch, R. *Phys. Letts. A* **157**, 399 (1991).
5. Van Eester, D., *Plasma Phys. Controlled Fusion* **35**, 441-451 (1993).
6. Batchelor, D.B., Jaeger, E.F., and Colestock, P.L., *Phys. Fluids* **B1**, 1174 (1989).
7. Sauter, O. and Vaclavik, J., *Nucl. Fusion* **32**, 1455 (1992).
8. Dumont, R.J., Phillips, C.K., and Smithe, D.N., "Effects of non-Maxwellian Plasma Species on ICRF Propagation and Absorption in Toroidal Magnetic Confinement Devices," this conference.
9. Stix, T.H., "Susceptibilities for a Hot Plasma in a Magnetic Field," in *Waves in Plasmas*, New York: American Institute of Physics, 1992, pp. 237-264.

Gabor Wave Packet Method to Solve Plasma Wave Equations

A. Pletzer*, C. K. Phillips* and D. N. Smithe[†]

Princeton Plasma Physics Lab, Princeton NJ 08543
[†]*Mission Research Corporation*

Abstract. A numerical method for solving plasma wave equations arising in the context of mode conversion between the fast magnetosonic and the slow (e.g ion Bernstein) wave is presented. The numerical algorithm relies on the expansion of the solution in Gaussian wave packets known as Gabor functions, which have good resolution properties in both real and Fourier space. The wave packets are ideally suited to capture both the large and small wavelength features that characterize mode conversion problems. The accuracy of the scheme is compared with a standard finite element approach.

INTRODUCTION

The problem of computing RF wave dynamics in fusion plasmas is numerically challenging due to locally fine-scale resonance and short-wavelength mode conversion effects [1]. The conversion of fast magnetosonic into ion Bernstein waves, for instance, requires the resolution of waves with dramatically different wavelengths.

Many codes written (e.g. the Mets code [2]) rely on a Fourier decomposition of the waves, requiring many modes to capture short-wavelength phenomena. Thus, there is a need to explore other more efficient numerical approaches, which provide better local resolution, and can take better advantage of windowing or multiple scale-length aspects of the problem. Due to the constraint that the dielectric tensor is most easily expressed analytically for sinusoidal waves, the extension of Fourier to using wave packets with a Gaussian envelop (Gabor functions) is most natural.

In this article, we explore a novel numerical approach based on expanding the solution in Gabor functions, which combines the advantages of the Fourier and finite element methods. The method, which we refer to as the Gabor element method (GEM), allows for a high degree of flexibility in the specification of boundary conditions. The process of discretization leads to, effectively, a sparse matrix due to the limited support of the Gabor functions. Therefore, GEM shares many similarities with the finite element method (FEM). However, GEM differs from FEM in that a single set of basis functions can be used to solve differential equations, in principle, of arbitrary order.

To validate GEM, we focus on two test problems. First, GEM is applied to solve a second order, Airy-type equation with a linear turning point (cut-off). The solution is then compared to that obtained using linear FEM. Next, a fourth-order, Wasow-type model equation describing the mode coupling between fast and slow waves, with two well separated wavelengths, is solved.

THE GABOR ELEMENT METHOD

Our aim is to solve ordinary differential equations of arbitrary order $2N$ (an even integer),

$$\sum_{i,j=0}^{N} (-)^i \frac{d^i}{dx^i} \left[f_{ij}(x) \frac{d^j y(x)}{dx^j} \right] = s(x) \;;\; x \in [0,1], \tag{1}$$

subject to N boundary conditions at $x = 0$ and 1

$$\frac{d^i}{dx^i} \left[\sum_{j=0}^{N} f_{N,j}(x) \frac{d^j y(x)}{dx^j} \right]_0^1 = C_i + \sum_{j=0}^{N-1} B_{ij} \frac{d^j y(x)}{dx^j} \;;\; i = 0 \cdots N-1. \tag{2}$$

In (1), f_{ij} and s are user supplied functions of the independent variable x. Note that conditions (2) are flexible enough to accommodate Dirichlet, Neumann, or Robin boundary conditions by allowing, if required, the C_i's and B_{ij}'s to be infinite.

The Gabor element method is now presented. Following a Galerkin approach, (1) is multiplied by a test function $h(x)$ and integrated over the domain $[0,1]$ to yield

$$a(h,y) = b(h) \tag{3}$$

where

$$a(h,y) \equiv \int_0^1 dx \sum_{ij} \frac{d^i h}{dx^i} f_{ij} \frac{d^j y}{dx^j} + \sum_{i=1}^{N-1} \sum_{\ell=0}^{i} (-)^{i+\ell} \left[\frac{d^\ell h}{dx^\ell} \frac{d^{i-\ell-1}}{dx^{i-\ell-1}} \left(\sum_j f_{ij} \frac{d^j y(x)}{dx^j} \right) \right]_0^1 \tag{4}$$

$$+ \sum_{\ell=0}^{N} (-)^{\ell+N} \left[\frac{d^\ell h}{dx^\ell} \sum_j B_{N-\ell-1,j} \frac{d^j y}{dx^j} \right]_0^1$$

and

$$b(h) \equiv \int_0^1 dx\, h\, s - \sum_{\ell=0}^{N} (-)^{\ell+N} \left[\frac{d^\ell h}{dx^\ell} C_{N-\ell-1} \right]_0^1. \tag{5}$$

The first term in (4) represents the energy functional while the two subsequent terms arise after integrating by parts i times the term of (1) in $[\,]$. Next, the solution

$$y(x) = \sum_\gamma g_\gamma(x) y_\gamma \;;\; g_\gamma(x) \equiv e^{i 2\pi u_j x} e^{-(x-x_i)^2/(2w_i^2)} \tag{6}$$

is expanded in Gabor wave packets $g_\gamma(x)$ where $\gamma = (i,j)$. Equation (6) is a double expansion in wave-numbers $2\pi u_j$, $j = -(N_F-1)/2 \cdots (N_F-1)/2$ and envelop positions $x_i, i = 0, \cdots N_x - 1$. Upon inserting (6) into (3) and choosing $h(x) = g_{\gamma'}^*$ we then get $N_x N_F$ linear coupled equations

$$\sum_\gamma A_{\gamma',\gamma} y_\gamma = b_{\gamma'}$$

for the unknowns y_γ, where $A_{\gamma',\gamma} \equiv a(g_{\gamma'}, g_\gamma)$ and $b_{\gamma'} \equiv b(g_{\gamma'})$.

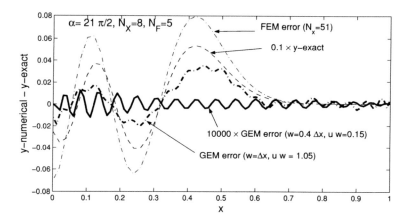

FIGURE 1. Error of the GEM solution for the Airy type equation using two combinations of u and w parameters. Notice the reduction factor of 10 000 used to plot the GEM solution obtained using $w = 0.4\Delta x$ and $uw = 0.15$. The FEM error obtained using linear hat elements (same number of degrees of freedom) is shown for comparison.

RESULTS

For simplicity, we will assume in the following that the phase-space lattice is uniform: $x_i = i\Delta x$ and $u_j = ju$. The accuracy of GEM will depend on the values of the grid spacing Δx, the fundamental frequency u and the Gaussian (half) width w. It can be proved that the Gabors form a frame only under the condition that $u\Delta x < 1$ [3]. Moreover, in order for the Gabors to overlap, we must have $2w \sim \Delta x$.

To determine more precisely the optimal u and w parameters, we solve equation $y'' + \alpha^2(1-2x)y = 0$ whose solution, the Airy function $\text{Ai}[(\alpha/2)^{2/3}(2x-1)]$ is a propagating wave for $x < 1/2$ but evanescent for $x > 1/2$. Figure 1 shows the pointwise error of the Gabor solution for $\alpha = 21\pi/2$, using 8 envelops and 5 Fourier modes $(-2\cdots+2)$. The exact solution (reduced by a factor of 0.1) is shown as a dashed line. The GEM error (dash-dotted line) compares favorably with the FEM error obtained using the same number of degrees of freedom. Notice that the FEM error is proportional to the second derivative y'', as expected for linear hat elements. However, the choice of $w = \Delta x$ and $uw = 1.05$ is suboptimal; changing these parameters to $w/\Delta x = 0.4$ and $uw = 0.15$ suppresses the error by a factor > 10000. This emphasizes the ability of GEM to capture the solution more accurately than FEM with a small number of degrees of freedom.

To model the coupling of fast to slow waves, we solve the Wasow equation

$$\left(\frac{d^2}{dx^2} + k^2[1 - 0.5(x-0.5)]\right)\left(\frac{d^2}{dx^2} + k^2[1 - 160(x-0.5)]\right)y + \alpha y = 0$$

subject to boundary conditions $y(0) = 0$, $y(1) = 1$, and $y'(0) = y'(1) = 0$, with $k^2 = 2 \times 10^3$ and $\alpha = 8 \times 10^6$. To the right of $x = 0.506$, the slow wave is evanescent while to the left it is propagating with $\lambda \to 0.01$. The wavelength of the fast wave ranges

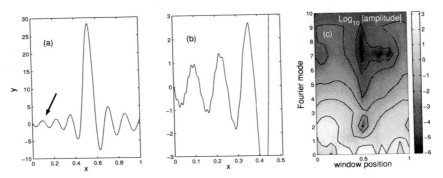

FIGURE 2. Solution of the fourth order equation obtained using $w/\Delta x = 1.0$, $u\Delta x = 0.9$, $N_x = 21$ and $N_F = 21$: (a) solution, (b) blow-up of $x < 0.5$ region, and (c) spectrogram (log of amplitude).

from $0.13 - 0.16$ across. Figure 2 shows the solution (a) with the short wavelength contribution from the slow wave clearly noticeable in (b). Picture (c) shows the mode structure in phase space, which depends on the choice of u and w parameters.

CONCLUSIONS

Good accuracy was achieved with the Gabor element method (GEM) when $u\Delta x < 1$ and $w/\Delta x \sim 0.4 - 1$. A small Gaussian width w yields a sparser matrix system but requires more Fourier modes. A larger $w/\Delta x \sim 1$ can be more efficient but yields a residual error that cannot be suppressed by increasing the resolution in phase space.

When optimally chosen, u and w yield an error that is insensitive to high order derivatives of the solution. This confers to GEM the capability to extract small and large features equally well. For problems with an oscillatory solution, GEM typically outperforms the finite element method error by several orders of magnitudes.

ACKNOWLEDGMENTS

Work supported by US DOE contract DE-AC02-76CH03073.

REFERENCES

1. Swanson, D. G., *Theory of Mode Conversion and Tunneling in Inhomogeneous Plasmas*, John Wiley & Sons, Inc., 1998.
2. Smithe, D. N., Phillips, C. K., Hosea, J. C., Majeski, R. P., and Wilson, J. R., "Investigation of RF Absorption by Fast Ions and High Temperature Plasmas using the METS95 Wave Analysis Tool," in *12th Topical Conference on RF Power in Plasmas, Savannah*, American Institute of Physics, New York, 1997, p. 367.
3. Daubechie, I., *Ten Lectures on Wavelets*, Society for Industrial and apllied Mathematics, 1992, chap. 3, pp. 53–63.

O-X Mode Conversion in an Axisymmetric Plasma at Electron Cyclotron Frequencies

Kaya Imre and Harold Weitzner

Courant Institute of Mathematical Sciences, New York University
251 Mercer St., New York, NY 10012

Abstract

Results from the study of the O-X mode conversion at electron cyclotron frequencies in an axisymmetric plasma are made explicit. A precise description of the possible incident and transmitted waves is given. A proof of a result, previoiusly verified numerically, is given.

In a paper of the same title submitted for publication elsewhere[1] by one of us (H.W.) the reflection and transmission with mode conversion in an axisymmetric plasma at frequencies on the order of the electron cyclotron frequencies is considered. The cold plasma dielectric function modeled the plasma response, and wave equations valid near the mode conversion region were found. An integral representation of a family of solutions was found. The validity of the asymptotic expansion of the solutions depended on, among other matters, properties of solutions of a set of transcendental equations, and numerical calculations justified the result. This note is an addendum to that paper in which a proof of the properties is given and a thorough description of the solution is given.

The steady magnetic field is $\mathbf{B} = -Q(\psi)\nabla\psi \times \nabla\theta + \nabla\psi \times \nabla\phi$ where ψ is the flux function, θ and ϕ are the poloidal and the ignorable toroidal angle and Q is the safety factor. The major radius coordinate is r and the elements of the dielectric tensor are $\kappa_\perp$, $\kappa_\wedge$, and $\kappa_\|$. The incident wave number is k and the condition for O and X mode cut-offs are $\kappa_\| = 0$ and $\kappa_\perp - \kappa_\wedge - k_\|^2 = 0$, respectively. Mode conversion occurs when the surfaces intersect, which is along a curve we identify as $\psi = \theta = 0$. We expand the relevant functions near these values and define

$$-2\kappa_\| r B_p/N = c_1\psi + \ldots \tag{1}$$

$$Y = -2\kappa_\|(rB_p/N)^2 + (\kappa_\perp - \kappa_\wedge - k_p^2)rB_T/(Nk_p) = c_2\psi + c_3\theta + \ldots \quad (2)$$

$$d = [\nabla\psi \cdot \nabla\psi/Y_{,\theta} + iB_T^2/(QB)](N/rB_p)^2/(c_1\nabla\psi \cdot \nabla\psi) \quad (3)$$

$$d\mu^2 - 2\mu - 1 = 0 \quad (4)$$

and

$$1 + 1/\mu = |1 + 1/\mu|(\exp(i\omega)) \quad (5)$$

For the solutions given by the integral representations there is only total mode transmission with total mode conversion. These results are for incident O modes or incident X modes. The regions in which the waves propagate are given in terms of a parameter ζ, where

$$\zeta = [(\theta/\psi)(c_3) + (c_2/c_1)][N/(rB_p)]^2 \quad (6)$$

There are infinitely many wave solutions found each of which is characterized by two values ζ_1 and ζ_2, such that the wave propagates for $\zeta_1 \leq \zeta \leq \zeta_2 \leq 0$ and is exponentially damped for other values of ζ. The values for ζ_1 and ζ_2 are obtained from a spiral in a complex plane. We introduce complex variables τ, z and w related by the conditions

$$z = r_0 \exp(\theta \tan\omega + i\theta) = |z|\exp(i\theta) \quad (7)$$

where r_0 is real and positive, but arbitrary and is a property of the incident wave

$$\tau = (d^*\mu^* + z/\mu^*)/(1-z) \quad (8)$$

and

$$w = (\tau d\mu + \zeta)/(\zeta - \tau/\mu). \quad (9)$$

We allow θ to cover all values in (7) so that (7) and (8) define a double spiral in the τ plane, spiralling around the center at $(2 + 1/\mu^*)$ at one end and $-1/\mu^*$ at the other end. We consider any segment of the spiral of largest extent such that $|\tau|$ is monotone on the segment and Im τ does not change sign. We take $|\tau_1| \leq |\tau| \leq |\tau_2|$ on that segment. It is easy to see that there are infinitely many such segments. We then have

$$-|\tau_2|^2 \leq \zeta \leq -|\tau_1|^2. \quad (10)$$

We see that there is a rich family of solutions which propagate with total mode conversion.

The proof of this result depended on the following property. The condition for a stationary point of the phase is just

$$w = \exp[(1 + e^{2i\omega})\log r_0 + e^{-2i\omega} \log z] \tag{11}$$

where w, τ, z and ζ are related by (8) and (9). We require that for any stationary point of the phase if τ is real then z must also satisfy (7), or z must lie on the spiral. We prove the result as follows. From (5) it follows that $\omega \neq N\pi$ for any N, and we exclude the special cases $\omega = \pi/2 + N\pi$, which must be treated separately, see (7). We set

$$w = |w|\exp(i\phi) \tag{12}$$

and τ and ζ are real in (8) and (9) only if

$$\sin\omega \sinh(\log|z|) + \cos\omega \sin\theta = 0 \tag{13}$$

$$\sin\omega \sinh(\log|w|) - \cos\omega \sin\phi = 0. \tag{14}$$

From (11) we find if

$$X = \log|z| - 2F\cos\omega \tag{15}$$

$$Y = -\theta - 2F\sin\omega \tag{16}$$

and

$$F = \cos\omega \log(|z|/r_0) - \theta \sin\omega \tag{17}$$

then

$$(X - \log|z|)\sin\omega = (Y + \theta)\cos\omega \tag{18}$$

If $F \neq 0$, $X - \log|z|$ and $Y + \theta$ are non-zero and

$$(\sin\omega)(X - \log|z|)\left[\frac{\sinh X - \sinh\log|z|}{X - \log|z|}\right] = (Y + \theta)\cos\omega \left(\frac{\sin Y + \sin\theta}{Y + \theta}\right) \tag{19}$$

Now

$$\left|\frac{\sin Y + \sin\theta}{Y + \theta}\right| = \left|\frac{\sin(\frac{Y+\theta}{2})}{\frac{Y+\theta}{2}}\right| \left|\cos\left(\frac{Y-\theta}{2}\right)\right| \leq 1, \tag{20}$$

while

$$\left|\frac{\sinh\alpha - \sinh\beta}{\alpha - \beta}\right| = \left|\frac{\sinh(\frac{\alpha-\beta}{2})}{(\frac{\alpha-\beta}{2})}\right| \cosh\left(\frac{\alpha+\beta}{2}\right) > 1. \tag{21}$$

Thus, from (18), (20) and (21), it follows that (19) cannot be true, and hence $F = 0$, so that z satifies (7) and lies on the spiral, and an essential element has been shown.

[1]Weitzner, H., "O-X Mode Conversion in an Axisymmetric Plasma at Electron Cyclotron Frequencies", submitted to *Phys. Plasmas*.

Ultrahigh Resolution Simulations of Mode Converted ICRF and LH Waves with a Spectral Full Wave Code [1,2]

J. C. Wright*, P. T. Bonoli*, E. D'Azevedo† and M. Brambilla**

MIT - Plasma Science and Fusion Center Cambridge, MA 02139
†*Computer Science and Mathematics Division, Oak Ridge National Lab Oak Ridge, TN*
**Institute für Plasma Physik Garching, Germany*

Abstract. Full Wave (FW) studies of mode conversion (MC) processes in toroidal plasmas have required prohibitive amounts of computer resources in the past. The TORIC code solves the linear fourth order reduced wave equation in the ion cyclotron range of frequencies (ICRF), in toroidal geometry using a Fourier representation for the poloidal and toroidal dimensions and finite elements in the flux dimension. It has been parallelized to increase available computer processing and memory. Mode conversion scenarios may now be done routinely in a few hours of elapsed time with 32 processors. We present examples of fully converged studies of MC on Asdex-Upgrade and Alcator C-Mod. The new capabilities of the code extend beyond what is required for ICRF MC and permit studies of the very small wavelengths of lower hybrid (LH) waves. We will show preliminary results of FWLH calculations in C-Mod geometry.

INTRODUCTION

Modeling of fast magnetosonic waves (FW), in two dimensional toroidal plasmas has been possible for some years[1], but those scenarios involved simple electron Landau damping (ELD) of fast waves for on-axis heating and current drive. New physics algorithms have allowed finite Larmor radius (FLR) full wave codes such as TORIC[2] to accurately capture the damping of mode converted ion Bernstein waves (IBW)[3]. Recent experimental and analytic work[4] have created renewed interest in the presence of the ion cyclotron wave (ICW)[5] in toroidal plasmas. All three of these waves can be present at the ion-hybrid mode conversion layer in plasmas.

Resolving the disparate scales involved in mode-conversion (MC) physics requires resolution far beyond what is available in the memory of serial processor machines. To this end, we have introduced out-of-core memory management and parallel algorithms to greatly extend the range and geometry of problems that can be done with the TORIC code. We show demonstrably converged cases of mode conversion that agree with first principle estimates on required resolution. Finally, We present applications of the speed

---

[1] Work supported by SciDAC and USDOE Contract No. DE-FC02-01ER54648.
[2] Research sponsored by the Mathematical, Information, and Computational Sciences Division; Office of Advanced Scientific Computing Research; U.S. Department of Energy, under Contract No. DE-AC05-00OR22725 with UT-Battelle, LLC.

and resolution increases to Alcator C-Mod and Asdex Upgrade machines, including preliminary runs in the lower hybrid range of frequencies.

THE TORIC CODE

TORIC is a Finite Larmor Radius (FLR) full wave code. "Full wave", means that it solves Maxwell's equations in the presence of a plasma and wave antenna. It does this for a fixed frequency with a linear plasma response (Eq. (1a)) in a mixed spectral-finite element basis (Eq. (1b)). Where $\mathbf{J}^A$ and $\mathbf{J}^P$ are the antenna and plasma currents. Note that $\mathbf{J}^P$ is the oscillating current driven by the plasma wave and is distinct from the current induced in the electrons during current drive experiments that contributes towards the confining magnetic field.

$$\nabla \times \nabla \times \mathbf{E} = \frac{\omega^2}{c^2} \left\{ \mathbf{E} + \frac{4\pi i}{\omega} \left(\mathbf{J}^P + \mathbf{J}^A \right) \right\} \quad (1a)$$

$$\mathbf{E}(\mathbf{x}) = \sum_m \mathbf{E}_m(r) \exp(im\theta + in\phi) \quad (1b)$$

The FLR approximation retains the second harmonic wave frequency and the second order in the ion gyro-radius, ($\rho_i = v_{ti}/\Omega_{ci}$), for plasma interactions with the wave. It also retains the physics of the three ICRF waves: ICW, IBW, and FW. Near the mode conversion region, the FLR approximation breaks down and $k_\perp \rho_i \sim 1$, where ($\rho_i = v_{ti}/\Omega_{ci}$), and $k_\perp$ is the perpendicular wave number. In these regions, damping from a WKB approximation is used to modify the anti-Hermitian part of the conductivity operator in TORIC to capture the proper damping, while the real part describing propagation remains unchanged[2].

Using the spectral representation in Eq. (1b) and from the condition that $k_\perp \rho_i \simeq 1$ we estimate the maximum poloidal mode number needed for mode conversion studies. Taking $k_\perp \sim \frac{m}{r}$, the required resolution is approximately $M_{\max} \equiv r/\rho_i \equiv \rho^*$. In Alcator C-Mod, $m_{\max} \equiv M \simeq 173$, given a mode conversion layer at $r/a = 0.5$ in a deuterium plasma with a central field of 7.85 T and ion temperature of 1.8 keV. In larger, next-step devices, the requirements will be significantly higher. In LHRF, ($\Omega_e \gg \omega \gg \Omega_i$), dispersion yields: $\frac{\omega_{pe}}{\omega} k_\| \sim k_\perp \sim \frac{m}{r} \rightsquigarrow M \simeq 1000$, which is far in excess of what has been possible in the past.

To handle the large resolution requirements in these wave regimes, several serial and parallel modifications have been made to the code. The stiffness matrix created by the linear system in Eq. (1a) is a block tridiagonal system with $3N_\psi$ blocks. There is no toroidal dependence in the system, so the toroidal mode number enters only algebraically. The blocks are dense and have $(6N_m)^2$ elements each and represent the poloidal coupling. The system quickly exceeds the memory of a serial system even for small problems ($N_\psi = 100, N_m = 100$, 64 bit reals $\rightarrow 0.8GB$.) By writing blocks to disk during the inversion process, the problem size becomes limited by available disk space, and the memory requirements of a single block. This still imposes a constraint around

$N_m \simeq 511$ for 1 GB of RAM. To have the resolutions needed to for LHRF, we must distribute the problem and memory across multiple processors.

The dense blocks of poloidal coupling coefficients are distributed across nproc processors and the ScaLAPACK [6] library of parallelized linear algebra routines is used to carry out the matrix inversion and solution for the electric field vector, $\mathbf{E}(\mathbf{x})$. During the postprocessing phase when power and current deposition profiles are calculated, direct message passing interface (MPI) calls are used to redistribute the calculations along the radial flux variable dimension. This results in an acceleration of a factor of nproc in speed for the post-processing phase. The parallelization is nearly ideal, since these calculations are not coupled in that direction. This is significant, since for problems at large resolution, the post-processing calculations accounted for about half of the processing time. With these changes, it is reduced to an insignificant amount when the number of processor is large (> 10).

RESULTS

If insufficient resolution is used in mode-conversion studies, it is reflected in the power spectrum as high amplitudes in the largest modes. In Fig. (1a) we see that even at the largest radius shown ($\equiv r/a = 0.9$), the amplitude is below 1%, and so the case is well resolved. In Fig. (1b), simulations of the dependence of mode conversion efficiency show improved agreement with experiment as the resolution is raised from $N_m = 15$ to $N_m = 255$. We believe that the overestimation of electron power absorption at higher concentrations is due to the presence of a shear Alfvén resonance on the highfield side of the experiment that is not measured. In Fig. (1c), three separate wave scales may be seen. The ICRF propagates in from the right. On axis, the IBW continues to the left at r=-10cm. Off axis, particularly below the midplane, the ICW is clearly seen propagating to the right to the low field side. Also, the evanescent region to the right of the MC layer at r = -6cm is unmistakable. Figure (1d) shows a first attempt at full-wave simulations in the LHRF. It is run at a resolution sufficient to resolve the LH wavelength, $N = 480, N_m = 1023$. The antenna polarization was rotated to couple to the mainly electrostatic LH wave, however the grill antenna model presently in the code cannot effectively couple to the LH branch. The result is we couple to the FW in the LHRF. The mode's power deposition dependence on the plasma , and the measured perpendicular wavenumber are consistent with the FW and not the LHW. A new wave-guide model is currently being implemented in TORIC that should improve the coupling to the LHW.

The algorithms for solving full wave dispersion in realistic toroidal geometries have been developed by the RF community over the past few decades. The availability first, of more powerful serial processors and later of parallel architectures has opened up a broader range of physics regimes to these codes. The traditional problems of central heating and current drive are now done so fast and routinely that large parameter scans, coupling to antenna and Fokker-Planck codes, and the integration of more detailed dispersion models such as parallel gradient effects and non-Maxwellian particle distributions is now feasible. The increases in resolution capabilities allows very fine

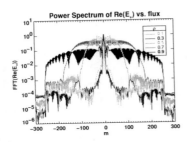
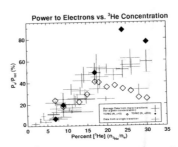
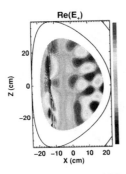
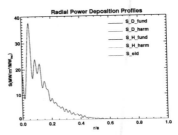

(a) Converged poloidal power spectrum demonstrating sufficient resolution in mode conversion simulation.

(b) D(^{3}He) mode conversion electron heating in Alcator C-Mod [$B_0 = 7.85$ fT, $f_0 = 80$MHz]. Volume integrated IBW electron absorption versus electron density.

(c) 2-D fields at 480x1023 resolution, show FW, IBW, and ICW together.

(d) Plot of FW power in LHRF. FW antenna model in code is not appropriate for coupling to LH waves.

FIGURE 1.

scale study of mode conversion processes in present-day machines and in future next-step machines where the gyro-radius scaling will only increase the needed resolution. For the first time we can consider lower hybrid frequency regime calculations and the wide range of physics issues involved, such as the role of wave focusing and diffraction in LH spectral broadening.

REFERENCES

1. Brambilla, M., and Krücken, T., *Nucl. Fusion*, **28**, 1813–1833 (1988).
2. Brambilla, M., *Nucl. Fusion*, **38**, 1805–1817 (1998).
3. Bernstein, I. B., *Phys. Rev.*, **109**, 10–21 (1958).
4. Lin, Y., et al., *Plasma Phys. Controlled Fusion* (2003).
5. Perkins, F. W., *Nucl. Fusion*, **17**, 1197 (1977).
6. Choi, J., Dongarra, J. J., Ostrouchov, L. S., Petitet, P., A., Walker, D. W., and Whaley, R. C., *Scientific Programming*, **5**, 173–184 (1996).

AUTHOR INDEX

A

Agarici, G., 98
Agnew, S., 403
Akhtar, M. K., 411, 427
Alberti, S., 297, 344
Almagri, A. F., 331
Amarante, G., 98
Anderegg, F., 403
Anderson, D. T., 331
Anderson, F. S., 331
Anderson, J. K., 367
Andre, R., 185
Angelini, B., 74
Angioni, C., 297
Apicella, M. L., 74
Apruzzese, G., 74
Arends, E., 359
Arnoux, G., 297
Artaud, J. F., 275, 455
Assas, S., 41
Austin, M. E., 348

B

Baity, F. W., 98, 102, 423
Baker, D. R., 313
Baranov, Y., 138, 227, 359
Barbato, E., 24, 74
Barnsley, R., 138
Basiuk, V., 275, 455
Basse, N. P., 243
Batchelor, D. B., 463, 471, 475, 487
Baxi, C. B., 325
Baylor, L. R., 313
Beaumont, B., 98, 251, 259
Beaumont, P., 138, 197
Becker, W., 154
Bécoulet, A., 219, 455
Behn, R., 297
Bell, G. L., 372, 396
Bell, R. E., 193, 201
Bengtson, R. D., 423
Berger-By, G., 275
Bergkvist, T., 197, 459, 479
Bering, III, E. A., 423

Bernabei, S. I., 74, 201, 209, 283
Berry, L. A., 463, 467, 471, 475, 487
Bers, A., 392, 396
Bertalot, L., 74
Bertocchi, A., 74
Beunas, A., 251
Bibet, P., 251, 259
Biewer, T. M., 193
Bigelow, T. S., 372, 396
Bilato, R., 106
Birus, D., 106
Blackman, T., 146, 150
Blanchard, P., 297
Bobkov, V., 154
Boivin, R. L., 166
Bonoli, P. T., 24, 50, 158, 162, 178, 205, 209, 471, 511
Borra, M., 74
Bosco, J., 243
Bosia, G., 82, 247
Bouquey, F., 275
Bourdelle, C., 275
Brambilla, M., 106, 511
Braun, F., 106, 154
Brémond, S., 82, 90, 98, 259
Brons, S., 94
Brzozowski, J., 41
Buceti, G., 74
Budny, R., 41
Burrell, K. H., 313

C

Cairns, R. A., 392
Callis, R. W., 325, 335
Camenen, Y., 297
Cardinali, A., 74, 255
Carter, M. D., 205, 209, 367, 372, 396, 423
Cary, J. R., 380, 467
Castaldo, C., 74, 255
Cengher, M., 367
Centioli, C., 74
Cesario, R., 41, 74, 255
Challis, C., 197, 255
Chan, V. S., 41, 86, 352

Chang-Diaz, F. R., 423
Chappuis, P., 98
Chattopadhyay, P. K., 263, 367
Chiu, S. C., 86
Choi, M., 86, 305
Clary, J., 275
Cluggish, B., 403
Cocilovo, V., 74
Coda, S., 297, 344
Colas, L., 90, 259
Condrea, I., 297
Condren, B., 419
Conroy, S., 138
Crisanti, F., 74, 279
Cunningham, G., 359

D

Damiani, C., 98
Darbos, C., 275
Darrow, D. S., 185, 193
D'Azevedo, E., 471, 475, 511
De Angelis, R., 74
DeBaar, M., 41, 66
Decker, J., 392, 447, 495
deGrassie, J. S., 41, 313
De Marco, F., 74
Deng, C., 331
Denning, C. M., 411, 427
De Vries, P., 66, 138, 279
D'Ippolito, D. A., 463, 487
Doane, J. L., 335
Dominguez, R., 380
Dumont, R. J., 185, 213, 439, 499
Dumortier, P., 94, 142
Durodié, F., 94, 98, 102, 118, 122, 142, 150

E

Efthimion, P. C., 384, 396
Egedal, J., 185
Ekedahl, A., 227, 259, 267
Ellis, R. A., 335
Erents, K., 227
Eriksson, L.-G., 41, 66, 271, 455, 479
Ershov, N., 471
Esposito, B., 74

F

Fadnek, A., 102
Fanthome, J., 98
Felton, R., 66, 279
Fenzi-Bonizec, C., 41
Figueiredo, A., 66
Fiore, C., 162, 178
Forest, C. B., 367
Fredd, E., 283
Fredian, T. W., 243
Freeman, R., 403
Fuchs, V., 267
Fukuda, T., 321

G

Gabellieri, L., 74
Gantenbein, G., 317
Gatti, G., 74
Gerhardt, S. P., 331
Gilleland, J., 403
Giovannozzi, E., 74, 74
Giroud, C., 41, 138
Giround, C., 197
Giruzzi, G., 275
Glover, T. W., 423
Goetz, J. A., 166, 263
Gondhalekar, A., 41
Goniche, M., 227, 267
Goode, B., 467
Goodman, T. P., 297, 344
Gorelov, Y. A., 325, 335
Gormezano, C., 74
Goulding, R. H., 98, 102, 118, 122, 423
Gowers, C., 138
Granucci, G., 227
Graves, J., 297
Green, M. T., 325
Greenough, N., 283
Griffin, S. T., 415
Grimes, M., 243
Grolli, M., 74
Gryaznevich, M., 359

H

Harling, J., 66
Hartmann, D. A., 106, 154

Harvey, R. W., 24, 185, 205, 305, 344, 348, 367, 380, 396, 471, 495
Hawkes, N., 41
Hellsten, T., 41, 110, 126, 138, 197, 459, 479
Henderson, M. A., 297
Heuraux, S., 90
Higaki, H., 114
Higaki, K., 340
Hoang, G. T., 41, 275, 455
Hogge, J-P., 297
Hojo, H., 114
Horinouchi, K., 114
Hosea, J. C., 98, 102, 166, 209, 283
Hubbard, A., 162
Huber, A., 197
Hughes, J. W., 158
Humphreys, D. A., 348
Hutchinson, I. H., 50, 178
Huysmans, G., 275, 455

I

Iannone, F., 74
Ichimura, M., 114
Ide, K., 114
Ide, S., 321
Idei, H., 289
Igami, H., 340, 376
Ikeda, Y., 321
Imbeaux, F., 24, 66, 271, 275, 455
Imre, K., 507
Indireshkumar, K., 201
Ingesson, L. C., 138
Inoue, D., 114
Irby, J. H., 162, 166
Isayama, A., 321
Isler, R., 403
Isobe, M., 58
Ito, S., 289

J

Jacobson, V. T., 423
Jaeger, E. F., 185, 205, 209, 463, 471, 475, 487
Joffrin, E., 66, 197, 227, 279
Johnson, T., 41, 126, 138, 197, 459, 479

Jones, B., 396
Jones, G. H., 98, 102

K

Kaita, R., 213
Kajiwara, K., 325, 335
Kakimoto, S., 114
Kambouchev, N. D., 243
Kang, Y., 431
Kato, A., 58
Kaufman, A. N., 483
Kaufman, M. C., 263
Kaye, A., 98
Kaye, S. M., 201, 205, 209
Kazarian, F., 251, 259
Keller, A., 317
Kim, J. Y., 491
Kim, S. S., 491
Kiptily, V. G., 41, 66, 138, 197
Kirov, K., 41
Kobayashi, S., 289
Koch, R., 130
Koone, N., 419
Kroegler, H., 74
Kubo, S., 289
Kumazawa, R., 58
Kung, C. C., 283
Kuzikov, S., 247
Kwon, M., 491
Kyrytsya, V., 134

L

Labombard, B., 158
Laborde, L., 279
La Haye, R. J., 305, 348
Lamalle, P. U., 41, 98, 118, 122, 146, 150
Lancellotti, V., 134
Laqua, H. P., 15
Lashmore-Davies, C. N., 392
Laugier, J., 267
Lawson, K., 66
Laxåback, M., 110, 126, 138, 197, 459, 479
LeBlanc, B. P., 185, 201, 205, 209
Leggate, H., 66, 138

Leigheb, M., 74, 74
Lennholm, M., 275, 279
Leuterer, F., 317
Likin, K. M., 331
Lin, L., 162
Lin, Y., 50, 158, 162, 166, 178, 243
Lin-Liu, Y.-R., 305
Litaudon, X., 66, 279, 455
Litvak, A., 403
Lloyd, B., 359
Loesser, G. D., 98, 102, 283
Lohr, J., 305, 313, 325, 335, 348
Lomas, P. J., 227
Lorenz, A., 98
Luce, T. C., 305, 348
Lyssoivan, A., 130

M

Maddaluno, G., 74
Maekawa, T., 340, 376
Maget, P., 275
Maggiora, R., 134
Magne, R., 275
Mailloux, J., 197, 227, 255, 259
Maingi, R., 201
Majeski, R., 213
Mantica, P., 66
Mantsinen, M. J., 41, 66, 126, 130, 138, 197, 227
Maraschek, M., 317
Marinucci, M., 74
Marmar, E. S., 50, 162, 178
Matthews, G., 197
Mau, T. K., 185, 201, 205, 209
Mayoral, M.-L., 138, 146, 197, 259
Mazon, D., 255, 275, 279, 455
Mazurenko, A., 50
Mazzitelli, G., 74
McDonald, D., 227
Mead, M., 98
Medley, S., 185
Meigs, A., 41
Menard, J. E., 185
Meo, F., 41, 197
Merlino, R. L., 3
Messiaen, A., 94, 142
Messineo, M. A., 98
Micozzi, P., 74

Miller, R., 403
Mizuno, Y., 289
Monakhov, I., 98, 118, 130, 138, 146, 150, 197
Monakhov, R. H., 118
Monari, G., 74
Moreau, D., 279
Moriyama, S., 321
Murakami, M., 305
Murari, A., 279
Mutoh, T., 58
Myra, J. R., 463, 475, 487

N

Nagasaki, K., 321
Nelson, B. E., 98, 102
Nelson-Melby, E., 50, 297, 344
Nguyen, F., 41, 197
Nieter, C., 380
Nightingale, M., 98, 118, 146, 150
Nikkola, P., 297, 344
Nomura, G., 58
Notake, T., 289
Noterdaeme, J.-M., 41, 66, 106, 130, 138, 154, 197, 227, 259

O

O'Brien, M., 359
O'Connell, R., 263
Ohkawa, T., 403
Ohkubo, K., 289
Oliva, S. P., 263
Omelchenko, Y. A., 86
O'Neill, R., 403
Orsitto, P., 74
Osakabe, M., 58
Ozaki, T., 58

P

Pacella, D., 74
Paméla, J., 98
Panaccione, L., 74
Panella, M., 74
Paoletti, F., 74, 255

Papitto, P., 74
Parisot, A., 158, 166, 178
Park, B. H., 491
Parker, R., 158, 162
Parker, R. R., 243
Passeron, C., 455
Pavlo, P., 388
Pericoli, V., 66, 227
Pericoli-Ridolfini, V., 74, 279
Petrzilka, V., 227, 267
Petty, C. C., 305, 348, 352
Peysson, Y., 267, 271, 275, 455, 495
Phillips, C. K., 50, 185, 205, 209, 213, 439, 499, 503
Phillips, D. W., 205
Phillips, P., 50
Pieroni, L., 74
Piliya, A. D., 359
Pinsker, R. I., 313, 367, 384
Pletzer, A., 499, 503
Pochelon, A., 297
Podda, S., 138
Ponce, D., 325, 335
Porkolab, M., 158, 162, 166, 178
Portafaix, C., 98
Porte, L., 297
Prager, S. C., 174
Prater, R., 305, 313, 335, 348
Preinhaelter, J., 388
Pronko, S., 403
Putvinski, S., 403

R

Rachlew, E., 66, 197
Ram, A. K., 158, 392, 396
Rantamäki, K., 227
Rasmussen, D. A., 102, 209, 372, 396
Ravera, G., 74
Redi, M. H., 185
Riccardo, V., 98
Rice, J., 162, 178
Righetti, G. B., 74
Righi, E., 41
Riva, M., 279
Rokhman, Y. I., 243
Romanelli, F., 74
Rosenberg, A. L., 185, 205

Ryan, P. M., 102, 122, 185, 201, 205, 209

S

Saida, T., 58
Saito, K., 58
Salmi, A., 130, 138
Sartori, F., 197
Sartori, R., 227
Sasao, M., 58
Sauter, O., 297, 344
Saveliev, A. N., 359
Scarabosio, A., 297
Scharer, J. E., 170, 411, 427
Schilling, G., 50, 162, 166, 178
Ségui, J. L., 275
Seki, M., 321
Seki, T., 58
Semeraro, L., 98
Sentoku, Y., 86
Sevier, L., 403
Sharapov, S., 138
Shevchenko, V., 359, 388
Shimozuma, T., 289
Shimpo, F., 58
Shukla, S., 431
Sibley, A., 403
Silva, C., 227
Smirnov, A. P., 367, 380, 396, 471
Smithe, D. N., 439, 471, 499, 503
Soukhanovskii, V., 213
Spaleta, J., 213
Sparks, D. O., 102
Squire, J. P., 423
Staebler, A., 41, 197
Stamp, M., 227
Sternini, E., 74
Stillerman, J. A., 243
St John, H. E., 86
Stockli, M. P., 431
Sund, R., 170
Suttrop, W., 317
Suzuki, T., 321
Svidzinski, V. A., 174
Swain, D. W., 102, 185, 201, 205, 209
Sykes, A., 359

T

Takase, Y., 235
Takeuchi, N., 58
Takita, Y., 289
Tala, T., 279
Talarico, C., 98
Talmadge, J. N., 331
Tanaka, H., 340
Tartoni, N., 74
Taylor, G., 201, 384, 396
Tennfors, E., 197
Terry, D. R., 243
Testa, D., 41, 138
Thio, Y. C. F., 415
Thomas, M. A., 263
Torii, Y., 58
Tracy, E. R., 483
Tran, M. Q., 297
Tregubova, E. N., 359
Tresset, G., 279
Tuccillo, A. A., 41, 66, 74, 138, 197, 227, 259
Tudisco, O., 74
Turker, E., 98
Tysk, S. M., 411, 427

U

Uchida, M., 340, 376
Umstadter, K., 403
Utsumi, H., 114

V

Vahala, G., 388
Vahala, L., 388
Valovic, M., 388
Van Eester, D., 66, 130, 138, 197
Vecchi, G., 134
Vitale, V., 74
Vlad, G., 74
Volpe, F., 359
Vulliez, K., 98

W

Wade, M. R., 305, 348
Wade, T., 403
Walden, A., 98, 118, 146, 150, 197

Walton, R., 98, 102
Watari, T., 58
Weisen, H., 41
Weitzner, H., 507
Weix, P. J., 263
Welton, R. F., 431
Wesner, F., 106, 154
White, B. O., 427
Whitehurst, A., 122, 150
Wilgen, J. B., 201, 209, 372, 396
Wilson, J. R., 98, 162, 166, 185, 193, 201, 205, 209, 283, 396
Winslow, D., 403
Wolfe, S., 50, 162
Wong, K.-L., 305, 352
Wouters, P., 98, 118, 146
Wright, J. C., 50, 471, 511
Wukitch, S. J., 50, 158, 162, 166, 178, 243

Y

Yamaguchi, Y., 114
Yamamoto, T., 58
Yatsu, K., 114
Yeom, J. H., 491
Yokota, M., 58
Yoshimura, Y., 289
Yoshinaga, T., 340
You, K. I., 491
Yuh, H., 162

Z

Zabeo, L., 279
Zacek, F., 227
Záèek, F., 267
Zastrow, K.-D., 41, 66, 279
Zerbini, M., 74
Zhai, K., 331
Zhuang, G., 297
Zhurovich, K., 162
Zohm, H., 317
Zonca, F., 74
Zou, X. L., 275